METHODS IN MOLECULAR BIOLOGY™

Series Editor
John M. Walker
School of Life Sciences
University of Hertfordshire
Hatfield, Hertfordshire, AL10 9AB, UK

For further volumes:
http://www.springer.com/series/7651

Microfluidic Diagnostics

Methods and Protocols

Edited by

Gareth Jenkins

Key Laboratory for Organic Electronics and Information Displays (KLOEID), Nanjing University of Technology, Institute of Advanced Materials (IAM), Jiangsu, China

Colin D. Mansfield

Scientific and Medical Communications Consultant, Lyon, France

Humana Press

Editors
Gareth Jenkins
Institute of Advanced Materials (IAM)
Key Laboratory for Organic Electronics and Information Displays (KLOEID)
Nanjing University of Technology
Jiangsu, China

Colin D. Mansfield
Scientific and Medical Communications Consultant
Lyon, France

ISSN 1064-3745 ISSN 1940-6029 (electronic)
ISBN 978-1-4939-6212-9 ISBN 978-1-62703-134-9 (eBook)
DOI 10.1007/978-1-62703-134-9
Springer New York Heidelberg Dordrecht London

Printed on acid-free paper

Humana Press is a brand of Springer
Springer is part of Springer Science+Business Media (www.springer.com)

Preface

Microfluidic techniques are becoming widely incorporated into medical diagnostic systems due to the inherent advantages of miniaturization. In particular, the application of microfluidics to point-of-care testing (POCT) devices and high-throughput screening is predicted to become increasingly important, and consequently, the interest in microfluidic diagnostics is rapidly growing. The inherent advantages of scaling down include increased speed, efficiency, a reduction in the demand for sample and reagents, and the potential for multiplexing and parallelization. Other often cited advantages, yet by no means universally achievable, include increased portability, lower device costs (through mass production), and more highly integrated and automated systems leading to powerful yet easy to use devices. Such potential has led to widespread predictions that such technologies will help revolutionize health care provision at a particularly timely moment. When faced with the multiple challenges of increased costs, an aging population, bringing health care to developing countries, and the need to shift the business models of pharmaceutical companies away from palliative care to more responsive and personalized therapeutics, it is easy to see that microfluidic diagnostics is well placed to take a centrally important role.

The sheer number of different methods and applications available has, however, led to a diffuse and fragmented field with little standardization. From a practical and commercial perspective, microfluidic diagnostics have not yet had as much of an impact in "real-world" applications as had been widely predicted although steady progress has been made. This may be partially attributed to the difficulty in translating academic research into practical solutions. In particular, the highly interdisciplinary nature of the field can be daunting to new researchers, especially those coming from more established and well-defined disciplines who seek to apply the benefits of microfluidics to their own work. Many challenges are faced in order to convert promising concepts from the lab bench through to practical and commercially viable devices. As well as technological challenges, regulatory hurdles and issues relating to intellectual property (IP) and other commercial concerns further complicate the routes for technology transfer.

This book seeks to partly address some of these problems by providing a set of protocols necessary for the development of a variety of microfluidic diagnostic technologies. It pulls together a range of methods from leading researchers in the field, covering subjects such as microfluidic device fabrication, on-chip sample preparation, diagnostic applications and detection methodologies. The protocols described range from cutting-edge developments to established techniques and basic demonstrations suitable for education and training; from basic fabrication methods to commercializing research.

What you need to know and how to do it: each protocol offers step-by-step instructions, including an introductory overview of the technique, a list of materials and reagents required, as well as helpful tips and troubleshooting advice. Insightful reviews along with advice on how to successfully develop and commercialize microfluidic diagnostic technologies makes this volume indispensable reading for scientists entering the field as well as providing a reference text for those already established. Due to the multidisciplinary nature of the field, little background knowledge is assumed, providing an accessible text for scientists from a range of disciplines including biomedical researchers, engineers, biochemists, and clinicians.

This book is organized into three parts: "Microfluidic Diagnostics: From the Classroom to the Boardroom" contains a number of protocols suitable for the educational demonstration of microfluidic techniques, as well as chapters relating to commercialization issues, such as the microfluidic device market, patent filing, and regulatory affairs. In addition, the opening chapter provides an overview of present technology and future trends in point-of-care microfluidic diagnostics. "Fabrication and Manipulation Protocols" contains a number of protocol and review chapters detailing various microfluidic fabrication methods for the manipulation of fluidic samples on the microscale. "Application Protocols" contains protocols and reviews for various applications of microfluidic diagnostics and a range of detection methodologies.

In preparing this book we would first and foremost like to express our gratitude to all the authors whose hard work and excellent contributions will, we hope, form a useful and informative text for many other researchers in the field. We appreciate their time and especially their patience during a long and arduous review process. Dr Jenkins would like to express his thanks to his friends and colleagues at Imperial College London, in Nanjing, and at Xiamen University and also to the members of the European Consortium of Microfluidics (hosted by the Centre for Business Innovation) for many useful discussions during the preparation of this work. He would also like to acknowledge financial support from the UK Department for Business Innovation and Skills and from the State Key Laboratory of Physical Chemistry of Solid Surfaces at Xiamen University, China for the support of his UK-China Fellowship. Special thanks are due to his wife and newborn son whose support has been unending and indispensable throughout. Dr Mansfield would like to thank the series editor, John Walker, for inviting him to participate in this project, as well as for his guidance during preparation of the book. He would especially like to express gratitude to his wife, Fidji, and sons, James and Ryan, for their support and patience while he spent countless weekends and evenings away from them working on this book.

Jiangsu, China — ***Gareth Jenkins***
Lyon, France — ***Colin D. Mansfield***

Contents

Contributors

Amit Asthana • *Centre for Cellular and Molecular Biology, Council of Scientific and Industrial Research, Hyderabad, Andhra Pradesh, India*
Emanuele Barborini • *Tethis SPA, Milan, Italy*
Fiona Bessoth • *Patent Attorneys Ter Meer Steinmeister & Partner, Munich, Germany*
Hugh A. Bruck • *Department of Mechanical Engineering, University of Maryland, College Park, MD, USA*
Samantha Byrnes • *Department of Biomedical Engineering, Boston University, Boston, MA, USA*
Roberta Carbone • *Tethis SPA, Milan, Italy*
Anthony Cass • *Institute of Biomedical Engineering & Department of Chemistry, Imperial College, London, UK*
Michael P. Cuchiara • *Department of Bioengineering, MS-142, BRC, Rice University, Houston, TX, USA*
Jonathan Day • *DNA Electronics Ltd, Institute of Biomedical Engineering, Imperial College, London, UK*
Don L. DeVoe • *Department of Mechanical Engineering, University of Maryland, College Park, MD, USA*
Andy Fan • *Department of Biomedical Engineering, Boston University, Boston, MA, USA*
Bruce K. Gale • *Department of Mechanical Engineering, State of Utah Center of Excellence for Biomedical Microfluidics, University of Utah, Salt Lake City, UT, USA*
John Greenman • *Postgraduate Medical Institute, University of Hull, Hull, UK*
Friederike J. Gruhl • *Karlsruhe Institute of Technology (KIT), Institute for Microstructure Technology (IMT), Eggenstein-Leopoldshafen, Germany*
Song-I Han • *School of Nano Engineering, Inje University, Gimhae, Republic of Korea*
Ki-Ho Han • *School of Nano Engineering, Inje University, Gimhae, Republic of Korea*
Stephen J. Haswell • *Department of Chemistry, University of Hull, Hull, UK*
Samantha M. Hattersley • *Postgraduate Medical Institute, University of Hull, Hull, UK*
Harikrishnan Jayamohan • *Department of Mechanical Engineering, State of Utah Center of Excellence for Biomedical Microfluidics, University of Utah, Salt Lake City, UT, USA*
Gareth Jenkins • *Institute of Advanced Materials (IAM), Key Laboratory for Organic Electronics and Information Displays, Nanjing University of Posts and Telecommunication, China; Institute of Advanced Materials, Nanjing University of Technology, China; Institute of Biomedical Engineering, Imperial College, London, UK*
Robert C. Jones • *Fluidigm Corporation, South San Francisco, CA, USA*
Lily Kim • *Wyss Institute of Biologically Inspired Engineering at Harvard University, Brookline, MA, USA*
Minseok S. Kim • *Department of Bio and Brain Engineering, Korea Advanced Institute of Science and Technology (KAIST), Yuseong-gu, Daejeon, Republic of Korea*
Catherine Klapperich • *Department of Biomedical Engineering, Boston University, Boston, MA, USA*

YORDAN KOSTOV • *Steven Sun Division of Biology Office of Science and Engineering, FDA Center for Devices and Radiological Health (CDRH), Silver Spring, MD, USA; University of Maryland Baltimore County, Baltimore County, MD, USA*
LAWRENCE KULINSKY • *Department of Mechanical and Aerospace Engineering, University of California, Irvine, CA, USA*
SEYONG KWON • *Department of Bio and Brain Engineering, Korea Advanced Institute of Science and Technology (KAIST), Yuseong-gu, Daejeon, Republic of Korea*
ERIC T. LAGALLY • *Michael Smith Laboratories, University of British Columbia, Vancouver, BC, Canada; Department of Chemical and Biological Engineering, University of British Columbia, Vancouver, BC, Canada*
KERSTIN LÄNGE • *Karlsruhe Institute of Technology (KIT), Institute for Microstructure Technology (IMT), Eggenstein-Leopoldshafen, Germany*
XUEFEI LENG • *Department of Chemical Biology, College of Chemistry and Chemical Engineering, Xiamen University, Xiamen, People's Republic of China*
PENG LI • *Department of Mechanical, Industrial and Systems Engineering, University of Rhode Island, Kingston, RI, USA*
LOUISE LUND • *Michael Smith Laboratories, University of British Columbia, Vancouver, BC, Canada; Center for High Throughput Biology, University of British Columbia, Vancouver, BC, Canada*
MARCIN LYGAN • *Institut des Nanotechnologies de Lyon (INL), Ecully Cedex, France*
MARC MADOU • *Department of Mechanical and Aerospace Engineering, University of California, Irvine, CA, USA; Department of Biomedical Engineering, University of California, Irvine, CA, USA; Ulsan National Institute of Science and Technology (UNIST), Ulsan, South Korea*
COLIN D. MANSFIELD • *Institut des Nanotechnologies de Lyon (INL), UMR CNRS, Ecully Cedex, France*
RADOSLAW MAZURCZYK • *Institut des Nanotechnologies de Lyon (INL), UMR CNRS 5270, Ecully Cedex, France*
ZAHRA NOROOZI • *Department of Mechanical and Aerospace Engineering, University of California, Irvine, CA, USA*
JUNJIE OU • *Department of Mechanical and Mechatronics Engineering, University of Waterloo, Waterloo, ON, Canada*
ERIC OUELLET • *Michael Smith Laboratories, University of British Columbia, Vancouver, BC, Canada; Department of Chemical and Biological Engineering, University of British Columbia, Vancouver, BC, Canada; Biomedical Engineering Program, University of British Columbia, Vancouver, BC, Canada*
JE-KYUN PARK • *Department of Bio and Brain Engineering, Korea Advanced Institute of Science and Technology (KAIST), Yuseong-gu, Daejeon, Republic of Korea; KAIST Institute for the NanoCentury, Yuseong-gu, Daejeon, Republic of Korea*
WALTER PECK • *Whitney Point High School, Whitney Point, NY, USA*
SCOTT T. PHILLIPS • *The Pennsylvania State University, University Park, PA, USA*
JIAN QIN • *Fluidigm Corporation, South San Francisco, CA, USA*
RAMESH RAMAKRISHNAN • *Fluidigm Corporation, South San Francisco, CA, USA*
MICHAEL RAPP • *Karlsruhe Institute of Technology (KIT), Institute for Microstructure Technology (IMT), Eggenstein-Leopoldshafen, Germany*
AVRAHAM RASOOLY • *Division of Biology, Office of Science and Engineering, FDA Center for Devices and Radiological Health (CDRH), Silver Spring, MD, USA; National Cancer Institute, Rockville, MD, USA*

MEITAL RECHES • *Institute of Chemistry and Center for Nanoscience and Nanotechnology, The Hebrew University of Jerusalem, Jerusalem, Israel*
CAROLYN L. REN • *Department of Mechanical and Mechatronics Engineering, University of Waterloo, Waterloo, ON, Canada*
P. ROACH • *Institute for Science and Technology in Medicine, Keele University, Staffordshire, UK*
HIMANSHU J. SANT • *Department of Mechanical Engineering, State of Utah Center of Excellence for Biomedical Microfluidics, University of Utah, Salt Lake City, UT, USA*
STEVEN SCOTT • *Institute of Biomedical Engineering & Department of Chemistry, Imperial College, London, UK*
CHENREN SHAO • *Department of Mechanical Engineering, University of Maryland, College Park, MD, USA*
SANJIV SHARMA • *Institute of Biomedical Engineering & Department of Chemistry, Imperial College, London, UK*
N.J. SHIRTCLIFFE • *Biomimetic Materials, Hochschule Rhein-Waal, Rhine-Waal University of Applied Sciences, Kleve, Germany*
MONPICHAR SRISA-ART • *Department of Chemistry, Faculty of Science, Chulalongkorn University, Bangkok, Thailand*
VISHAL TANDON • *Department of Biomedical Engineering, Cornell University, Ithaca, NY, USA*
NICOLE K. THOM • *The Pennsylvania State University, University Park, PA, USA*
R. TOON • *Nemaura Pharma Limited, Loughborough, Leicestershire, UK*
L. SUZANNE WEAVER • *Fluidigm Corporation, South San Francisco, CA, USA*
CHENG WEI T. YANG • *Michael Smith Laboratories & Department of Chemical and Biological Engineering, University of British Columbia, Vancouver, BC, Canada*
JENNIFER L. WEST • *Department of Bioengineering, MS-142, BRC, Rice University, Houston, TX, USA*
SHUANG YANG • *Department of Mechanical Engineering, University of Maryland, College Park, MD, USA*
CHAOYONG JAMES YANG • *Department of Chemical Biology, College of Chemistry and Chemical Engineering, Xiamen University, Xiamen, People's Republic of China*
ANDREA ZANARDI • *Tethis SPA, Milan, Italy*

Section I

Microfluidic Diagnostics: From the Classroom to the Boardroom

Chapter 1

Present Technology and Future Trends in Point-of-Care Microfluidic Diagnostics

Lawrence Kulinsky, Zahra Noroozi, and Marc Madou

Abstract

This work reviews present technologies and developing trends in Point-of-Care (POC) microfluidic diagnostics platforms. First, various fluidics technologies such as pressure-driven flows, capillary flows, electromagnetically driven flows, centrifugal fluidics, acoustically driven flows, and droplet fluidics are categorized. Then three broad categories of POC microfluidic testing devices are considered: lateral flow devices, desktop and handheld POC diagnostic platforms, and emergent molecular diagnostic POC systems. Such evolving trends as miniaturization, multiplexing, networking, new more sensitive detection schemes, and the importance of sample processing are discussed. It is concluded that POC microfluidic diagnostics has a potential to improve patient treatment outcome and bring substantial savings in overall healthcare costs.

Key words: Point of care, POC, POCT, Microfluidics, Diagnostics, Lateral flow, Molecular diagnostics, Immunoassay

1. Introduction

Microfluidic diagnostics had an explosive growth in the last 20 years spurred by the convergence of clinical diagnostic techniques (such as blood gas analysis, immunoassays, and molecular biology testing) and mature microfabrication technology (1) that allowed production of submillimeter-size fluidic channels and reservoirs in a variety of material systems (for example: silicon, polydimethylsiloxane (PDMS), poly(methyl methacrylate) (PMMA), etc.). Miniaturization of a chemical lab has apparent immediate benefits: dramatically smaller amount of sample and reagents needed for the analysis; lower test costs; exponential reduction in test times due to the fact that diffusion distances in microfluidic systems are very small compared to the macroscopic lab tests; multiplexing—performing multiple types of tests from the same sample; possibility of integration and automation of all process steps on the same

Gareth Jenkins and Colin D. Mansfield (eds.), *Microfluidic Diagnostics: Methods and Protocols*, Methods in Molecular Biology, vol. 949, DOI 10.1007/978-1-62703-134-9_1, © Springer Science+Business Media, LLC 2013

platform; and development of a wide variety of Point-of-Care (POC) testing devices.

It is our belief that POC diagnostic systems will revolutionize the practice of medicine and have a strong potential to dramatically reduce healthcare costs. In some cases where tests should be performed immediately POC testing, often referred to as POCT, is the only option—for example in critical care and operating rooms or threatening infectious diseases. In other cases, when all the required tests can be done in a physician's office (as opposed to the patient's sample being sent to a central laboratory for processing and then the patient visiting the doctor's office again when test results are ready) POC testing offers the following benefits: (a) substantial savings in overall healthcare costs (as a result of reduced number of patient visits to medical offices); (b) an improvement in patient quality of life (avoidance of psychological stress caused by the uncertainty of his/her possible state of health or prognosis while waiting for results); (c) a possibility to start treatment earlier, which in some cases can affect the outcome of the treatment; and (d) a reduction of errors related to mixing up the results of various patients as compared to large test laboratories. It seems to us that one of the possible future trends will be an emerging support for POC testing by health insurance companies and government insurance programs (such as Medicare), which will benefit from the lowering of medical costs. Eventually some of these cost savings will be passed on to consumers in the form of lower health insurance premiums.

Before we commence our review of the state of microfluidic POC development it is necessary to clarify the terminology that we use. We consider that POC test platforms are self-contained diagnostic Lab-on-a-Chip (LOC) platforms (2), micro Total Analysis Systems (3), and fluidic cartridges or Lateral Flow (LF) strips (4)—all either with integrated or dedicated readout systems with footprints ranging from the small chip to a desktop system that can be placed in doctors' offices, hospitals, or mobile first responders' vehicles and designed to be operated by minimally trained personnel. In other words, POC diagnostic devices should require minimal manual operations (other than sample collection) and thus should contain all the necessary reagents, and it is highly desirable that all process steps (including sample preparation and pretreatment) are automated and integrated within the system. It is understandable that due to the complex processing steps not every clinical test can be performed outside of the laboratory, but we believe that whenever possible, there is a great benefit to be derived from the development of a wider array of POC tests.

As the size of a system shrinks into the micrometer range, the surface area-to-volume ratio increases, and surface forces (rather than body forces such as gravitational forces that dominate physics at the everyday "macro"scale) become the most significant influence for operation of microdevices (5, 6). For example, capillary and electrostatic forces play dominant roles at microscopic dimensions, and there are also challenges to mix fluids in reservoirs and microfluidic channels where convection is limited and diffusion becomes a key transport mode. We will first briefly review various types of microfluidic devices according to the predominant propulsion forces in each system (i.e., capillary-driven devices, pressure-driven devices, etc.). The details of fluidic functions (including metering, gating, separation, etc.) that control the microfluidics of each system type are not however elaborated on in this work as they are discussed in several comprehensive reviews (2, 7, 8).

We can categorize microfluidic diagnostic devices into three subsets of POC testing platforms: (a) LF test devices; (b) handheld or benchtop POC testing systems for blood gas, electrolytes, blood chemistry, and detection of certain protein markers—these systems typically use cartridges, tubes, and other non-LF platforms; and (c) molecular biology-based POC testing (such as nucleic acid testing)—the newest and fastest-growing sector of the POC testing market. In each case we will consider the examples of POC testing devices that are available commercially and will observe some emerging technology and trends for each class of diagnostic platforms. The newly released TriMark Publications' report (9) indicates that the global POC testing market reached $7.7 billion in 2010 and is growing at 7% per year. During 2010, 950 million POC tests were carried in US hospitals and the annual POC test number will grow to 1.5 billion by 2012. Presently, the vast majority of commercial POC testing is focused on blood glucose testing and other LF POC testing platforms. During the same period, academic research has been very active in developing other (non-capillary) fluidic platforms leading to a wide arsenal of fluidic handling techniques for non-LF platforms, including molecular diagnostic techniques. We will conclude our review with a summary of the observed present and evolving trends in the development and commercialization of POC microfluidic diagnostic devices.

2. Microfluidic Technologies

Microfluidic diagnostics uses microfluidic technologies to accomplish a predetermined set of operations (i.e., to bring the sample and reagents together, to add buffer, to implement wash, to facilitate the readout, etc.) required by the specific biochemistry of the tests and detection techniques. This section lists an arsenal of available microfluidic techniques and the next section describes commercial and developing POC diagnostic platforms followed by analyses on how microfluidic technologies are used in these tests. We categorize most of the microfluidic techniques according to the force employed for fluid propulsion (such as pressure-driven flow, electromagnetically driven flow, etc.). It should be noted that besides the main force used for fluid propulsion, other forces might also be utilized for specific fluidic operations on the same platform (gating, separation, etc.). For instance, while centrifugal fluidic platforms rely on a centrifugal pseudo-force as the principle means for fluid propulsion, other forces are employed as well, for example, capillary forces are used in valving, and electromagnetic forces are employed in the cell lysis process on the same platform. In addition to the classification of fluidic devices based on the predominant fluid propulsion force, it is also possible to classify fluidic devices according to the type of flow employed on the platform—i.e., whether there is a continuous flow or the so-called segmented flow (where fluid is advanced in discrete packets or droplets). We will see that segmented flows (also called droplet microfluidics), an extremely important emerging technology, can be achieved on a variety of platforms, for example, centrifugal platforms, pressure-driven platforms, electromagnetically driven flows, etc. Thus, we will consider "droplet microfluidic devices" as a distinct category of microfluidic devices.

2.1. Capillary Flow Devices

When the size of the fluidic channels is reduced to hundreds of microns and below, surface forces (rather than body forces such as gravitational forces) start to dominate the behavior of fluid systems. For example, aqueous solutions whisk along hydrophilic capillary walls (such as a piece of fleece) with the fluid advancing through the hollow capillary (or along the interfiber spaces) without any applied pressure. This technology is very appealing as it does not require external pumps. Capillary flow devices such as LF immunoassays or blood glucose test strips are the most successful commercial microfluidic diagnostic platforms existing today. While capillary diagnostic test devices tend to be inexpensive and widely accepted as POC testing platforms, complex tests where multiple steps such as mixing, dilutions, washing, etc. are required are difficult to adapt to them. However, as we will see later, there are some efforts under way to introduce multiplexing on capillary platforms. It is expected that there will be a strong continued push to increase the variety of POC tests performed using capillary fluidic diagnostic platforms.

2.2. Platforms with Applied Lateral Pressure

In this type of pressure-driven fluidic device, external pumps (or various ingenious built-in micro-pumps) (10, 11) are used to drive fluids (samples, reagents) through the system. This type of platform is very flexible as many fluidic operations such as mixing, valving, metering, separation, etc. have been developed over the years (12). Flow of liquids in microchannels is a low-Reynolds number process, with a consequence that flow is laminar and mixing between two liquid streams coming together happens by diffusion across the liquid/liquid interface. This process is relatively slow and many types of mixers (both active and passive) have been developed to promote more effective mixing (13, 14), including mixers designed to promote chaotic advection to create "folds" in the fluid in order to decrease the effective diffusion distances (15–17). Other fluidic operations utilize a laminar flow profile to separate and sort cells in microchannels (18, 19). The advantage of this type of platform is that a wide variety of fluidic operations are available and this platform is used extensively in academic research. Disadvantages include the presence of pumps, the need for fluidic connections to the pump, and relatively large dead volumes.

2.3. Platforms with Applied Transverse Pressure

This type of platform uses transverse pressure, to propel, or to stop the fluid flow. The microchannels and reservoirs are made of soft plastic that can be squeezed or pinched off by the external pressure caused by fluid flowing in the adjacent channels (see Fig. 1) or by other transverse forces. The material choice for this platform is limited to soft elastomers (20). This platform makes it possible to create very large fluidic networks (21) and thus might be an attractive choice for drug screening and other high-throughput applications.

2.4. Centrifugal Fluidics Platforms

On the centrifuge-based fluidics platforms (often manufactured in the shape of a disk and called Lab-on-a-CD (22)), the centrifugal pseudo-force directs the flow of the fluid from the reservoirs placed close to the center of the disk to the reservoirs located near the edge of the disk. A motor is required for the disk rotation, but there is no need for external pumps, thus no need for a fluidic connection between the pump and the fluidic platform. The valving on the CD platform is accomplished either based on the so-called passive valving, relying on the interplay between centrifugal forces (dependent on the rotation speed) and the capillary forces (dependent on channels' material and geometry), or based on active valving, where some form of external actuation is utilized (for example, the use of an infrared focused light source to melt wax plugs (23)). A wide variety of fluidic functions are available (including valving, mixing, aliquoting, blood fractionation, and cell lysis), making CD devices a very appealing sample-to-answer fluidic platform for diagnostic applications (24, 25).

2.5. Electrically and Magnetically Driven Flows

Electrically and magnetically driven flows are governed by such processes as electrophoresis, electro-osmosis, dielectrophoresis, electrowetting, and ferrohydrodynamics. An *electro-osmotic flow* is produced as a result of fixed charges present on the surface of microfluidic channel walls. These surface charges cause a charge separation within the solution near the channel walls and the formation of an electrical double layer (EDL). When an electric field is then applied along the length of the microchannel, the mobile charges within the EDL will be swept towards the oppositely

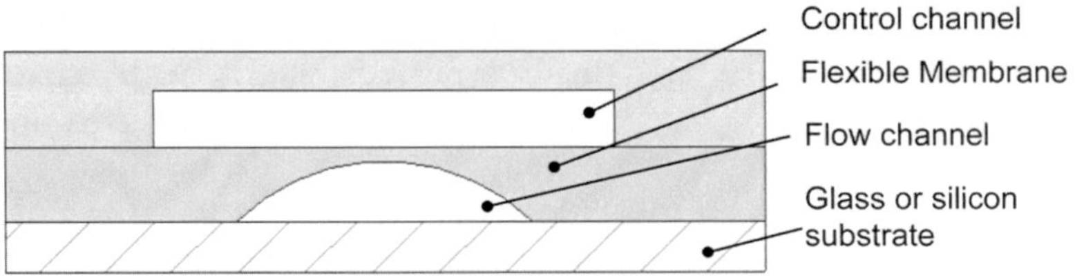

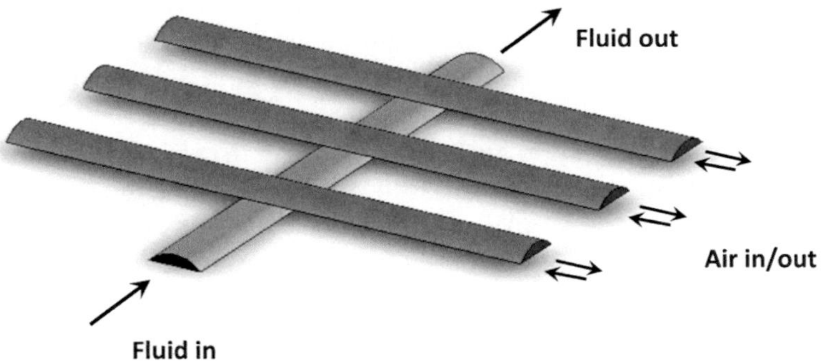

Fig. 1. A two-layer polydimethylsiloxane (PDMS) push-down microfluidic valve. An elastomeric membrane is formed where the flow channel is positioned orthogonal to the control channel directly above. Adapted from ref. 20.

charged electrode, moving fluid with them. Electro-osmosis is most pronounced in channels of several hundred microns in diameter and smaller, since in these geometries the EDL occupies a considerable part of the channel's cross section (26). *Electrophoresis* is the motion of charged molecules and particles under the influence of a spatially uniform electric field—it is used extensively to separate and purify charged macromolecules such as proteins and nucleic acids (27). Usually electrophoretic separation is accompanied by electro-osmosis, and these two processes are summarized under the heading of *electrokinetic flow* (28). One of the advantages of the electrokinetic flow is the plug-like (non-parabolic) velocity profile that helps to avoid dispersion of analytes or reagents within the fluid. *Dielectrophoresis* can be used to transport, trap, separate, and sort different types of particles and cells based upon their polarizability in nonuniform electric fields (29). *Electrowetting* is a modification of the wetting properties (contact angle) of dielectric surfaces caused by an applied electric field—i.e., a hydrophobic surface can become hydrophilic and vice versa, with the effect being reversible with a change of the applied voltage. A coordinated change in the hydrophobicity of several adjacent pads can cause a droplet to move from one pad to another and thus, it is possible to program the movement of a fluid in a complex pattern by applying potential to (and thus changing the contact angle of) individually addressable pads (30). Some of the newest fluidic devices employ *ferrofluids*—a suspension of magnetic particles that is moved around by switching on and off electromagnets underneath different regions of the fluidic platform. Magnetic particles drag fluid with them and thus, similarly to the electrowetting application, droplets can be moved, merged, and separated with ease (31).

2.6. Acoustically Driven Flows

Surface acoustic waves (SAWs) that travel over the surface of a substrate and propel a liquid droplet in the direction of the wave propagation have been used in microfluidic applications to generate, propel, mix, and break up liquid droplets (32). An attractive feature of acoustically driven flow is that in the MHz frequency range, sessile drops on a piezoelectric surface can be propelled at velocities as high as 1–10 cm/s. Enhanced mixing and microcentrifugation can be achieved within individual droplets to separate or mix different phases on a microscale (33). However, open architecture of SAWs might not work well for molecular diagnostic POC applications as the heating step during the polymerase chain reaction (PCR) step will cause evaporation and therefore, additional measures should be implemented to prevent evaporation, further complicating the setup.

2.7. Segmented Flow (Droplet Microfluidic[1]) Devices

In chemical and biochemical processes it can be advantageous to have a fluid travel in separate packets or droplets, either as single droplets traveling on a surface or between two planes (as for example, in the electrowetting processes described above) or as droplets of one phase separated and carried by another fluid (of different phase) that can be generated on centrifugal or pressure-driven platforms. These droplets can carry samples and they can be combined with other droplets, carrying buffers, washes, reagents, etc. to implement very complex multistep chemical or biochemical processes (such as molecular diagnostics) on a compact platform. This is also a powerful way to multiplex the analysis as each droplet (out of virtually millions) can carry a different type of reagent—thus, for example, it would be a very flexible platform for drug screening or process optimization. The dispersion of the sample in the microchannel can be minimized if the sample is moved within a droplet. Droplets can also serve as protective vehicles for drug delivery (34), as microreactors where reaction speeds will be very fast due to the short diffusion distances (35), and in many other applications (36). The toolset for creating droplets, transporting droplets, combining droplets, splitting droplets, putting cells within droplets, or droplets within droplets has been developed on a wide variety of platforms (including pressure driven, centrifugal, electrokinetic, etc.) (37, 38).

[1]The term "digital microfluidics" is also used.

3. Types of Point-of-Care Testing Platforms

Microfluidic diagnostic technologies can be separated into lab-based testing technologies/devices and POC testing platforms. Both types of platforms utilize similar microfluidic handling processes as outlined above, but there are also significant differences. First, lab-based testing does not need to be completely integrated (i.e., sample preparation steps can be performed on separate platforms and then pipetting robots can transfer metered sample between different stations) and second, there is no need to limit the overall weight/size of the testing platform and readout equipment. In contrast, POC testing platforms require a significant integration, where sample preparation steps as well as compact detection technologies can present additional problems for the creation of compact, inexpensive, portable (or desktop) diagnostic systems. There are a number of different tests that can be conducted on POC diagnostic platforms: blood glucose monitoring, blood gas and electrolytes, rapid coagulation tests, drugs of abuse screening, pregnancy and fertility tests, fecal occult blood testing, hemoglobin testing, cancer marker testing, cholesterol testing, infectious disease testing, etc. We will consider these tests from the microfluidic (not biological or biochemical) perspective. Depending on the complexity of the biochemical steps, some of these tests are performed without dilution and washing steps and are adapted directly to LF platforms—the most widespread and commercially successful type of POC testing to date. Another type of platform for POC testing is the so-called "handheld" and "desktop" POC test stations that can perform multiple assays from the same sample (multiplex) or do quantitative automated multiple tests (such as blood gases and blood chemistry in addition to immunoassays). The final type of the POC diagnostic platform is molecular biology-based testing, which represents the fastest-growing POC testing segment, and even though there are currently very few molecular diagnostic POC tests available commercially it is a subject of very active academic interest and we should expect new products on the market in the near future. This review is not exhaustive and it does not cover some POC testing devices such as in vivo diagnostic sensors.

3.1. Strip-Based POC Testing Platforms: Lateral Flow Tests

3.1.1. Lateral Flow POC Diagnostic Technologies

Strip-based tests that use capillary forces are among the most ubiquitous and commercially successful POC tests. These tests can roughly be separated into two categories—qualitative tests that identify the presence or absence of a particular analyte in the sample and quantitative (or semiquantitative) tests that typically also use readout devices.

An example of a qualitative LF test is the pregnancy test. A typical over-the-counter pregnancy test is an immunoassay where the presence or absence of a specific protein (such as human chorionic

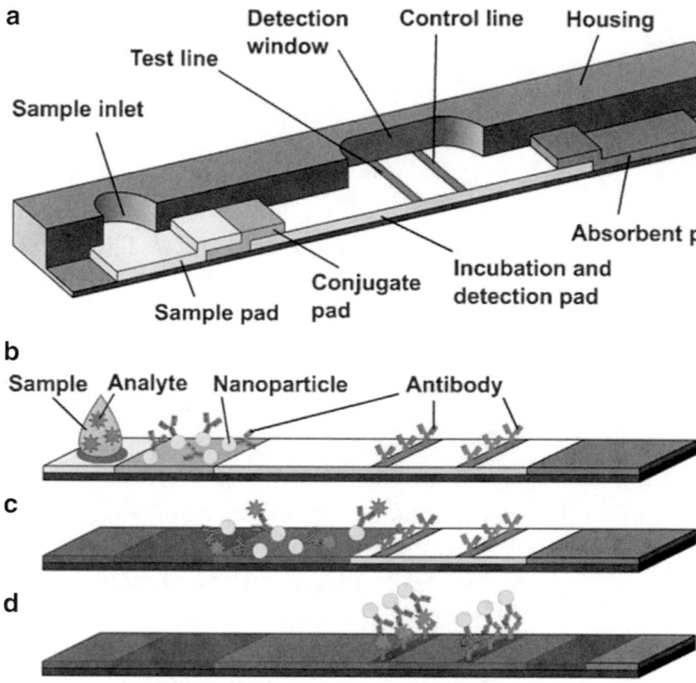

Fig. 2. Schematic design of a lateral flow test: (**a**) sample pad (sample inlet and filtering), conjugate pad (reactive agents and detection molecules), incubation and detection zone with test and control lines (analyte detection and functionality test), and final absorbent pad (liquid actuation); (**b**) start of assay by adding liquid sample; (**c**) antibodies conjugated to colored nanoparticles bind the antigen; (**d**) particles with antigens bind to test line (positive result), particles without antigens bind to the control line (proof of validity). (Reproduced from ref. 2 with permission from RCS Publishing).

gonadotrophin (hCG)) is detected based on the interaction of antibodies immobilized on a substrate with antigens in the sample. Figure 2 (2) presents a schematic design of one type of an LF immunoassay test that utilizes immunochromatography for the detection of a specific protein. While the sample passes over the conjugate pad, the antigens in the analyte form complexes with monoclonal antibodies conjugated to latex or gold nanoparticles or to an enzyme. The enzyme by reacting with multiple substrate molecules provides for optical signal amplification (absorption or fluorescence) by activating multiple dye molecules in a substrate. Capillary action continues to carry antibody–antigen pairs (with the attached tags) over the detection pad where a *test line* contains an array of immobilized polyclonal antibodies specific to the same antigen. The antigen (that is already paired with the tagged antibody) attaches itself to the immobilized polyclonal antibody, producing an antibody–antigen–antibody sandwich that is anchored on the test line and thus change of the test line can be observed directly or with fluorescent microscope (depending on the type of tag). The *control line* contains anchored immunoglobulin G (IgG) that binds to a nonspecific region of the antibodies and therefore, tagged antibodies will also be captured on the control line in a properly functioning LF immunoassay test unit.

The most popular *quantitative* LF test is utilized in POC systems for home blood glucose level measurements. Typically, a drop of blood is placed on a test strip where capillary action carries the sample to a region of the strip where an enzyme (such as glucose oxidase, GOx) is embedded. The reaction of glucose with oxygen in the presence of GOx can be detected electrochemically in a handheld reader, e.g., by amperometric detection of the peroxide reaction product, where the current is proportional to the patient's blood sugar level.

Quantification in other LF tests is accomplished with automated detection using reflectometry. For example, Metrika, Inc. (now with Bayer HealthCare) has introduced such a portable device for measurement of HbA1c that was approved by the FDA for home use (39). Similarly, the RAMP™ System (from Response Biomedical Corp.) provides a quantitative measurement of cardiac markers to assist in the diagnosis of a heart attack (40). In the latter LF device immunochromatography is used with fluorescently labeled latex particles for detection as well as for internal calibration to measure the levels of Troponin I and CK-MB in a whole blood sample in less than 15 min.

3.1.2. Evolving Trends in Lateral Flow Testing Platforms

Evolving trends in LF testing include the use of more sensitive and selective recognition elements. Examples include: nucleic acid hybridization-based LF devices (41) or a combination of antibody–antigen recognition with nucleic acid hybridization in nucleic acid LF immunoassays (42); and utilization of advanced labels such as resonance-enhanced absorption (43), chemiluminescence (44), up-converting phosphors (45), silver-enhanced gold nanoparticle labels (46), etc. Other trends involve developing a larger number of immunoassays (47), advancing quantification of the detection (48), and quality control in manufacturing to increase reproducibility of tests (49). In order to further increase the test sensitivity for LF assays some manufacturers have introduced microfabricated posts and grooves in the fluidic microchannels to better control fluid flow through the system (see Fig. 3) (50, 51). Some of the immunoassays incorporate wash steps and thus cannot be in the format of LF (immuno) assays. Implementation of more sensitive detection techniques leads to the creation of more expensive and complex automated POC desktop platforms that we consider below.

In addition to the goal of developing more sensitive tests, there is another pressing need—that of developing inexpensive, easy-to-use, disposable POC diagnostic tests for the resource-limited settings of developing countries (52, 53). The latest area of intense research activity is in the simplification of LF devices to their bare

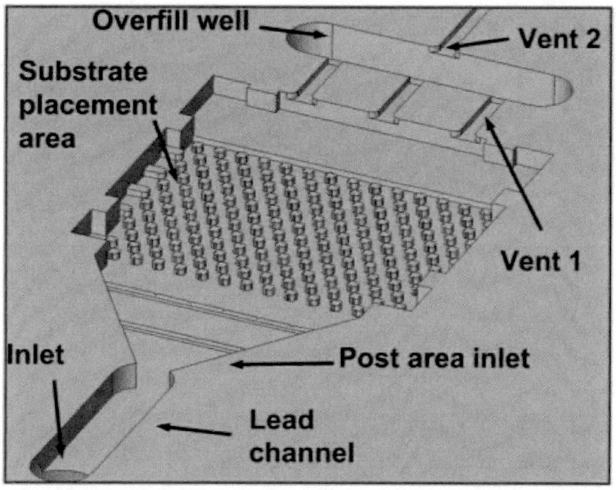

Fig. 3. An array of microfabricated features such as 50 μm posts and micromachined grooves controls fluid flow on the LF immunoassay platform. (Reproduced from ref. 50 with permission from American Association for Clinical Chemistry, Inc.).

ingredients—patterned filter paper impregnated with reagents (the so-called bioactive paper (54, 55)) forms a basis of simple microfluidic paper-based analytical devices (μPADs) (56–58). This extremely simple test platform was used in proof-of-concept demonstration for urinalysis, for quantitative colorimetric detection, for multiplexing, and even in constructing a simple filter paper and adhesive tape-based, multilayer three-dimensional fluidic network (see Fig. 4) (59). In addition, fast prototyping and production techniques for paper-based diagnostic devices have been proposed (60).

3.2. Automated Handheld and Desktop POC Platforms

3.2.1. Handheld and Desktop POC Diagnostic Technologies

So far we have considered some fairly simple one-step assays (mostly LF immunoassays and glucose test strips) that are either qualitative and do not require a readout device or semiquantitative and only need a simple readout device (e.g., a reflectometer).

Now we will discuss multistep POC tests that are somewhat more complex and require a desktop station (or, in a few cases, a handheld device) that will automate sample preprocessing, assay steps, and detection. While LF tests mostly use capillary forces for fluid transport, the handheld/desktop POC platforms employ a full range of microfluidic technologies—from centrifugal fluidics to electrokinetic flows.

Many desktop POC platforms have a set of microfluidic cartridges for various types of tests (e.g., blood gases and electrolytes, protein markers, etc.). For example, one of the most versatile platforms, namely, the i-STAT Analyzer (Abbott Point-of-Care) (61, 62), is a handheld device that has separate fluidic cartridges for blood gases and blood electrolytes, lactate, coagulation, hematology, and cardiac markers. Several drops of whole blood are dispensed

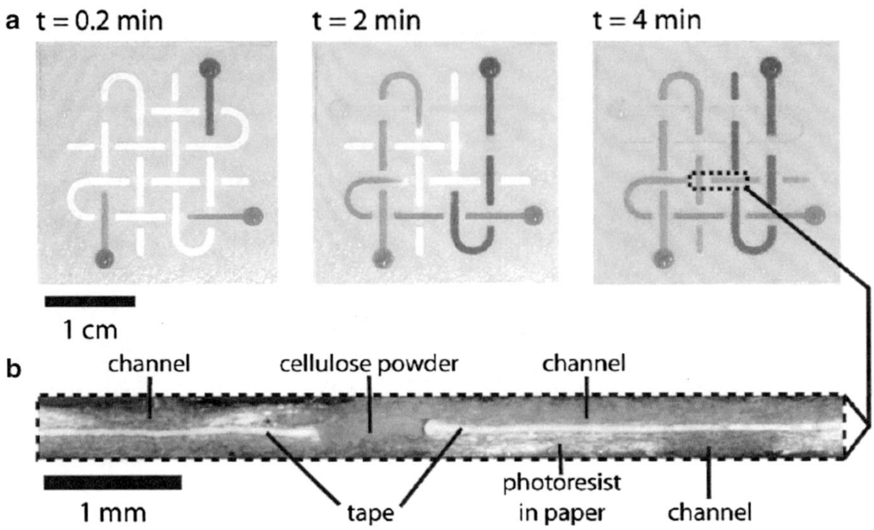

Fig. 4. Three-dimensional paper-tape stack demonstrating possibility for fluidic networks in paper-based analytical devices. (Reproduced from ref. 59 with permission from National Academy of Sciences, USA).

into the cartridge, which is then sealed and inserted into the analyzer. The self-testing and calibration routine is initiated when a barb punctures a sealed reservoir containing the calibration solution, which washes over the sensor array. The fluidic control is based on an applied transverse pressure, when the analyzer device presses the internal air pouch that, in turn, displaces the calibrant into the waste reservoir and sends the blood sample into the sensing chamber. The diagnostic tests of the i-STAT Analyzer are performed via amperometric or potentiometric detection on the thin-film biosensor array. There are many other commercial handheld and desktop analyzers on the market, but in contrast with the i-STAT device, the majority of manufacturers produce stand-alone blood gas analyzers, stand-alone immunoassay testing platforms, or two (or more) separate instruments for various types of tests. For example, the Biosite Triage® MeterPlus specializes in immunoassay test panels for cardiac markers (63) (see Fig. 5), while Roche has a dedicated Cobas b221 POC blood analyzer and a separate handheld CoaguCheck coagulation monitor (64); Siemens produces POC RapidPoint blood gas analyzers, another POC test station Stratus® CS Acute Care™ Diagnostic System for cardio markers, and yet another CLINITEK platform for urinalysis (65).

Centrifugal fluidics serves as a basis for several POC diagnostics applications: Abaxis' Piccolo xPress provides whole blood analysis on a disk (66), Gyros' Bioaffy CD are designed for POC diagnostics and development of immunoassays (67, 68); Samsung has developed an immunoassay from whole blood on a disk with active valving executed by laser-irradiated ferrowax microvalves (69); IMTEK's researchers have designed an integrated disk with built-in

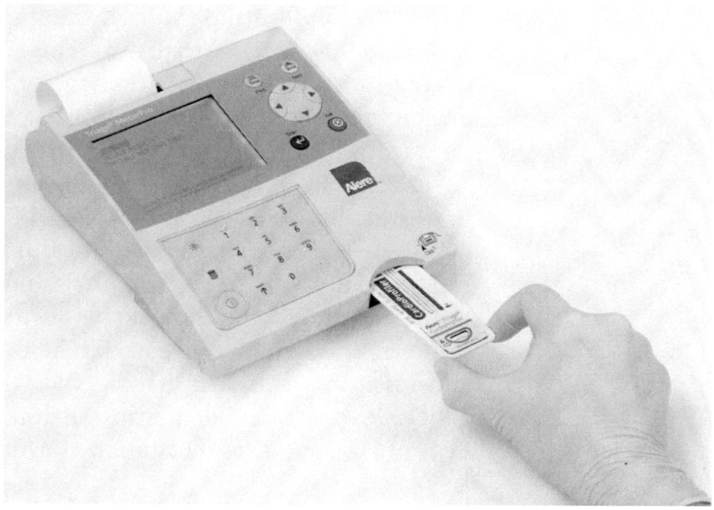

Fig. 5. Biosite's POCT Desktop Platform Triage® MeterPlus with a test cartridge. (Reproduced with permission from Alere San Diego, Inc.).

total internal reflection surfaces to optimize rapid determination of alcohol level in whole blood (70). One of the attractive features of the centrifugal platform for POC testing is that the sample preparation steps can be integrated seamlessly within the device—for example, whole blood fractionation can be achieved by centrifugation (for example, just the plasma fraction can be selected for subsequent processing) (71); also efficient mechanical cell lysis has been developed on CD platform (72, 73).

3.2.2. Evolving Trends in Handheld and Desktop POC Diagnostic Platforms

The field of micro Total Analysis Systems (μTAS) or LOC is quite mature with new detection techniques being developed constantly (74, 75). We envision incorporation and adaption of μTAS and LOC technologies to accelerate detection rates and sensitivities in POC diagnostic platforms, including a range of tools and processes such as employment of electrokinetic techniques (27, 76, 77), use of magnetic beads (78, 79), etc. A new generation of biosensors will be developed (80, 81) (especially those relying on electrochemical detection, as evidenced by the success of i-STAT platform discussed above) and novel detection technologies (such as optofluidics (82, 83) and label-free detection (84)) will be implemented on POC testing platforms. Presently "LOC" strongly resembles "Chip on a Lab" where all of the processing steps are conducted on a small chip with tiny amounts of sample and reagent, but to achieve a good sensitivity, powerful and bulky detection devices are used—thus the overall size of the POC desktop platform is often dominated by the detection device. The miniaturization trend will continue, especially for detection systems, in order to decrease the footprint of the POC diagnostic platforms. The POC testing fluidics propulsion systems will become even less power-demanding and multiplexing technology (such as barcodes) will become more widespread. This last set of trends is demonstrated by the recent development of an LF chip for multiplexed protein detection (85). The information collection, sharing, computerization of patient files, and integrity of test data are receiving an increased attention as an area for potential cost savings and improvement of health care. In order to store patient data to facilitate POC test data sharing with health providers, and to avoid possible mix-ups of the test results, almost all of the newest POC testing handheld and desktop platforms have built-in data storage, printers, and/or special barcode labels (with patient information and test data) and they have dedicated communication ports for Ethernet or wireless data transmission to Laboratory Information System (LIS) and/or Hospital Information System (HIS) (86).

3.3. Molecular Diagnostic POC Platforms

3.3.1. Molecular Diagnostic Technologies for POC Applications

There are two primary reasons why molecular diagnostic detection is growing in demand despite the fact that there is an extensive list of available immunoassays. The first reason is the so-called seroconversion—a period that is needed for our body to accumulate enough of the antibodies against a specific antigen to be detectable via immunoassays. This delay (for immunoassay detection) in the case of many infectious diseases has serious consequences such as permanent damage or even death and the possibility that an undiagnosed virus carrier will continue to infect other people. The second reason is amplification that increases test sensitivity—nucleic acids can be amplified via PCR technology, but proteins cannot be amplified.

The field of academic research in the area of molecular diagnostics systems is too extensive to be discussed here, even superficially, and we invite interested readers to look at the recent reviews on the subject (87–90). Instead of a survey of academic research on POC molecular diagnostic systems, we will describe several molecular diagnostic platforms that are commercially available or under development. These technologies use disposable units (cartridges or pouches) with pre-packed reagents for sample preparation as well as a non-disposable portable closed system for housing permanent components including electronics, pumps, and controls for processing assays.

Cepheid's GeneXpert® (91) (represented in Fig. 6) is a real-time PCR-based DNA analysis system incorporating sample preparation (DNA extraction), real-time DNA amplification, and detection. This modular system has multiple bays allowing it to simultaneously analyze multiple samples, with the degree of multiplexing limited to its five fluorescent channels. The system provides "sample-to-answer" results from an unprocessed sample in about 1 h. The disposable cartridge incorporates the body for housing preloaded reagents, a valve that moves the sample and reagents to the active area located in the base of the cartridge by rotating to different

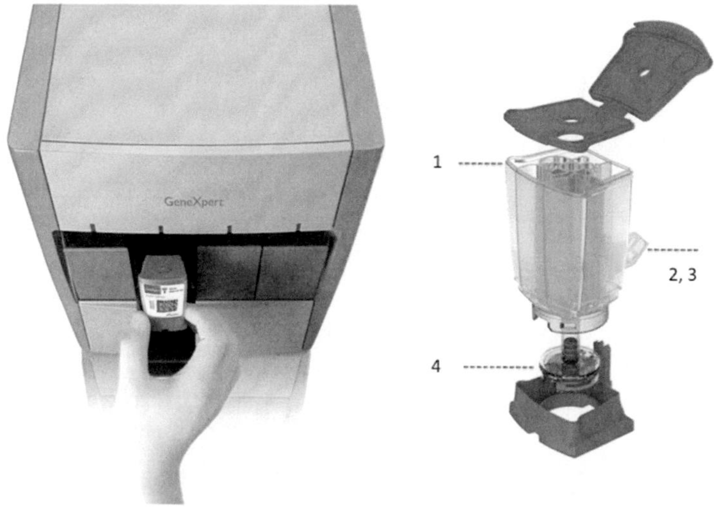

Fig. 6. GeneXpert® platform and disposable cartridge with components such as (*1*) processing chambers that contain reagents, filters, and capture technologies necessary to extract, purify, and amplify target nucleic acids; (*2*) optical window that enables real-time optical detection; (*3*) reaction tube for rapid thermal cycling; and (*4*) valve that enables fluid transfer from chamber to chamber and may contain nucleic acid lysis and filtration components. (Reproduced with permission from Cepheid).

positions, and a reaction tube for performing real-time PCR (RT-PCR). The cell lysis is carried out by ultrasonic energy generated by a sonic horn that acts upon the base of the cartridge where the active area is located. The extracted DNA solution is subsequently washed, purified, and concentrated by moving through different chambers, and then advanced to the reaction chamber where RT-PCR takes place. The platform utilizes an integrated fluorescence detection system. The final results are processed using dedicated software. A barcode system is used to store related information about each test. Other advantages of the GeneXpert® system are its small footprint and low power consumption.

Iquum's Liat (Lab-in-a-tube) *System* (92), composed of an analyzer and a disposable tube, processes a single sample with a complete "sample-to-answer" time (that includes a sample preparation step) of 30–60 min. Raw sample, such as whole blood, plasma, urine, or a swab, is placed within a tube that is inserted into the analyzer where the RT-PCR process takes place. The fluidic transport during assay processing is directed by actuators applying a transverse pressure to collapsible compartments of the tube to displace liquid in the adjoining reservoir. Magnetic beads (controlled by a built-in magnet) are used to bind and transport nucleic acid of the sample. One notable advantage of the Liat tube system is that all assay steps are performed inside the tube that contains all the reagents and a sample. After the test is complete, the tube containing the biohazardous waste is safely disposed of.

Enigma ML (93) or "minilab" is a modular multiplexed RT-PCR system for analyzing both DNA and RNA targets from different body liquids or swabs. The results are obtained in 30–45 min.

The Idaho Technology's FilmArray (94) is also a single sample fluorescence-based system for simultaneous detection of 18 viral and three bacterial pathogens. All reagents required to perform the assay are preloaded and lyophilized in a disposable pouch. No precise sample metering prior to testing is required; the disposable pouch with a built-in vacuum system draws in the required volume upon injection of a sample. Cell lysis is performed using a bead beating process. The released nucleic acids are then bound to magnetic beads and are transferred to the purification chamber where a wash buffer removes all the debris while the nucleic acids remain bound to the beads. An elution buffer separates the nucleic acids from the magnetic beads and flows them into the PCR chamber where they are amplified through a two-stage PCR process: during the first stage of PCR, the FilmArray performs a single, large-volume, multiplexed reaction; in the second stage PCR is performed in small wells, each of which contains a primer designed to detect one specific target. Image processing software is used for analysis of the fluorescence intensity in each well. Results from a raw sample are obtained in 1 h.

In all four systems described here, barcode scanning is used for recording of information about the test and the pouch ID. Table 1 presents a quick comparison between the features of the four discussed molecular diagnostic POC platforms.

3.3.2. Evolving Trends in POC Molecular Diagnostic Platforms

There are a growing number of companies developing POC Molecular Diagnostic Platforms with both PCR as well as RT-PCR technologies. The main difference between these two nucleic acid amplification technologies is that, even though implementation of RT-PCR is somewhat easier, at the moment there is a limit to the multiplexing capacity of solution-based RT-PCR because of the

Table 1
Comparison between four POC molecular diagnostic platforms

Company	System	Hands-on time (min)	Run time (min)	Multiplex capacity	Throughput	Status
Cepheid	GeneXpert	2	45	5	Modular (48 samples)	FDA cleared—available commercially
Iquum	Liat	5	30	6	One sample	In development
Enigma	Enigma ML	1	30	6–12	Modular (six samples)	In development
Idaho Technology	FilmArray	5	60	20+	One sample	In development

limited number of distinct types of fluorophores (five or six), while for PCR with microarrays there is almost unlimited multiplex capacity. Future developments will emphasize the adaptation of nucleic acid amplification and detection technologies demonstrated on LOC platforms (for example, digital fluidics (95), electrophoresis (96), and other technologies (97)) to POC molecular diagnostic devices. We anticipate more development and implementation of the isothermal amplification strategies (98) and of novel detection techniques such as bioluminescence (99). It is also expected that, as molecular diagnostic technologies develop higher sensitivity, much less intrusive (but also less concentrated) types of samples such as saliva, sweat, or tears will be used more often.

4. Conclusions

Microfluidic diagnostics for POC testing has the potential to reduce health care costs and improve patents' outcomes. It is likely that government as well as health insurance providers will help in the wider adaptation of POC tests, through more favorable coverage and reimbursement policies and the encouragement of widespread use in doctors' offices. We can summarize evolving general trends in POC diagnostics as miniaturization (especially needed for detection systems), multiplexing, and networking (collection, storage, and transfer of data). Present demands are in the areas of developing POC testing platforms for resource-limited environments (53); in automation, simplification, and integration of sample preparation and preprocessing steps (100–102); and also in transfer of the wide range of tests performed in the academic setting based upon LOC formats into commercial POC diagnostic platforms.

It was observed that the field of microfluidics, which initially showed a great promise to be used in a large number of unique and useful applications, has never quite found a commercially successful, ubiquitous "killer" application (103, 104). It is feasible that POC diagnostics might just be that "killer" application (or range of applications) for the microfluidic toolbox.

Acknowledgments

This work was supported by the National Science Foundation grants ECCS-0801792 and NIRT-0709085, National Institute of Health grant 1 R01 AIO89541-01, and UC Lab Fees Award 09-LR-09-117362 and sponsored by World Class University (WCU) program (R32-2008-000-20054-0) through the National Research Foundation of Korea funded by the Ministry of Education, Science and Technology.

References

1. Madou MJ (2002) Fundamentals of microfabrication: the science of miniaturization. CRC Press, Boca Raton, FL
2. Mark D, Haeberle S, Roth G, von Stetten F, Zengerle R (2010) Microfluidic lab-on-a-chip platforms: requirements, characteristics and applications. Chem Soc Rev 39:1153–1182
3. Tudos AJ, Besselink GAJ, Schasfoort RBM (2001) Trends in miniaturized total analysis systems for point-of-care testing in clinical chemistry. Lab Chip 1:83–95
4. Posthuma-Trumpie GA, Korf J, van Amerongen A (2009) Lateral flow (immuno) assay: its strengths, weaknesses, opportunities and threats. A literature survey. Anal Bioanal Chem 393:569–582
5. Nghe P, Terriac E, Schneider M, Li ZZ, Cloitre M, Abecassis B, Tabeling P (2011) Microfluidics and complex fluids. Lab Chip 11:788–794
6. Squires TM, Quake SR (2005) Microfluidics: fluid physics at the nanoliter scale. Rev Mod Phys 77:977
7. Haeberle S, Zengerle R (2007) Microfluidic platforms for lab-on-a-chip applications. Lab Chip 7:1094–1110
8. Dittrich PS, Tachikawa K, Manz A (2006) Micro total analysis systems. Latest advancements and trends. Anal Chem 78:3887–3907
9. Point of Care Diagnostic Testing World Markets (2012) TriMark Publications, LLC http://www.trimarkpublications.com/products/Point-of-Care-Diagnostic-Testing-World-Markets.html. Accessed 12 Nov 2012
10. Laser DJ, Santiago JG (2004) A review of micropumps. J Micromech Microeng 14:R35
11. Nisar A, Afzulpurkar N, Mahaisavariya B, Tuantranont A (2008) MEMS-based micropumps in drug delivery and biomedical applications. Sensor Actuat B Chem 130: 917–942
12. West J, Becker M, Tombrink S, Manz A (2008) Micro total analysis systems: latest achievements. Anal Chem 80:4403–4419
13. Mansur EA, Ye M, Wang Y, Dai Y (2008) A state-of-the-art review of mixing in microfluidic mixers. Chin J Chem Eng 16:503–516
14. Nam-Trung N, Zhigang W (2005) Micromixers—a review. J Micromech Microeng 15:R1
15. Tekin H, Sivagnanam V, Ciftlik A, Sayah A, Vandevyver C, Gijs M (2011) Chaotic mixing using source–sink microfluidic flows in a PDMS chip. Microfluid Nanofluid 10:749–759
16. Casadevall i Solvas X, Lambert RA, Kulinsky L, Rangel RH, Madou MJ (2009) Au/PPy actuators for active micromixing and mass transport enhancement. Micro Nanosyst 1:2–11
17. Stroock AD, Dertinger SKW, Ajdari A, Mezić I, Stone HA, Whitesides GM (2002) Chaotic mixer for microchannels. Science 295:647–651
18. Andersson H, van den Berg A (2003) Microfluidic devices for cellomics: a review. Sensor Actuat B Chem 92:315–325
19. Yi C, Li C-W, Ji S, Yang M (2006) Microfluidics technology for manipulation and analysis of biological cells. Anal Chim Acta 560:1–23
20. Unger MA, Chou HP, Thorsen T, Scherer A, Quake SR (2000) Monolithic microfabricated valves and pumps by multilayer soft lithography. Science 288:113–116
21. Thorsen T, Maerkl SJ, Quake SR (2002) Microfluidic large-scale integration. Science 298:580–584
22. Madou M, Zoval J, Jia G, Kido H, Kim J, Kim N (2006) Lab on a CD. Annu Rev Biomed Eng 8:601–628
23. Abi-Samra K, Hanson R, Madou M, Gorkin Iii RA (2011) Infrared controlled waxes for liquid handling and storage on a CD-microfluidic platform. Lab Chip 11:723–726
24. Gorkin R, Park J, Siegrist J, Amasia M, Lee BS, Park J-M, Kim J, Kim H, Madou M, Cho Y-K (2010) Centrifugal microfluidics for biomedical applications. Lab Chip 10:1758–1773
25. Jens D et al (2007) The centrifugal microfluidic bio-disk platform. J Micromech Microeng 17:S103
26. Gaudioso J, Craighead HG (2002) Characterizing electroosmotic flow in microfluidic devices. J Chromatogr A 971:249–253
27. Dolnik V, Liu S (2005) Applications of capillary electrophoresis on microchip. J Sep Sci 28:1994–2009
28. Bousse L, Cohen C, Nikiforov T, Chow A, Kopf-Sill AR, Dubrow R, Parce JW (2000) Electrokinetically controlled microfluidic analysis systems. Annu Rev Biophys Biomol Struct 29:155–181
29. Hunt TP, Issadore D, Westervelt RM (2008) Integrated circuit/microfluidic chip to programmably trap and move cells and droplets with dielectrophoresis. Lab Chip 8:81–87
30. Lee J, Moon H, Fowler J, Schoellhammer T, Kim C-J (2002) Electrowetting and electrowetting-on-dielectric for microscale liquid handling. Sensor Actuat A Phys 95:259–268
31. Ali B, Nam-Trung N (2010) Programmable two-dimensional actuation of ferrofluid droplet using planar microcoils. J Micromech Microeng 20:015018

32. Yeo LY, Friend JR (2009) Ultrafast microfluidics using surface acoustic waves. Biomicrofluidics 3:012002–012023
33. Wixforth A (2006) Acoustically driven programmable microfluidics for biological and chemical applications. J Assoc Lab Auto 11: 399–405
34. Xu Q, Hashimoto M, Dang TT, Hoare T, Kohane DS, Whitesides GM, Langer R, Anderson DG (2009) Preparation of monodisperse biodegradable polymer microparticles using a microfluidic flow-focusing device for controlled drug delivery. Small 5:1575–1581
35. Shum HC, Bandyopadhyay A, Bose S, Weitz DA (2009) Double emulsion droplets as microreactors for synthesis of mesoporous hydroxyapatite. Chem Mater 21:5548–5555
36. Abdelgawad M, Wheeler AR (2009) The digital revolution: a new paradigm for microfluidics. Adv Mater 21:920–925
37. Teh S-Y, Lin R, Hung L-H, Lee AP (2008) Droplet microfluidics. Lab Chip 8:198–220
38. Casadevall i Solvas X, deMello A (2011) Droplet microfluidics: recent developments and future applications. Chem Commun 47:1936–1942
39. Kennedy L, Herman WH (2005) Glycated hemoglobin assessment in clinical practice: comparison of the A1cNow™ point-of-care device with central laboratory testing (GOAL A1C study). Diabetes Technol Ther 7:907–912
40. Brooks DE, Devine DV, Harris PC, Harris JE, Miller ME, Olal AD, Spiller LJ, Xie ZC (1999) RAMPTM: a rapid quantitative whole blood immunochromatographic platform for point-of-care testing. Clin Chem 45:1676–1678
41. Mens PF, van Amerongen A, Sawa P, Kager PA, Schallig HDFH (2008) Molecular diagnosis of malaria in the field: development of a novel 1-step nucleic acid lateral flow immunoassay for the detection of all 4 human *Plasmodium* spp. and its evaluation in Mbita, Kenya. Diagn Microbiol Infect Dis 61:421–427
42. Blažková M, Koets M, Rauch P, van Amerongen A (2009) Development of a nucleic acid lateral flow immunoassay for simultaneous detection of *Listeria* spp. and Listeria monocytogenes in food. Eur Food Res Technol 229:867–874
43. Assadollahi S, Reininger C, Palkovits R, Pointl P, Schalkhammer T (2009) From lateral flow devices to a novel nano-color microfluidic assay. Sensors 9:6084–6100
44. Rudolf Seitz W (1984) Immunoassay labels based on chemiluminescence and bioluminescence. Clin Biochem 17:120–125
45. Corstjens P, Zuiderwijk M, Brink A, Li S, Feindt H, Niedbala RS, Tanke H (2001) Use of up-converting phosphor reporters in lateral-flow assays to detect specific nucleic acid sequences: a rapid, sensitive dna test to identify human papillomavirus type 16 infection. Clin Chem 47:1885–1893
46. Chu X, Fu X, Chen K, Shen G-L, Yu R-Q (2005) An electrochemical stripping metalloimmunoassay based on silver-enhanced gold nanoparticle label. Biosens Bioelectron 20: 1805–1812
47. von Lode P (2005) Point-of-care immunotesting: Approaching the analytical performance of central laboratory methods. Clin Biochem 38:591–606
48. Kupstat A, Kumke MU, Hildebrandt N (2011) Toward sensitive, quantitative point-of-care testing (POCT) of protein markers: miniaturization of a homogeneous time-resolved fluoroimmunoassay for prostate-specific antigen detection. Analyst 136:1029–1035
49. Tisone TC, O'Farrell B (2009) Manufacturing the next generation of highly sensitive and reproducible lateral flow immunoassay. In: Wong R, Tse H (eds) Lateral flow immunoassay. Humana Press, New York, pp 1–26
50. Pugia MJ, Blankenstein G, Peters R-P, Profitt JA, Kadel K, Willms T, Sommer R, Kuo HH, Schulman LS (2005) Microfluidic tool box as technology platform for handheld diagnostics. Clin Chem 51:1923–1932
51. Schulte TH, Bardell RL, Weigl BH (2002) Microfluidic technologies in clinical diagnostics. Clin Chim Acta 321:1–10
52. Urdea M, Penny LA, Olmsted SS, Giovanni MY, Kaspar P, Shepherd A, Wilson P, Dahl CA, Buchsbaum S, Moeller G, Hay Burgess DC (2006) Requirements for high impact diagnostics in the developing world. Nature 444:3–8
53. Yager P, Edwards T, Fu E, Helton K, Nelson K, Tam MR, Weigl BH (2006) Microfluidic diagnostic technologies for global public health. Nature 442:412–418
54. Pelton R (2009) Bioactive paper provides a low-cost platform for diagnostics. Trends Anal Chem TRAC 28:925–942
55. Martinez AW, Phillips ST, Whitesides GM (2008) Three-dimensional microfluidic devices fabricated in layered paper and tape. Proc Natl Acad Sci USA 105:19606–19611
56. Martinez AW, Phillips ST, Whitesides GM, Carrilho E (2009) Diagnostics for the developing world: microfluidic paper-based analytical devices. Anal Chem 82:3–10

57. Abe K, Suzuki K, Citterio D (2008) Inkjet-printed microfluidic multianalyte chemical sensing paper. Anal Chem 80:6928–6934
58. Chitnis G, Ding Z, Chang C-L, Savran CA, Ziaie B (2011) Laser-treated hydrophobic paper: an inexpensive microfluidic platform. Lab Chip 11:1161–1165
59. Martinez AW, Phillips ST, Whitesides GM (2008) Three-dimensional microfluidic devices fabricated in layered paper and tape. Proc Natl Acad Sci 105:19606–19611
60. Martinez AW, Phillips ST, Wiley BJ, Gupta M, Whitesides GM (2008) FLASH: a rapid method for prototyping paper-based microfluidic devices. Lab Chip 8:2146–2150
61. Connelly NR, Magee M, Kiessling B (1996) The use of the iSTAT portable analyzer in patients undergoing cardiopulmonary bypass. J Clin Monit 12:311–315
62. Papadea C, Foster J, Grant S, Ballard SA, Cate JC IV, Southgate WM, Purohit DM (2002) Evaluation of the i-STAT portable clinical analyzer for point-of-care blood testing in the intensive care units of a University Children's Hospital. Ann Clin Lab Sci 32: 231–243
63. Sykes E, Karcher RE, Eisenstadt J, Tushman DA, Balasubramaniam M, Gusway J, Peterson VJ (2005) Analytical relationships among biosite, Bayer, and Roche methods for BNP and NT-proBNP. Am J Clin Pathol 123:584–590
64. Cosmi B, Palareti G, Carpanedo M, Pengo V, Biasiolo A, Rampazzo P, Morstabilini G, Testa S (2000) Assessment of patient capability to self-adjust oral anticoagulant dose: a multicenter study on home use of portable prothrombin time monitor (COAGUCHECK). Haematologica, 85: 826 http://www.roche.com/products/product-list.htm?region=us&type=divdia&id=Diagnostics. Accessed 12 Nov 2012
65. Guy M, Newall R, Borzomato J, Kalra PA, Price C (2009) Diagnostic accuracy of the urinary albumin: creatinine ratio determined by the CLINITEK Microalbumin and DCA 2000+ for the rule-out of albuminuria in chronic kidney disease. Clinica Chimica Acta, 399: 54-58 http://healthcare.siemens.com/point-of-care/urinalysis/clinitek-status-analyzer. Accessed 12 Nov 2012
66. Schembri C, Ostoich V, Lingane P, Burd T, Buhl S (1992) Portable simultaneous multiple analyte whole-blood analyzer for point-of-care testing. Clin Chem 38:1665–1670
67. Inganas M, Derand H, Eckersten A, Ekstrand G, Honerud A-K, Jesson G, Thorsen G, Soderman T, Andersson P (2005) Integrated microfluidic compact disc device with potential use in both centralized and point-of-care laboratory settings. Clin Chem 51:1985–1987
68. Honda N, Lindberg U, Andersson P, Hoffmann S, Takei H (2005) Simultaneous multiple immunoassays in a compact disc-shaped microfluidic device based on centrifugal force. Clin Chem 51:1955–1961
69. Lee BS, Lee JN, Park JM, Lee JG, Kim S, Cho YK, Ko C (2009) A fully automated immunoassay from whole blood on a disc. Lab Chip 9:1548–1555
70. Steigert J, Grumann M, Brenner T, Riegger L, Harter J, Zengerle R, Ducree J (2006) Fully integrated whole blood testing by real-time absorption measurement on a centrifugal platform. Lab Chip 6:1040–1044
71. Haeberle S, Brenner T, Zengerle R, Ducre J (2006) Centrifugal extraction of plasma from whole blood on a rotating disk. Lab Chip 6:776
72. Kido H, Micic M, Smith D, Zoval J, Norton J, Madou M (2007) A novel, compact disk-like centrifugal microfluidics system for cell lysis and sample homogenization. Colloids Surf B Biointerfaces 58:44–51
73. Siegrist J, Gorkin R, Bastien M, Stewart G, Peytavi R, Kido H, Bergeron M, Madou M (2010) Validation of a centrifugal microfluidic sample lysis and homogenization platform for nucleic acid extraction with clinical samples. Lab Chip 10:363–371
74. Arora A, Simone G, Salieb-Beugelaar GB, Kim JT, Manz A (2010) Latest developments in micro total analysis systems. Anal Chem 82:4830–4847
75. Abgrall P, Gué AM (2007) Lab-on-chip technologies: making a microfluidic network and coupling it into a complete microsystem—a review. J Micromech Microeng 17:R15
76. Vandaveer WR, Pasas-Farmer SA, Fischer DJ, Frankenfeld CN, Lunte SM (2004) Recent developments in electrochemical detection for microchip capillary electrophoresis. Electrophoresis 25:3528–3549
77. Vrouwe EX, Luttge R, Vermes I, van den Berg A (2007) Microchip capillary electrophoresis for point-of-care analysis of lithium. Clin Chem 53:117–123
78. Shinkai M (2002) Functional magnetic particles for medical application. J Biosci Bioeng 94:606–613
79. Zhang H, Meyerhoff ME (2005) Gold-coated magnetic particles for solid-phase immunoassays: enhancing immobilized antibody binding efficiency and analytical performance. Anal Chem 78:609–616
80. Nicu L, Leïchlé T (2008) Biosensors and tools for surface functionalization from the

macro- to the nanoscale: the way forward. J Appl Phys 104:111101
81. Leca-Bouvier B, Blum LJ (2005) Biosensors for protein detection: a review. Anal Lett 38:1491–1517
82. Monat C, Domachuk P, Eggleton BJ (2007) Integrated optofluidics: a new river of light. Nat Photon 1:106–114
83. Erickson D (2010) Optofluidics. In: Kakaç S, Kosoy B, Li D, Pramuanjaroenkij A (eds) Microfluidics based microsystems. Springer, Netherlands, pp 529–551
84. Yu X, Xu D, Cheng Q (2006) Label-free detection methods for protein microarrays. Proteomics 6:5493–5503
85. Wang J, Ahmad H, Ma C, Shi Q, Vermesh O, Vermesh U, Heath J (2010) A self-powered, one-step chip for rapid, quantitative and multiplexed detection of proteins from pinpricks of whole blood. Lab Chip 10:3157–3162
86. Blick KE (2001) The essential role of information management in point-of-care/critical care testing. Clin Chim Acta 307:159–168
87. Zhang Y, Ozdemir P (2009) Microfluidic DNA amplification – a review. Anal Chim Acta 638:115–125
88. Lee HH, Dineva MA, Chua YL, Ritchie AV, Ushiro-Lumb I, Wisniewski CA (2010) Simple amplification-based assay: a nucleic acid-based point-of-care platform for HIV-1 testing. J Infect Dis 201(Suppl 1):S65–S72
89. Lee WG, Kim Y-G, Chung BG, Demirci U, Khademhosseini A (2010) Nano/microfluidics for diagnosis of infectious diseases in developing countries. Adv Drug Deliv Rev 62:449–457
90. Lien K-Y, Lee G-B (2010) Miniaturization of molecular biological techniques for gene assay. Analyst 135:1499–1518
91. McNiven M, Talaulikar D (2012) Establishment of a conversion factor for the Cepheid GeneXpert BCR-ABL assay. Pathology, 44: 55–57 http://www.cepheid.com/systems-and-software/genexpert-system/. Accessed 12 Nov 2012
92. Coombs R, Dragavon J, Harb S (2011) Validation of a novel lab-in-a-tube analyzer and single-tube system for simple/rapid HIV-1 RNA quantification. 18th conference on retroviruses and opportunistic infections http://www.iquum.com/about/about.shtml. Accessed 12 Nov 2012
93. Tayo A, Ellis J, Phillips LL, Simpson S, Ward DJ (2012) Emerging point of care tests for influenza: innovation or status quo. Influenza and Other Respiratory Viruses, 6: 291-298 http://www.enigmadiagnostics.com/template2.php?page=instruments.php&m=5. Accessed 12 Nov 2012
94. Hammond SP, Gagne LS, Stock SR, Marty FM, Gelman RS, Marasco WA, Poritz MA, Baden LR (2012) Respiratory Virus Detection in Immunocompromised Patients with FilmArray Respiratory Panel Compared to Conventional Methods. Journal of Clinical Microbiology, 50: 3216-3221 http://www.idahotech.com/FilmArray/. Accessed 12 Nov 2012
95. Hua Z, Rouse JL, Eckhardt AE, Srinivasan V, Pamula VK, Schell WA, Benton JL, Mitchell TG, Pollack MG (2010) Multiplexed real-time polymerase chain reaction on a digital microfluidic platform. Anal Chem 82:2310–2316
96. Thaitrong N, Liu P, Briese T, Lipkin WI, Chiesl TN, Higa Y, Mathies RA (2010) Integrated capillary electrophoresis microsystem for multiplex analysis of human respiratory viruses. Anal Chem 82:10102–10109
97. Chen L, Manz A, Day PJ (2007) Total nucleic acid analysis integrated on microfluidic devices. Lab Chip 7:1413–1423
98. Notomi T, Okayama H, Masubuchi H, Yonekawa T, Watanabe K, Amino N, Hase T (2000) Loop-mediated isothermal amplification of DNA. Nucleic Acids Res 28:e63
99. Holland CA, Kiechle FL (2005) Point-of-care molecular diagnostic systems – past, present and future. Curr Opin Microbiol 8:504–509
100. Dineva MA, Mahilum-Tapay L, Lee H (2007) Sample preparation: a challenge in the development of Point-of-Care nucleic acid-based assays for resource-limited settings. Analyst 132:1193
101. Huang Y, Mather EL, Bell JL, Madou M (2002) MEMS-based sample preparation for molecular diagnostics. Anal Bioanal Chem 372:49–65
102. Kulinski MD, Mahalanabis M, Gillers S, Zhang JY, Singh S, Klapperich CM (2009) Sample preparation module for bacterial lysis and isolation of DNA from human urine. Biomed Microdevices 11:671–678
103. Blow N (2007) Microfluidics: in search of a killer application. Nat Methods 4:665–670
104. Becker H (2009) Hype, hope and hubris: the quest for the killer application in microfluidics. Lab Chip 9:2119–2122

Chapter 2

Teaching Microfluidic Diagnostics Using Jell-O® Chips

Cheng Wei T. Yang and Eric T. Lagally

Abstract

Microfluidics has emerged as a versatile technology that has found many applications, including DNA chips, fuel cells, and diagnostics. As the field of microfluidic diagnostics grows, it is important to introduce the principles of this technology to young students and the general public. The objective of this project was to create a simple and effective method that could be used to teach key microfluidics concepts using easily accessible materials. Similar to the poly(dimethylsiloxane) soft lithography technique, a Jell-O® "chip" is produced by pouring a mixture of Jell-O® and gelatine solution into a mold, which is constructed using foam plate, coffee stirrers, and double-sided tape. The plate is transferred to a 4°C refrigerator for curing, and then the Jell-O® chip is peeled off for experimental demonstrations. Three types of chips have been fabricated with different molds: a JELLO mold, a Y-channel mold, and a pH-sensor mold. Using these devices, the basics of microfluidic diagnostics can be demonstrated in one or two class periods. The method described in this chapter provides teachers with a fast and inexpensive way to introduce this technology, and students with a fun and hands-on way to understand the basics of microfluidic diagnostics.

Key words: Microfluidics, Microfluidic diagnostics, Lab-on-a-chip, Microfluidics education, Teaching methods, Jell-O microfluidics

1. Introduction

Microfluidics is a multidisciplinary field that utilizes fundamentals of physics, biology, chemistry, and engineering to create miniaturized and integrated devices for various applications, including DNA chips, biological assays, and chemical synthesis (1). Because it uses small volumes of fluid samples, microfluidics has the potential to revolutionize modern biology and medicine by significantly reducing costs and reaction times associated with an analysis (2). Many types of materials have been explored for creating microfluidic channels and chips. Because it is inexpensive, optically transparent, and biocompatible, poly(dimethylsiloxane) (PDMS) elastomer has been extensively used in microfluidics (3). Soft lithography is the common technique for fabricating PDMS microfluidic chips (4).

Gareth Jenkins and Colin D. Mansfield (eds.), *Microfluidic Diagnostics: Methods and Protocols*, Methods in Molecular Biology, vol. 949, DOI 10.1007/978-1-62703-134-9_2,

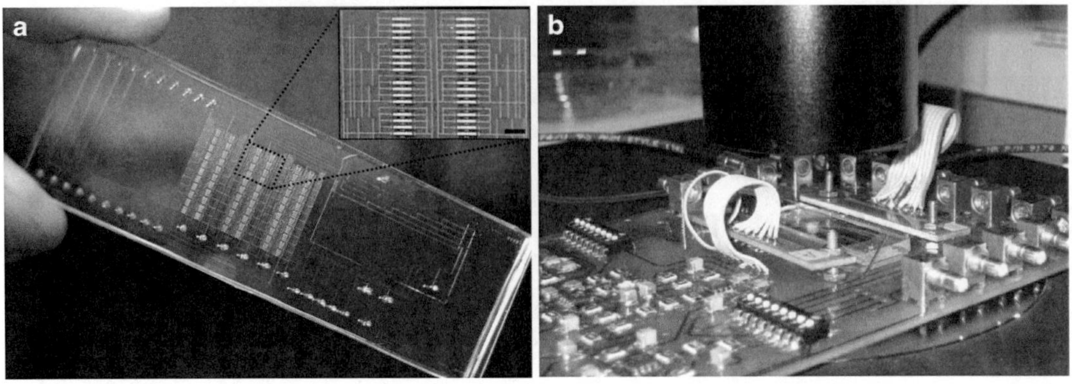

Fig. 1. (**a**) Photograph of a PDMS chip for affinity reagent isolation. (**b**) Photograph of a PDMS chip platform for bacterial pathogen detection.

In our laboratory, PDMS soft lithography is being used to create microfluidic chips for affinity reagent isolation (Fig. 1a) and bacterial pathogen detection (Fig. 1b). A general workflow of the soft lithography fabrication process is presented in Fig. 2.

Other extensions of microfluidics are being explored in other materials as well. For example, much effort is currently focused on producing low-cost microfluidic diagnostics for addressing the issue of global public health using both paper- and thread-based microfluidic devices (5–7). As the field of microfluidic diagnostics continues to grow and becomes an integral part of our daily lives, it is important to transfer the current research efforts and applications of this technology to young students and the general public. Recently, we have devised a set of demonstrations to illustrate the use of Jell-O® and other inexpensive materials for teaching microfluidics (8). Using these experiments, people can easily and effectively learn concepts such as microfluidic chip fabrication, laminar flow, dimensionless numbers, pH sensing, and diagnostics. These demonstrations can also serve as a bridge between nonscientists and scientists by creating a platform for discussing current microfluidics research. Moreover, these educational endeavors can help to inspire the next generation of young scientists into the field of microfluidics. This chapter describes the use of Jell-O® chips for teaching microfluidics and microfluidic diagnostics to young students and the general public.

2. Materials

2.1. General Mold Construction

1. Six 6 in. foam plates, round (see Note 1).
2. Several flat wooden coffee stirrers.
3. Single- and double-sided tape.

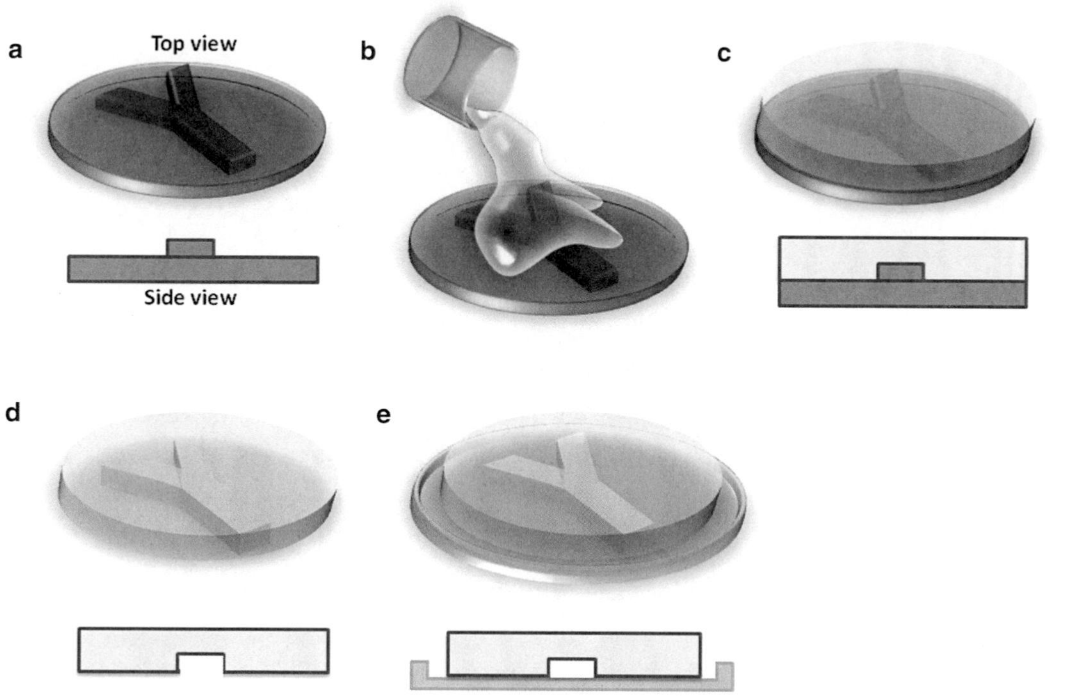

Fig. 2. Scheme for producing Jell-O® chips using soft lithography. (**a**) A negative mold is made with desired features. (**b**) Liquid chip material is poured onto the mold. (**c**) Mold with liquid material is cured. (**d**) Solidified chip is peeled off and (**e**) placed on a rigid substrate for experiments. (Reproduced from ref. 8 with permission from American Chemical Society).

4. Scissors.
5. Personal protective equipment (gloves, lab coat, and safety goggles).

2.2. General Jell-O® Chip Fabrication

1. Two 85 g boxes of lemon-flavored Jell-O® jelly powder (see Note 2).
2. Two 7 g pouches of unflavored (the Original) Knox® Gelatine (see Note 3).
3. Two beakers of 120 mL of purified water for dissolving Jell-O® and Knox Gelatine.
4. One metal stirrer.
5. Hot plate.
6. Six premade molds with specific patterns.
7. PAM® Original no-stick cooking spray.
8. Some tissue paper.
9. Refrigerator with temperature of 4°C.
10. Flat 5 in. aluminum pans.
11. Personal protective equipment (gloves, lab coat, and safety goggles).

2.3. Module 1: "JELLO Chip" Demonstration

1. Jell-O® microfluidic chips, each with a continuous channel depicting the letters "JELLO."
2. Round drinking straws.
3. One disposable transfer pipet per Jell-O® chip.
4. Food-grade color dye, green.
5. Small vials of water with a few drops of green food coloring dye (~30 mL each).
6. Personal protective equipment (gloves, lab coat, and safety goggles).

2.4. Module 2: "Y-Channel Chip" Demonstration

1. Jell-O® microfluidic chips, each with a Y-shaped channel.
2. Round drinking straws.
3. Two disposable transfer pipets per Jell-O® chip.
4. Food-grade color dye, blue.
5. Small vials of clear water (~30 mL each).
6. Small vials of water with a few drops of blue food coloring dye (~30 mL each).
7. Personal protective equipment (gloves, lab coat, and safety goggles).

2.5. Module 3: "pH-Sensor Chip" Demonstration

1. Jell-O® microfluidic chips, each with two straight channels.
2. Round drinking straws.
3. Two disposable transfer pipets per Jell-O® chip.
4. Two small pieces of acid-sensing pH paper and two small pieces of base-sensing pH paper.
5. Double-sided tape.
6. Small vial of 1 M hydrochloric acid (or cooking vinegar).
7. Small vial of 1 M sodium hydroxide (or dissolved antacid solution).
8. Personal protective equipment (gloves, lab coat, and safety goggles).

3. Methods

In general, the molds are created using foam plates, wooden coffee stirrers, and double- and single-sided tape. The coffee stirrers are first cut into various shapes and sizes, depending on the purpose of the mold, using a pair of scissors. These pieces of coffee stirrers are taped onto a foam plate using double-sided tape to create the desired mold pattern. Single-sided tape is then adhered to the junctions of the wooden sticks to reduce the gap. Three types of

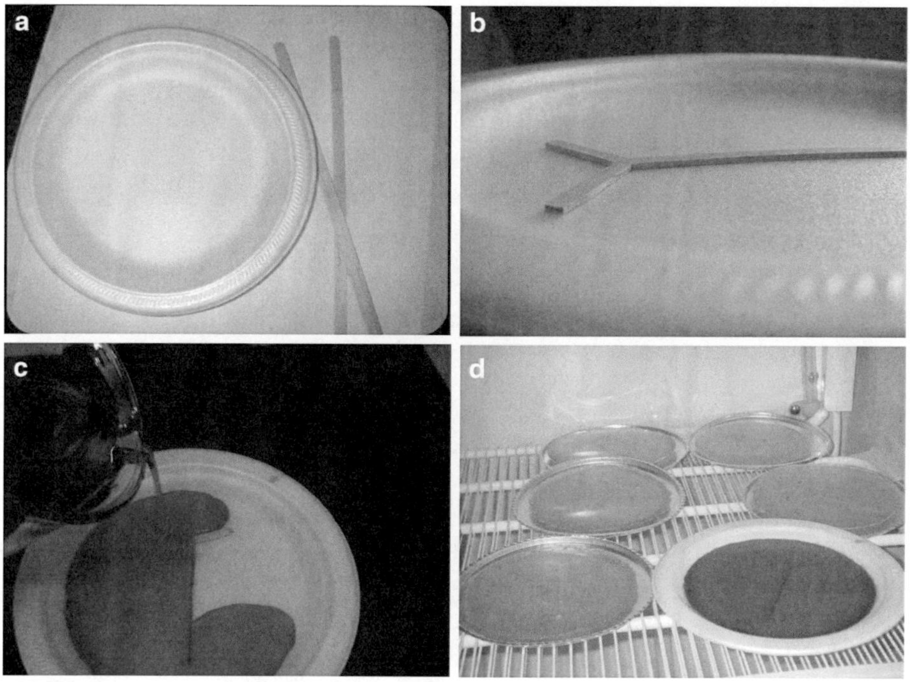

Fig. 3. General workflow for producing Jell-O® chips using soft lithography approach. (**a**) Foam plate and wooden coffee stirrers are starting materials for making the mold. (**b**) A negative mold is made with desired features using double-sided tape. (**c**) Jell-O® and gelatine liquid mixture is poured onto the mold. (**d**) The molds with liquid material are left to cure in a 4°C refrigerator. Solidified chips are peeled off and placed on aluminum pans for experiments at room temperature (reproduced from ref. 8 with permission from American Chemical Society).

molds have been constructed to illustrate the diverse concepts that can be taught using this teaching method: a "JELLO" mold, a Y-channel mold, and a pH-sensor mold. In general, the Jell-O® chips are made by pouring a liquid mixture of both Jell-O® and gelatine into the molds. These plates are left to cure in a 4°C refrigerator for about 2 days. When ready, the Jell-O® chips are removed from the refrigerator, peeled from the molds, and placed on aluminum dishes for demonstrations. The high sugar content from the Jell-O® and gelatine mixture provides a natural seal on the aluminum dishes, and the strength of the seal is suitable for the low-pressure applications presented here. A general workflow for fabricating these Jell-O® chips is shown in Fig. 3.

Instructors should allocate two 1-h class periods to conduct the demonstration(s). The first class period is dedicated to introducing microfluidics and soft-lithography, constructing the molds, preparing the Jell-O® and gelatine mixture, pouring the mixture into the molds, and moving the plates to the refrigerator (see Note 4). The second class period is focused on conducting the hands-on experiments, observing the microfluidics phenomena, elucidating the accompanying theory, and discussing some current and relevant applications. For more mature audiences (high-school, university, general public), the learners should conduct both chip fabrication and experimentation. For younger audiences (grade-school and middle-school), the mold and the chips should be made in advance by instructors; these students would participate by manipulating the chips to form a seal on an aluminum pan, and conducting the experiments. The main learning outcomes are summarized in Table 1 below.

3.1. General Mold Construction

1. "JELLO" Chip: Cut the coffee stirrers into rectangular shapes of various lengths, according to the letters "JELLO." Using double-sided tape, attach these small pieces of wooden sticks onto a foam plate to form a continuous channel depicting the letters "JELLO." Use small pieces of single-sided tape to cover

Table 1
A summary of learning outcomes for Module I: "JELLO Chip" demonstration, Module II: "Y-Channel Chip" demonstration, and Module III: "pH-Sensor Chip" demonstration (reproduced from ref. 8 with permission from American Chemical Society)

Parameter	Module I	Module II	Module III
Target learners	Grade-school science students	High school science students	High school science students
Mold fabrication difficulty	Medium	Medium	Easy
Experimental difficulty	Easy	Medium	Medium
Learning objectives	Basics of microfluidic fabrication Soft lithography Concept of pressure-driven flow Diversity, complexity, and flexibility of designs	Visualization of laminar flow Differences between laminar flow and turbulent flow Significance of dimensionless parameters Current microfluidic applications of laminar flow	Differences between acids and bases Fundamentals of pH sensing Concept of parallelization Current microfluidic parallelization applications
Questions to be answered	What is microfluidics and how are microfluidic chips made? How are channels formed in microfluidic chips? How do liquids flow in microfluidic chips? Can fluid be passed through the chip with only one inlet and no outlet?	Why do the two solutions not mix in this Jell-O chip? What is the difference between turbulent flow and laminar flow? What are dimensionless numbers? How can dimensionless numbers help us to build our devices?	What are acids and bases? How can we determine the pH of solutions using pH papers? What is parallelization? What are current microfluidic applications of parallelization?

the junctions of coffee sticks to ensure a smooth overall mold surface (see Note 5).

2. Y-Channel Chip: Two pieces of coffee stirrers are needed for forming one Y-channel mold. Cut the first coffee stirrer at both ends using a pair of scissors to obtain a long rectangular-shaped wooden stick of ~3 in. long. One end of this stick should be flat (outlet) and the other end should be further cut into a dagger shape. Cut the second coffee stirrer to obtain two smaller rectangular-shaped sticks of the same length (~1 in. long) (see Note 6). Using double-sided tape, tape the longer stick near the bottom half of a foam plate. Similarly, tape the two smaller sticks at the dagger-shaped end of the longer stick to form a mold with the letter "Y" (see Note 5). Use small pieces of single-sided tape to cover the junctions of coffee sticks to ensure a smooth overall mold surface.
3. pH-Sensor Chip: Two pieces of coffee stirrers are required for forming the pH-sensor mold. Cut both of the coffee stirrers to obtain two long rectangular-shaped wooden sticks of the same length (~3 in. long). Using double-sided tape, attach these sticks to the middle of the foam plate at ~1 in. apart.

3.2. General Jell-O® Chip Fabrication

1. After constructing the mold plates (see Note 7), mix two pouches of Jell-O® jelly powder in 120 mL of purified water in one beaker (using a metal stirrer). Mix two pouches of Knox Gelatine powder in another beaker with the same amount of water (see Note 8).
2. Place the first beaker (containing partially dissolved Jell-O®) on a hot plate and heat the solution to a boil (see Note 9). Remove beaker from the heat and pour the content of the second beaker (containing partially dissolved gelatine) into the first beaker. Reheat the mixture of Jell-O® and gelatine solution to a boil on the hot plate, and finally remove this beaker from the heat.
3. Apply a small amount of cooking spray onto the inside rim of the foam plate (with tissue paper) to facilitate the peeling of the Jell-O® chips after curing. Pour the mixture of Jell-O® and gelatine solution into the molds (an amount that can adequately cover the wooden sticks). Approximately six mold plates can be filled with the amount of solution prepared.
4. Transfer the molds with liquid mixture to a 4°C refrigerator for curing (see Note 10) and cure the chips for about 2 days to obtain more robust Jell-O® chips.
5. When ready for demonstration, carefully peel the cured Jell-O® chips off of the mold. Bending the foam plate at the rim may help with the peeling process. Be careful when peeling the Jell-O® chip near the wooden sticks to prevent any tears, which can result in leakage in the chip.

Fig. 4. A chip made into the letters "JELLO" and filled with green colored water demonstrates the ease of making complex patterns with the Jell-O® technique (reproduced from ref. 8 with permission from American Chemical Society).

6. Determine the side of the chip with hollow channel(s). Place this side against an aluminum pan to create a natural and reversible seal, and to form an enclosed channel. Make sure to eliminate all visible air bubbles between the chip and aluminum pan (see Note 11).

3.3. "JELLO Chip" Demonstration

1. These instructions assume that chips with a continuous channel depicting the letters "JELLO" are cured and ready to be used for demonstration.
2. To motivate this demonstration, questions including "What is microfluidics?," "How are microfluidic chips made?," and "How are channels formed in microfluidic chips?" can be presented to the class. It may be helpful to use Figs. 1 and 2 to facilitate the discussion.
3. After class discussion, peel the "JELLO" chips from their foam plates and place them on aluminum pans as described in Subheading 3.2, then proceed with experimental demonstration. These instructions explain the procedure for working with one chip (see Note 7).
4. Using a round drinking straw, puncture an inlet hole at the tip of letter "J" and an outlet hole at the end of letter "O" with a gentle twisting motion.
5. Add a few drops of green food coloring dye into a small vial of water, and load the green water into a disposable transfer pipet.
6. By gently squeezing the pipet bulb, inject the green water into the channel via the inlet hole. The resulting fluid flow can be directly visualized without the use of any imaging apparatus. An example of the results produced is shown in Fig. 4.
7. Learning Objectives including "Basics of microfluidic fabrication" and "Soft lithography" can be easily taught using the "JELLO" demonstration presented above. After conducting the hands-on experiment, the similarities and differences between

Jell-O® chip and PDMS chip fabrication can be highlighted. For example, negative molds for PDMS chip fabrication are produced using photolithography; the PDMS pre-polymer is cured in an oven at 60°C; and the feature sizes of PDMS microfluidic chips are usually a few micrometers in width (4). However, the general fabrication concepts and PDMS soft lithography can be easily explained using this demonstration.

8. The concept of pressure-driven flow can also be explained using this demonstration. Questions including "How do liquids flow in microfluidic chips?" and "Can fluid be passed through the chip with only one inlet and no outlet?" can be used to motivate the class discussion. When pressure is applied to the pipet bulb, a pressure is applied to the colored water in the transfer pipet, and a fluid flow is observed. The fluid flow stops as soon as the pressure from the pipet bulb is released. In contrast, if there is only one inlet and no outlet, then the fluid cannot flow through the channel. This phenomenon occurs because the air present in the channel has no place to escape. Furthermore, if a large enough pressure is exerted on the fluid in this inlet-only system, then the reversible seal between chip and aluminum pan would break. The outlet provides a path for the air inside the channel to escape, therefore allowing the fluid to flow.
9. Finally, this demonstration can be used to illustrate the level of creativity that can occur in designing microfluidic chips. Depending on our specific needs, we can fabricate molds and chips with varying flexibility, diversity, and complexity. Currently, microfluidic chips are being designed to address specific problems in microfluidic diagnostics.

3.4. "Y-Channel Chip" Demonstration

1. These instructions assume that chips with a continuous channel depicting the letter "Y" are cured and ready to be used for demonstration.
2. To motivate this demonstration, pose the question "What would happen when you pour clear water and blue water in a cup?" to the class. Evidently, the students would expect mixing of the two solutions. Subsequently, pose the question "What would happen when you inject clear water and blue water in the 'Y-channel' chip?." The majority of the students would most likely answer that mixing of the fluids would occur. Without revealing the answer, obtain the aluminum pans with "Y-channel" chips and proceed with experimental demonstration. These instructions explain the procedure for working with one chip (see Note 7).
3. Using a round drinking straw, puncture two inlet holes at the top of letter "Y" and an outlet hole at the bottom of letter "Y" with a gentle twisting motion.

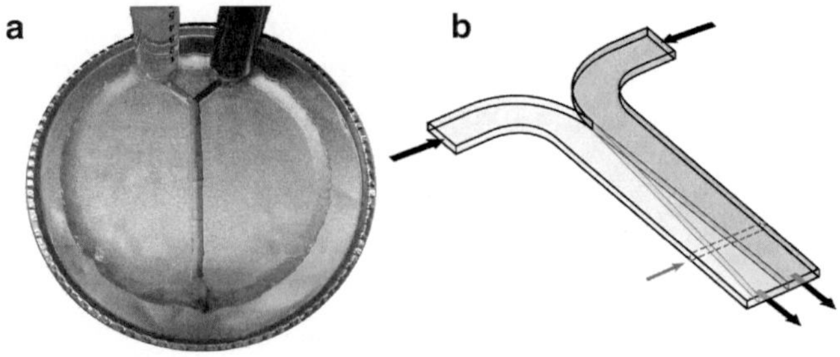

Fig. 5. (**a**) A Jell-O® Y-channel chip with a Reynolds number of 30. The injection of colored water to one inlet and clear water to the second results in the classic laminar flow profile, in which both streams remain separate and mix solely by diffusion along the length of the channel. (**b**) Diagram of laminar flow diffusive mixing occurring at the interface between two different fluids along the channel length. This phenomenon is governed by the Péclet number (reproduced from ref. 8 with permission from American Chemical Society).

4. Prepare two small vials of purified water, and add a few drops of blue food coloring dye to one of the vials. Obtain two disposable transfer pipets: load one pipet with blue water and the other one with clear water.
5. Simultaneously inject clear water and blue water into left channel and right channel, respectively (see Note 12). The resulting laminar fluid profile can be directly visualized without the use of a microscope (see Note 13). An example of the results produced is shown in Fig. 5.
6. Furthermore, the flow rate of one solution can be changed (by changing the pressure applied to the pipet bulb), and the shifting of the interface between clear and colored water can be observed. It is counterintuitive to see that the two fluids do not mix in the Y-channel chip, so this demonstration provides a convenient starting point for discussing the differences between turbulent flow and laminar flow.
7. Students who are more mathematically advanced can be introduced to dimensionless parameters or numbers for a more comprehensive understanding (see Note 14). For examples, dimensionless parameters including the Reynolds number (*Re*) (see Note 15) and Péclet number (*Pe*) (see Note 16) can be discussed and calculated in class. To gain a better understanding of *Pe* numbers, diffusion and diffusion coefficient (*D*) may also need to be discussed (see Note 17).
8. The significance of dimensionless numbers can also be highlighted, because learning how to use dimensionless numbers is an important skill for scientists and engineers. For example, dominant forces in the fabricated microfluidic devices can be calculated using dimensionless numbers. Reynolds number can be used to determine whether laminar or turbulent flow dominates; Péclet number can be used to determine whether

convective mass transfer or diffusion dominates. Conversely, the value of a parameter can also be changed to switch between the analytical regimes, and dimensionless numbers can facilitate in the designing of microfluidic chips.

9. In these Jell-O® Y-channel chips, reliable separation of analytes based solely on diffusion or molecular size cannot be easily achieved (Fig. 5a); however, this result can be achieved in smaller microfluidic systems (Fig. 5b). For example, diffusive mixing has been used in microfluidic T-sensors for chemical concentrations measurements (9) and for rapid determination of diffusion coefficients for molecules of varying sizes (10).
10. After discussing the theory behind the Y-channel chip, some current applications of this device can be highlighted. In addition to the two T-sensor examples discussed above, laminar flow can also be used to separate the anode and cathode streams in Y-shaped fuel cells, without the use of a polymer electrolyte membrane (11–13).

3.5. "pH-Sensor Chip" Demonstration

1. These instructions assume that "pH-sensor" chips are cured and ready to be used for demonstration.
2. To motivate this demonstration, questions such as "What are acids and bases?" and "How can we determine the pH of solutions using pH papers?" can be presented to the class. Obtain aluminum pans with "pH-sensor" chips and proceed with experimental demonstration. These instructions explain the procedure for working with one chip (see Note 7).
3. Using a round drinking straw, puncture two inlet holes at the top of both channels, and two outlet holes at the bottom of both channels with a gentle twisting motion (to form two separate channels).
4. Use double-sided tape to attach the small pieces of pH paper onto the aluminum pan, within the boundaries of the channels. In both of the channels, tape a piece of acid-sensing pH paper near the inlet and tape a piece of base-sensing pH paper near the outlet.
5. Carefully prepare small vials of 1 M HCl and 1 M NaOH (each with volume of ~30 mL) (see Note 18). Obtain two disposable transfer pipets, and load one with NaOH and the other with HCl.
6. Simultaneously inject NaOH into the left channel and HCl into the right channel, and directly visualize the resulting color changes in the pH papers. An example of the results produced is shown in Fig. 6.
7. Alternatively, safer acidic or basic solutions, such as cooking vinegar and dissolved antacid solution, could be used for this experiment (see Note 19).

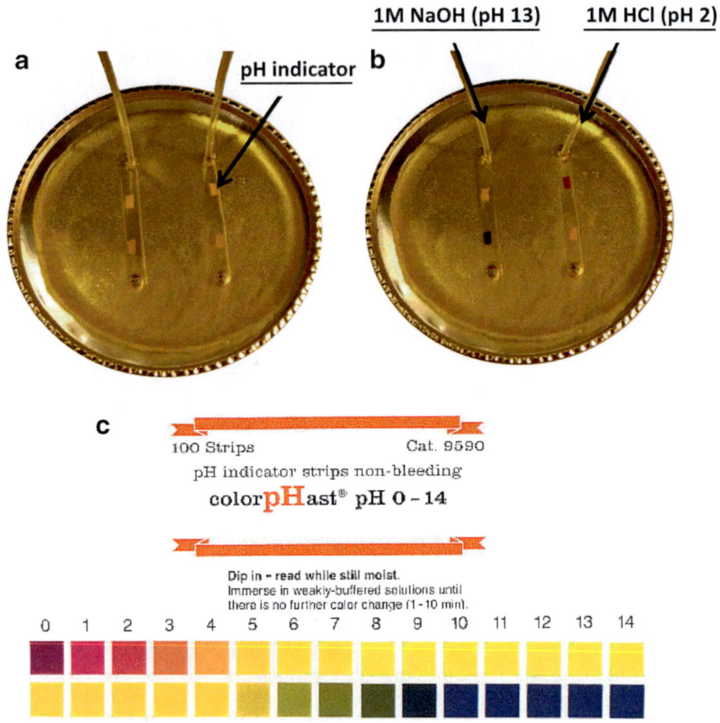

Fig. 6. Dual Jell-O® chip pH sensor. (**a**) Two different pH-sensing regions per channel can be used to detect different solutions. (**b**) The addition of either acidic or basic solutions to each channel results in a distinct pattern of color change visible to the naked eye. (**c**) pH indicator chart for reference (adapted from EM-Reagents) (reproduced from ref. 8 with permission from American Chemical Society).

8. When 1 M NaOH is injected into the channel on the left, the base-sensing pH paper nearest the channel outlet turns blue. In contrast, when 1 M HCl is injected into the channel on the right, the acid-sensing pH paper nearest the channel inlet turns purple. As a result, the differences between acids and bases, and the fundamentals of pH sensing can be explained.
9. Optionally, this demonstration can be combined with the Y-channel demonstration to form a more integrated learning activity. For example, a weak acid and a weak base can be injected over a pH paper placed within the Y-channel chip. Therefore, distinct pattern of color change can be observed on the same pH-sensing strip due to laminar flow, diffusive mixing, and neutralization.
10. The concepts of parallel analysis and microfluidic diagnostics can also be introduced with this demonstration. As an example of parallel analysis, first assume that the indicator strips near the channel inlets represent areas of the microfluidic chip that contain a type of ligand (capturing agent) that detects analyte A. Similarly, assume that the indicator strips near the channel outlets

represent a second type of ligand that detects analyte B. When two unknown fluid samples are introduced into the channels, the presence of analyte A or B in each solution can be determined. Therefore, we are running two separate tests in parallel.

11. This concept of parallel analysis can be further extended. For example, first assume that the channels have been modified with four different ligands. Consequently, four independent tests can be conducted on one fluid sample in parallel. Therefore, current efforts in microfluidic diagnostics to conduct hundreds of tests in parallel can be easily explained using this simple demonstration (14, 15).

4. Notes

1. To be more environmentally sustainable, flat paper plates may be used instead of foam plates.
2. Lemon-flavored Jell-O® jelly powder was used because it created chips with the best optical transparency. Other flavors could also be used, as long as fluid flow can still be seen through the chips.
3. This is a modification of the initial protocol, which used only one pouch of Knox® Gelatine. It has been found that this new approach shortened the curing time and also increased the robustness of the chips.
4. Since 2 days of curing time is recommended, it is ideal to make these chips on Friday and continue with the demonstration on Monday.
5. During the taping of the sticks onto the foam plate, the sticks need to be pressed against each other well to reduce the gaps between them. It may also help to wear gloves during the taping process since tape becomes less sticky when oils from the fingers are transferred to the tape.
6. The length for the two sides of the dagger-shaped end should be the same as the width of the smaller coffee sticks to ensure a continuous and smooth Y-channel geometry.
7. The amount of Jell-O® and gelatine mixture produced in this protocol is enough to completely cover six mold plates. Thicker chips can be made by reducing the number of molds made (to three or four). However, increasing the thickness of the chips will most likely affect and lengthen the curing time. The number of Jell-O® chips needed for a presentation or a workshop depends on the number students present. From our experience, a ratio of 2–3 students per chip produced the best learning experience.

8. Jell-O® and gelatine powder are not readily soluble in cold water. Therefore, the solution in this beaker will remain opaque until heated to a boil in the following steps.
9. Before using the hot plate, it is helpful to cover the hot plate with some aluminum foil to prevent any spills from drying up on hot plate surface and to facilitate the cleanup procedure. During the heating process, the solution should be stirred often.
10. If a large quantity of Jell-O® chips is required for the demonstration/workshop, it is helpful to place the foam plates on a cart with wheels, and then pour the Jell-O® and gelatine mixture into the plates. Subsequently, the cart can be wheeled into and stored in a 4°C cold room for curing the chips.
11. From experience, it is important to store the Jell-O® in the refrigerator until immediately prior to the demonstrations. If left at room temperature for too long, the chips may lose their robustness and the peeling process may fail.
12. To prevent leakage from the inlets, the transfer pipets should be held at a 45° angle to the surface, and pointed into and parallel to the inlets of the Y-channel. Both clear and dyed water should be injected evenly and slowly into the channels in order to create the laminar flow profile.
13. If Y-channel chips are too difficult to fabricate, simpler T-channel chips can also produce similar results. Two rectangular-shaped coffee stirrers can be combined to make the letter "T" on a foam plate. Puncture the two inlet holes and one outlet hole as before. Laminar flow profile can still be observed using the T-channel chips.
14. Dimensionless parameters do not have a physical unit, and they are usually defined as a ratio of two properties (these properties may have units individually, but the units cancel out when they are combined) (16).
15. Reynolds number (Re) is a dimensionless parameter that relates inertial to viscous forces in fluid flows: $Re = \rho U_0 L_0 / \eta$, where ρ is the fluid density, U_0 is the characteristic velocity, L_0 is the typical length scale, and η is the shear viscosity (16, 17). A high Re implies that inertial forces are dominant; and a low Re implies that viscous forces are dominant. Therefore, Re is used to differentiate between laminar and turbulent flows. Using water as the working fluid ($\rho = 1.0$ g/cm^3 and $\eta = 0.010$ g/cm s) with a velocity of 1.0 cm/s and a channel radius of 0.30 cm yields $Re = 30$. Typically, the transition from laminar to turbulent flow in round pipes occurs at $Re = 2{,}000$–$3{,}000$ (16), so $Re = 30$ indicates that the flow is within the laminar flow regime.

16. Péclet number (*Pe*) is another dimensionless parameter that is defined as $Pe = U_0L_0/D$, where U_0 is the characteristic velocity, L_0 is the typical length scale, and D is the diffusion coefficient.

17. Understanding the implications of *Pe* requires an understanding of diffusion. Elementary diffusion explanations can be initiated using the Jell-O® chips described here, as laminar flow is an ideal flow regime in which to demonstrate the effects of diffusion. For a molecule with an unknown diffusivity, the diffusion coefficient can be modeled as a spherical solute: $D \approx k_B T/6\pi\eta a$, where k_B is the Boltzmann constant, T is the temperature, η is the shear viscosity, and a is the molecule size (16). The diffusion coefficient of typical food coloring dye is about 200 μm²/s (18); therefore, calculation yields $Pe = 150{,}000$ in our Jell-O® chips. This large *Pe* indicates that convective mass transfer dominates and little diffusion along the length of the channel occurs (Fig. 5a). However, as the typical length scale decreases (such as in the case of research microfluidic systems), intermediate *Pe* is achieved, and the differences in solute diffusion rates will become more pronounced (Fig. 5b).

18. Be extremely careful when working with high concentrations of acids and bases: Proper protective clothing including lab coat, gloves, and goggles must be worn at all times. This step may not be suitable for younger students.

19. Household solutions that are safer than 1 M HCl and NaOH may be used. Common acidic solutions include vinegar and lemon juice; common basic solutions include dissolved antacid solutions and soapy water.

Acknowledgment

Jake Abbot and Cameron Lawson, two Grade-10 high school students from Prince of Wales Mini School in Vancouver (working through a mentorship program at the Michael Smith Laboratories), helped with developing the initial protocol and testing the first Jell-O® chips. An undergraduate student working in our laboratory, Adrian Lee, created new experiments based on their techniques. The authors would like to thank Drs. David Ng and Joanne Fox for the initial idea and the 2009 UBC iGEM Team and Lagally lab members for their technical support. Financial support of this work by the Michael Smith Laboratories' start-up funding to ETL is gratefully acknowledged.

References

1. Whitesides GM (2006) The origins and the future of microfluidics. Nature 442:368–373
2. Paguirigan AL, Beebe DJ (2008) Microfluidics meet cell biology: bridging the gap by validation and application of microscale techniques for cell biological assays. Bioessays 30:811–821
3. Liu C (2007) Recent developments in polymer MEMS. Adv Mater 19:3783–3790
4. McDonald JC, Duffy DC, Anderson JR, Chiu DT, Wu H, Schueller OJA, Whitesides GM (2000) Fabrication of microfluidic systems in poly(dimethylsiloxane). Electrophoresis 21: 27–40
5. Li X, Tian J, Nguyen T, Shen W (2008) Paper-based microfluidic devices by plasma treatment. Anal Chem 80:9131–9134
6. Li X, Tian J, Shen W (2009) Thread as a versatile material for low-cost microfluidic diagnostics. ACS Appl Mater Interfaces 2:1–6
7. Martinez AW, Phillips ST, Whitesides GM, Carrilho E (2010) Diagnostics for the developing world: microfluidic paper-based analytical devices. Anal Chem 82:3–10
8. Yang CWT, Ouellet E, Lagally ET (2010). Using inexpensive Jell-O chips for hands-on microfluidics education. Anal Chem. 82:5408–5414
9. Kamholz AE, Weigl BH, Finlayson BA, Yager P (1999) Quantitative analysis of molecular interaction in a microfluidic channel: the T-sensor. Anal Chem 71:5340–5347
10. Kamholz AE, Schilling EA, Yager P (2001) Optical measurement of transverse molecular diffusion in a microchannel. Biophys J 80: 1967–1972
11. Choban ER, Markoski LJ, Wieckowski A, Kenis PJA (2004) Microfluidic fuel cell based on laminar flow. J Power Sources 128:54–60
12. Kjeang E, Djilali N, Sinton D (2009) Microfluidic fuel cells: a review. J Power Sources 186:353–369
13. Phirani J, Basu S (2008) Analyses of fuel utilization in microfluidic fuel cell. J Power Sources 175:261–265
14. Kerschgens J, Egener-Kuhn T, Mermod N (2009) Protein-binding microarrays: probing disease markers at the interface of proteomics and genomics. Trends Mol Med 15:352–358
15. Phillips K, Cheng Q (2007) Recent advances in surface plasmon resonance based techniques for bioanalysis. Anal Bioanal Chem 387:1831–1840
16. Squires T, Quake, S (2005) Microfluidics: fluid physics at the nanoliter scale. Reviews of Modern Physics 77 (3),977–1026, 2005
17. Beebe DJ, Mensing GA, Walker GM (2002) Physics and applications of microfluidics in biology. Annu Rev Biomed Eng 4:261–286
18. Inglesby MK, Zeronian SH (2001) Diffusion coefficients for direct dyes in aqueous and polar aprotic solvents by the NMR pulsed-field gradient technique. Dyes Pigments 50:3–11

Chapter 3

Fundamentals of Microfluidics for High School Students with No Prior Knowledge of Fluid Mechanics

Vishal Tandon and Walter Peck

Abstract

Three microfluidics-based laboratory exercises were developed and implemented in a high school science classroom setting. The first exercise demonstrated ways in which flows are characterized, including viscosity, turbulence, shear stress, reversibility, compressibility, and hydrodynamic resistance. Students characterized flows in poly(dimethylsiloxane) microfluidic devices in the other two exercises, where they observed the mixing characteristics of laminar flows, and conservation of volumetric flow rate for incompressible flows. In surveys, the students self-reported increased knowledge of microfluidics, and an improved attitude toward science and nanotechnology.

Key words: Microfluidics, Fluid mechanics, High school physics, Reynolds number, Hydraulic resistance, Laminar flow, Continuity, Dynamic similarity, Lab-on-a-chip, Microfluidics education, Teaching methods

1. Introduction

1.1. Learning Objectives

Inquiry-based learning, learning by posing questions and designing experiments to answer them, has the potential to both improve science education and to motivate previously uninterested students when incorporated alongside more traditional pedagogy (1, 2). A positive attitude toward science, in particular, has been shown to correlate with achievement, and incorporation of hands-on activities tends to improve attitudes toward science (1). In addition to incorporating more inquiry-based laboratory exercises, the inclusion of cutting-edge research topics has also been proposed as a way of improving attitudes toward science (3).

One example of a rapidly growing field of modern research is microfluidics, which has applications in bioanalysis, chemical synthesis, and point-of-care medical diagnostic devices (4). Microfluidics is an interdisciplinary area of research that tradition-

Gareth Jenkins and Colin D. Mansfield (eds.), *Microfluidic Diagnostics: Methods and Protocols*, Methods in Molecular Biology, vol. 949, DOI 10.1007/978-1-62703-134-9_3, © Springer Science+Business Media, LLC 2013

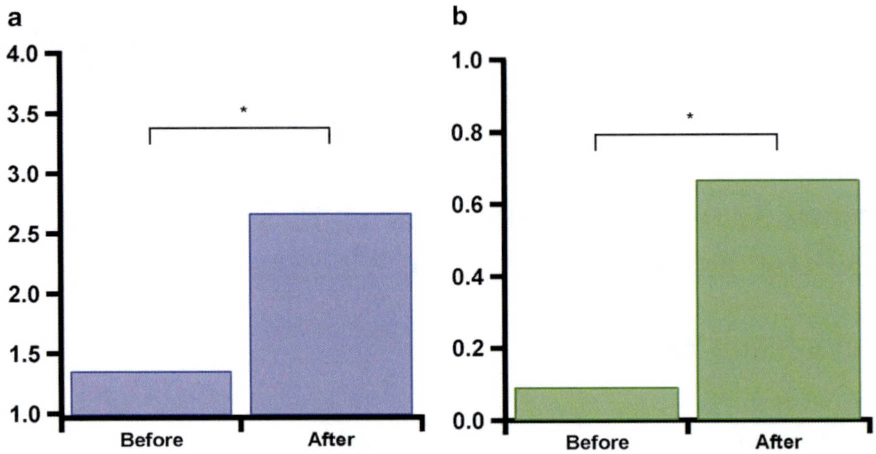

Fig. 1. (**a**) Self-reported understanding of what microfluidics is, on a scale of 1–4 (1 indicates no understanding, whereas 4 indicates a strong understanding). (**b**) Fraction of students that claim the ability to give an example of microfluidics in everyday life. The data sets in (**a**) and (**b**) were analyzed with one-tailed *t*-tests. $p < 10^{-9}$ and $n = 54$ for both data sets, indicating a statistically significant improvement in microfluidics knowledge after participation in these lab exercises.

ally would be considered too specialized to be taught at the K-12 level (i.e., high school or secondary school level). Many high school physics students have had little or no training in fluid mechanics at all, as it is often not emphasized in standard, state-sanctioned curricula (see, for example, (5)). Microfluidics-based laboratory exercises have the potential to increase student education in several areas, including fluids, mechanics, chemistry, biology, and electromagnetism.

Students initially reported little or no understanding of microfluidics, but indicated an improvement in their knowledge of the field after participating in the exercises (Fig. 1). In addition, 9 out of 22 students (41%) in the honors class indicated that they would be more interested in doing research in science after the activity, and 6 out of the 22 students (27%) indicated that they would be more interested in doing research in nanotechnology specifically. The activities described here were also integrated as curricula as part of the Cornell's Learning Initiative in Medicine and Bioengineering (CLIMB) program (6).

The laboratory activities described in the following sections were successfully implemented in a physics classroom at Whitney Point High School (Whitney Point, NY), led by Walter Peck (physics teacher) and Vishal Tandon (Ph.D. student in biomedical engineering, Cornell University). Students engaged in three lab activities: one designed to teach basic fluid mechanics (Subheading 3.1), and two that allowed students to characterize flows in poly(dimethylsiloxane) (PDMS) microfluidic devices (Subheadings 3.2 and 3.3). Honors students also designed their own method for measuring average fluid velocities in microfluidic devices.

1.2. Basic Fluid Mechanics Activity

This laboratory exercise is intended to give students a basic understanding of different ways in which fluids are characterized, and how flows can change dramatically depending on the nature of the fluid involved, the speed of flow, and the size of the system. In particular it focuses on giving students an intuitive understanding of viscosity, turbulence, reversibility of Stokes Flow, compressibility, and hydraulic resistance.

1.3. Flows in PDMS Microfluidic Devices

Devices fabricated from PDMS are ubiquitous in microfluidics for bioanalytical applications (4, 7). Protocols for inexpensive Jell-O-based devices that demonstrate fabrication and laminar flow have been developed previously (8). Here we present experiments that utilize actual, small-scale PDMS devices, and do not focus on fabrication. The challenges in implementing such devices in a high school classroom include the following: (a) fabrication is expensive and will likely require assistance/donation from a local university; (b) working with them requires microscopes; and (c) working with these devices is nontrivial, and may frustrate less motivated students. The advantages of using real PDMS devices include the following: (a) students get a visual sense of how small microfluidic devices are; (b) students will learn about challenges in interfacing microfluidic devices with the macroscopic world; and (c) using prefabricated devices shortens the length of the activity to one class period (approximately 1 h).

Fabrication protocols for PDMS devices are difficult to implement in a high school setting and are beyond the scope of this teaching protocol; therefore, we do not present details here. Instead, collaboration with a local university is recommended, with protocols for fabricating PDMS devices available in the literature (6, 7, 9), or suitable devices obtained from a commercial supplier. Briefly, for fabrication of the devices used in this protocol, masters were patterned in SU-8 photoresist on Si wafers using UV lithography, developed, and cured. PDMS was poured over the masters, cured with a cross linker, peeled off the master, and bonded to glass slides to make the completed devices.

The laminar flow PDMS device (Fig. 2) consists of two inlets and two outlets. The inlet channels converge into a single channel, and then diverge again at the outlets. It is designed to demonstrate how low Reynolds number (*Re*) flow may differ from students' intuitions, which are likely based on high Re flow. Students will expect the fluids to mix in the single channel, but the fluids will remain mostly separated since convective transport from the inlet to the outlet dominates over diffusive transport in the transverse direction.

The step-down flow PDMS device (Fig. 2) consists of one channel that reduces in size in three stages. As fluid travels from a wider channel to a smaller one, the average fluid velocity must increase (for an incompressible flow) in order to maintain conservation of mass. This device is designed to allow students to observe and potentially quantify these velocity changes.

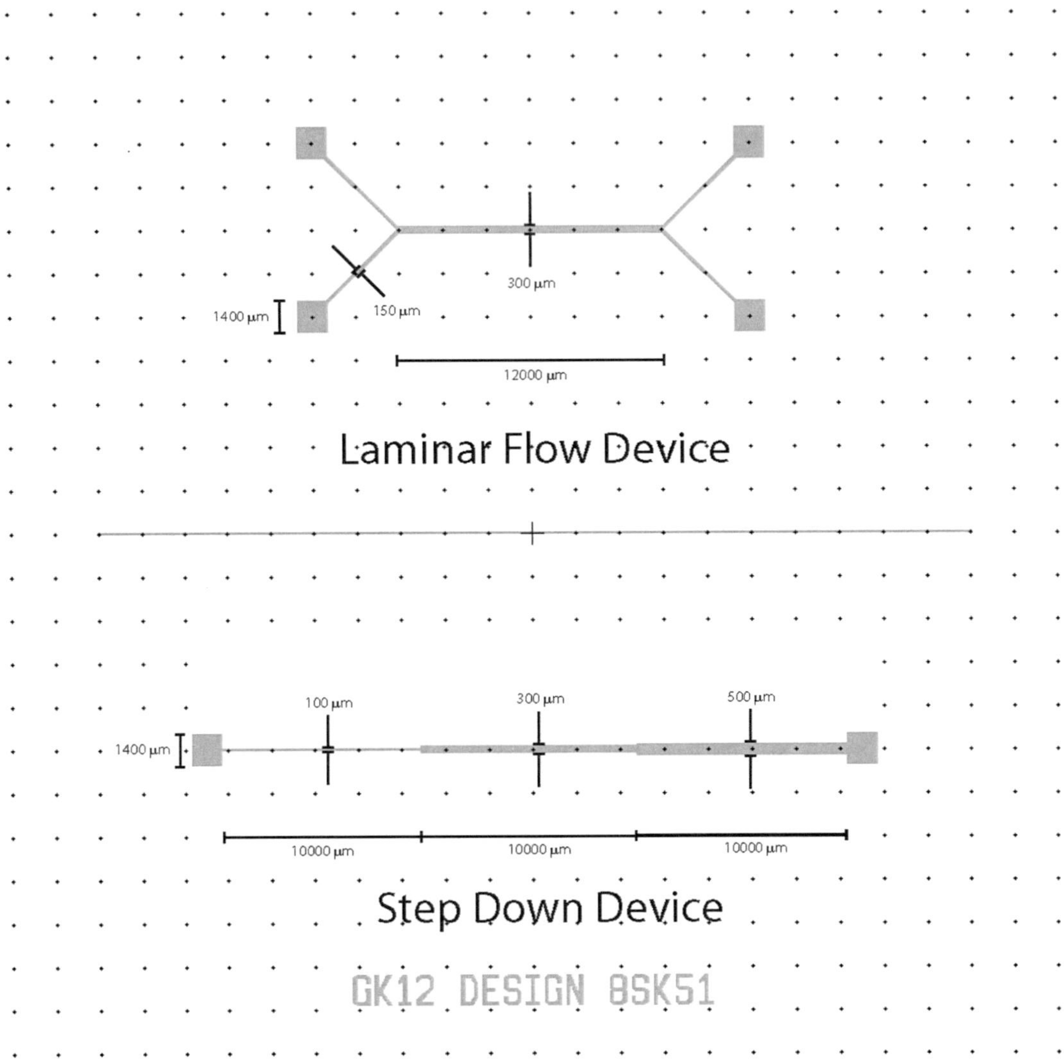

Fig. 2. CAD drawings of two PDMS devices: one that demonstrates laminar flow (*top*), and one that demonstrates constant flow rate for an incompressible fluid (*bottom*). These devices were designed and fabricated by Sowmya Kondapalli at Cornell University.

1.4. Safety Note

The instructor should appropriately caution the students as to possible safety issues involved with any of the experimental steps. Some basic safety guidance is provided below. Gloves should be worn to prevent staining from dye solutions. Safety goggles should also be worn to protect the eyes. The capillaries used are sharp enough to prick skin (but not cut it). If real syringe needles are used, they should be used with great care and should not be used to pierce skin at any time. Blunt syringe needles may be used instead of standard needles. None of the chemicals involved are dangerous, but the instructor should be familiar with the information provided on any material safety data sheet (MSDS) supplied with a particular product.

2. Materials

2.1. General Material Common to All Experiments

1. Deionized water (see Note 1).
2. Dye (e.g., food coloring).
3. Disposable gloves.
4. 5 ml BD Luer-Lok plastic syringes.

2.2. Basic Fluid Mechanics

1. Glycerin (see Note 2).
2. Transfer pipettes.
3. Stirring sticks.
4. Luer-Lok Adapters (Labsmith, Livermore, CA, Catalog #C360-300).
5. Plugs (Labsmith, Livermore, CA, Catalog #C360-101).
6. PEEK fittings for 360 μm capillaries (Labsmith, Livermore, CA, Catalog #C360-100).
7. 60 mm Petri dishes.
8. 100 mm Petri dishes.
9. Silica capillary—360 μm OD/100 μm ID (Polymicro Technologies, Phoenix, AZ, Catalog #TSP100375) (see Note 3).

2.3. Laminar Flow PDMS Device

1. Laminar flow PDMS device (see Note 4).
2. Tygon tubing (23 gauge) fit with stainless steel pins (see Note 5).
3. Syringe needles (23 gauge) (see Note 6).
4. Microscope.
5. Syringe pump (with space for at least two syringes, see Note 7).
6. Clamp stand and clamps.
7. Small weights (200–500 g).
8. Waste container, cotton swabs, or napkins.

2.4. Straight Channel Step-Down Device

1. Straight channel step-down PDMS device (see Note 5).
2. Tygon tubing (23 gauge) fit with stainless steel pins.
3. Syringe needles (23 gauge) (see Note 6).
4. Fluorinert solution (3M, St. Paul, MN).
5. Microscope.
6. Syringe pump (see Note 4).
7. Ruler.
8. Stopwatch.
9. Clamp stand and clamps.
10. Small weights (200–500 g).
11. Waste container, cotton swabs, or napkins.

3. Methods

3.1. Basic Fluid Mechanics

3.1.1. Teacher Preparation

1. Photocopy printed materials (lab instructions for the students) (see Note 8).
2. For each group of students, prepare four beakers of solution:
   (a) Glycerin
   (b) Water
   (c) Dyed Glycerin
   (d) Dyed Water
3. For each group of students, place the lid of a 60 mm Petri dish into the base of a 100 mm dish. Pour glycerin into the 100 mm dish so that it surrounds but does not cover the 60 mm lid.
4. Fill one 5 ml syringe with air, and one with water. Plug the ends of the syringes using Labsmith Luer-Lok adapters and plugs (see Note 9).
5. Fill two 5 ml syringes with water. Using Labsmith Luer-Lok adapters and PEEK fittings, attach a 3 cm length of silica capillary to each syringe, one with ID 100 μm, and one with ID 25 μm.
6. Allocate stirring sticks and transfer pipettes for each group.

3.1.2. Student Experimental Procedure

1. Give students a few minutes to observe the water and glycerin utilizing the stirring sticks. Point out that the glycerin is more viscous than water, and ask them to write down what viscosity means based on their observations.
2. Using a transfer pipette, students add a drop of dye to the water beaker without stirring. They should write down their observations of the resulting flow, and they should pay close attention as to whether or not the fluid is mixing. Ask students if they think the fluids can be "unmixed." Students can stir the fluid, and estimate the time it takes for the fluids to fully mix.
3. Repeat step 2 with the glycerin beaker. In addition to asking students the same questions as in step 2, point out the differences between turbulent (water) and laminar (glycerin) flow (Fig. 3), and ask students which they think is more difficult to model mathematically.
4. Using a transfer pipette, students add dye to the glycerin in the 100 mm petri dish. They should draw a simple, recognizable pattern, such as a square or a circle (Fig. 4).
5. Have students rotate the 60 mm petri dish (keeping its position fixed concentric with the 100 mm dish) slowly, about a half turn. Ask students to describe the deformation pattern of the dye, noting how it changes as a function of distance from the rotating dish.

Fig. 3. Comparison of relatively low viscosity flow of dye in water to relatively high viscosity flow of dye in glycerin. (**a**) A drop of dye is added to water, and photographed after about 5 s. (**b**) A drop of dye is added to glycerin, and photographed after about 5 s. Some of the dye was pushed down with a transfer pipette to make it visible in the picture. With stirring, this is also a demonstration of high (water) vs. low (glycerin) Reynolds flow.

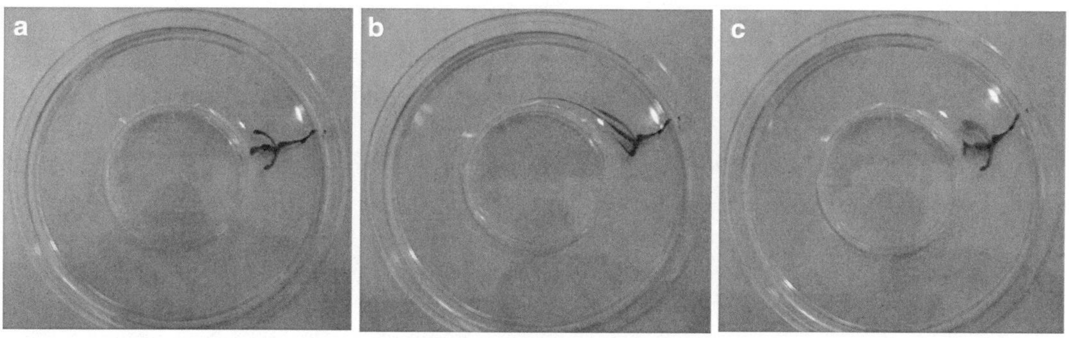

Fig. 4. Experiment that demonstrates the reversibility of Stokes Flow. (**a**) A pattern is drawn with dye in glycerin. (**b**) The center Petri dish is rotated counterclockwise, causing shear and deforming the pattern. (**c**) The center Petri dish is rotated clockwise, reversing the deformation and (nearly) restoring the original pattern.

6. Ask students to rotate the 60 mm dish back to the starting position by rotating slowly in the opposite direction. Students should comment on the resulting pattern (it should be very similar to the starting pattern). This is a crude demonstration of the reversibility of Stokes flow.
7. Give students a set of plugged syringes: one filled with air, and one filled with water. Ask students which fluid they think they will be able to compress by pressing down on the plunger of the syringe. Students should then attempt to compress each of the fluids; they can write down a rough quantitative estimate of compression by noting the change in volume as measured

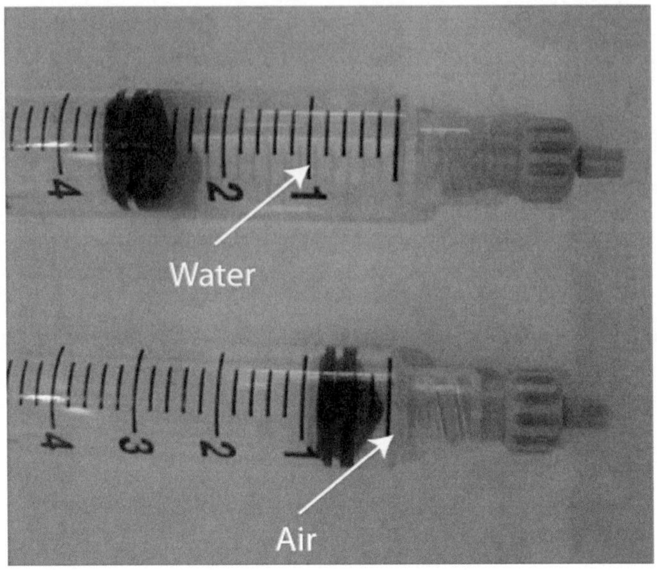

Fig. 5. Demonstration of fluid compressibility. The top syringe was initially filled with 3 ml of water, while the bottom syringe was initially filled with 3 ml of air. Both syringes are sealed at their ends. After depressing the plunger of each syringe (by hand), the air reduces in volume, while the water does not.

by the graduations on the syringes (Fig. 5). It should be clear that the air is compressible, while the water is not. Ask students how they know that the air is actually being compressed, and not leaking out of the syringe (the answer is that the plunger rebounds to its original position when released).

8. Give students a set of water-filled syringes with silica capillaries attached: one with a 25 μm inner diameter (ID) and one with a 100 μm ID. Students should try to force water to flow out of a given capillary by pushing on the syringe plunger. Upon comparing the two capillaries, students will find that it is much easier to push fluid through the larger capillary (Fig. 6). Ask students whether this difference is proportional to the size difference between the capillaries (factor of 4), or whether it is larger or smaller (it takes a pressure gradient 16 times larger to drive fluid at the same flow rate in a capillary that has a diameter a factor of 4 smaller). Make analogies to electrical circuits and resistivity, as they are much more likely to have been covered in a standard high school physics curriculum.

3.2. Laminar Flow PDMS Device

3.2.1. Teacher Preparation

1. Photocopy printed materials (lab instructions for the students) (see Note 8).
2. For each group of students, set up a microscope, a laminar flow device (see Note 10), inlet and outlet Tygon tubing, two syringes, two syringe needles, and beakers with dye and with

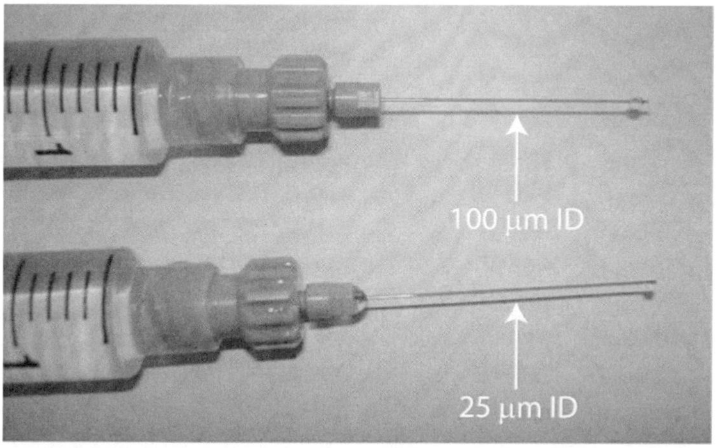

Fig. 6. Demonstration of hydraulic resistance. For the same applied pressure, flow rates are much larger in the capillary with 100 μm ID than for the capillary with 25 μm ID. This can be seen by the size of the droplets at the outlets pictured above. Students applied pressure by hand, giving them a tactile demonstration of hydraulic resistance as well (they found it harder to push fluid through the smaller capillary).

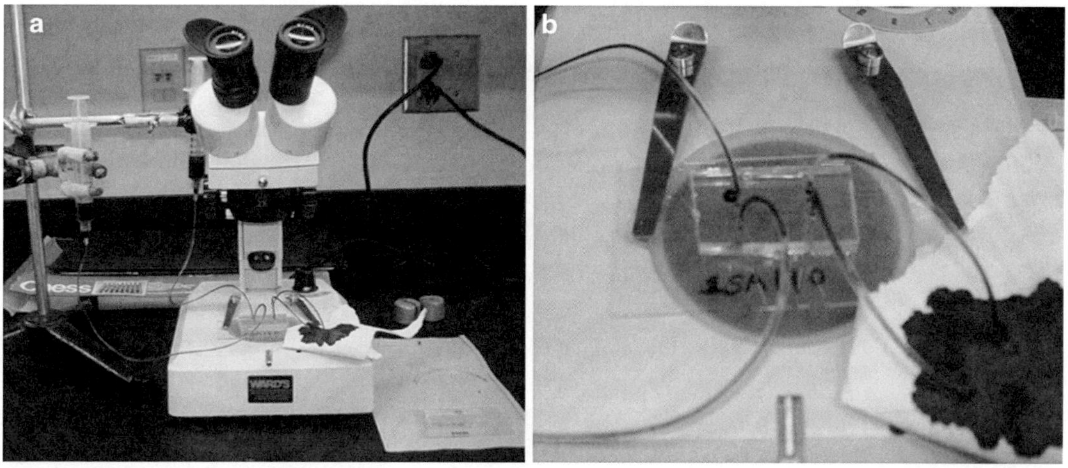

Fig. 7. (**a**) Overview of the experimental setup for studying the laminar flow PDMS device on a microscope. (**b**) Close-up picture of the laminar flow PDMS device setup on a microscope.

clear water (or dye of a different color). The dye should be concentrated; otherwise it is difficult to see at small scales.

3. Set up a syringe pump or clamp stands with clamps and weights at each station (Fig. 7) depending on whether syringe pumps are available (see Note 7).

3.2.2. Student Experimental Procedure

1. Draw 2 ml of the dye into one syringe, and 2 ml of water (or dye of a different color) into another syringe.
2. Attach syringe needles to both syringes. Leave the caps on the needles while you are doing this.

3. Mount the syringes on the syringe pump.
4. Remove the caps from the needles and slide Tygon tubes onto them. Use the end of the tube that does not have a pin.
5. Place the laminar flow PDMS device onto the microscope stand, and insert the pins from the Tygon tubes connected to the syringes into the inlets of the device (the device is symmetrical, so either side is fine). There are holes punctured in the PDMS where you should put the pins.
6. Connect two Tygon tubes to the outlets of the device, pushing the pin ends of the tubes into the holes in the PDMS. These tubes should go to a waste container.
7. Bring the center part of the device into focus in the microscope.
8. Lightly press on one syringe and fill the channel with dye solution; this may temporarily require removing it from the syringe pump (see Note 11).
9. Set the syringe pump to infusion mode at 10 μl/h. Adjust the flow rate as necessary (see Note 12).
10. Observe the flow with the microscope, and record your observations (see Note 13).
11. Move the chip so that you are looking at the branch point near the outlets. Record your observations.

3.3. Step-Down PDMS Device

3.3.1. Teacher Preparation

1. Photocopy printed materials (lab instructions for the students) (see Note 8).
2. For each group of students, set up a microscope, a step-down flow device, inlet and outlet Tygon tubing, two syringes, two syringe needles, and beakers with dye and Fluorinert solution. The dye should be concentrated; otherwise it is difficult to see at small scales.
3. Set up a syringe pump or clamp stands with clamps and weights at each station (Fig. 8) depending on whether syringe pumps are available.
4. Have rulers and stopwatches available to allow students to come up with a way to quantify velocity changes.

3.3.2. Student Experimental Procedure

1. Draw 2 ml of food dye solution into one 5 ml syringe. Try to remove all the air bubbles from the syringe.
2. Draw 2 ml of Fluorinert into another 5 ml syringe. Fluorinert is a liquid that is immiscible with water, but has similar viscosity. Again, remove the air bubbles.
3. Connect both syringes to the two-way valve using the Luer-Lok connection. Screw them in tightly so that they do not leak.
4. Screw a syringe needle onto the two-way valve. Leave the cap on the syringe needle while you do this.

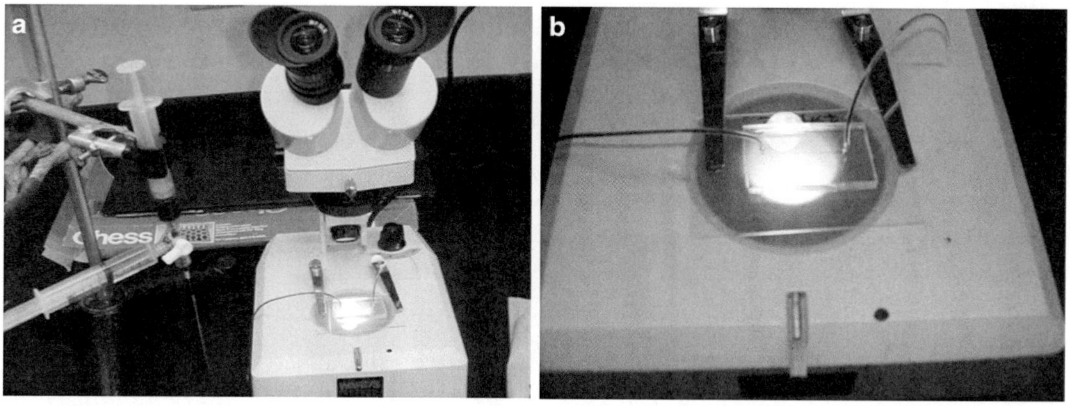

Fig. 8. (**a**) Overview of the experimental setup for studying the step-down PDMS device on a microscope. (**b**) Close-up picture of the step-down PDMS device setup on a microscope.

5. Mount the food dye syringe onto the syringe pump (see Note 14). The Fluorinert syringe will be at a right angle to the dye syringe. Make sure that it is accessible, and not obstructing anything.
6. Remove the cap from the syringe needle, and slide a Tygon tube over it (the end that does not have a pin in it).
7. Place the step-down PDMS device on the microscope stage, and insert the pin from the Tygon tube connected to the syringe needle into the inlet of the device (the side with the widest channel).
8. Place a pin from another Tygon tube into the outlet of the device. The other end of the tube should go into a waste container.
9. The valve stops flow in the direction it is pointing. Turn the valve toward the food dye syringe. Gently press on the Fluorinert syringe to fill the device. Fluorinert is transparent, so it may be difficult to see it in the channel. Watch for fluid coming out of the outlet (see Note 11).
10. Turn the valve to the Fluorinert syringe. Manually apply pressure from the syringe pump until dye solution fills the inlet tube and is about to enter the device. Set the syringe pump infusion rate to 10 μl/h. The infusion rate can be adjusted as necessary (see Note 15).
11. Observe the interface between the dye and the Fluorinert solution moving through the channel. Pay attention to how the flow velocity changes as the interface moves through the smaller sections of the channel. If the interface is lost, the valve may be turned and plugs of Fluorinert can be injected. An optical

interface is necessary in order to observe the fluid velocity (see Note 16).

12. Design an experiment to quantitatively measure the speed with which the interface moves through the step-down device (see Note 17).

4. Notes

1. Some standard laboratory equipment is listed in the materials section, since it may not be commonly found in a high school classroom.
2. Corn syrup may be substituted, but is messier to work with.
3. Silica capillary may be too expensive to buy for the classroom, though it may be possible to obtain samples from Polymicro or other capillary manufacturers. Only a few cm of capillary are necessary, and they can be reused.
4. Protocols for fabricating PDMS devices are available (6, 7, 9), though they may be difficult to implement in a high school setting. Collaboration with a local university is recommended.
5. Approximately 0.5 cm pins are cut from stainless steel tubing with an outer diameter of 0.025″ (McMaster-Carr, Princeton, NJ). The steel tubing should be cut with a saw or a Dremel tool to ensure that the ends are not pinched off. It is easier to insert a long stainless steel tube into a Tygon tube, and cut the steel tube off to the desired length.
6. For safety concerns, blunt syringe needles may be used instead of standard needles, though standard needles are easier to work with.
7. A syringe pump enables precise control of flow rates, allowing students to make quantitative measurements. Syringe pumps are expensive, and may not be available, however. If syringe pumps are not available, an alternative method using clamp stands, clamps, and weights may be used. Syringes are clamped in a vertical position with the plunger on top and the needle below. A mass may be balanced on the plunger to apply pressure, and students can estimate the pressure by dividing the weight of the mass by the area of the plunger. If m is the mass of the weight, g is the acceleration due to gravity, and r is the radius of the plunger, then the pressure is $mg/\pi r^2$.
8. Time in a high school classroom tends to be extremely limited. Any preparation for the activity that is not an explicit part of the experiment and does not aid in teaching the students should be done ahead of time.

9. Labsmith fittings are expensive, so the syringes can be plugged in some other way (e.g., with epoxy) if necessary.
10. It may also be possible to purchase and use capillary electrophoresis devices, though they may be expensive.
11. The device may leak at the inlet if the pressure is too high. If this happens, clean up the leak with a cotton swab or a napkin.
12. If using the clamp stand setup, use the following procedure instead: Try to press on both syringes equally. Once you get both fluids into the channel, balance the weights on top of the plungers so that you apply a roughly equal pressure to both syringes. Use what you know about force and pressure to estimate the pressure applied.
13. Students may get frustrated if they have trouble setting up the device. Try to give them new things to try if this happens, e.g., a new device, reversing the direction of the device, different inlet/outlet tubing, etc. Avoid setting the device up for them, though since time is limited, it may become necessary.
14. If no syringe pump is available, mount the food dye syringe to a clamp stand using a clamp such that the plunger is facing vertically upward. The Fluorinert syringe will be at a right angle (parallel to the lab bench) to the dye syringe.
15. If no syringe pump is available, apply pressure to the food dye syringe either by hand or by carefully balancing a weight on top of the plunger for the syringe. Using the weight gives you a quantitative estimate of the applied pressure.
16. Tracer particles, such as polystyrene beads, are often used to observe the velocity field. In order for the particles to be visible, they either need to be large enough, or fluorescently labeled and observed in a fluorescence microscope. For the former, challenges arise from particle sedimentation, and for the latter, an expensive fluorescence microscope is necessary, something not commonly found in high schools. In addition, the tracer particles are expensive as compared to Fluorinert solution.
17. Give students stopwatches and rulers at this point. It may also be useful to give students a diagram showing the dimensions of the device, and an equation for the hydraulic resistance of a rectangular channel: $R = 1 - 0.63\left(\frac{h}{w}\right)\frac{12\mu L}{h^2 w}$, where R is the hydraulic resistance; L, h, and w are the length, height, and width of the channel, respectively; and μ is the viscosity of the fluid.

References

1. Freedman MP (1997) Relationship among laboratory instruction, attitude toward science, and achievement in science knowledge. J Res Sci Teach 34(4):343–357
2. Anderson RD (2002) Reforming science teaching: what research says about inquiry. J Sci Teach Educ 13(1):1–12
3. http://www.gk12.org/. Accessed 12 Nov 2012
4. Stone HA, Stroock AD, Ajdari A (2004) Engineering flows in small devices: microfluidics toward a lab-on-a-chip. Annu Rev Fluid Mech 36:381–411
5. http://www.p12.nysed.gov/ciai/mst/sci/ls. Accessed 12 Nov 2012
6. http://climb.bme.cornell.edu/flow.php. Accessed 12 Nov 2012
7. McDonald JC, Duffy DC, Anderson JR, Chiu DT, Wu H, Schueller OJA, Whitesides GM (2000) Fabrication of microfluidic systems in poly(dimethylsiloxane). Electrophoresis 21: 27–40
8. Yang CWT, Ouellet E, Lagally ET (2010) Using inexpensive jell-o chips for hands-on microfluidics education. Anal Chem 82:5408–5414
9. Kondapalli S, Kirby BJ (2009) Refolding of beta-galactosidase: microfluidic device for reagent metering and mixing and quantification of refolding yield. Microfluid Nanofluid 7:275–281

Chapter 4

Measuring Microchannel Electroosmotic Mobility and Zeta Potential by the Current Monitoring Method

Chenren Shao and Don L. DeVoe

Abstract

Electroosmotic flow (EOF) is an electrokinetic flow control technique widely used in microfluidic systems for applications including direct electrokinetic pumping, hydrodynamic pressure generation, and counterflow for microfluidic separations. During EOF, an electric field is applied along the length of a microchannel containing an electrolyte, with mobile ions near the charged microchannel walls experiencing a Coulomb force due to electrostatic interactions with the applied electric field that leads to bulk solution movement. The goal of this laboratory is to experimentally determine the fixed channel surface charge (zeta potential) and electroosmotic mobility associated with a given microchannel substrate material and buffer solution, using a simple current monitoring method to measure the average flow velocity within the microchannel. It is a straightforward experiment designed to help students understand EOF physics while gaining hands-on experience with basic world-to-chip interfacing. It is well suited to a 90-min laboratory session for up to 12 students with minimal infrastructure requirements.

Key words: Electroosmotic flow, Zeta potential, Electrokinetics, Surface charge, Microfluidics, Teaching methods

1. Introduction

Applications of microfluidic devices commonly involve the flow of ionic solutions (electrolytes) within channels with dimensions measured in microns or nanometers. When the pH of the electrolyte solution differs from the isoelectric point of the substrate material, the channel sidewalls possess a nonzero surface charge density and thus exhibit a surface potential termed the zeta potential (ζ). As described by electrical double layer theory (1), counterions within the bulk electrolyte are attracted to the channel walls through Coulomb charge interactions, leading to a net excess of mobile

Gareth Jenkins and Colin D. Mansfield (eds.), *Microfluidic Diagnostics: Methods and Protocols*, Methods in Molecular Biology, vol. 949, DOI 10.1007/978-1-62703-134-9_4,

counterions near the fixed surface, with an exponential decrease in magnitude of electrical potential (φ) from the channel wall to the bulk solution. The characteristic dimension of this layer of mobile ions is defined by the Debye length (λ_D), given by

$$\lambda_D = \left(\frac{\varepsilon k_B T}{2q^2 c_0} \right)^{1/2}, \tag{1}$$

where ε is the electrical permittivity of fluid, k_B is the Boltzmann constant, T is the solution temperature, c_o is the bulk electrolyte ion concentration, and q is the charge of a single electrolyte ion (defined by the product of ionic valence and electron charge). The Debye length scales inversely with the square root of the electrolyte ionic strength, with typical values on the order of 1–100 nm depending on ion concentration. When an external electrical field is applied along the length of a microchannel longitudinal Coulomb forces acting on mobile ions within the electrical double layer, it results in drag forces acting on the fluid. This leads to a plug-like flow of the bulk fluid within the microchannel, with uniform velocity far from the channel walls. When the channel radius is significantly larger than λ_D, the bulk velocity (u) can be determined by solving the Navier–Stokes equations with the addition of an electrostatic body force. Under the assumption of steady one-dimensional longitudinal flow under isobaric boundary conditions, the Navier–Stokes equation may be combined with the Poisson equation relating electrical charge and potential to yield the *inner solution* for EOF velocity:

$$u(y) = \frac{\varepsilon E}{\eta}(\varphi(y) - \varphi_o), \tag{2}$$

where η is the dynamic viscosity of fluid, E is the external electrical field applied along the channel length, φ_o is the potential at the channel wall, and $\varphi(y)$ is the electrical potential within the double layer at a distance y from the wall. Because the potential in the bulk solution far beyond the Debye length approaches zero ($\varphi(y) \sim 0$ for $y \gg \lambda_D$), the bulk flow velocity is given by

$$u = \frac{\varepsilon \zeta}{\eta} E, \tag{3}$$

where the wall potential, φ, has been explicitly replaced by the zeta potential, ζ. By defining the electroosmotic mobility (μ_{eo}) as the linear term relating applied electric field and bulk flow velocity ($\mu_{eo} = \varepsilon\zeta/\eta$), this relationship may also be expressed as

$$u = \mu_{eo} E. \tag{4}$$

For a more detailed derivation of the EOF inner solution and electroosmotic mobility, students may be referred to a number of good references such as (2, 3).

Since zeta potential is a material-dependent property, it is important to characterize this parameter for any given microfluidic substrate material. In this laboratory session, a current monitoring method (4) is used to experimentally determine u, and by extension μ_{eo}, for a set of chosen parameters (solution pH, ionic concentration, and channel material). In the current monitoring method, a simple microchannel connecting two open reservoirs is filled with a known ionic solution, followed by the introduction of a second solution with different ionic strength into one of the reservoirs. When an electric field is applied along the channel length, the first electrolyte is displaced by the second solution, resulting in a change in resistance within the channel due to the difference in ionic strength. The average flow velocity can be determined by monitoring the time required for the current to plateau, indicating that the second electrolyte has fully displaced the initial solution within the microchannel.

This experiment may be carried out using a simple single-channel microfluidic chip design fabricated using any substrate material such as glass, PDMS, or thermoplastics. Glass is a particularly desirable material to use, because zeta potential as a function of pH is well defined for silica surfaces, allowing students to compare their estimates of ζ with published data over a range of pH values. Alternatively, the instructor may find it valuable to ask the students to evaluate multiple chip materials using buffer at a single pH value. Regardless, it is recommended that the chip consists of a single channel with hydraulic diameter below 100 μm and a length of at least 2 cm. Shorter or wider channels may not present sufficient hydrodynamic resistance during buffer introduction, leading to spurious current measurements due to unwanted hydrodynamic flow. Each channel end should terminate at an open reservoir, preferably with a volume on the order of 10 μL.

During the lab, each student receives a handout describing the basics of EOF physics, measurement theory, safety precautions, lab procedures, and lab report requirements. It is important to emphasize the safety precautions before any experiment is performed. Depending on the students' backgrounds, it may also be necessary to train them on the use of a pipette to transfer buffer solutions from storage vials to the on-chip reservoirs. During the lab, each student will make his or her own independent measurements. The instructor may also choose to ask each student to collect data from all other students' experiments to improve statistical analysis of the resulting data.

2. Materials

1. Glass, elastomer, or thermoplastic microfluidic chip containing a single channel connecting two open reservoirs.
2. 10 and 20 mM Tris buffer solutions, pH 8.1 (Sigma Aldrich Corp., St. Louis, MO, USA).
3. High voltage power supply (Bertan Series 230; Spellman, Hauppauge, NY, USA).
4. Platinum electrodes (SN 765000; A-M Systems, Sequim, WA, USA).
5. Resistor (4.6 kΩ).
6. Electrical prototyping breadboard.
7. Precision digital multimeter (34410A; Agilent Technologies, Santa Clara, CA, USA).
8. Micropositioning 3-axis probe manipulators with magnetic bases (DCM 200 series; Cascade Microtech, Inc., Beaverton, OR, USA).
9. 10–100 μL adjustable pipette (Eppendorf, Hauppauge, NY, USA).
10. 200 μL disposable pipette tips (Fisher Scientific, Pittsburg, PA, USA).
11. Small vacuum pump (DOA-P704-AA; Gast Manufacturing, Inc., Benton Harbor, MI, USA).
12. Timer.

3. Methods

3.1. Preparations

3.1.1. Electronics Setup

1. *Warning*: Improper use of high-voltage equipment can result in significant injury or death. Although the power supply recommended for use in this lab has an internal current limiter, there remains the danger of electrocution. Always ensure that the power supply chassis as well as any nearby conductive surfaces that could come into contact with high voltage are properly grounded. Students should be instructed on appropriate use of the power supply, and local safety protocols should be reviewed for any additional institutional requirements. Proper safety precautions must be observed during both preparation and execution of the lab session.
2. As shown in Fig. 1, the voltage supply is connected to two microchannel reservoirs, with a 4.6 kΩ resistor attached between one reservoir and the power supply ground terminal.

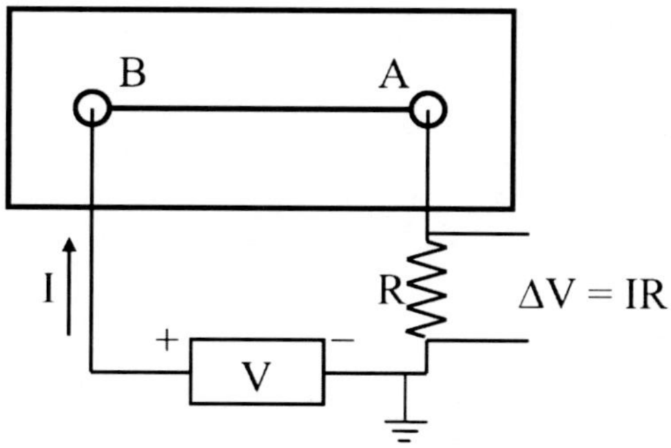

Fig. 1. Schematic of the electrical circuit used for current monitoring. A and B are reservoirs of a single channel chip. *R* is resistance; *V* is voltage; *I* is current.

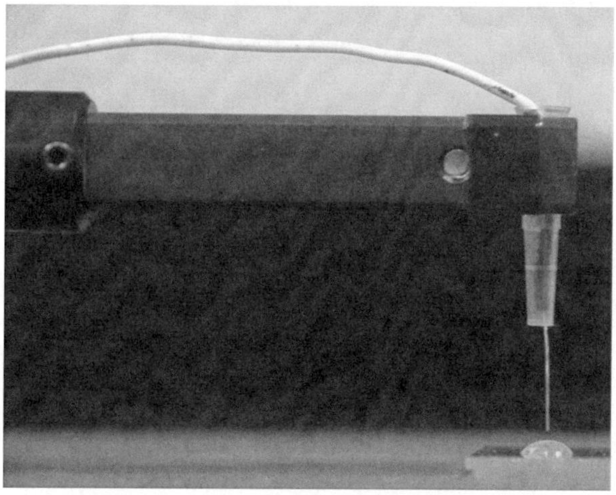

Fig. 2. View of a platinum electrode inserted through a 200 μL pipette tip and secured to a precision positioner before being lowered into buffer solution within an on-chip reservoir.

This particular resistance value is not critical, but the resistor should be large enough to provide good signal-to-noise ratio for measuring microampere currents while being small enough to prevent significant voltage drops that could reduce the electric field within the microchannel.

3. A voltage meter is attached across the resistor to monitor voltage drop, with a prototyping breadboard used to simplify the connections.
4. The microfluidic chip is placed on a stainless steel plate that is co-grounded with the voltage supply.
5. The platinum electrodes are held above the chip reservoirs by precision positioners that are magnetically fixed on the steel plate. To electrically isolate the electrodes, each wire is placed through the tip of a 200 μL disposable pipette tip (see Fig. 2) before being clamped to the micropositioner arms. The use of

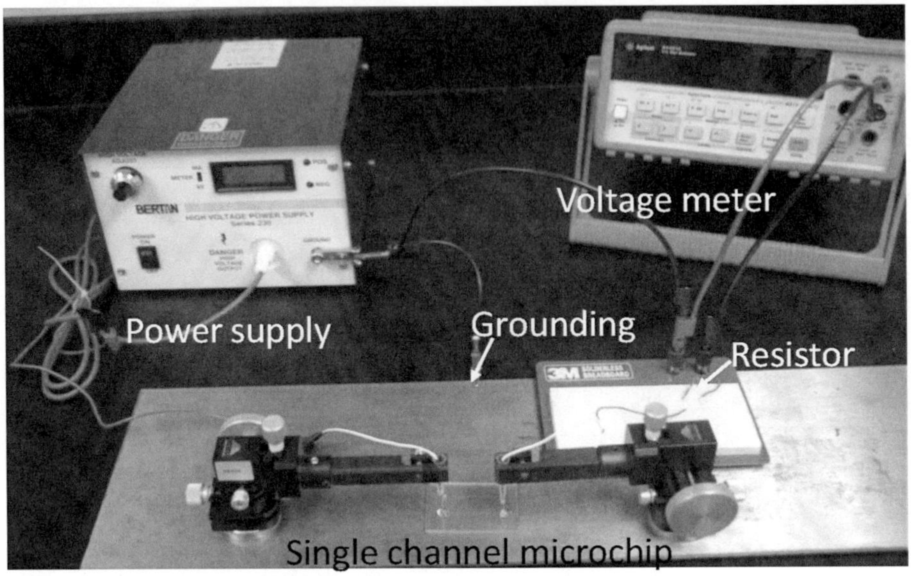

Fig. 3. Overview of the current monitoring experimental apparatus.

micropositioners allows the electrodes to be readily adjusted to contact buffer solution in the reservoirs of each new chip tested during the laboratory session. A photograph of the assembled experimental system is shown in Fig. 3.

3.1.2. Safety Precautions

In addition to being instructed on the potential dangers and safe practices when operating a high-voltage power supply, the students are required to respect the following rules and should be reminded of them before any experiment is performed.
Before conducting the experiments:

1. Verify that the power supply knob is turned to 0 V before approaching the station.
2. Ensure that all personnel are out of reach of the experimental apparatus.
3. Inform the lab instructor and other students before turning on the power supply and performing the experiment.

After completing the experiments:

1. Turn the voltage knob on the power supply to 0 V.
2. Turn off the power supply.

3.2. Channel Conditioning

1. Before performing a test using a chip that has been exposed to air for a long period, it is important to properly condition the microchannel surfaces by applying a high electric field along the channel length (see Note 1).
2. Pipette several drops of 20 mM Tris buffer into reservoir A (see Fig. 1). Using a vacuum pump attached to a disposable

pipette tip through a flexible tube, gently apply vacuum to reservoir B to fill the channel with buffer solution (see Note 2). If bubbles appear in the channel during this process, vacuum out the buffer and repeat this step (see Note 3).

3. Pipette a large drop of 20 mM Tris buffer into reservoir B, making sure that no bubbles are trapped in reservoir. If bubbles are found, flush the chip with vacuum and repeat steps 1 and 2.
4. Adjust the precision positioners to bring the electrodes into contact with the fluid within each reservoir. When using Tris as the electrolyte solution, the positive electrode should be positioned to reservoir B containing the large drop of buffer. The Tris base counterion is H^+, and thus EOF will be directed from the positive electrode to the negative electrode. If a different electrolyte is employed, the proper electrode configuration should be determined before conducting the experiment.
5. Turn on the voltage supply and increase the voltage to define an electric field of 300 V/cm. Monitor the current by reading the voltage meter across the resistor. If the current is negligible or very noisy, it is likely that bubbles are trapped either inside the channel or in the reservoirs. In this case steps 1 through 4 must be repeated.
6. Condition the chip for approximately 3 min, with the students recording the voltage meter readings at 10-s time intervals. Depending on the chip material, longer or shorter conditioning periods may be appropriate. Turn off the voltage supply. Withdraw the electrodes from reservoirs and vacuum out the buffer.

3.3. EOF Measurements

1. Flush the chip with vacuum after the initial conditioning step or following a prior EOF test.
2. Pipette 20 mM Tris buffer into reservoir A and gently vacuum the buffer through the channel from reservoir B so that the buffer will fill the channel without bubbles (see Notes 4 and 5).
3. Pipette a large drop of 10 mM Tris buffer into reservoir B.
4. Adjust the precision positioners to bring the electrodes into contact with the fluid within each reservoir.
5. Increase the voltage to define an electric field of 150 V/cm along the channel length, and monitor the voltage across the resistor, with measurement recorded by the students every 5 s (see Note 6). Note that for large channels, Joule heating may limit the maximum electric field to a lower value than 150 V/cm (see Note 7).
6. As 10 mM buffer replaces the 20 mM buffer that was initially in the channel, the current will fall to nearly half of its original value within several minutes, depending on channel length and

electric field. Once the voltage readings plateau and remain stable for 20–30 s stop recording data.

7. Reduce the applied voltage to 0 V, turn off the power supply, withdraw the electrodes from the reservoirs, and vacuum out the buffer (see Note 8).

   If sufficient time is available, several simple variations may also be performed to extend the laboratory:

   (a) Reverse the buffers, initially filling the channel with 10 mM buffer and replacing it with 20 mM by EOF.

   (b) Employ several different buffer solutions with varying pH.

   (c) Repeat the experiments using chips fabricated from a different substrate material.

3.4. Data Analysis

1. Students are asked to perform the following data analysis for presentation in a laboratory report.
2. Using the measured voltage data collected during the initial channel conditioning step, together with the known resistor value, plot the channel current $I(t)$ versus time (t) measured during channel conditioning. Discuss the shape of the data, paying particular attention to any observed reductions in current fluctuations over time, and suggest one or more possible reasons why the overall slope may be nonzero.
3. For each set of measured EOF data, plot $I(t)$ versus t, and determine the time at which the current plateaus. Note any unexpected trends in the data, and discuss any reasons that could explain these trends.
4. From the measured data, determine the average linear flow velocity (u), and calculate the electroosmotic mobility (μ_{eo}) and channel wall zeta potential (ζ) using Eqs. 3 and 4. Compare the resulting ζ value(s) to published data for the same material(s) used in the experiment.
5. From the known electrolyte molarity, determine the ionic concentration of the bulk solution and calculate the Debye length (λ_D) from Eq. 1. Discuss the channel dimensions used in the experiment relative to the double layer thickness, and how the channels width and height are sufficiently large to avoid overlap of the electrical double layer.

4. Notes

1. Conditioning the microchannel surfaces by applying a high electric field along the channel length serves to stabilize the channel surface charge, and may also remove low-molecular-weight contaminants that could otherwise mask the native surface charge of the channel sidewalls, leading to irreproducible

experimental measurements. Poor experimental results during the laboratory session can often be directly attributed to insufficient channel conditioning.

2. Depending on the diameter of the reservoir, it may be necessary to cut the pipette tip to ensure a sufficient seal with the chip.
3. Note that for hydrophilic chip materials such as glass, this step may be unnecessary due to self-priming of the channels by capillary action.
4. When using vacuum to fill the microchannel, it is common for students to over-vacuum the buffer and unintentionally empty reservoir A. Students practice the filling process several times using a test chip to learn how to gently fill the channel without exhausting buffer in the inlet reservoir. If a vacuum pump is not readily available for channel filling, an alternative method is to use the pipette in withdraw mode to pull solution through the channel.
5. While vacuum filling can generally ensure that no bubbles are trapped in the microchannel, on occasion small bubbles may remain within the channel. If the current is unexpectedly low or unstable when the power supply is switched on, it is very likely that bubbles are trapped between the low-concentration and high-concentration buffers in reservoir B. In this event the chip should be flushed by vacuum and refilled before repeating the experiment.
6. If readings are performed manually (instead of using a digital recorder or data acquisition system), it is recommended that one student be tasked with reading the voltage values out loud for other students to write down.
7. Joule heating is implicated when voltage readings exhibit drift that is not consistent with EOF. In this case, reduce the field strength as needed to eliminate Joule heating.
8. After each experiment, students may try to adjust the electrode positioners without first switching off the power supply. Similarly, it is also common for students to remove their chips from the test system while the electrodes are still submerged in the reservoirs. The lab instructor must pay attention to these points and remind the students to properly shut down the equipment when finishing their measurements.

References

1. Stern OZ (1924) The theory of the electrolytic double-layer. Electrochem 30:9
2. Kirby BJ (2010) Micro- and nanoscale fluid mechanics: transport in microfluidic devices. Cambridge University Press, UK
3. Kirby BJ, Hasselbrink EF (2004) Zeta potential of microfluidic substrates: 2. Data for polymers. Electrophoresis 25(2): 203–213
4. Huang XH, Gordon MJ, Zare RN (1988) Current-monitoring method for measuring the electroosmotic flow-rate in capillary zone electrophoresis. Anal Chem 60(17): 1837–1838

Chapter 5

Overview of the Microfluidic Diagnostics Commercial Landscape

Lily Kim

Abstract

Since its birth in the late 1980s, the field of microfluidics has continued to mature, with a growing number of companies pursuing diagnostic applications. In 2009 the worldwide in vitro diagnostics market was estimated at >$40 billion USD, and microfluidic diagnostics are poised to reap a significant part of this market across a range of areas including laboratory diagnostics, point-of-care diagnostics, cancer diagnostics, and others. The potential economic advantages of microfluidics are numerous and compelling: lower reagent and/or sample volumes, lower equipment costs, improved portability, increased automation, and increased measurement speed. All of these factors may help put more information in the hands of doctors and patients sooner, enabling earlier disease detection and more tailored, effective treatments. This chapter reviews the microfluidic diagnostics commercial landscape and discusses potential commercialization challenges and opportunities.

Key words: Microfluidic diagnostics, Commercialization, Market, Industry, Immunoassays, Clinical chemistry, Nucleic acid detection, Point-of-care diagnostics

1. Introduction

Microfluidic diagnostics have the potential to transform medicine, giving more information to the physician and patient that may allow faster, more effective treatment. Despite significant progress in academia and industry over the last two decades, fewer microfluidic products than expected have reached the market, and none to date has emerged as a clear "killer application"—yet. Though challenges exist, momentum is building with a growing number of start-ups, suppliers, foundries, and consortiums to support successful commercialization.

When considering the commercial perspective, one should be aware that the term "microfluidic" may be used more broadly in industry than in academia and could include many types of devices

Gareth Jenkins and Colin D. Mansfield (eds.), *Microfluidic Diagnostics: Methods and Protocols*, Methods in Molecular Biology, vol. 949, DOI 10.1007/978-1-62703-134-9_5, © Springer Science+Business Media, LLC 2013

that manipulate various fluid volumes (e.g., a consumable cartridge handling relatively large milliliter-scale volumes may be considered a microfluidic device). Because many medical diagnostics are testing some type of body fluid, diagnostics are a natural fit for microfluidic technology. In some devices, microfluidic properties and principles lie at the heart of the diagnostic technology, whereas other devices use microfluidic channels to simply route the sample to a sensor. In keeping with this commercial view, in this chapter we broadly define microfluidic diagnostics to include devices that manipulate and process small quantities of liquids, typically in the micro- and nano-liter volume range.

As seen in the examples in this chapter, there are many potential clinical and economic advantages of applying microfluidic technology to diagnostics. Microfluidics can offer lower reagent volumes, faster results, higher throughput, miniaturized instrument formats, and/or increased automation, all of which can lower costs. In some cases microfluidics may enable reduced sample volumes (e.g., low-volume blood samples that eliminate the need for a separate phlebotomy visit by reducing the skill level needed to collect blood). Finally, the microscale aspect of microfluidics can enable new types of diagnostics that are not possible or practical using other methods. The widespread popularity of lateral flow diagnostics (e.g., home pregnancy tests) is an example of the potential power of microfluidic diagnostics to change medicine.

Diagnostics may be used not only for initial diagnosis but also for monitoring (potentially increasing the market size). Trends toward personalized medicine, more frequent monitoring, and a tighter link between diagnostics and therapeutics are also motivating the development of microfluidic diagnostic systems. Linking diagnostics to therapeutics is fundamental to personalized medicine, using specific information about the individual patient to create a more effective, targeted therapeutic approach. For biotech and pharmaceutical companies facing increasing challenges (e.g., maturing markets increasingly dominated by generic drugs), personalized medicine and linking diagnostics to therapeutics may offer new business directions and opportunities.

With the prospect of personalized medicine and low-cost genome sequencing, the line between research tool vs. diagnostic is becoming blurred. While the needs of the research and medical markets are different, several companies developing microfluidic devices are applying them to both markets. Because there are no clinical regulatory requirements for lab instruments, some companies are generating initial revenue by first selling machines to researchers, but simultaneously planning clinical trials for eventual launch into medical markets.

To begin exploring these issues, this chapter first reviews the concept of a market, then gives an overview of current commercial development, and finally concludes with a brief discussion of factors that may impact commercialization.

2. Commercial Perspective: What is a Market?

First, to understand the commercial perspective it is important to understand the concept of a market. In business, a market is often thought of as the group of potential customers for a product. For example, lung cancer physicians and patients have related needs and problems—they form a market. These patients have needs that may be addressed by a corresponding array of products. A specific technology like microfluidic diagnostics is just one of many technologies that may address these needs.

Because a specific technology may have applications in multiple markets, it is important to identify which target markets to focus on since this can affect design decisions. For example, when developing a cancer diagnostic, a commercial team may need to first decide which target market(s) within cancer (e.g., lung, breast, prostate) is most attractive for initial development and launch. In parallel, estimating the size of the market(s) is critical for understanding the commercial viability of a new diagnostic. For example, how many patients are there with lung cancer? What kind of diagnostic tests do they need, and how often do they need them? How much money is currently spent on diagnostics for lung cancer? In 2009 the in vitro diagnostic market was estimated at >$40 billion USD, and microfluidic technology is well positioned to capture a substantial portion of this market in the future (1).

Thinking about the diagnostics market means thinking about not only patients but also the healthcare providers who will be delivering care. In many cases, healthcare providers would actually be the direct customers for a new diagnostic. What are the needs of the physicians, nurses, and technicians who may be using the device? What are the clinical and economic advantages of the new diagnostic, and who reaps these benefits? Who will pay for the new instruments and tests?

Also, note that the choice of target market(s) helps determine the competitive landscape. For example, when thinking about academic competition, one may focus only on competing academic groups developing similar technologies. However, when thinking about the competition in a market, it is important to consider *all* current and emerging competition, no matter what technology is employed. To succeed, microfluidic diagnostics will have to compete against (and offer significant advantages over) conventional tests as well as emerging competitors. The main competitors may not be microfluidic at all.

All of these market-related questions must be considered when assessing the potential value of a new microfluidic diagnostic.

3. Overview of Commercial Development

The next section gives an overview of some of the main types of microfluidic diagnostics currently under commercial development. While some devices mentioned here are already on the market, many have not yet launched at the time of publication. With the large number of products in development and new start-up companies constantly forming (Table 1), this situation is likely to change quickly in the next few years as more devices launch. The following summary is organized into some of the main types of diagnostics, including those for cell capture, clinical chemistry tests, tests based on immunoassays, and molecular diagnostics.

3.1. Microfluidic Devices for Cell Capture

While conventional flow cytometers can sort individual cells, because they operate in a serial fashion they are not efficient enough for use in capturing very rare cells in the bloodstream. The ability of microfluidics to operate at the scale of individual cells has encouraged the commercial development of devices for capturing and isolating specific cell types.

3.1.1. Rare Cell Capture in Cancer

Cancer detection is one application where identification of rare cells is important, and a handful of companies (e.g., On-Q-ity, Biocept, Cytoscale, and Superior Nanobiosystems) are developing microfluidic diagnostics to attack the problem.

Circulating tumor cells (CTCs) are cells that have separated from the primary tumor to circulate in the bloodstream. CTCs are critical because they enable cancer to spread throughout the body, eventually leading to death. The ability to detect rare CTCs may enable earlier diagnosis and allow the physician to predict a patient's response to therapy. However, CTCs are difficult to detect using conventional techniques because they are so rare; in the bloodstream CTCs occur at a frequency of one in a billion cells. Conventional flow cytometers (which operate serially) can sort up to ~50,000 cells per second (2). Thus, using conventional flow cytometry it could take over 5 h to capture a single rare cell. Clinical trials are under way investigating CTCs in many types of cancer, including breast, prostate, colorectal, and lung cancer (3, 4).

On-Q-ity is a start-up developing cancer diagnostics for detecting rare CTCs using microfluidic technology (the "CTC-chip") commercialized out of the Massachusetts General Hospital (5, 6). Whole blood flows through the microfluidic device, which contains a field of anti-Ep-CAM-coated pillars. The pillars capture live tumor cells of epithelial origin as they travel through the device (Fig. 1a). Since some tumor cells are non-epithelial, On-Q-ity also uses size selectivity (by controlling the size and spacing of the pillars) to capture Ep-CAM-negative CTCs. Once captured, the cells can be counted and analyzed using fluorescent labeling, genetic

Table 1
Key companies pursuing microfluidic diagnostics (listed alphabetically)

Company	Website (www.)	Company	Website (www.)
➢ Abaxis	abaxis.com	➢ LeukoDx	leukodx.com
➢ Abbott Point of Care (i-STAT)	abbottpointofcare.com	➢ Maxwell Sensors	maxwellsensors.com
➢ Achira	achiralabs.com	➢ Micro2Gen	micro2gen.com
➢ Akonni Biosystems	akonni.com	➢ MicroCHIPS	mchips.com
➢ Axis-Shield	axis-shield.com/Afinion	➢ Micronics	micronics.net
➢ Biosite	biosite.com	➢ Micropointbio	micropointbio.com
➢ Biosurfit	biosurfit.com	➢ Mode Diagnostics	modedx.com
➢ Boehringer Ingelheim	boehringer-ingelheim.de	➢ Molecular Vision	molecularvision.co.uk
➢ Boston Microfluidics	bostonmicrofluidics.com	➢ Mycrolab	mycrolab.com
➢ CapitalBio Corporation	capitalbio.com	➢ Nanobiosym	nanobiosym.com
➢ Celula	celula-inc.com	➢ Nanomix	nano.com
➢ Cepheid	cepheid.com	➢ Nanosphere	nanosphere.us
➢ Claros Diagnostics	clarosdx.com	➢ On-Q-ity	on-q-ity.com
➢ Cleveland Biosensors	clevelandbiosensors.com	➢ Pathogenetix	pathogenetix.com
➢ Clondiag	clondiag.com	➢ PerkinElmer	perkinelmer.com
➢ Crospon	crospon.com	➢ Philips Applied Technologies	apptech.philips.com
➢ Daktari Diagnostics	daktaridx.com	➢ Rheonix	rheonix.com
➢ DEOS Labs	deoslabs.com	➢ Samsung	samsung.com
➢ Diagnostics for All	dfa.org	➢ Sensivida	sensividamedical.com
➢ Diagnoswiss	diagnoswiss.com	➢ SensLab	senslab.de
➢ DNA Electronics	dnae.co.uk	➢ SFC Fluidics	sfc-fluidics.com
➢ Epocal	epocal.com	➢ Siemens	siemens.com
➢ Fluidigm	fluigent.com	➢ Siloam Biosciences	siloambio.com
➢ Fluidmedix	fluimedix.com	➢ Smart holograms	smartholograms.com
➢ FocusDx	focusdx.com	➢ T2 Biosystems	t2biosystems.com
➢ Helicos Bioscences	helicosbio.com	➢ TearLab	tearlab.com
➢ Ingeneron	ingeneron.com	➢ TECAN	tecan.com
➢ IonTorrent (Life Technologies)	iontorrent.com	➢ Veridex	veridex.com
➢ Kumetrix	kumetrix.com	➢ Verinata Health	verinata.com

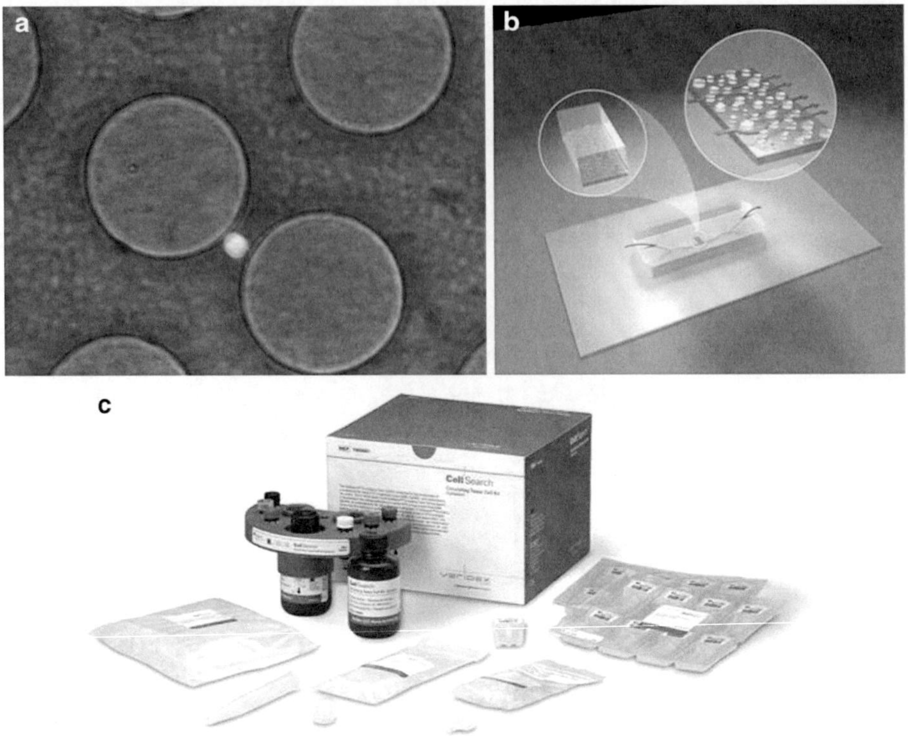

Fig. 1. Competing devices for rare CTC capture. (**a**) On-Q-ity device operates based on posts that capture cells due to antibody labeling and size selectivity. Photo shows an intact cell captured between two posts. (**b**) Biocept device design has antibody-labeled circular posts of a different geometric configuration. (**c**) The Veridex CellSearch system uses immunomagnetic beads rather than microfluidic structures to capture cells. Veridex recently announced a collaborative effort to also develop a microfluidic approach to CTC capture.

assays, or other measurements. The potential for further analysis is key to extracting information that would inform personalized cancer treatment. Cancer is currently one of the main focuses of personalized medicine due to the development of targeted therapies and progress in the cancer genomics field.

A competing start-up, Biocept, is also developing a microfluidic-based cancer diagnostic to detect CTCs. Biocept's proprietary approach also contains a field of microfabricated posts functionalized with antibodies (Fig. 1b) (7). Similar to On-Q-ity, Biocept claims that captured cells may be analyzed using fluorescent tags or also processed via molecular analysis. It remains to be seen how the On-Q-ity and Biocept devices will differentiate themselves, both technologically and by their target markets within cancer.

In addition, CytoScale and Superior NanoBiosystems also report developing devices for rare cell capture, and there are many academic groups pursuing research in the field (3, 8, 9). As of 2006, Micronics was also reported to be developing a microfluidic device for CTC collection (10, 11). However, the current state of their rare-cell-capture program is uncertain as it is no longer actively mentioned on their Web site (12).

All of these microfluidic approaches will have to contend with a non-microfluidic competitor, Veridex, which has already launched a CTC-counting system for the clinic (Fig. 1c). The Veridex CellSearch system uses a different technical approach based on immunogenic beads (13). Anticipating this competition, On-Q-ity has used data to claim that their technology is more sensitive. In comparison with the Veridex CellSearch, On-Q-ity's technology allows capture of viable cancer cells for characterization. The ability to perform further analysis on the captured viable cells could offer a significant advantage against Veridex's early lead in the market (14). However, Veridex recently announced a collaboration with a group at the Massachusetts General Hospital (the same group that invented the On-Q-ity technology) to develop their own microfluidic approach (15). As can be seen, the competitive dynamic is constantly shifting.

3.1.2. Rare Cell Capture for Prenatal Diagnosis

Prenatal diagnosis is another area where the ability to capture rare cells in the bloodstream would be valuable. Currently fetal cells are sampled using a variety of invasive approaches, including chorionic villus sampling and amniocentesis, which both present a small but finite risk of miscarriage—an especial concern to pregnant women over 35 years old who have higher rates of miscarriage but are also the same patients recommended to undergo such procedures due to higher risk of birth defects. Simply counting the fetal cells is not enough—they must be processed further to yield test results.

Artemis Health is working on a noninvasive microfluidic approach to capture fetal cells circulating in maternal blood, enabling access to the entire fetal genome. The Artemis system uses size selectivity to distinguish between maternal and fetal cells, since fetal cells are larger than maternal red blood cells and platelets. The fetal cell-enriched sample is then collected for further analysis (16).

Superior NanoBiosystems is also considering the prenatal market for a microfluidic rare cell capture device, although there is little information available on its approach (9).

3.1.3. Counting CD4 Cells in HIV

Another application where cell counting is important is HIV therapy. Because CD4-positive lymphocytes are the main targets of HIV, CD4 counts are used to aid in diagnosis, assessment of the stage of HIV, evaluation of treatment efficacy, and informing treatment decisions. In the developed world CD4 counts are typically performed using flow cytometry, which may not be available in resource-limited settings due to high instrument cost (typically over $25,000 per system) but also limited portability. Several companies, including Daktari Diagnostics and Alere, are developing microfluidic products to enable CD4 counts at the point of care (POC). In addition, others such as Zyomyx are developing alternate non-microfluidic approaches. In 2009 the market for CD4 testing in

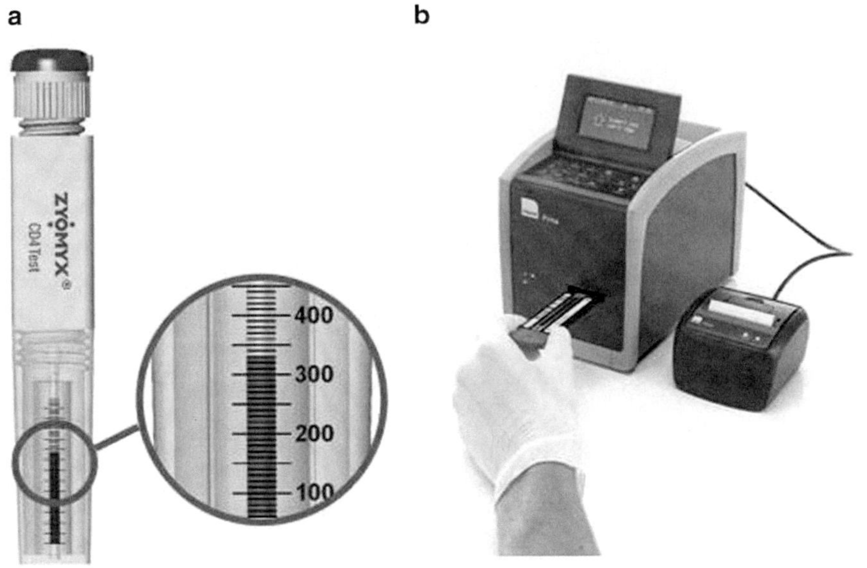

Fig. 2. Competing devices for CD4 counting. (**a**) Readerless design from Zyomyx involves a thermometer-like tube where the CD4 count can be read with the naked eye. (**b**) The benchtop Alere Pima™ CD4 Test counts CD4 cells optically and uses a consumable cartridge with a viewing window.

sub-Saharan Africa was estimated at $150 million USD (17), smaller than many US diagnostics markets but still a significant sum.

Daktari Diagnostics is developing a microfluidic CD4 counting system for HIV patients in the developing world, where design goals include lowered costs, ease of use, stability under high temperatures, and portability. A disposable cartridge employs microfluidic cell chromatography to capture and isolate CD4 cells from the rest of the material in the blood, automating what might otherwise be complex sample preparation steps. The captured CD4 cells are then counted using an impedance-based electrical measurement that helps lower the cost and size of the device when compared with bulky and expensive optics (18, 19). The resulting instrument is a small, handheld system that is designed to minimize energy consumption and prolong battery life while producing rapid CD4 counts within minutes.

While the Daktari approach is compelling, Zyomyx has developed a competing thermometer-like, readerless device that uses antibody-based methods to pull CD4 cells from a blood sample. The process is performed in a closed, disposable cartridge providing ease of use and cost benefits, and requires no instruments to count the cells (Fig. 2a). In 2009 Imperial College's CD4 Initiative (funded by the Gates Foundation) chose Zyomyx as the most promising approach among three teams it had founded in 2007; competing teams were from Beckman Coulter and the Macfarlane Burnet Institute (Australia) (20). The Gates Foundation is also helping to fund Daktari Diagnostics (21) (and many other microfluidic diagnostic efforts in global health).

Using yet another method, Alere has launched a benchtop CD4 counting system, the Alere Pima™ CD4 Test, with an emphasis on POC use. The system uses static image analysis for cell counting and can deliver an absolute CD4 cell count in whole blood in 20 min (Fig. 2b). Alere is also investigating the development of tests using whole blood samples for measuring viral load in HIV (22). The Pima™ system has been tested by nurses and technicians in Zimbabwe and found comparable to standard flow cytometry (23).

Since many of these devices have not yet launched, it remains to be seen how the specific combination of cost, sensitivity, ease of use, portability, and other factors will translate into eventual market share for these competing technologies.

3.2. Devices that Enable Point of Care Using Clinical Chemistry

While many conventional diagnostics are currently available in a hospital's central lab, physicians and patients must often wait hours or days to receive results. POC diagnostics enable physicians to make more informed diagnosis and treatment decisions, helping to converge on a correct diagnosis and appropriate treatment as early as possible. To address these needs, there are many devices on the market and under development which have adapted current assays to the POC setting. Further enabling POC, many of the tests based on more conventional clinical chemistry have electrical readouts, simplifying instrumentation.

The Abbott i-STAT is one of the oldest POC devices on the market, having launched in 1992 (24). The i-STAT consists of a consumable microfluidic cartridge which delivers a small-volume blood sample to bioelectronic sensors that are read when the cartridge is plugged into a handheld reader (Fig. 3a). Tests cover a wide range of measurements, including blood gases, electrolytes, glucose, measures of coagulation, and cardiac markers including troponins and B-type Natriuretic Peptide (BNP) (different cartridges are required for different tests). According to Abbott, the i-STAT is now used in one out of three hospitals in the USA; however it has taken a long time to reach success. Before acquisition by Abbott in 2004 for $392 million USD, i-STAT was not profitable despite revenues of >$55 million USD (25), illustrating some of the financial challenges of creating a commercially successful device.

Several other companies have since entered the space for POC clinical chemistry tests, including competitor Abaxis with their Piccolo system (26). In 2009 Abaxis was chosen by Fortune Small Business Magazine as one of the top 100 fastest-growing small businesses with a reported revenue of $106 million USD in 2008 (27). Interestingly, in 2010 Abaxis began selling an Abaxis-branded version of the i-STAT for the veterinary market.

Roche Diagnostics is also a key player in the diagnostics space, with conventional-scale products for the clinical lab market as well

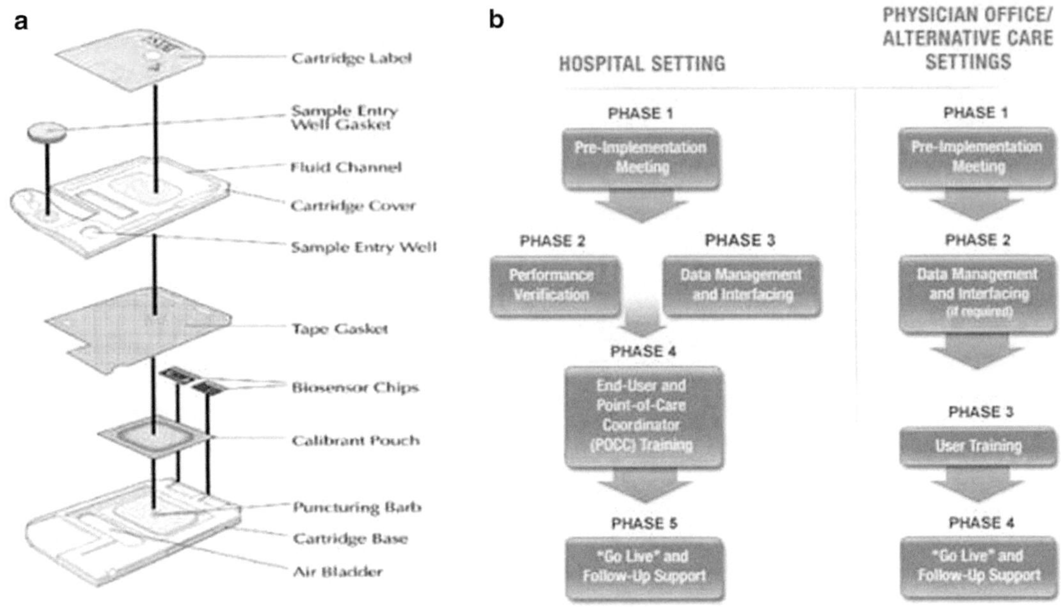

Fig. 3. (**a**) Exploded view of consumable microfluidic cartridge in the Abbott i-STAT device. Simple microfluidic channels lead the sample and a calibrant solution to electronic sensors on the edge of the cartridge that interface with a handheld reader. (**b**) Abbott's educational program for healthcare providers using the i-STAT is carefully planned to start with preparation and continue with follow-up support after users have been trained.

as newer, miniaturized products, such as the desktop Reflotron instrument capable of POC blood analysis. Another major player in the microfluidic diagnostics market is Siemens, having a series of launched POC products, including blood gas and urinalysis instruments (28).

Even electronics manufacturer Samsung has entered the field with a countertop device that performs many of the same measurements as the devices mentioned above, including cholesterol, glucose, and 17 others (29). It is not clear which markets Samsung will be targeting; however it appears that their device may have a significantly lower price point than competitors (30).

Also note that it is not only large companies who have entered the market: recent start-up Claros Diagnostics has developed a benchtop system for POC prostate-specific antigen (PSA) detection for monitoring of prostate cancer and will be marketing the system in Europe in 2011. The device will allow patients to get their results within minutes during the doctor's office visit, allowing easier monitoring of prostate cancer treatment and recurrence (31, 32).

Finally, in the realm of global health, nonprofit Diagnostics for All has taken an innovative approach toward clinical chemistry tests by creating ultracheap paper-based microfluidic chips targeted at <$1 US per chip (33, 34). Reagents are dried onto channels patterned in the paper device. When a liquid sample wicks along the paper, conventional enzyme-based assays are performed with a readerless colorimetric output. The hope is that the paper-based microfluidic format may make diagnostic testing more affordable in the developing world (33).

Some may argue whether some of these devices are truly microfluidic, since the microfluidic component may simply be a channel delivering the sample to the sensor. However, most agree that the market for these devices will continue to grow. Some of the big players, such as Roche and Abbott, have already established a significant presence through acquisitions, but many smaller companies continue to enter the field.

3.3. Devices Using Immunoassays

A number of companies are pursuing an immunoassay approach to microfluidic diagnostics, where the principle of detection is based on antibody–antigen binding. For the purposes of this section, these devices differ from the antibody-based cell capture described above because they trap only the analyte molecules, not cells. Arguments for taking immunoassays down to the microscale include reduced reagent volumes (and thus reduced costs), smaller sample sizes (in the case of blood samples), automation of assays that would otherwise require several labor-intensive steps (typical of conventional ELISAs), higher sensitivity compared with conventional immunoassays, smaller POC instrument size, and faster results due to POC diagnosis as well as shorter reaction times. The potential applications for immunoassay-based methods span a wide range of disease areas, from infectious disease to prostate cancer to Alzheimer's disease.

Lateral flow tests are some of the most widespread examples of microfluidic immunoassay devices and have been on the market for many years. Unlike typical academic microfluidics, lateral flow tests do not contain a microfluidic flow channel; instead fluid moves along a wicking strip that contains embedded reagents for a colorimetric immunoassay (35). Tests are currently available for detecting pregnancy, diseases such as HIV, and biomarkers such as troponin. These devices have had a huge impact because of their low cost, quick results, ease of use, and portability. However, most lateral flow tests have limited ability to provide quantitative results.

In addition to lateral flow tests, there is much commercial activity in other types of immunoassays. Companies like Alere have been steadily acquiring microfluidic diagnostics, including immunoassay-based instruments developed at Biosite (36). Acquired by Inverness/Alere in 2007 for $1.6 billion USD, Biosite's immunodiagnostics were among the first microfluidic diagnostics on the market, dating back to 1990 (37–39). Over the past 20 years Biosite has developed and launched tests in a variety of disease areas, including infectious disease, cardiovascular disease, and substance abuse (40). At the time of acquisition, Biosite reported revenues of $300 million USD. Although the microfluidic components enable a disposable cartridge model, microfluidics is not a primary focus for Biosite, whose key proprietary technology includes protein biomarkers and fluorescent dyes for immunoassays.

In contrast, Quanterix is an example of an immunoassay where the microsystem component plays a large role. Quanterix is developing a microfluidic platform to measure very small quantities of proteins (down to single molecules) using high-sensitivity immunodiagnostics (41, 42). Detection of small numbers of protein molecules is significant, since there is no direct amplification process for proteins as there is for DNA (although reverse translation is possible). Though not employing typical microfluidic channels, the Quanterix approach involves thousands of femtoliter-volume microfabricated wells. An ELISA measurement takes place inside each well; the small well volume allows detectable amplification of the signal from very small numbers of analyte molecules. Quanterix is interested in applying this technology to disease areas where a highly sensitive immunodiagnostic may be a good fit, for example measuring PSA to monitor for recurrence of prostate cancer, or in Alzheimer's disease.

Using yet another approach, Boston Microfluidics is developing an ELISA-based diagnostic focusing on sexually transmitted illnesses. In this case, the microfluidic aspect enables automation of an assay (ELISA) that would normally require several steps (43). The goal is to create a simple device where unskilled users could quickly and easily determine their infection status during their doctor's office visit, instead of having to wait for lab results. The small volume of blood required would also make a trip to a phlebotomist for a standard-volume blood sample unnecessary.

Finally, a number of companies are employing microfluidic methods for high-throughput, improved immunoassays. For example, Akonni Biosystems is developing microfluidic gel-drop microarrays on chip (which can use antibody–antigen binding but also nucleic-acid binding). Instead of a conventional microarray where the probes are attached to a 2D substrate, Akonni's system involves probes distributed throughout a 3D microdroplet, offering potential advantages of increased probes per area compared with a typical 2D array. Plexigen and Randox have also developed immunoassay arrays for the clinical laboratory to increase throughput of traditional ELISA-type assays (44–46). Siloam Biosciences already has a microfluidic immunoassay product for the research market and is developing a microfluidic POC diagnostic; however it is not clear what specific markets it is targeting (47).

While microfluidic diagnostics based on immunoassays have already been successful commercially and offer considerable promise, as genetic-based tests mature there may be competition in certain areas where molecular diagnostics could offer higher sensitivity (depending on the situation) due to PCR-based amplification.

3.4. Devices Incorporating Genetic Tests (Molecular Diagnostics)

Although the human genome project was completed in 2003, we are still awaiting significant impact of the genetic revolution on daily medical care. Part of this delay results from lack of instrumentation that would allow affordable sequencing. Recognizing that the ability to sequence vast amounts of DNA will be in our futures, many companies are developing molecular diagnostics, ranging from those detecting a limited set of genes to whole genome sequencing.

The holy grail of molecular diagnostics is an instrument that can perform whole genome sequencing for under $1,000 USD (a threshold seen as enabling access and reimbursement for widespread use). This extremely demanding requirement has made miniaturized, massively parallel assays an essential component. The potentially huge genome-sequencing market, estimated at over $1 billion USD, has drawn a large number of companies (48). The competition is fierce and some companies have already begun to withdraw from the field. Key contenders (past and present) include Illumina, Life Technologies, Pacific Biosciences, 454 Life Sciences (acquired by Roche), Helicos Biosciences, Complete Genomics, Applied Biosystems, Solexa (acquired by Illumina), and U.S. Genomics (now Pathogenetix). While these diagnostics do not all fall in the microfluidic realm, manipulation of small fluid volumes is important in many of these systems.

One example of a company where microfluidic technology is at the core of cheaper sequencing is GnuBio, a start-up out of the Weitz lab at Harvard. GnuBio is still in early-stage development, yet has made the bold claim that its picoliter-droplet technology may enable a $30 genome. Each droplet would contain sample DNA as well as a barcode and probe. The concept is that by using massively parallel picodroplets (e.g., 1,000 channels per device, one million droplets per second), lower reagent costs would be achieved (49, 50). It is a tall order, and it will be interesting to see what GnuBio can demonstrate in the next few years. As of June 2011, GnuBio press releases indicated it was ahead of schedule and hoped to be shipping production models as early as 2012 (51).

Whichever companies emerge as the winners will have some share of the human genome sequencing and cancer genomics markets. However, clinicians will still have to contend with a lack of scientific understanding of what to do with these newly acquired, vast amounts of data. Depending on how quickly medical knowledge advances, actual use and adoption of genome sequencing may be initially limited to niche markets. In addition, a transition to personalized medicine may require a new way of conducting clinical trials. Regulatory hurdles are already significant, even for traditional drug development. Establishing a regulatory framework for personalized medicine could introduce further delays that indirectly affect the uptake of genome sequencing for the masses.

In addition to genome sequencing of humans, molecular diagnostics for infectious disease is another large market of interest. It has been well established which antibiotics can kill specific bacterial strains. However, because conventional tests are slow, physicians currently treat infections using antibiotics empirically with their best guess antibiotic (usually a broad-spectrum antibiotic). In the short term, this practice can harm patients with rare infections not covered by standard antibiotics, because it delays the start of effective treatment. In the long term, empiric antibiotic prescription leads to overuse of antibiotics and contributes to the development of resistant bacterial strains. Rapid POC diagnostics could transform infectious disease, allowing rapid treatment with an antibiotic known to kill the identified infecting organism.

Cepheid, formed in 1996, has developed and launched a molecular diagnostic system called GeneXpert, which automates PCR-based DNA testing (52). The system uses microfluidic cartridges, with each cartridge detecting a single pathogen (current offerings include MRSA, Vancomycin Resistant Enterococcus, Group B Streptococcus, and others). A recent study published on the use of GeneXpert in diagnosing tuberculosis has created much interest, because of its potential to significantly reduce tuberculosis in the developing world (53). Unlike conventional smear tests which are only ~50% accurate, the Cepheid test can produce ~98% accurate results in 90 min, a potentially game-changing advance. However, it remains to be seen whether the cost of the instrument and consumables will allow widespread use in global health markets (54, 55). Cost constraints are less stringent in the G7, and Cepheid has already started seeing promising revenues, reported at $44.8 million USD in 2008 (56).

In addition to Cepheid, many other companies are developing nucleic-acid-based approaches for infectious disease diagnostics. For example, Wave80 Biosciences is developing molecular diagnostics for viral infections (e.g., hepatitis C, hepatitis B, HIV/AIDS). It is using nucleic acid tests to directly measure the amount of virus circulating in the blood. Norchip is developing a diagnostic for human papillomavirus (HPV) based on nucleic-acid-based sequence amplification of RNA (57, 58). Finally, Biolithic is developing a handheld POC molecular diagnostic device to rapidly detect sexually transmitted diseases using microfluidics highly integrated with electronics (59).

However, just as there are certain markets where PCR may prove superior to immunoassay, there are others where antibody-based approaches might hold an advantage over PCR. For example, in infectious disease the amplification of PCR can cause unwanted false positives (low specificity) if the sample contains a small amount of contaminating pathogen/DNA. In addition, some physicians dislike PCR-based approaches to detecting pathogens in the bloodstream, because pathogenic nucleic acid in the bloodstream may not necessarily indicate the presence of live pathogens. Therefore, there may be some cases where direct detection of pathogens themselves (e.g., via antibody capture) may be desired. For example, Magna Diagnostics is a start-up using a combined approach with immunomagnetic pathogen-cell-capture followed by nucleic acid analysis for rapid pathogen identification in the sepsis market (60).

Given the large number of companies vying for a share of the infectious disease diagnostics market, it is likely that improved pathogen detection will be available soon across a variety of disease areas.

4. Commercialization Challenges and Opportunities

Over the past several years, many articles and opinion pieces have bemoaned the slow progress in commercialization of microfluidics. The typical refrain is that commercialization has been much slower than anticipated (61–65). In 2009 one analyst even went so far as to blame the economic recession on the failure of innovation (specifically mentioning MEMS technology) (66).

However, is any of this unusual for a developing technology? The Gartner hype curve is a well-publicized analysis of the typical time course for adoption of new technologies, starting with initially overblown excitement that leads to a trough of disillusionment, followed by a period of slow but steady growth. While the Gartner curve itself is controversial, many technologies, not just microfluidics, have experienced similar hurdles to adoption—for example, traditional MEMS. Holger Becker of microfluidic ChipShop believes that a period of steady growth may indeed be starting (65). The increasing number of start-ups and successful acquisitions, with products finally hitting the market, is a positive sign. Most new ventures will still fail due to the risky nature of start-ups; however as the field matures companies will begin consolidating and making incremental improvements as market leaders are established.

Many questions remain on exactly how and why adoption of these new tests will occur. For example, when is immunoassay a better fit, and when is genetic testing required or superior? How much would payers (whether they are national healthcare systems or private insurers) be willing to pay per test? How will improved availability of diagnostics change the way the diagnostics are used? Will diagnostics save money on healthcare in the long run? In terms of miniaturization, is there some limit beyond which smaller size is no longer beneficial? On the other hand, as society as a whole adapts to smartphones as a carry-everywhere multi-device, will there be a smartphone equivalent in the world of diagnostics? To what extent will medical diagnostics be incorporated into everyday life?

Additionally, one of the biggest hurdles may be physicians who are reluctant to adopt new diagnostics. A 2008 Medco market research study found that most physicians do not feel confident enough to use genetic testing with their patients (although usage improved significantly when outreach programs were put in place) (67, 68).

It is also important to be aware that there may be considerable institutional resistance to adopting POC tests in the hospital. Ironically, at the same time that academics and start-ups have been developing POC tests, over the past decade hospitals have actually been moving away from POC diagnostics. Motivations for centralized testing include improving standardization of results, enabling centralized billing and record keeping, and improving efficiency. Thus, adoption of new, decentralized, POC devices within hospitals may require significant institutional changes. Diagnostics companies (especially those with POC devices) must prepare for extensive educational efforts and an uphill battle to encourage uptake (e.g., see Fig. 3b for an example of the educational program related to the i-STAT).

Although regulatory approval for diagnostics has historically been less burdensome than for pharmaceuticals, regulatory hurdles are becoming a growing challenge for diagnostics as well. In addition, new types of diagnostics such as genetic testing are motivating the FDA to reconsider its regulatory processes in a scramble to keep up with emerging technologies. For example, in June of 2010 the FDA cracked down on five companies selling direct-to-consumer genetic tests, saying that these needed to pass regulatory approval first (69).

On the positive side, microfluidic diagnostics may enable new business models because they can potentially deliver results at low price points appropriate for emerging markets. Several microfluidic diagnostics companies (e.g., Claros diagnostics, Advanced Liquid Logic) are simultaneously pursuing projects both in the developing world as well as in the G7. While some of these cheaper tests offer lower resolution/sensitivity than typically found in the G7, in many cases technologies developed for low-resources settings could find additional customers in the G7 if they can offer significant cost savings or enable use in new settings (e.g., in the home) (70, 71).

And companies are already starting to see commercial success with microfluidic diagnostics. Alere's combined portfolio (which contains several microfluidic diagnostic technologies) brought in a reported $1.9B in 2009 (72). The companies discussed in this chapter are only a fraction of those developing new microfluidic diagnostics; by the time of publication many new companies will have started, and some will have closed due to the challenges of starting a company. As more devices are commercialized, microfluidic diagnostics could help deliver on the promises of increased prevention, evidence-based medicine, and personalized treatment to improve lives.

5. Further Reading

For more on the commercialization of microfluidic diagnostics, see Holger Becker's excellent "Focus" series in Lab on a Chip (73). In addition, the LinkedIn group "Lab on a chip and Microfluidic Devices (a.k.a., Microfluidics)" contains many helpful discussions, often with contributions from leaders in the field (74). Finally, Kevin Davies's book, "The $1000 Genome," gives an overview (including the commercial perspective) on the race for the genome sequencing market (48).

References

1. Kalorama Information: In Vitro Diagnostics Market Doubled Over Decade: Kalorama Reports. http://www.kaloramainformation.com/about/release.asp?id=1771. Accessed 12 Nov 2012
2. Sklar LA, Carter MB, Edwards BS (2007) Flow cytometry for drug discovery, receptor pharmacology and high throughput screening. Curr Opin Pharmacol 7:527–534
3. Maheswaran S, Haber DA (2010) Circulating tumor cells: a window into cancer biology and metastasis. Curr Opin Genet Dev 20:96–99
4. Home—ClinicalTrials.gov. http://clinicaltrials.gov/. Accessed 12 Nov 2012
5. On-Q-ity: Diagnose, Monitor, Save. http://www.on-q-ity.com/publications/. Accessed 12 Nov 2012
6. Nagrath S et al (2007) Isolation of rare circulating tumour cells in cancer patients by microchip technology. Nature 450:1235–1239
7. Company Description—Biocept Website. http://www.biocept.com/. Accessed 12 Nov 2012
8. Cytoscale Website. http://cytoscale.com/. Accessed 25 Oct 2010
9. Superior NanoBioSystems—Home. http://www.superiornanobiosystems.com/index.html. Accessed 12 Nov 2012
10. Near patient microfluidic rare cell detection: a paradigm shift in flow cytometry. http://www.cli-online.com/index.php?id=1024&tx_ttproducts_pi1(product)=2714&swords=near%20patient%20gerdes. Accessed 12 Nov 2012
11. Lancaster C et al (2005) Rare cancer cell analyzer for whole blood applications: microcytometer cell counting and sorting subcircuits. Methods 37:120–127
12. Micronics Microfluidics: Lab Automation & Nanotechnology. http://www.micronics.net/technology. Accessed 12 Nov 2012
13. Veridex, LLC, a Johnson & Johnson company in vitro diagnostics oncology. http://www.veridex.com/cellsearch/CellSearchHCP.aspx.Accessed 12 Nov 2012
14. Sequist LV et al (2009) The CTC-chip: an exciting new tool to detect circulating tumor cells in lung cancer patients. J Thorac Oncol 4:281–283
15. Veridex, LLC Announces Collaboration to Develop Next-Generation Circulating Tumor Cell (CTC) Technology with Massachusetts General Hospital. http://www.jnj.com/connect/news/all/veridex-llc-announces-collaboration-to-develop-next-generation-circulating-tumor-cell-technology-with-massachusetts-general-hospital. Accessed 12 Nov 2012
16. Welcome to Artemis Health—Our mission is to develop tests that provide accurate and timely information about fetal and maternal health. http://www.verinata.com/. Accessed 12 Nov 2012
17. StreetInsider.com—Canaccord Adams Raises Price Target On Inverness Medical (IMA) to $49, Citing Pima System. http://www.streetinsider.com/Analyst+Comments/Canaccord+Adams+Raises+Price+Target+On+Inverness+Medical+(IMA)+to+$49,+Citing+Pima+System/4974389.html. Accessed 12 Nov 2012
18. Daktari Diagnostics, Inc. http://www.daktaridx.com/. Accessed 12 Nov 2012
19. Daktari Diagnostics, Backed by Gates Foundation, Raises Funds for HIV Test Study | Xconomy. http://www.xconomy.com/boston/2010/06/17/daktari-diagnostics-backed-by-gates-foundation-raises-funds-for-hiv-test-study/. Accessed 12 Nov 2012
20. University Funding and Grants | Scientific Frontline Communication Center. HIV treatment test closer to manufacture with new $7.3 million grant. http://www.sflorg.com/comm_center/unv_funding/. 12 Nov 2012

21. Daktari Diagnostics, Backed by Gates Foundation, Raises Funds for HIV Test Study | Xconomy. http://www.xconomy.com/boston/2010/06/17/daktari-diagnostics-backed-by-gates-foundation-raises-funds-for-hiv-test-study/. Accessed 12 Nov 2012
22. PIMA: Home. http://pimatest.com/en/home.html. Accessed 12 Nov 2012
23. Mtapuri-Zinyowera S et al (2010) Evaluation of the PIMA Point-of-Care CD4 Analyzer in VCT Clinics in Zimbabwe. J Acquir Immune Defic Syndr 55:1–7
24. i-STAT® System | Point-of-Care Testing | Handheld Blood Analyzer | Abbott Point of Care.com. http://www.abbottpointofcare.com/. Accessed 12 Nov 2012
25. SEC Info—I Stat Corporation/DE—10-K—For 12/31/02. http://www.secinfo.com/dqtWy.29.htm#ecri. Accessed 12 Nov 2012
26. Abaxis—Home. http://www.abaxis.com/index.asp. Accessed 12 Nov 2012
27. 2009 FSB 100: Top Small Businesses—Abaxis Inc. (ABAX)—FORTUNE Small Business. http://money.cnn.com/magazines/fsb/fsb100/2009/snapshots/45.html. Accessed 12 Nov 2012
28. Siemens Global Website. http://www.siemens.com/entry/cc/en/. Accessed 12 Nov 2012
29. I-Micronews—MICROFLUIDICS: Samsung ventures into medical device market. http://www.i-micronews.com/lectureArticle.asp?id=5168. Accessed 12 Nov 2012
30. Welcome to the Medical Device Space Samsung. http://www.medgadget.com/archives/2010/07/welcome_to_the_medical_device_space_samsung_1.html. Accessed 12 Nov 2012
31. Claros Diagnostics—Corporate Overview. http://www.opko.com/. Accessed 22 Aug 2010
32. Prostate Cancer Results While You Wait—Technology Review. http://www.technologyreview.com/printer_friendly_article.aspx?id=24932. Accessed 12 Nov 2012
33. BVGH Reports | Global Health R&D | BIO Ventures for Global Health (BVGH). http://www.bvgh.org/Biopharmaceutical-Solutions/Data-Center/BVGH-Reports.aspx. Accessed 14 June 2011
34. Diagnostics For All. http://www.dfa.org/. Accessed 12 Nov 2012
35. Posthuma-Trumpie GA, Korf J, Amerongen A (2008) Lateral flow (immuno)assay: its strengths, weaknesses, opportunities and threats. A literature survey. Anal Bioanal Chem 393:569–582
36. Inverness stays in acquisition mode | IVD Technology. http://www.ivdtechnology.com/article/inverness-stays-acquisition-mode. Accessed 12 Nov 2012
37. Halperin A (2007, March 26) A Hefty Price for Biosite. BusinessWeek: http://www.businessweek.com/stories/2007-03-26/a-hefty-price-for-biositebusinessweek-business-news-stock-market-and-financial-advice Accessed 06 Oct 2012
38. Inverness completes Biosite acquisition—Daily Business Update—The Boston Globe. http://www.boston.com/business/ticker/2007/06/inverness_compl.html. Accessed 12 Nov 2012
39. Biosite agrees to Inverness acquisition for $92.5 per share—Mass High Tech Business News. http://www.masshightech.com/stories/2007/05/14/daily39-Biosite-agrees-to-Inverness-acquisition-for-925-per-share.html. Accessed 12 Nov 2012
40. Biosite® Inc. Web Site—Products. http://www.alere.com/us/en.html. Accessed 29 Nov 2010
41. Quanterix—The Leader in Single Molecule Analysis. http://www.quanterix.com/. Accessed 12 Nov 2012
42. Rissin DM et al (2010) Single-molecule enzyme-linked immunosorbent assay detects serum proteins at subfemtomolar concentrations. Nat Biotechnol 28:595–599
43. Boston Microfluidics. http://www.bostonmicrofluidics.com/. Accessed 12 Nov 2012
44. Plexigen. http://www.plexigen.com/. Accessed 30 Nov 2010
45. Fitzgerald SP et al (2005) Development of a high-throughput automated analyzer using biochip array technology. Clin Chem 51:1165–1176
46. Randox Laboratories. http://www.randox.com/index.php. Accessed 12 Nov 2012
47. Better ImmunoAssays through Innovative Microfluidics—Siloam Biosciences. https://www.siloambio.com/. Accessed 12 Nov 2012
48. Davies K (2010) The $1,000 Genome: The revolution in DNA sequencing and the new era of personalized medicine simon and schuster
49. The $30 Genome?—Technology Review. http://technologyreview.com/biomedicine/25481/?mod=related. Accessed 12 Nov 2012
50. GnuBio Enters Next-Generation Sequencing Sweepstakes—Bio-IT World. http://www.bio-itworld.com/news/06/02/10/Gnubio-sequencing-debut-consumer-genetics.html. Accessed 12 Nov 2012
51. GnuBIO ahead of schedule with DNA sequencing plans—Mass High Tech Business News.

http://www.masshightech.com/stories/2011/06/06/daily22-GnuBIO-ahead-of-schedule-with-DNA-sequencing-plans.html. Accessed 12 Nov 2012

52. Cepheid | GeneXpert System. http://www.cepheid.com/systems-and-software/genexpert-system/. Accessed 12 Nov 2012
53. Boehme CC et al (2010) Rapid molecular detection of tuberculosis and rifampin resistance. N Engl J Med 363:1005–1015
54. Small PM, Pai M (2010) Tuberculosis diagnosis—time for a game change. N Engl J Med 363:1070–1071
55. Automated Test for Drug-Resistant Tuberculosis Gives Results in Hours—NYTimes.com. http://www.nytimes.com/2010/09/07/health/07global.html. Accessed 12 Nov 2012
56. Cepheid Reports Record Revenue of $44.8 Million—FierceHealthcare. http://www.fiercehealthcare.com/press-releases/cepheid-reports-record-revenue-44-8-million. Accessed 12 Nov 2012
57. NORCHIP. http://www.norchip.com/. Accessed 12 Nov 2012
58. Future Science News and Analysis. Cancer Screening Made Simple, Thanks to Micro-Fluidic Technology —Science Daily. http://www.sciencedaily.com/releases/2010/10/101014113914.html. Accessed 06 Oct 2012
59. Biolithic | Consumer Diagnostic Devices. http://www.biolithic.com/. Accessed 12 Nov 2012
60. Magna Diagnostics. http://magnadiagnostics.com/. Accessed 12 Nov 2012
61. Shoveling Water—Technology Review. http://www.technologyreview.com/biomedicine/24086/. Accessed 12 Nov 2012
62. Abe Lee tells Kathleen Too about the fundamentals of micro- and nanofluidics and lab-on-a-chipdevices.http://www.rsc.org/Publishing/ChemTech/Volume/2009/05/Abe_Lee_interview.asp. Accessed 12 Nov 2012
63. Blow N (2007) Microfluidics: in search of a killer application. Nat Meth 4:665–670
64. Blow N (2009) Microfluidics: the great divide. Nat Meth 6:683–686
65. Becker H (2009) Hype, hope and hubris: the quest for the killer application in microfluidics. Lab Chip 9:2119–2122
66. The Failed Promise of Innovation in the U.S.—BusinessWeek. http://www.businessweek.com/magazine/content/09_24/b4135000953288.htm. Accessed 12 Nov 2012
67. medco.com®: About Medco, latest news. http://www.medcohealth.com/medco/corporate/home.jsp?ltSess=y&articleID=CorpAlertMedco_physician_use_gen_tests. Accessed 01 Dec 2010
68. Stanek EJ et al (2012) Adoption of pharmacogenomic testing by US physicians: results of a nationwide survey. Clin Pharmacol Ther. 91(3):450–458
69. F.D.A. Faults 5 Companies on Genetic Tests—NYTimes.com. http://www.nytimes.com/2010/06/12/health/12genome.html. Accessed 12 Nov 2012
70. Daktari Diagnostics Closes $2.8M Series A Round to Combat Global HIV Crisis | Xconomy. http://www.xconomy.com/boston/2009/09/04/daktari-diagnostics-closes-28m-series-a-round-to-combat-global-hiv-crisis/. Accessed 12 Nov 2012
71. The Boomerang: Healthcare Innovation Goes Where it Must, To the Developing World « Boston Biotech Watch. http://bostonbiotechwatch.com/2010/05/27/the-boomerang-healthcare-innovation-goes-where-it-must-to-the-developing-world/. Accessed 12 Nov 2012
72. Russell J (2010) Alere Investor Presentation, BMO Capital Markets 2010 Healthcare Conference. http://investor.alere.com/news/upload/Alere_Investor_presentation_5-Aug-2010.pdf. Accessed 01 Dec 2010
73. Becker H (2010) Collective wisdom. Lab Chip 10:1351–1354
74. Lab on a chip and Microfluidic Devices (a.k.a., Microfluidics) | LinkedIn. http://www.linkedin.com/groups?mostPopular=&gid=713657. Accessed 12 Nov 2012

Chapter 6

Practical Aspects of the Preparation and Filing of Patent Applications

Fiona Bessoth

Abstract

This chapter is intended to give a brief introduction into the most important patent basics. More importantly, it provides some practical advice to scientists on various considerations when contemplating and preparing the filing of a patent application, irrespective of whether this is assisted by a patent professional or not.

Key words: Intellectual property, Invention, European Patent Convention, Patent Cooperation Treaty

1. Introduction to Patents

The most important step in the preparation of a patent application is a thorough analysis of the invention itself. "The invention" is typically a particular product or process that has been developed. The primary objective of the analysis is to identify whether there is a broader concept or a general principle that makes this particular product or process work, and that is applicable in a broader field. Once a concept is identified, the patent application can be drafted on the basis of the concept, rather than just the particular product or process that was actually developed, thus providing a greater scope of protection and more adequate reward for the inventor. A diligent analysis of the invention does not only provide a better basis for a patent application but can also provide valuable insights and incentives for future research. It is not a rare occurrence that discussions about inventions lead to further developments and inventions.

Gareth Jenkins and Colin D. Mansfield (eds.), *Microfluidic Diagnostics: Methods and Protocols*, Methods in Molecular Biology, vol. 949, DOI 10.1007/978-1-62703-134-9_6, © Springer Science+Business Media, LLC 2013

2. Some Patent Basics

Before entering into the more practical aspects, some basic aspects of the patent system should be outlined. The fundamental idea behind the creation of a patent system was to reward an inventor by an exclusive right to the invention over a number of years (usually 20) in return for a disclosure of the invention to the public. Allowing the public access to information about new inventions was supposed to stimulate and promote further technological progress and innovation.

Patents are exclusive rights for inventions, granted by an authority of a country or a group of countries. Patents entitle their owners to prevent or stop others from making, using, offering for sale, selling, or importing a product or a process without their permission (a license). It is important to note that a patent is a right to keep others from making use of the patented invention. It is not automatically a right to use one's own invention. Rather, it is necessary to respect other people's protective rights when practicing the invention. For instance, a patented speaker of a mobile phone may make use of a material that is protected by the patent of a different owner. Therefore, a license for the use of said material would be necessary to avoid infringement of the patent on the material. It is also important to bear in mind that a patent is a territorial right, which is limited to the boundaries of the country or the group of countries granting it. There is no such thing as a "world patent." Therefore, the filing of a patent application necessarily involves a choice of the territory or the territories where patent protection is desired. This choice could be based on a country's market potential, a competitor's place of manufacture, sites of competing research centers, and various other factors. It is an integral part of the filing strategy.

While some degree of harmonization of standards has been reached in patent systems worldwide, it is important to understand that different countries may have quite different requirements when it comes to the question of what is principally regarded as being patentable, as well as the patent granting procedure itself.

3. What is Patentable?

Patents are granted for inventions that are novel, inventive, and capable of industrial applicability. This principle is ingrained in practically all patenet systems. However, there are significant variations in what these terms are regarded as entailing and in the list of exemptions from this general principle. Within the framework of the various patent laws, an invention is generally understood as a solution to a technical problem. Most patent laws comprise a list of exemptions from what is regarded as an invention for the purpose of patenting, but these lists vary from one country or convention to another.

Under the European Patent Convention (EPC), for instance, discoveries, scientific theories, and mathematical methods; playing games or doing business; as well as presentations of information are generally barred from patent protection, to name but a few examples. However, while the discovery of a physical phenomenon, such as centrifugal forces, for instance, cannot be patented, its application for a particular purpose may well be patentable. In the case of centrifugal forces, for instance, their use to drive fluids from one reservoir through a system of channels to another on a spinning disc-shaped microchip may be patentable. A similar principle applies to the discovery of substances that occur naturally. Whilst a compound naturally occurring in the leaf of a certain tree may not be patentable, its use in a medicament would principally be.

It is worth noting that the patent world uses the word "invention" merely as a technical term, which does not say anything about the significance of the invention or its appreciation by the scientific community. It is not uncommon for scientists to be under the impression that patents are granted only for groundbreaking inventions—which is not true. An invention may be based on just an incremental improvement of a known device or a small "trick," which has a certain technical or economic benefit. On the other hand, the technical or scientific significance of the invention may have (or even should have) an impact on the obtainable scope of protection, with more far-reaching inventions generally entitling to broader protection. It may also be worth noting that it is principally irrelevant to patentability whether the invention was made by chance or whether it is the outcome of years of painstaking research. Post-it® sticky notes are but one example of accidental inventions that turned out to be a huge commercial success.

As mentioned above, a fundamental patentability requirement for an invention is its susceptibility of "industrial application." This requirement demands that the invention must be capable of being used for an industrial or a business purpose, as opposed to a merely private, purely intellectual, or aesthetic activity. "Industrial" is to be interpreted broadly and covers anything from agriculture to sanitary products (the latter not being regarded as private from a patent view at all). Put simply, an invention should have some practical use. It is one of the criteria that rarely present a stumbling stone to the patenting of inventions. One of the more intuitive requirements for an invention to be patentable is that it must be novel. This seems a simple enough statement, but it is quite intriguing how differently an assessment of novelty may play out in the patent practices of the various countries.

One of the common denominators of the various patent laws is that an invention is regarded as novel if it does not form part of the prior art. In general, the prior art encompasses all relevant technical knowledge available to the public anywhere in the world prior to the effective filing date of the patent application. The availability to the public may be in the form of a written document (most common scenario), an oral presentation or other verbal communication, or through prior public use. To give one of the less frequently occurring examples, a Donald Duck cartoon turned out to be a highly significant piece of prior art. The cartoon's illustration of Donald's nephews filling a sunken ship with table tennis balls to make it resurface meant that a patent application directed to a method of salvaging a submerged vessel by filling it with flotation devices was not regarded as being novel.

One of the basic differences between patent systems when it comes to novelty is the point in time that is taken as the starting point for the assessment of novelty, i.e., which decides what constitutes part of the prior art and what does not. As indicated above, in most countries, the filing date of a patent application constitutes the relevant date (the so-called first-to-file principle). The USA, however, currently still works on a first-to-invent principle, where the actual date when the invention was made is decisive. This is somewhat of an oversimplification, as there is a distinction between conception of an invention and its reduction to practice, and there are various requirements to be met, but this goes beyond the scope of this chapter. It was therefore common practice in the USA to have laboratory notebooks signed on a regular basis, so as to be able to furnish evidence of when an invention was made, if need be. In any case, there has been a recent change. The Leahy-Smith America Invents Act (AIA) converts the United States patent system from a "first to invent" system to a "first to file" system. The "first-inventor-to-file" provision will take effect March 16, 2013. The importance of the filing date cannot be emphasized enough. It is particularly important in highly competitive areas where the lead in a certain field is fought over by several research groups. For instance, it is not uncommon for a large number of patent applications to be filed in the wake of a conference, where a new finding was presented, with patent applications being directed to practical uses to which this new finding can be applied. The disastrous 2010 oil spillage in the Gulf of Mexico, for instance, provoked a surge in the filing of applications directed to devices and processes to solve such leakage problems.

It is this novelty requirement that makes it necessary (in most countries) for the invention to be kept secret before a patent application is filed. In practice, this means that a patent application should be filed before publishing a scientific paper, or presenting a new product, giving a presentation to potential investors or even talking about it down the pub (which generally does not involve a secrecy agreement). The prior art one accidentally creates oneself may well turn out to be the biggest obstacle to a patent application. It is one of those acts that are sometimes designated to be one of the "seven deadly sins of the inventor" (http://www.epo.org/topics/patent-system/seven-sins.html). When submitting abstracts or posters to conferences, it should be borne in mind that at least some of their contents may be published before the actual opening of the conference, which could present a bar to their patentability. Therefore, the message is simple: Silence is golden.

Some countries are kinder to their inventors than others, at least concerning this aspect. The USA and Japan, for instance, provide for a novelty grace period, meaning that disclosure of the invention by the inventor does not count as prior art against the inventor's own patent application directed to that invention, as long as this disclosure has occurred within a certain time frame of the filing of that patent application (usually 6 or 12 months). However, it is important to note that this novelty grace period can turn out to be a false friend on an international level, as a disclosure prior to filing constitutes a bar to patentability in all those countries where this rule does not apply. In particular US inventors, who are used to this generous rule, should be aware of this. They might be well advised to file a patent application before disclosing anything in public, at least if patent protection outside the USA is not completely ruled out. On a plus side, this rule does not only concern inventions made in the USA. Thus, if an invention is accidentally disclosed before a patent application was filed, all is not lost; obtaining patent protection in the USA and Japan would principally still be an option.

A further requirement for an invention to be patentable is that it must be inventive. This requirement is intended to eliminate changes to the prior art that would have been obvious or even trivial. Even though it may already be apparent from the above, it is noted that novelty and inventive step are assessed on an objective basis, namely, on the basis of all available prior art. The assessment does not take into account the actual starting point of the inventor, i.e., the knowledge the inventor may or may not have had of a certain piece of prior art. So, while an invention may be creative and nonobvious when considering the inventor's knowledge and starting point, it may still fail the obviousness test when compared to a different piece of prior art.

4. Anatomy of a Patent Application

It has been mentioned above what the patentability requirements are and that patent applications must be filed with the competent authority for a certain country or a group of countries. Before outlining what exactly happens to the application after its filing, a quick look is taken at the necessary contents of a patent application.

A patent application usually comprises the following parts:

- A description of the invention
- Claims
- Drawings (where appropriate)
- An abstract

The claims are the heart of the patent application/patent. A claim is a very short and concise description of the essential elements of an invention. The claims define the desired scope of protection. Therefore, the claims are of key importance both in patent prosecution (from the filing of an application to the grant of a patent) and patent litigation (enforcing a patent).

The importance of careful claim drafting and careful review of the claims by the inventor cannot be emphasized enough. Practical advice concerning this aspect will be given further below. Generally, claims may be directed to the following subject-matter:

- Products (such as a device, compound, apparatus).
- Methods/processes (e.g., a method for manufacturing, analytical methods, methods for altering/treating, etc.).
- The use of a product or a process for a certain purpose (e.g., a new medical indication).

Often, an invention comprises several aspects that belong in different categories. An invention residing in a particular microchip-based immunoassay could include a claim directed to the microchip itself and a claim to the immunoassay (i.e., analytical method, on chip if the immunoassay itself had been known beforehand)—always presuming both are novel, of course. If the microchip itself had been known beforehand, for instance, it would principally still be possible to claim the immunoassay (on chip) and the use of the chip for this particular immunoassay.

There is a further differentiation of claims, that of independent claims and dependent claims. An independent claim is the broadest claim to any specific subject-matter, that is, the one that defines that particular aspect of the invention most broadly (e.g., microchip, analytical method, method of manufacture, fluorescent marker, etc.). It is usually followed by one or more dependent claims, which refer to the independent claim and optionally one or more preceding dependent claims, and are typically increasingly specific.

The subject-matter of a dependent claim is also defined by the subject-matter of the independent claim to which it refers. The reference to another claim just avoids the necessity to repeat in its entirety the subject-matter defined therein for any subsequent claim. Dependent claims are therefore directed to specific embodiments of the invention. They may strengthen a patent infringement case, when more than one claim is infringed, rendering the infringement more obvious. If, for instance, an independent claim is directed to an electrophoretic microchip comprising a separation channel and an injector, dependent claims could define the dimensions of the separation channel, the particular design of the injector, the material the chip is made out of, any surface coating, and the like.

The description, together with drawings, if present, must sufficiently disclose/enable the claimed subject-matter. Expressed differently, the description must put a person of average skill in the field in a position to put the invention into practice. For example, if a claim is directed to a hitherto unknown compound, the description must set out how to obtain, for instance synthesize, the compound. Equally, if a claim is directed to a microchip, the description must describe how the microchip can be made, unless this belongs to the common general knowledge in the field. The description generally contains different subsections:

- An introductory sentence indicating the field of the invention, that is to say, the general category to which the invention belongs (e.g., "The present invention relates to a microchip for chemical analysis and a method for its manufacture"). This introductory sentence typically lists the beginning of the independent claims.
- An introduction in which related prior art and its drawbacks are typically mentioned and the task the inventors have sought to achieve by the invention is described (e.g., "The detection limits that current microchip-based analytical systems are able to provide are still unsatisfactory, in particular for the detection of absorption of UV/visible light … It is therefore an object of the present invention to provide a microchip for chemical analysis that provides better detection limits …")
- Care should be taken in formulating the problem to be solved or task to be achieved. It should be free from hindsight, that is to say, free from the inventive concept. In the above example, any reference to an "increased absorption path" in the problem would already contain one step toward the solution to the problem (other solutions could have been to go in an entirely different direction, i.e., choosing fluorescence, using a signal enhancer, etc.). The object or the task should be as general and abstract as the description of the problems associated to the prior art will permit, for example, "easier to manufacture," "better resolution," "higher speed," etc.

- The introduction is usually followed by a description of the invention itself, starting with the general and subsequently increasing in specificity (e.g., "The object has been solved by a microchip comprising a detection cell that has at least one reflective surface arranged such that … The inventive flow cell has the advantage that a light beam passing through has a longer absorption path as a result of the beam being reflected back … thus passing twice through the analyte … The reflective surface can be made of … by … be disposed such that …").
- A patent application should contain at least one specific example, otherwise referred to as "exemplary embodiment" in patent jargon. In the USA, there is a requirement to describe the best mode of carrying out the invention.
- Both the general description and the description of the exemplary embodiment can be illustrated by one or more figures, if appropriate. These figures are typically schematic line drawings with reference numerals that are explained in the description, but could also be photos (e.g., microscope images), amino acid sequences, electropherograms, or other graphical representations of analytical data. Inventions relating to a method, such as a method of manufacture, for instance, could be illustrated by a flowchart or a sequence of schematic drawings illustrating the individual steps or the effect they are intended to achieve (e.g., a sequence of etching steps describing the fabrication of a microstructure).
- Patent applications also typically contain an abstract, which is for information purposes only, i.e., its purpose is to give an idea of the contents of the patent application. It plays no part in patent prosecution or patent infringement. Often, the abstract just reiterates the independent claim considered as reflecting the invention best.

One of the most important things to bear in mind is that a patent application may not be supplemented by more information at a later date. There appear to be some minor exceptions to this rule in some countries, but even then the room for additions is minimal and certainly insufficient to remedy fatal flaws.

5. Patent Prosecution

Before going into more details concerning the drafting or checking of a patent application, an outline of the patent examination procedure is given below, as this will help clarify the role of the claims during the patent prosecution stage.

The most important steps of a typical patent prosecution process are explained below by way of reference to the proceedings before the European Patent Office (EPO). There are some variations between different countries/conventions, but the following main elements are generally implemented in most countries at least roughly in the same manner.

- The applicant files a patent application with the competent authority (paying attention to the filing requirements, such as a completed official request form, etc.).
- In a first step of the patent grant procedure before the EPO a formalities examination takes place, involving checking that the application documents are in order, the necessary fees are paid, etc.
- In a second step, a search for related prior art is carried out order for the assessment of whether the subject-matter claimed in the application meets the patentability requirements, i.e., is novel, inventive, and liable to industrial application. The applicant may amend the claims, if desired, in order to better distinguish the claims over the related prior art.
- The patent application is published 18 months after its filing date. Usually, the prior art search will be completed by then and the publication of the application also contains the Search Report, which provides a list of what prior art has been identified in relation to which claims. This publication of the patent application, in turn, makes the application prior art for later patent applications.
- The next stage is that of substantive examination. An official communication is issued by an examiner, setting out any objections to novelty, inventive step, industrial susceptibility, as well as clarity and sufficiency of disclosure, if necessary. The applicant can amend the claims in order to meet the examiner's objections. The claims can be amended, for instance, by combining an independent and a dependent claim, thus limiting the claimed subject-matter and thus the scope of protection. It is also possible to add limitations from the description.
- If the examiner is then (after one or more exchanges with the applicant) satisfied that the independent claims are limited to patentable subject-matter and that the claims and the description are generally in order, a patent is granted. Otherwise, the patent application is refused.

As evident from the above, the dependent claims can provide fallback positions in case it should turn out that the subject-matter of the independent claim(s) does not meet the relevant requirements. Now that the claim structure and the purpose of the claims in patent prosecution have been explained, some considerations for the reviewing (and drafting) of patent applications, and in particular the claims, are outlined below. For completeness sake, it is noted that in some patent systems, the filing of claims together with the application is not necessary in order to be accorded a filing date (for instance, under the EPC or in a US provisional application). However, in order to be granted a patent eventually, claims must be furnished at some stage of the procedure. These claims must then be taken from within the description, thus generally severely limiting the options for claim drafting, because there is then no more room for generalization. Therefore, it is generally strongly advisable to include claims from the outset.

6. Do the Claims Truly Reflect the Invention?

As already mentioned before, it is usually a particular product or process ("embodiment" in patent jargon) rather than an abstract concept that provides the incentive for the filing of a patent application. It may be tempting to claim just that product, or possibly just a little bit around it; however, that might not do the invention justice. Therefore, in the preparation of a patent application and in particular in the drafting and reviewing of claims, it is important to understand what exactly the invention is. This understanding may be aided by answering the following questions:

- What does the invention achieve? What problem or task is solved by the invention? Think of all the advantages associated with the invention as well as the perimeter within which the invention was supposed to operate. For example, "A micromixer is capable of delivering fast and efficient mixing, even in high pressure applications, even having a volume of only …, allowing on-chip detection." The advantages do not need to be scientific or technical in a narrow sense, but can also encompass concepts such as being more economical, more user friendly, better suited for application in the field, etc.
- How is the problem solved or the task achieved? Is there a physical phenomenon that is taken advantage of? Is a correlation or interplay between two factors/aspects required? Is there a synergy of multiple factors/aspects? In the micromixer example above, the general idea might be to split the flows of the liquids up and arrange them in an alternating fashion, with the underlying fundamental explanation being the decrease of the diffusion distance.
- Are there alternative ways to solve the problem by taking advantage of the same solution, for instance taking advantage of the same physical phenomenon? It may help to take a competitor's perspective by asking yourself, "What would I have to do to avoid the patent claim and still achieve the same or similar effect (some drawbacks being acceptable)?" One might even think of utterly unrealistic alternatives, just to work out

what all the alternatives that are assumed to provide the same or similar effect have in common. Are there common functions or structures in these alternatives?

- Are there other applications/fields where the invention could be put to good use?
- Are the claim features (contents of the claims, such as parts of a product or steps of a process) absolutely essential? Could they be defined in a more general/functional manner, or even deleted from the independent claim altogether?

Working out all the above questions may seem to require much effort. However, identifying the heart of the invention does not only provide the best basis for a patent application and thus adequate patent protection but often also furnishes valuable insights that might spark further development and inventions. So the time and energy spent is generally a good investment. Once the above questions are answered, it is worth thinking about the prior art known to you. Is the concept novel over the prior art? If not, which part of the concept is novel and can be claimed?

Sometimes, inventors receiving a draft set of claims from an attorney for review have difficulty understanding the claim and wonder whether it is actually still their invention defined therein. No need to panic! The attorney may just have done an excellent job and taken the invention to a fairly high conceptual level that even the inventors had not recognized before. It is important that the inventor should understand the claim so as to be able to assess whether it does indeed reflect the invention/the inventive concept and excludes the known prior art.

Claims are, by their very nature, verbal definitions of inventions. Therefore, it should also be questioned whether this definition is sufficiently clear and unambiguous: That is, are the claim terms clear or at least clearly defined within the description of the patent application? Are they used in accordance with their usual meaning? Physical parameters present in the claims warrant particular attention. An average particle size, for instance, may have different values, depending on how it is measured. Therefore, the exact measurement method used should be indicated at least in the description.

7. Patent Filing Strategy and Tools Designed to Make Life Easier

The preparation of the patent application also involves a decision where patent protection would be desirable. As mentioned before, patents are territorial rights, such that, principally, the filing of a patent application is necessary for each country or region where grant of a patent and thus patent protection is sought. Of course, the simultaneous filing of patent applications in numerous countries would be

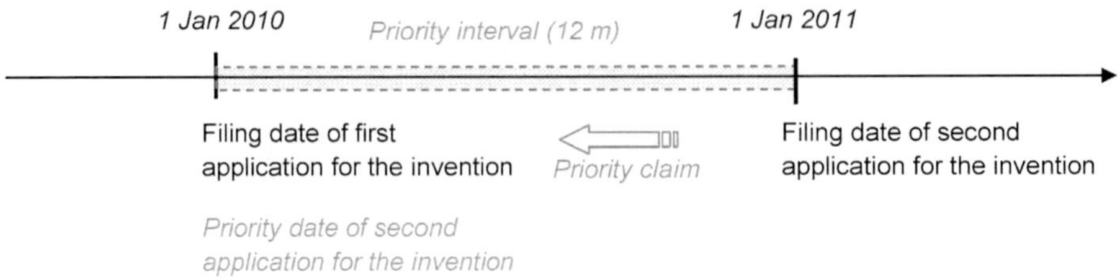

Fig. 1. Priority claim timeline.

associated with a high investment of organizational and financial resources, which may be rendered even more difficult by a possible uncertainty of the potential success of the invention underlying the patent application. Several instruments are available to facilitate the decision concerning which countries to seek patent protection, on the one hand, and to facilitate the practical filing of patent applications with effect in numerous countries, on the other hand.

The first instrument is the so-called priority claim (see Fig. 1). If an applicant (or its predecessor in title) has filed an application for a patent (or utility model/certificate) in or for any state that is party to the Paris Convention (PC) for the Protection of Industrial Property or any member of the World Trade Organization (WTO), then this applicant may claim priority when filing a patent application for the same invention within 12 months after the filing date of the first application (in any of the PC/WTO member countries/conventions). If the claim to priority is valid, i.e., if the formal requirements were fulfilled, the later application was filed in time, and the inventions in the two applications are indeed the same, then the priority claim has the effect that the date of the priority application (rather than the filing date of the later application) determines the prior art that can be cited against the later application. Thus, by filing a patent application in one country, the applicant effectively gains one year without any negative impact on patentability (because it is the filing date of that priority application that determines the prior art). During that year, the applicant has the chance to gather more information on the potential value of the invention, find potential investors, check market potentials, etc. to assist the decision as to if and where to pursue further patent applications. Depending on where the priority application was filed, the applicant may have a prior art search carried out and be in receipt of the search results. These search results give an idea whether and to what extent the claimed subject-matter is patentable, thus providing a further piece of information to assist in taking the decision if and where to pursue the patent application further. Thus, a bigger financial investment can be postponed until there is some indication of the prospects of success of the patent application on the one hand, and the invention itself on the other. Also, the priority application does not affect the lifetime of the later patent application, because the filing date of the later application is still the starting date for the patent life, irrespective of any priority claim. It may be added that if a priority is claimed, publication of the (later) application will be effected 18 months after the priority date, not the filing date of the later application.

The later application is an independent application; thus it is principally possible to add subject-matter, such as examples and experimental details, etc., and to remedy any deficiencies in the earlier (priority) application. However, care must be taken, because the claim to priority can only be valid if the later application is directed to the same invention. Expressed differently, the subject-matter claimed in the later application must have already been contained (in the same degree of generality or specificity) in the earlier (priority) application. The standards, according to which the priority claim is scrutinized, are rather strict. After all, it would be unfair to award an earlier effective filing date to an invention (and exclude prior art based on that assumption) if it was not absolutely certain that the invention had indeed already been made (and disclosed) at the filing date of the priority application.

Secondly, a number of patent conventions are in place, which facilitates the patent prosecution in more than one country. The EPC, for instance, allows the grant of a patent by the EPO for all or any of its Member States (consisting of 38 Contracting States as of 1st January 2011). The filing, search, and examination of the patent application all take place centrally with/by one authority, the EPO. However, once granted, a European patent is a "bundle" of individual national patents, rather than a patent that would be automatically valid for all member states. Currently, a patent must be enforced in each member state of the EPC individually, just like a national patent (i.e., a patent granted by the national authority of that member state). A true Community patent, which would automatically take effect across the entire European Community, has been planned for years but has so far failed to be implemented.

Thus, in order to obtain a patent in France, the United Kingdom, Italy, and Turkey, for instance, one may file patent applications with the national authorities of each of those states to obtain patents via the national route (potentially involving the need to provide translations into the national language for the filing or shortly after the filing), thus going through the patent search and examination process in each of these countries. In the alternative, one may file one patent application with the EPO designating France, the United Kingdom, Italy, and Turkey as the states in which the patent should apply. This has the advantage of only one central search and examination procedure, which can lead to the grant of a patent for all of these countries. For a European patent to take effect in the various states, the individual states may have specific requirements, such as the filing of a translation of the granted patent into the national language, the payment of a fee, etc.

Thus, in addition to an administrative simplification of the procedure by having only one central search and examination procedure, costs for translations arise only after grant.

There is a further tool that provides for a significant simplification, especially when seeking patent protection in various countries worldwide: the Patent Cooperation Treaty (PCT). The PCT makes it possible to seek patent protection for an invention simultaneously in each of a large number of countries (142 Contracting States as of 1st January 2011) by filing an "international" patent application. Such an international patent application may be filed by anyone who is a national or a resident of a Contracting State of the PCT. It may generally be filed with the national patent office of the Contracting State in which the applicant is a national or a resident. If that Contracting State is party to a Convention (e.g., EPC, Eurasian Patent Convention) it may also be filed with the competent patent office for that Convention, or with the International Bureau of the World Intellectual Property Organization (WIPO) in Geneva. What then follows is first the so-called international phase including:

- Formalities examination
- "International search": Search for prior art and issuance of an "international search report," as well as a written opinion on patentability
- Publication of the application by the International Bureau
- Optionally, a supplementary search or a preliminary examination of the application

The search and optional preliminary examination are carried out by one of the major patent offices for the PCT, and are therefore no different from any search carried out by a national office. ("International" stands for the search being carried out during the international phase, not the nature of the search—every prior art search is international in the sense that prior art from all over the world is searched.)

After this "international phase," the applicant may enter into one or more "national phases." The national (or regional) offices then examine the application and grant or refuse the national (or regional) patent on the basis of their national laws. Further action is required before proceeding into the "national phase," including payment of the required national (or regional) fees, and possibly translation of documents and appointment of a local representative, to each of the national (or regional) offices from which the applicant wishes to be granted a patent on the basis of his or her international application. These actions must be completed within a certain time limit, which varies from country to country, of between 20 and 42 months after the priority/filing date, but amounts to 30–31 months for most countries.

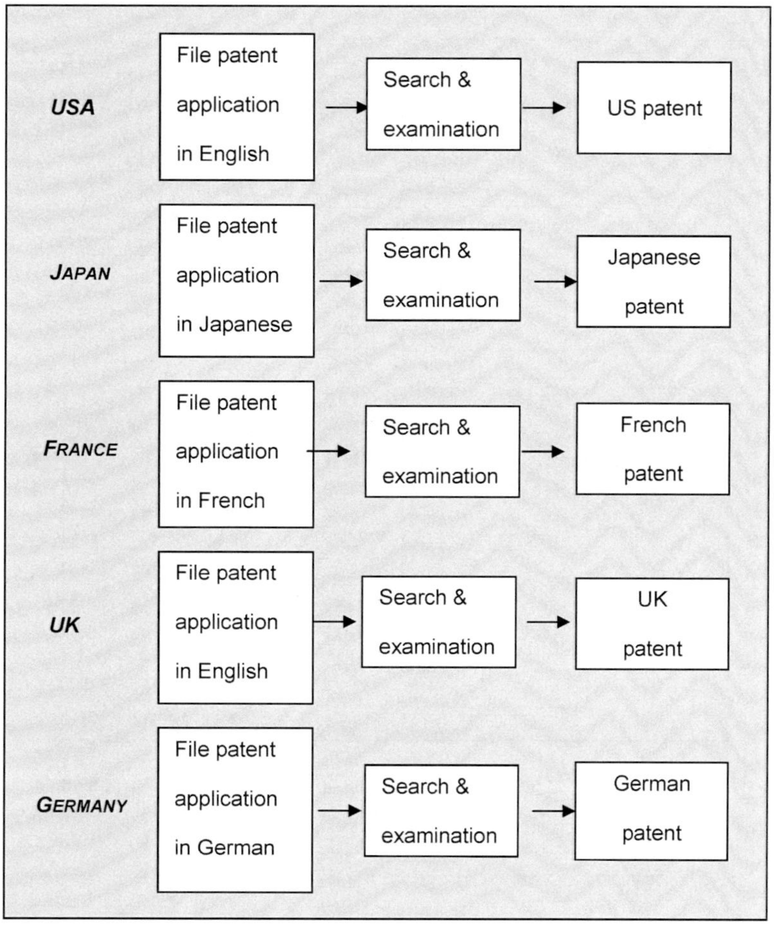

Fig. 2. Patent protection via national patent offices.

The PCT offers a number of advantages, compared to individually filing patent applications in various countries, amongst which the following are to be mentioned:

- Only one application in one language must be filed with one patent office (even if patent protection in numerous countries is desired).
- The applicant can consider the outcome of the prior art search and thus evaluate the chances of his or her invention being patented before taking a decision in which countries to pursue the application further (thus potentially saving expenses and/or postponing the associated expenses until—on average—30 months after the priority date).

Fig. 2 compares the different ways to obtain patent protection in the USA, Japan, France, the United Kingdom, and Germany via national patent offices. Figures 3 and 4 illustrate the routes via the national patent offices/EPC, and via the PCT, respectively.

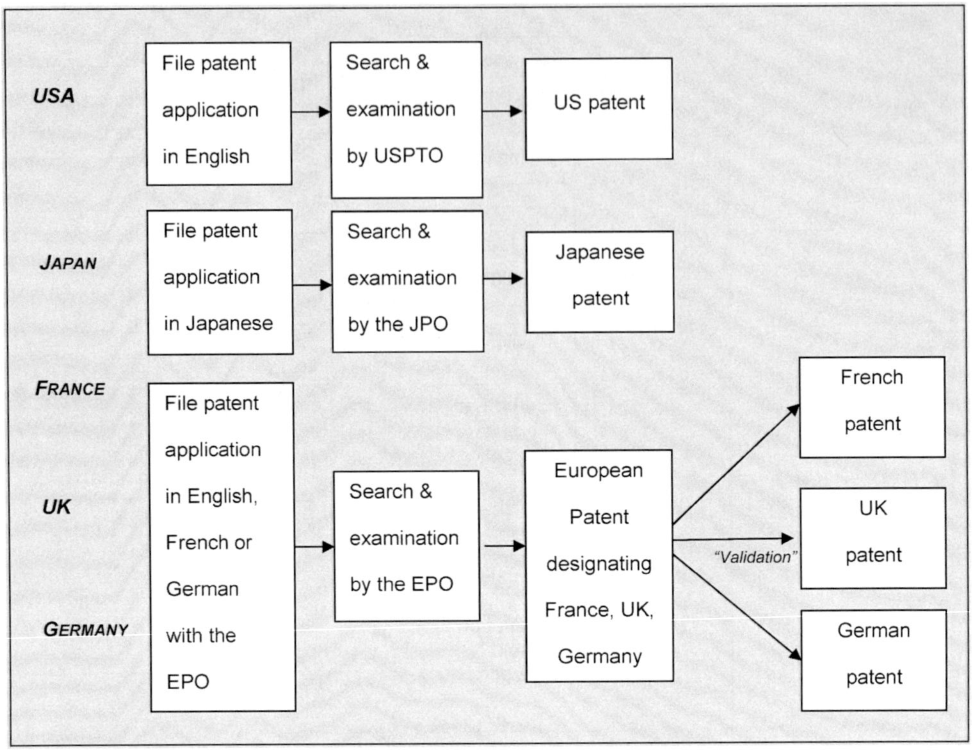

Fig. 3. Patent protection via national patent offices or European Patent Convention.

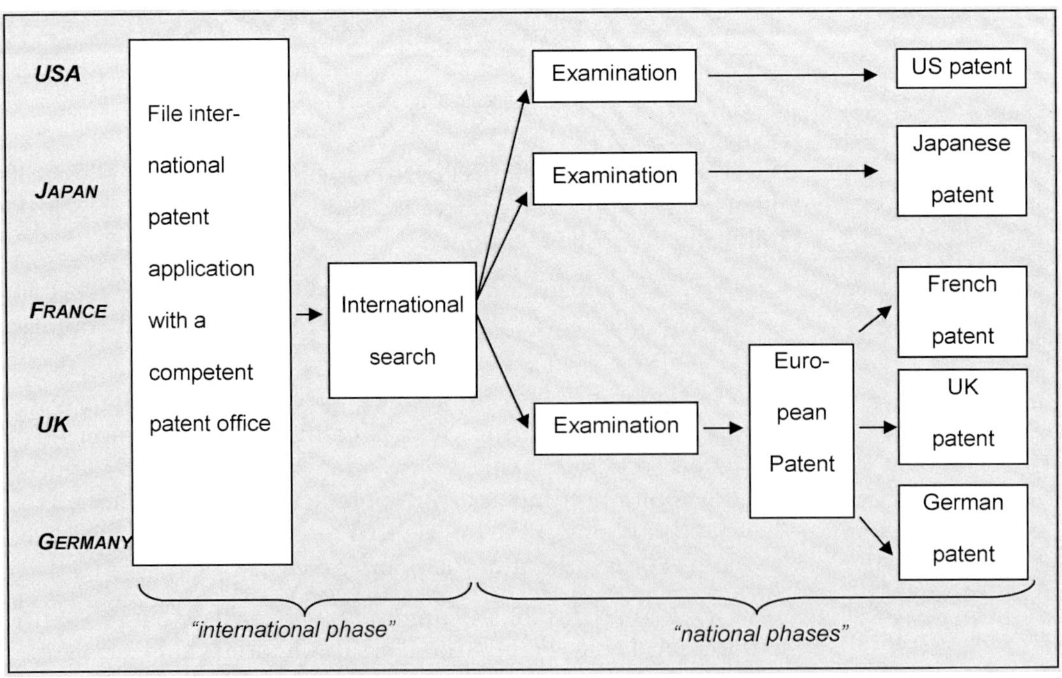

Fig. 4. Patent protection via the Patent Cooperation Treaty.

8. Summary

A thorough analysis of the invention provides an invaluable basis for the drafting of a patent application and the scope of patent protection to be gained from it. This is where the scientist or the engineer can provide the most significant input.

This chapter could only provide a rough outline of the various cornerstones of the patent system and mention some practical considerations when reviewing (and possibly even drafting) patent applications and particularly claims. More detailed information on the filing of patent applications, including the particular requirements in the various countries, can be gathered from various online sources, some of which are mentioned below.

9. Exemplary Online Sources of Information

- The *EPO* provides the guides "How to get a European patent. Guide for applicants. Part 1" (including information on exactly what must be filed in what form with whom and by when, as well as examples of patent applications and claims) and "How to get a European patent EURO-PCT. Guide for applicants. Part 2" and a plethora of further information material online:
- www.epo.org (for the applicant's guides, go via "Quick access," section "Grant procedure" and choose "Guide for applicants")
- The WIPO operates a Web site with information on the *PCT*, such as a very detailed Applicant's Guide, a document providing answers to frequently asked questions, and a site with links to national patent offices: www.wipo.int (for the applicant's guide: http://www.wipo.int/pct/en/appguide/index.jsp or choose "IP Services," "Patents," "PCT Resources," "Filing," and then the "PCT Applicant's Guide"; the directory of IP offices is accessible via www.wipo.int/directory/en/urls.jsp)
- A list of the members of the Paris Convention can be accessed via http://www.wipo.int/treaties/en/ip/paris/
- The members of the WTO can be accessed via http://www.wto.org/
- The *USPTO* equally provides a host of information in the section on "Patents": www.uspto.gov
- "*Espacenet*" is a European network of databases that provides a facility (free of charge) to search patent applications and patents (i.e., carry out a prior art search or search for patents that could be relevant for the freedom to operate one's own invention): www.espacenet.com

- The *Japanese Patent Office* provides information on procedures to obtain a patent and other aspects also in the English language: www.jpo.go.jp
- There is also English information available on the Web site of the *State Intellectual Property Office of the People's Republic of China*: www.sipo.gov.cn

Chapter 7

Introduction to In Vitro Diagnostic Device Regulatory Requirements

Jonathan Day

Abstract

A common application for microfluidics can be found in medical devices where the advantages of small volume measuring equipment can be exploited for In Vitro Diagnostics. This chapter focuses on the US and the EU regulations, explaining the broad landscape and regulatory pathways of each market.

Key words: In vitro diagnostic device, FDA regulations, 510K, EU regulations, CE marking, Quality management systems

1. US Regulatory Structure

The USA regulates in vitro diagnostic devices (IVD) through the Food and Drug Administration (FDA) which includes, but is not limited to, medical, veterinarian, cosmetic, and food products. IVDs are regulated as medical devices which fall under the jurisdiction of the Office for In Vitro Diagnostics (OIVD), which in turn comes under the control of the Center for Devices and Radiological Health (CDRH), all of which fall under the vast umbrella of the FDA. It is important to note that many IVDs may be combined with nondiagnostic products; therefore, an awareness of other offices/departments within the FDA that may be consulted as part of the approval process for IVDs is highly desirable. An example of a product that may require review from a different FDA department could be a diagnostic test intended for HIV diagnosis regulated by the FDA's Center for Biological Evaluation and Research (CBER). Although beyond the scope of a microfluidic IVD developer it is prudent for the would-be manufacturer to be aware that the regulatory control of IVDs is also covered by the Clinical Laboratory Improvement Amendments.

Gareth Jenkins and Colin D. Mansfield (eds.), *Microfluidic Diagnostics: Methods and Protocols*, Methods in Molecular Biology, vol. 949, DOI 10.1007/978-1-62703-134-9_7, © Springer Science+Business Media, LLC 2013

The FDA defines an IVD as: "… those reagents, instruments, and systems intended for use in diagnosis of disease or other conditions, including a determination of the state of health, in order to cure, mitigate, treat, or prevent disease or its sequelae. Such products are intended for use in the collection, preparation, and examination of specimens taken from the human body (Code of Federal Regulations, Title 21, Section 809.3) (21 CFR 809.3)."

The FDA also defines a regulatory authority as: IVDs are *medical devices* as defined in section 210(h) of the Federal Food, Drug, and Cosmetic Act, and may also be *biological products* subject to section 351 of the Public Health Service Act. Like other medical devices, IVDs are subject to premarket and postmarket controls. IVDs are also subject to the Clinical Laboratory Improvement Amendments (CLIA '88) of 1988.

2. US FDA Regulations; 510K and PMA

The FDA's IVD regulations are defined in the Code of Federal Regulations (21 CFR), which categorizes IVDs on a risk basis with three categories, namely, Class I, Class II, and Class III, with a higher class related to higher risk. The CFR is a codification of the general and permanent rules published in the Federal Register by the Executive departments and agencies of the Federal Government. Title 21 of the CFR is reserved for rules of the FDA. Each title (or volume) of the CFR is revised once each calendar year. In order to legally market your device in the USA it may require either clearance or approval from the FDA. In the case of class III devices this will require approval through the Pre-Market Application (PMA) process, whereas for most class II and some class I devices a pre-market notification (also known as a 510K after the section 510(k) of Federal Food Drug and Cosmetics Act) will be required.

2.1. Low-Risk Devices

Many class I devices and some class II devices are 510K exempt, meaning that subject to limitations on exemptions covered under parts of 21 CFR, a pre-market notification is not required before placing your device on the market. For IVDs this can be found in 21 CFR 862.9, although this does not include near-patient testing devices of an IVD that would otherwise be exempt if performed in a laboratory. In the section "Limitations of exemptions from section 510(k) of the Federal Food, Drug, and Cosmetic Act" it is stated, "The exemption from the requirement of premarket notification (section 510(k) of the act) for a generic type of class I or II device is only to the extent that the device has existing or reasonably foreseeable characteristics of commercially distributed devices within that generic type or, in the case of in vitro diagnostic devices, only to the extent that misdiagnosis as a result of using the device would not be associated with high morbidity or mortality."

2.2. Medium-Risk Devices

Medium-risk devices currently require pre-market notification following section 510(k) of the Food Drug and Cosmetics act. A 510K submission is required for the following three scenarios:

1. Introducing a device to the market for the first time.
2. Change in indications for use for a previously cleared device.
3. Making significant modification to a previously cleared device.

A 510K itself is not a form but evidence to prove substantial equivalence to a device already legally marketed in the USA (also known as a predicate) where evidence is created by the sponsor/manufacturer of the device and presented to the FDA to allow the determination of whether their substantial equivalence to the predicate. Some devices may go through a pre-market notification by using special controls prescribed by the FDA. Special controls may include special labeling and/or performance requirements that can be seen as prescribed design controls (part of 21 CFR 820), described later in the chapter.

2.3. High-Risk Devices

High-risk devices require a PMA from the FDA before legal marketing of the product can occur in the USA. If an IVD is not 510K exempt and is also not deemed to be a medium-risk device it will require a PMA. For IVDs this may be due to a high risk intended use such as for HIV/AIDS diagnosis (note that HIV diagnostic assays are types of devices that are also regulated by the Center for Biologics Evaluation and Research). PMAs are expensive and lengthy processes, prompting the FDA to promote the modular PMA, a module-based submission process suitable for some, although not all, sponsors/applicants.

2.4. Research Use Only and Investigational Use Only

IVDs have a unique characteristic compared to other medical devices in that they can be made available to the market under the labeling and controls of the "Investigational Use Only" device. Requirements for this include, but are not limited to, that tests must be confirmed by an approved test (i.e., gold standard) and that specific labeling accompanies the product. "Research Use Only" devices by definition cannot be considered to be medical devices as their intended use is beyond the scope of IVD and as long as the intended use does not fall under the IVD medical device definition they are not regulated by the FDA.

3. European Regulations and CE Marking

The European Union developed the CE marking process for many groups of products including toys, measuring devices, and medical devices. Medical devices are regulated by three separate directives: the medical device directive (MDD 93/42/EEC), the Active Implantable Medical Device Directive (AIMD 90/385/EEC), and the In Vitro Diagnostic Directive (IVDD 89/79/EC). The CE marking process is based upon the Conformity Assessment procedure which is defined in the Council Decision 93/465/EEC with eight conformity assessment modules:

Module A	Internal production control
Module Aa	Internal production control with supplementary requirements
Module B	EC type-examination
Module C	Conformity to type
Module D	Production Quality Assurance
Module E	Product quality assurance
Module F	Product verification
Module G	Unit verification
Module H	Full quality assurance

The EU regulating structure is different to that of the USA where the FDA has the ability to oversee not only the approval of submissions but also the compliance of manufacturers and developers through auditing and other activities. In comparison the EU could appear to be fragmented with the legislation coming from a central European commission and individual member states maintaining Competent Authorities (e.g., Medicines Healthcare Products Regulatory Agency or MHRA in the UK) to ensure compliance with the EU regulations/directives through the approval of Notified Bodies, who in turn are able to certify and where applicable execute the relevant CE marking modules to approve products for CE marking. A recent change to the directives requires a manufacturer who does not have a place of business within the EU to appoint a legal representative or Authorized Representative to make available information requested by the competent authority when required.

The IVD Directive has 24 Articles and 10 Annexes; combined they provide seven possible routes for CE marking depending on the device. As with the US regulations the intended use of the device determines which risk category the IVD is likely to come under.

3.1. Annex II List A

Annex II list A defines the highest risk category IVDs as either: "Reagents and reagent products, including related calibrators and control materials, for determining blood groups: ABO system, rhesus (C,c,D,E,e) anti Kell" or as "Reagents and reagent products, including related calibrators and control materials, for the detection, confirmation and quantification in human specimens of markers of HIV infection (HIV 1 and 2), HTLV I and II and hepatitis B, C and D." As this is the highest risk category the regulatory pathway for approval requires the intervention of a notified body. Additionally, as this classification offers significant risk there is a requirement to prepare a technical file, known as the Design Dossier, to ensure that the product meets essential common technical specifications as outlined by the Commission Decision of the 7 May 2002.

Regulatory approval for this classification of device can be in the form of either:

- EC Declaration of conformity/full quality assurance, design examination, verification of manufactured products following Annex IV of the directive (Module H) or
- EC type-examination (Annex V, Module B) plus EC declaration of conformity/production quality assurance/verification of manufactured products (Annex VII, Module D).

3.2. Annex II List B

A full list of Annex II products can be found in Directive 98/79/EC, which includes reagents and reagent products for the detection and quantification in human samples of rubella and toxoplasma, or determining Chlamydia or the tumoral marker PSA, as a few examples. Annex II List B devices also require the intervention of a notified body.

Regulatory approval for this classification of device can be any of the following:

- EC declaration of conformity/full quality assurance (Annex IV minus points 4 and 6, Module H),
- EC type-examination (Annex V, Module B) plus EC verification (Annex VI, Module F), or
- EC type-examination (Annex V, Module B) plus EC declaration of conformity/production quality assurance (Annex VII minus point 5, Module D).

3.3. General IVD

This IVD group covers the vast majority of IVDs. Providing the manufacturer complies with the requirements set out in Annex III of the IVD Directive and maintains a technical file showing compliance with the essential requirements of the directive, the manufacturer or authorized representative can "self-declare" conformity. This process is also known in the industry as self-certification, and although this sounds like an attractive route to launch a first product it also provides the most opportunities for errors and misinterpretation of the directive, as there is no requirement for the involvement of a notified body. Regulatory approval for this classification of device entails EC declaration of conformity (Annex III minus point 6, Module A).

3.4. Home Use IVD

When an IVD which does not fall under the Annex II list A or B categories and is intended for self-testing it requires the assessment of a notified body and cannot go through the self-certification EC declaration of conformity route. Regulatory approval for this classification of device is as follows:

- EC declaration of conformity (Annex III including point 6, Module Aa).
- EC declaration of conformity/full quality assurance (Annex IV minus points 4 and 6, Module H).
- EC type examination (Annex V, Module B) plus EC verification (Annex VI, Module F)
- EC type examination (Annex V, Module B) plus EC declaration of conformity/production quality assurance (Annex VII minus point 5, Module D).

4. Quality Management Systems

Both the USA and the EU require certain criteria to be met when developing, manufacturing, marketing, and postmarketing a device. The USA defines its regulations in 21 CFR 820 whereas the EU requires certain essential requirements to be satisfied. Both markets require high-risk devices to be tested by a third party either concurrently with a batch release or for the quality assurance system to be examined and approved before and while a device is marketed.

4.1. Medical Devices 21 CFR 820; Current Good Manufacturing Practice

Quality system regulation (QSR) requires a developer and manufacturer to provide a Quality Management System (QMS) that provides transparency and traceability of its design, manufacturer, and postmarket vigilance (complaint handling) activities, the main criteria of which are listed in Table 1.

4.2. International Standard ISO 13485

Although not a prerequisite in the EU but widely employed, the international standard ISO 13485 can be used to ensure that essential requirements are met. The standard is based on ISO 9000, a common QMS for many businesses and has many parallels to the US QSR; however, while there is a voluntary accreditation to the QSR (i.e., there is no requirement to be assessed before claiming a company is operational under a QSR management system) there is a requirement to be ISO 13485 certified by a notified body, to say that

Table 1
Main criteria for quality system regulation

Subpart	Criteria
Subpart A General Provisions	820.1 Scope (of requirements) 820.3 Definitions 820.5 Quality system
Subpart B Quality System Requirements	820.20 Management responsibility 820.22 Quality audit 820.25 Personnel
Subpart C Design Controls	820.30 Design controls
Subpart D Document Controls	820.40 Document controls
Subpart E Purchasing Controls	820.50 Purchasing controls
Subpart F Identification and Traceability	820.60 Identification 820.65 Traceability
Subpart G Production and Process Controls	820.70 Production and process controls 820.72 Inspection, measuring, and test equipment 820.75 Process validation
Subpart H Acceptance Activities	820.80 Receiving, in-process, and finished device acceptance 820.86 Acceptance status
Subpart I Nonconforming Product	820.90 Nonconforming product
Subpart J Corrective and Preventative Action	820.100 Corrective and preventative action
Subpart K Labeling and Packaging Control	820.120 Device labeling 820.130 Device packaging
Subpart L Handling, Storage, and Distribution	820.140 Handling 820.150 Storage 820.160 Distribution 820.170 Installation
Subpart M Records	820.180 General requirements 820.181 Device master record 820.184 Device history record 820.186 Quality system record 820.198 Complaint files
Subpart N Servicing	820.200 Servicing
Subpart O Statistical Techniques	820.250 Statistical techniques

a company is operating under a 13485 quality system. Similar to the QSR, the main sections of the standard are summarized in Table 2.

4.3. Design Controls

Although compliance to each part of the standards or regulations is necessary to have compliant QMS, in the context of this publication product development is worth expanding upon. Product development in both the US QSR and ISO 13485 is very similar

Table 2
Main sections of the international standard ISO 13485

Section	Criteria
Quality Management System	General requirements Documentation requirements
Management Responsibility	Management commitment Customer focus Quality policy Planning Responsibility, authority, and communication Management review
Resource Management	Provision of resources Human resources Infrastructure Work environment
Product Realization	Panning of product realization Customer-related processes Design and development Purchasing Production and servicing provision Control of monitoring and measuring devices
Measurement, Analysis, and Improvement	Monitoring and measurement Control of nonconforming product Analysis of data Improvement

and can be broken down into several phases often described by the US QSR terminology "Design Controls."

4.3.1. Design and Development Planning

A documented plan that describes or references the design and development activities, and personnel responsible for implementation of the development. The plan must be reviewed, updated, and approved as the design and development procedure evolves.

4.3.2. Design Inputs

Design requirements that are appropriate for both the intended use of the device and the end user requirements. These must be reviewed and approved by appropriate individuals. Examples of design inputs for an IVD may include the sensitivity and specificity of Enzyme-Linked Immuno assay and may go so far as to include the incubation times.

4.3.3. Design Outputs

Outputs are measures to ensure that your inputs have been met which include acceptance criteria for determining the proper functioning of the device.

4.3.4. Design Review

Documented reviews are required which involve participants from all functions involved in the development process and an individual who does not have direct responsibility for the development process. Each documented review must be signed by all persons performing the review with the date of the review and shall be documented in the Design History File (DHF).

4.3.5. Design Verification

Design verification can be described as making sure that the design outputs meet the design inputs and is usually performed during the documented design review, i.e., the device has been developed according to the plan; this should not be confused with design validation.

4.3.6. Design Validation

Under initial production lots (or simulated production lots) product may be validated against its intended use and end user requirements; this includes software validation and a risk analysis; again this may be covered in a documented design review.

4.3.7. Design Transfer

Transfer of the product can be viewed as the scale-up of the product into full manufacturing, ensuring that the product design is effectively transferred to the manufacturing process.

4.3.8. Design Changes

Process for documenting changes to the design (may occur after product release).

4.3.9. Design History File

The documented history of all design control activity.

5. Conclusions

When planning for commercialization of a microfluidic IVD medical device, wherever possible you must keep in mind the regulatory implications in relation to the intended use and end user requirements. Risk analysis is now a major part of regulatory requirements and outlining the risk and cost benefit when in early development phases can greatly increase the likelihood of a project's success. Many of the regulations require an effective quality system, which requires the thorough examination of the design of a product to guarantee successful validation of a product to end user requirements. As with all legislation change is inevitable; therefore the landscape above is likely to be occasionally reshuffled. At the time this chapter is being written there are plans to modify the In Vitro Diagnostic Directive and also the US FDA pre-market notification, which has recently received criticism. Therefore, an awareness of the shifts in legislation and how it may affect a product already released to the market is as important as placing a product on the market itself.

6. Exemplary Online Sources of Information

- The CFR 21, both current and historical versions, can be downloaded from the files of the Government Printing Office and can also be searched directly at http://www.gpoaccess.gov/cfr/index.html1. A revised 21 CFR is issued on approximately 1 April of each year. There are three types of searches that can be done on the CFR Title 21 database: (a) Search by Part and Section Number; (b) Select a CFR Part Number; and (c) Full Text Search—you may enter a single word (example DEVICE), an exact phrase (example DEVICE TRACKING), or multiple words connected by AND (example import AND export).
- Link to 21 CFR 809.3: www.accessdata.fda.gov/scripts/cdrh/cfdocs/cfcfr/CFRSearch.cfm?FR=809.3.
- Link to the CLIA Database lists records of all commercially marketed in vitro test systems that FDA has categorized under the CLIA ('88) since 31 January 2000: www.fda.gov/MedicalDevices/DeviceRegulationandGuidance/IVDRegulatoryAssistance/ucm124103.htm.
- Link to 21 CFR 862.9: www.accessdata.fda.gov/scripts/cdrh/cfdocs/cfcfr/CFRSearch.cfm?fr=862.9.
- Link to 21 CFR 820: http://www.accessdata.fda.gov/scripts/cdrh/cfdocs/cfcfr/CFRSearch.cfm?CFRPart=820&showFR=1.
- Link to directive 98/79/EC of the European parliament and of the council of 27 October 1998 on in vitro diagnostic medical devices: http://eur-lex.europa.eu/LexUriServ/LexUriServ.do?uri=OJ:L:1998:331:0001:0037:EN:PDF.

Section II

Microfluidic Diagnostics: Fabrication and Manipulation Protocols

Chapter 8

Microfluidic Device Fabrication by Thermoplastic Hot-Embossing

Shuang Yang and Don L. DeVoe

Abstract

Due to their low cost compatibility with replication-based fabrication methods, thermoplastics represent an exceptionally attractive family of materials for the fabrication of lab-on-a-chip platforms. A diverse range of thermoplastic materials suitable for microfluidic fabrication is available, offering a wide selection of mechanical and chemical properties that can be leveraged and further tailored for specific applications. While high-throughput embossing methods such as reel-to-reel processing of thermoplastics is an attractive method for industrial microfluidic chip production, the use of single chip hot embossing is a cost-effective technique for realizing high-quality microfluidic devices during the prototyping stage. Here we describe methods for the replication of microscale features in two thermoplastics, polymethylmethacrylate (PMMA) and polycarbonate (PC), using hot embossing from a silicon template fabricated by deep reactive-ion etching.

Key words: Hot embossing, Imprinting, Replication, Polymethylmethacrylate, Polycarbonate, Thermoplastics

1. Introduction

An oft-cited advantage of microfluidic technology is the ability to realize low-cost disposable platforms for chemical and biochemical analysis. Compared to traditional microfluidic platforms based on silicon or glass, devices fabricated from thermoplastics can be realized at significantly lower unit cost. This is due to a combination of lower material costs for thermoplastics, as well as compatibility with low-cost replication methods. A particularly effective replication method for low-volume chip fabrication is hot embossing. In a typical hot embossing process, a thermoplastic substrate is placed in contact with a rigid template containing the inverse features desired in the final device. Unlike elastomers, thermoplastics soften and flow above their characteristic glass transition temperature (T_g). By applying pressure between the template and substrate while

Gareth Jenkins and Colin D. Mansfield (eds.), *Microfluidic Diagnostics: Methods and Protocols*, Methods in Molecular Biology, vol. 949, DOI 10.1007/978-1-62703-134-9_8,

maintaining the substrate above glass transition temperature, the desired microchannel features can be transferred to the thermoplastic material.

Polymethylmethacrylate (PMMA) and polycarbonate (PC) are two thermoplastic materials commonly used for microfluidic applications. An advantage of PMMA is its high optical transparency into the ultraviolet range, while PC offers a compatibility with a wider range of solvents and a higher glass transition temperature well suited to applications such as polymerase chain reaction for nucleic acid amplification. The methods described here take advantage of a low-cost industrial hot press to achieve high-quality embossing results with the addition of a simple substrate holder. The methods may also be extended to other thermoplastic materials by suitably modifying the various process parameters.

2. Materials

1. Float glass, 125 mm^2 and 6 mm thick (Precision Glass and Optics, Santa Ana, CA, USA) (see Note 1).
2. Silicon wafers, 10 cm diameter, 550 μm thick (Silicon Quest International, Inc., Santa Clara, CA, USA).
3. PMMA sheet, T_g ~ 105°C, 1.6 mm thick (Acrylite FF; Evonik Rohm GmbH, Darmstadt, Germany).
4. Polycarbonate sheet, T_g ~ 150°C, 1.6 mm thick (McMaster Carr, Robbinsville, NJ, USA).
5. Deburring tool (WEL-DB-4, 90°; Reid Tool Supply, Muskegon, MI).
6. Stainless steel plate, Type 304, 2.0 mm thick, 200 mm × 200 mm (McMaster Carr, Robbinsville, NJ, USA).
7. Pyralux® FR0001 thermal adhesion film, 25 μm thick (DuPont Electronic Materials, Research Triangle Park, NC, USA).
8. Custom aluminum fixture (described in Subheading 3).
9. Acetone, methanol, isopropanol (IPA), deionized (DI) water, and dry nitrogen for rinsing sequences.
10. Hot press. Here we use a Carver Auto Four hot press (Carver, Inc. Wabash, IN, USA).
11. Bandsaw or circular coring tool for cutting PMMA and PC sheets.
12. Vacuum oven to dehydrate the PMMA and PC sheets and for storage of PMMA and PC chips.

3. Methods

3.1. Substrate Fixture Preparation

Effective embossing, particularly of high-aspect ratio features, requires that the template be rigidly fixed to the embossing tool. This is necessary to ensure that the template can be removed from the embossed substrate at an elevated temperature to prevent thermal expansion mismatch from breaking the template, and to ensure that the template and substrate are separated by normal rather than shear forces when the embosser platens are opened. There are several available methods to this end. For example, the template may be fixed to the platen by a vacuum plate or by using a temporary adhesive layer to bond the template directly to the platen. A recommended alternative is to use a simple mechanical fixture, such as the system depicted in Fig. 1. The fixture consists of two layers of aluminum, with one layer smaller than the other to provide space for small stainless steel clips to be mounted to the larger plate. These clips serve to hold in place the steel-mounted silicon template as in Fig. 1b, which is mounted to the lower platen of the hot press.

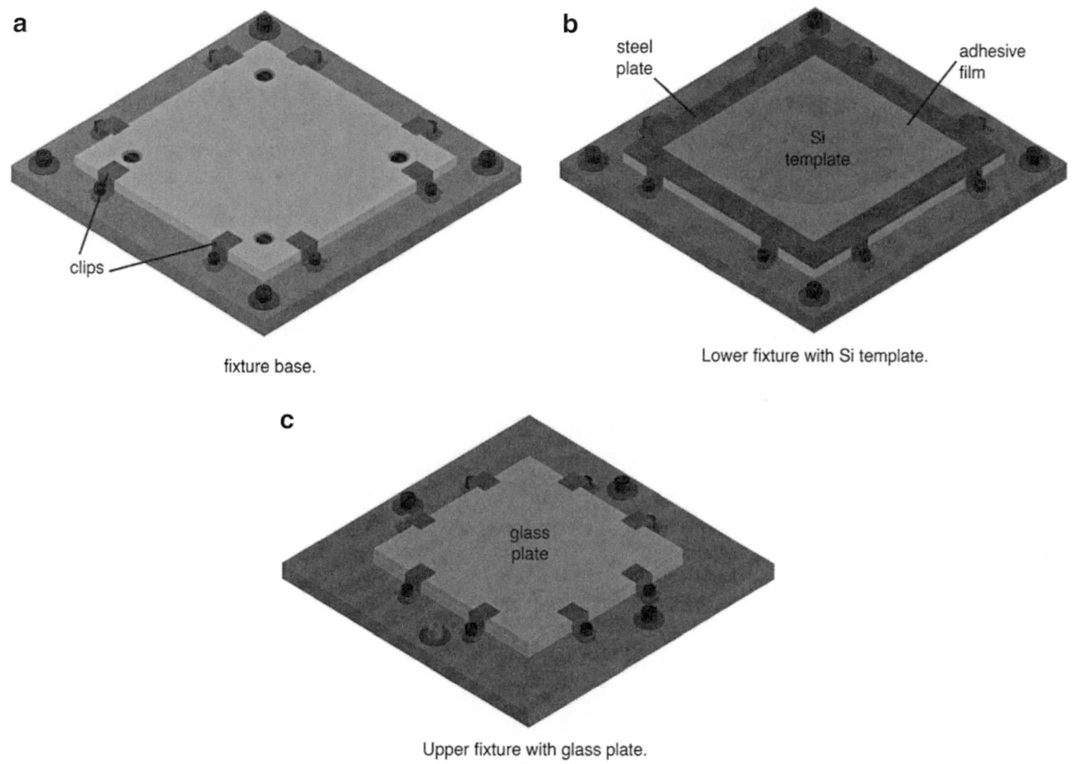

Fig. 1. Schematic diagram of a custom substrate fixture. (**a**) Fixture base: A 150 mm^2 aluminum plate is bolted to a 200 mm^2 aluminum base, with thin stainless steel clips around the periphery of the base. (**b**) Lower fixture with silicon template: Fixture holds the silicon template bonded to a steel plate. (**c**) Upper fixture with glass plate.

Fig. 2. Example silicon template fabricated by DRIE.

A second fixture is used to mount a glass plate to the upper platen. Details of fixture geometry depend on the specific process demands, but the schematic diagram shown in Fig. 1 serves as a suitable guideline for the basic design. Note that the aluminum plate must be thick enough to resist bending during demolding, but thin enough to avoid introducing a large thermal resistance and thermal sink to the embossing system. A total plate thickness of 6 mm is generally suitable. A set of custom fixtures can be made by a skilled machinist with several hours of labor.

3.2. Silicon Template Fabrication

A variety of silicon microfabrication techniques may be used for template fabrication, including both isotropic and anisotropic etching methods. A preferred method for template fabrication is deep reactive-ion etching (DRIE), a highly anisotropic silicon etching process capable of routinely generating aspect ratios up to ~20:1, and in some cases up to ~100:1 (1). DRIE template patterning follows standard microfabrication techniques including mask production, photolithography, DRIE etching, and photoresist removal. Details of these processes are beyond the scope of this chapter, but it is worth noting that researchers lacking access to appropriate in-house microfabrication facilities can take advantage of various commercial foundry services for this step. An example of a DRIE template designed for imprinting a chip used for multidimensional electrokinetic bioseparations (2) is shown in Fig. 2.

3.3. Thermoplastic Chip Preparation

1. Cut a PMMA or a PC sheet to the desired chip size, which must be smaller than the 10 cm diameter silicon wafer template. The sheet may be cut using a bandsaw or circular coring tool mounted on a drill press.

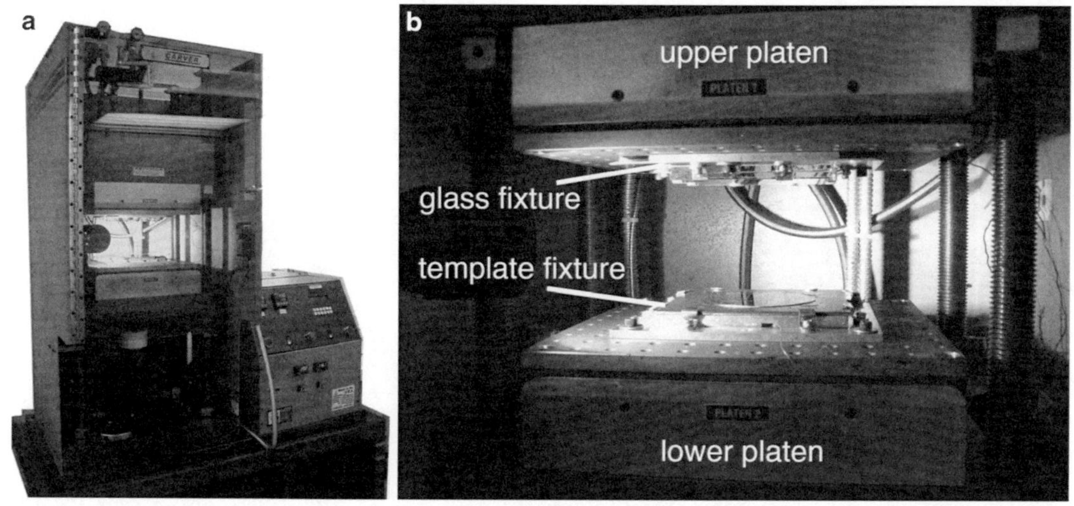

Fig. 3. (**a**) Carver Auto Four hot-press used for hot embossing. (**b**) Details of the platens showing configuration of the upper glass and lower template fixtures.

2. Remove all debris from the edges of the thermoplastic using a deburring tool.
3. For PMMA imprinting, clean the cut PMMA chip by rinsing with IPA and DI water prior to nitrogen drying, and dehydrate the PMMA sheet in a vacuum oven at 70°C overnight. Do not use methanol or acetone to clean PMMA (see Note 2). Storage in a vacuum oven is recommended (see Note 3).
4. For PC imprinting, clean the cut PC chip by methanol, IPA, and DI water in sequence, followed by nitrogen drying, and dehydrate the PC sheet in a vacuum oven at 100°C overnight. This dehydration step is particularly important for PC to minimize bubble formation during imprinting and to ease the demolding process.

3.4. Device Imprinting

While dedicated imprinting tools for thermoplastic embossing are commercially available, excellent imprinting results may be achieved using a low-cost industrial hot press. Here the use of a Carver Auto Four hot press is described. An image of the hot-press is shown in Fig. 3. The system consists of an adjustable upper platen, a lower platen positioned above a hydraulic actuator, and a digital controller allowing programming of force, temperature, and dwell time. Thermocouples in each platen monitor the working temperature.

3.4.1. Template Installation

1. Bond the silicon wafer to a 2 mm thick stainless steel plate using thermal adhesive film by pressing the materials together at 188°C for 3 h.
2. Fix the steel plate bonded to the template into a custom holding fixture using the fixture clips as shown in Fig. 1b. The silicon

Table 1
PMMA and PC hot embossing recipes

	PMMA	PC
Embossing temperature	128°C	168°C
Embossing pressure	1.38 MPa	1.72 MPa
Embossing time (nominal)	5 min	5 min
Demolding temperature	100°C	140°C

surface should be higher than the upper surface of the clips. Bolt the fixture onto the lower platen of the hot press.

3. Clean a 125 mm^2 and 6 mm thick float glass plate by sequential rinsing with acetone, methanol, IPA, and DI water, followed by nitrogen or air drying. Cleaning the glass is required to ensure good adhesion between the glass and thermoplastic during demolding from the template.
4. Fix the glass plate to the second holding fixture, Fig. 1c, and bolt this fixture to the upper platen of the hot press.
5. Blow each surface clean using a nitrogen gun before initiating the hot embossing process.

3.4.2. Hot Embossing Process

Table 1 describes embossing parameters for 1.6 mm thick PMMA and PC purchased from specific suppliers. Other thicknesses, grades, or brands of thermoplastics may have different properties such as glass transition temperature, thickness, stabilizers, etc. Therefore, the optimal imprinting recipe for specific materials should be experimentally determined, using these recipes as starting points. Images of various features imprinted into PMMA and PC chips following these recipes are shown in Fig. 4.

1. Set the target embossing temperature using the hot press control panel, with the temperature monitored by thermocouples mounted to each platen. The optimal embossing temperature is typically 20°C to 30°C higher than the glass transition temperature of the imprinted substrate materials. Note that the thermal resistance introduced by the various layers in the imprinting stack can result in temperature variations between the platen surfaces and thermoplastic chips, on the order of 4°C to 5°C. If higher variations are encountered for a particular arrangement, higher embossing temperatures than those reported here may be required.
2. Close the upper and lower platens of the hot press to bring the surfaces into contact, using the lowest force setting allowed by the embossing tool (~3 kN for a Carver Auto Four hot press).

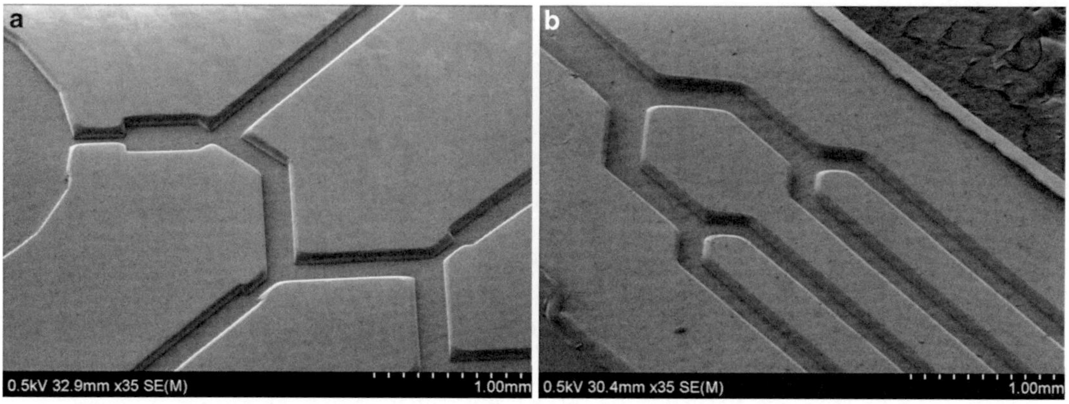

Fig. 4. SEM images of (**a**) PMMA and (**b**) PC microchannel structures imprinted from a DRIE silicon template using the reported recipes.

3. Hold the embossing force at the minimum level until the measured temperature reaches the desired embossing temperature (typically 10–15 min).
4. Apply the desired embossing pressure for a fixed embossing time as dictated by the appropriate thermoplastic recipe. Longer imprinting times are typically required for the embossing of large substrates, deep features, or complex patterns.
5. At the completion of the embossing time, adjust the temperature setting to room temperature and monitor the temperature readings.
6. When the hot press reaches the demolding temperature, open the platens. The thermoplastic should temporarily adhere to the upper glass plate and be smoothly separated from the silicon template.
7. Allow the platens to cool to the set end point temperature (see Notes 4 and 5).
8. After cooling, use a nitrogen gun to blow the imprinted thermoplastic chip away from the upper glass surface.
9. Store the imprinted chip in a vacuum environment prior to bonding to a cover plate, which may be performed by thermal bonding, solvent bonding, or other related techniques. For example, details for low-temperature bonding of thermoplastics including PMMA have been reported (3), together with a broader review of thermoplastic bonding techniques for microfluidic sealing (4).
10. A number of problems may occur during the hot embossing process including incomplete pattern transfer (see Notes 6 and 7), uneven pattern transfer (see Note 8), thermoplastic remaining on the template (see Note 9), and chip warping (see Note 10).

4. Notes

1. Large glass plates may be used for imprinting thermoplastic chips up to 10 cm^2. The thickness of glass is critical to transfer mechanical force and thermal heat to the imprinted materials. A thick glass will take a longer time for plastic heating-up and softening. The temperature difference between the plastic and the hot-press platen increases.
2. Do not use strong solvents to clean PMMA, such as acetone or methanol, which will dissolve PMMA and significantly damage its surface and decrease transparency.
3. Plastic should be stored in a vacuum oven prior to use. Absorbed moisture can result in bubble-like formations on the imprinted plastic surfaces. The storage temperature should be around 70% of the glass transition temperature, e.g., ~70°C for PMMA and ~100°C for PC.
4. Different end point temperature can be used for multiple cycle imprinting. Use a higher end point temperature, instead of room temperature, to reduce the time to heat the hot-press.
5. Many commercial hot presses allow water cooling of the templates. For imprinting temperatures of above 170°C, it is recommended to use only natural convective air cooling until the temperature drops below 170°C, at which point water cooling may be used to speed the process.
6. Incomplete pattern transfer: When the imprinting pressure or temperature is too low, template features cannot be completely transferred to the thermoplastic. In this instance, pressure should be increased while holding temperature constant. If a moderate (15–20%) increase in pressure does not improve pattern transfer, raise the temperature in 5% increments and repeat the pressure cycling to evaluate its impact on pattern transfer. Note that excessive pressure can damage the template, while excessive temperature can result in extremely difficult demolding. Note further that when adapting the reported recipes for other thermoplastics such as cyclic olefin copolymer (COC), different temperature and pressure values are generally required (see Note 7).
7. A recipe for imprinting COC previously reported in summary form refs. 2, 5 is similar to the PMMA recipe detailed here, although it must be noted that different grades of COC can have very different imprinting parameters due to variations in glass transition temperature and flow characteristics during embossing.

8. Uneven pattern transfer: When template patterns are not transferred evenly across the full wafer diameter, it is most likely due to poor parallelism of the hot-press platens. In this case the platens should be carefully adjusted using a dial gauge to make them as parallel as possible.
9. Thermoplastic remains on the template. If the demolding temperature is too low, or if the glass surface is not sufficiently clean, the thermoplastic chip may remain on the template following imprinting rather than sticking to the glass plate attached to the upper platen. In this event, immediately blow it off the template using a nitrogen gun. Failure to remove the chip may result in damage to the template.
10. Chip warping: If the demolding temperature is too high, the temperature gradient induced during removal of the chip from the glass by blowing with a nitrogen gun may result in warping of the chip. In this event, simply decrease the demolding temperature during the next fabrication cycle, keeping in mind that if the temperature is lowered excessively the thermoplastic may stick to the template.

References

1. Marty F, Rousseau L, Saadany B, Mercier B, Francais O, Mita Y, Bourouina T (2005) Advanced etching of silicon based on deep reactive ion etching for silicon high aspect ratio microstructures and three-dimensional micro- and nanostructures. Microelectron J 36:673–677
2. Yang S, Liu JK, Lee CS, DeVoe DL (2009) Microfluidic 2-D PAGE using multifunctional in situ polyacrylamide gels and discontinuous buffers. Lab Chip 9:592–599
3. Tsao CW, Hromada L, Liu J, Kumar P, DeVoe DL (2007) Low temperature bonding of PMMA and COC microfluidic substrates using UV/ ozone surface treatment. Lab Chip 7:499–505
4. Tsao CW, DeVoe DL (2008) Bonding of thermoplastic microfluidics. Microfluid Nanofluid 6:1–16
5. DeVoe DL, Lee CS (2006) Microfluidic technologies for MALDI-MS in proteomics. Electrophoresis 27:3559–3568

Chapter 9

Introduction to Glass Microstructuring Techniques

Radoslaw Mazurczyk and Colin D. Mansfield

Abstract

In this chapter an overview of manufacturing methods, leading to the fabrication of microstructures in glass substrates, is presented. Glass is a material of excellent optical properties, a very good electric insulator, biocompatible and chemically stable. In addition to its intrinsic qualities, glass can be processed with the use of manufacturing methods originating from the microelectronic industry. In this text two complete manufacturing protocols are described, each composed of standard microfabrication steps; namely, the deposition of masking layers, photolithographic patterning and pattern transfer via wet or dry etching. As a result, a set of building blocks is provided, allowing the manufacture of various microfluidic components that are frequently used in the domain of micro-total analysis system technology.

Key words: Micro-total analysis system, μ-TAS, Lab-on-a-chip, Glass etching, Microfluidics

1. Introduction

1.1. μ-TAS Fabrication Technologies

A microfabricated gas chromatography instrument (1) is widely considered as the earliest example of a micro-total analysis system (μ-TAS). The concept of miniaturization of bulk laboratory equipment into a compact, microchip format has grown and expanded enormously ever since this pioneering work and has been presented in a number of handbooks and review articles (2–10). This chapter describes the fabrication technologies that bring these devices to life.

Essentially, the μ-TAS manufacturing technologies may be grouped along three axes, defined with respect to the source material exploited, namely:

- Silicon and related materials (silicon oxide, silicon nitride etc.)
- "Plastics" e.g., polymers—PMMA, PDMS etc.
- Glasses (SiO_2-based only), quartz.

Gareth Jenkins and Colin D. Mansfield (eds.), *Microfluidic Diagnostics: Methods and Protocols*, Methods in Molecular Biology, vol. 949, DOI 10.1007/978-1-62703-134-9_9,

Each of the above options features its own particular advantages and disadvantages, which determines its field of applications. Silicon technology benefits, on the one hand, from the well-established know-how and huge fabrication base originating from the integrated circuit industry. On the other hand, however, being a semiconductor, silicon cannot easily be used in applications where high voltages are involved; for example, on-chip electrophoretic separations. "Plastic" microchips can be fabricated at low cost, employing methods such as injection molding or hot embossing. Unfortunately, polymer materials are usually hydrophobic by nature, which largely complicates handling of aqueous solutions in polymer-based devices. Glasses and quartz, being hydrophilic materials, are well-suited for use in microfluidic systems dedicated to biomedical applications, where aqueous solutions are ubiquitous. Excellent dielectric properties and the high optical transmission of this class of materials make them particularly attractive for wherever electroosmotic liquid manipulations coupled with photoluminescence excitation/detection are required. Essentially, glasses/quartz can be processed using standard microfabrication methods—layer deposition, photolithography and etching—originating from the semiconductor industry. Unfortunately, the transfer of technology is not straightforward and usually requires refinements, such as the development of special methods of surface treatment, the use of specific etching agents and masks. Nevertheless, on the whole glass technology is quite suitable for laboratory-scale fabrication and fast prototyping of a wide range of microsystems, including microfluidic devices. It should be noted, however, that in order to proceed with the experimental protocol described both access to microfabrication facilities and at least medium-level knowledge of the field is required. Readers who are novices in this domain are encouraged to refer to the appropriate textbooks, for instance, Introduction to Microfabrication by Franssila (11). In addition, subsequent hands-on training and/or support from personnel familiar with microfabrication is indispensable.

1.2. Glass Microstructuring

As mentioned earlier, to a certain extent the processing of glasses benefit from the technological base developed for silicon microtechnology. The common denominator for both is silicon dioxide, SiO_2, being the main chemical constituent of glasses, and ubiquitous insulating and masking material used in semiconductor processing. In the first approximation one can expect that the higher the SiO_2 content of a given type of glass, the simpler it becomes to exploit the "silicon technology pool." Let us then classify selected types of glasses, in the descending order with regard to the SiO_2 concentration. Quartz/fused silica are composed of pure SiO_2 by definition. Unfortunately, despite excellent optical and insulating properties, the prohibitive cost of substrates limits the use of these materials in microfluidics. This is due to the fact that in many applications microfluidic systems are conceived of as disposable devices; hence, a competitive price is one of the decisive factors. Apart from the economic aspect, high processing temperatures of quartz/fused silica (melting/softening point above 1,600°C) may pose an important technological obstacle. Borosilicate glass contains, apart from the SiO_2, at least 5% of boric oxide, B_2O_3—hence the name. For example, commercially available Pyrex 7740 glass is composed of: SiO_2 80.6%, B_2O_3 13%, Na_2O 4%, Al_2O_3 2.3% (percentage given by weight). The glass substrate least rich in SiO_2 is soda lime glass, which contains: SiO_2 60–75%, Na_2O (soda) 12–18%, CaO (lime) 5–12%, and Al_2O_3, SO_3, K_2O, MgO, Fe_2O_3, TiO_2 at less than 1% each.

Of the basic fabrication steps mentioned previously, the composition of substrate determines both the selection of masking layer and the etching agent. Whatever the etching method—dry or wet—the etching medium is supposed to remove the substrate material in the exposed areas only. Any damage to the masking material, leading to the penetration of etching medium through (or under) the masking layer, has a highly negative impact on the quality of the final device. Complex chemical composition of the substrate requires the use of multicomponent etching agents to target all the substrate's constituents. For example, the wet etching of quartz/fused silica follows the reaction: $SiO_2 + 6HF \rightarrow H_2SiOF_6$ (aqueous) $+ H_2O$, with hydrofluoric acid, HF, as the SiO_2 etching agent (ref. 11 p. 120).

However, if we look at the composition of soda lime glass, it becomes clear that such a scheme may not necessarily work in this case. Indeed, it turns out that non water-soluble deposits, originating from CaO, MgO, and Al_2O_3, are formed (12) during the etching in HF (or buffered HF). These deposits hinder the etching mixture from the full and uniform access to the structured substrate, hence resulting in a rough surface of the processed microdevice. Adding HCl to the etching mixture is a solution for this problem (12). A similar reasoning applies to dry, plasma etching, except that a gaseous environment is the etchant in this case. Plasma based on any fluorine-containing gas (CF_4, SF_6, CHF_3, or C_4F_8), usually mixed with Ar, or O_2, etches SiO_2 (11, 13). Once again, in the case of soda lime glass, nonvolatile species (NaF, AlF_3) are left behind, deteriorating the surface quality of etched features (14–16).

Obviously, finding the appropriate mask material, chemically resistant against all species in the etching environment, becomes more of a challenge as the number of these species increase. A number of strategies exist for both wet (17–25) and dry etching (13–16, 26–31), with masking layers made of photoresists, metals, polysilicon layers, silicon, and combinations of the above. Not only does the chemical resistance of the mask material need to be taken into account but also its internal mechanical stress and adhesion to the substrate. These two factors, if not properly tailored, lead to the deterioration of the mask during the course of etching, which results in numerous defects in the final device (23, 25).

In view of the above-mentioned problems, it is evident that the SiO_2 rich, borosilicate glasses are easier to process as compared with soda lime. Nevertheless, soda lime glasses have another, possibly crucial property in that it is possible to modify locally the optical properties of soda lime glass using the process of ion-exchange. This phenomenon paves the way for the fabrication of monolithically integrated, compact microfluidic–micro-optic devices (32–35). Bearing in mind the remarkable potential of this technology, the effort required to develop an etching method of the soda lime glass, will pay off. The two experimental protocols presented below are dedicated to realizing this task. Both methods share a common sample preparation step, after which the method bifurcates in two directions. The first one is wet etching in a liquid agent and the second method is dry etching in a reactive ion etching (RIE) system. In choosing which protocol is best suited to your needs it is worth mentioning that although wet etching has to date been better optimized for soda lime glass, dry etching is an interesting alternative because with no HF-based chemicals involved it is a much safer technique.

1.3. Inspection and Characterization of the Fabricated Devices

Although not explicitly covered by this protocol it will be necessary to inspect and characterize the dimensions of the final fabricated devices, and for best practice such inspection can also be performed during the process. In general, there are three basic questions addressed by this process: (1) What is the geometry of the channels? (2) Are there any solid precipitates or roughness of the etched features? (3) What does the device "look" like?

As a means to investigate the geometry of the channel, the use of an optical microscope (with Nomarsky effect if available) and a surface profiler are recommended. The optical microscope is sufficient to reveal the presence of any solid debris in the features. In addition, the surface profiler allows the characterization of the surface roughness at the channel's bottom. In certain cases (public presentations, journal papers) it is useful and attractive to present a perspective view of the fabricated device, i.e., a 2D surface image. Optical profilers (Wyko) and certain mechanical profilers (Dektak 150) allow the realization of such imagery; however, our preference is for SEM images. For this latter technique, the fabricated device must be metalized prior to the observation and connected to the ground potential of the SEM. This is to prevent accumulation of electric charge on the chip's surface, which will distort the image.

2. Materials

2.1. Materials Common for All Protocols

It is assumed that manufacturing will be carried out in a microfabrication laboratory, class 10000 (ISO Class 4) or better (see Note 1). The following processing and characterization tools are required:

1. Sputtering deposition system, equipped with a chrome (Cr) target.
2. UV lithography facility (see Note 2).
3. Wet processing post.
4. Characterization tools:
   (a) 1D surface profiler (for example, Alphastep from Kla Tencor, Dektak from Veeco or equivalent).
   (b) Optical microscope (resolution down to 1 μm).
   (c) Scanning electron microscope or 2D surface profiler (mechanical or optical, Dektak 150 or Wyko series from Veeco for example).
5. Small benchtop equipment and typical cleanroom consumables: oven, hot plate, stirring plate, laboratory glassware (made of Schott Duran or equivalent, PTFE, PP), wash bottles, PTFE (CTFE) tweezers, lint-free tissues, and cotton swabs.
6. Deionized water or double distilled water (henceforth named simply water).
7. Dry compressed nitrogen.
8. Soda lime glass slides 75 × 25 mm, thickness 0.9–1.1 mm (e.g., Corning product number 2947-75X25, or other suppliers).
9. Standard chemicals: acetone, ethanol, sulfuric acid 96–98%, hydrogen peroxide 30%, purity ≥95%.
10. Specialty chemicals:
    (a) AZ5214E photoresist (AZ Electronic Materials GmbH).
    (b) MIF 726A developer (AZ Electronic Materials GmbH).
    (c) Chromium etch: (e.g., 1020 series from Transgene Company Inc., etch rate 3.2–4 nm/s at 40°C, or Chrome etch 18 from Microresist Technology GmbH, etch rate 150 nm/min at 22°C).

2.2. Materials and Equipment for Wet Etching Method

1. Buffered oxide etch (BOE, also known as buffered HF, BHF) 7:1 (Transene Company Inc. or BASF for example).
2. Hydrochloric acid, 35–37%.
3. PTFE or PP beakers, size (diameter × height) 8 × 5 cm.

2.3. Materials and Equipment for Dry Etching Method

1. Reactive ion etching system (frequency 13.56 MHz), equipped with Ar and SF_6 process gases.

3. Methods

Figure 1 shows a generic technology flowchart for both wet and dry etching protocols. The following steps constitute the procedure: (a) substrate cleaning; (b) chromium (Cr) layer deposition; (c) UV lithography; (d) pattern transfer, i.e., exposing areas in the Cr layer; (e) etching (wet or dry); (f) mask removal; (g) final cleaning.

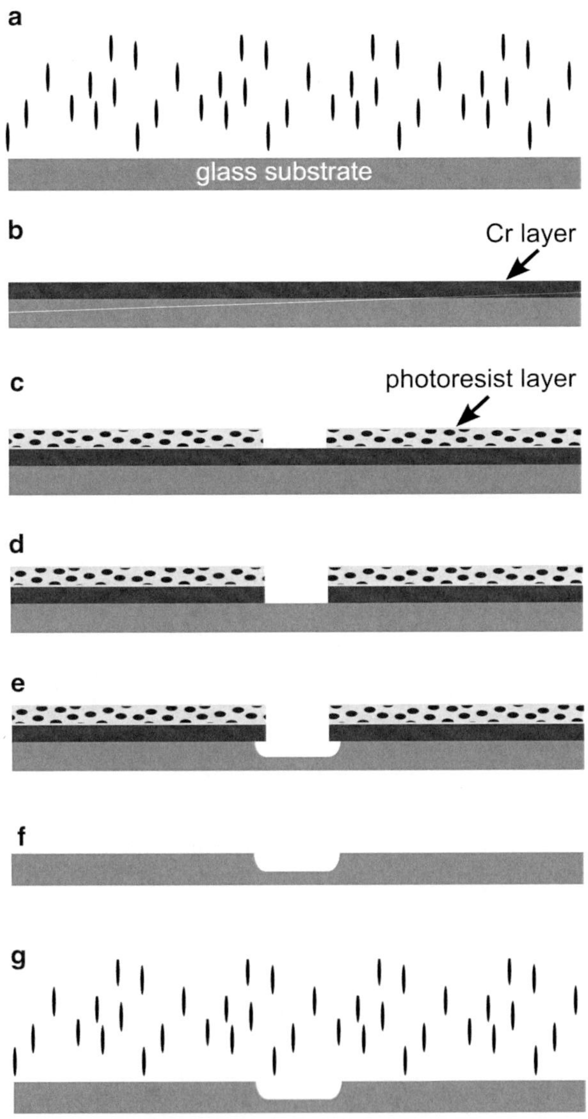

Fig. 1. Flowchart of the fabrication protocols: (**a**) initial substrate cleaning; (**b**) chromium layer deposition; (**c**) UV lithography; (**d**) pattern transfer; (**e**) etching; (**f**) mask removal; (**g**) final cleaning. The scheme applies directly to the wet etching method. In the case of the dry etching, the photoresist layer is removed before the etching step (explanation in text).

The scheme applies directly to the wet etching protocol. In case of the dry etching, however, the photoresist layer is removed between steps (d) and (e), i.e., only the Cr masking layer protects the substrate during the etching step (e). Both protocols share the initial/final cleaning steps, which are essentially the same.

3.1. Substrate Cleaning: Common to Both Protocols

The cleaning procedure should be sufficiently thorough as compared to the initial cleanliness of the slides. Here the worst case scenario is assumed, i.e., that the substrates are covered with solid particles (dust, fibers etc.), and stained with organic contaminants.

1. Remove any loose particles by blowing them off with nitrogen and washing the slides with ethanol or deionized water (simply referred to as water henceforth). If contamination is still noticeable to the naked eye the slides should be wiped with lint-free tissue or a swab soaked with acetone and then repeated with ethanol. Care must be taken not to scratch the surface with any hard solid particles, which might be left after the previous treatment. Already after such preliminary cleaning, the slides should have a glossy appearance, without visible smears or traces.
2. Wash the slides for 5 min in stirred acetone and subsequently for 5 min in stirred ethanol. Use two separate crystallization dishes for this operation.
3. Repeat step 2 by simply rinsing the slides using wash bottles, in the same sequence: acetone, then ethanol (see Note 3).
4. Dry the substrates under a stream of compressed nitrogen.
5. Wash the slides in water. For this, place them in a beaker of a shape and size, such that free access to both sides of the substrates is provided (see Note 4). Allow the water flow through the beaker for approximately 10 min.
6. Dry the substrates again under a stream of nitrogen.
7. Next the substrates are to be cleaned in the "piranha" mixture, which is very effective at removing organic contaminants. As this is one of the typical procedures used in microfabrication, only essential aspects will be explained here. In particular, please note that the mixture is hot, highly corrosive and potentially explosive. Hence, adequate safety precautions are essential, i.e., acid resistant gloves and apron, goggles or face mask should be worn at all times when handling the piranha solution.
8. Place the slide into a clean and dry crystallization dish.
9. Put the dish on a stirrer plate and pour 10 ml of hydrogen peroxide first and then, 30 ml of sulfuric acid. Depending on the total volume required, you can modify the amount of each product whilst maintaining the v/v ratio of 3:1 (see Note 5). Stir gently for approximately 10 min until the reaction is complete.

10. Prepare a beaker filled with 200 ml of water. Retrieve your slide from the piranha solution using the PTFE (CTFE) tweezers and dip it in water.
11. Take the slide out of the beaker and rinse both sides with water for approximately 5 min.
12. For the final rinsing, place the slide again into a beaker filled with water and let the water flow through the beaker for at least 15 min (see Note 6).
13. Dry the slide either under the nitrogen stream or by utilizing a spin-drying method (e.g., by spinning in a centrifuge with slide holders at $200 \times g$ for 5 min). The latter method is recommended as drying with a nitrogen gun is not always efficient on relatively large microscope slides. Please note that this drying step is essential to remove solid particles suspended on the surface layer of water (see Note 7).

3.2. Pattern Definition and Transfer via Wet Etching Method

According to the flowchart (Fig. 1b), a Cr masking layer is deposited at the first stage of the actual fabrication process. Chromium coated microscope slides can be purchased commercially although many research facilities choose to perform this step in-house, in which case the sputtering method is recommended as it allows uniform deposition of relatively dense layers of chromium. From experience, a layer of 150 nm provides good adhesion and integrity during further processing (see Note 8). In the next steps, a standard positive photolithography process is utilized with a dark-field mask to define features in the photoresist (see Note 2). In our case the features were two intersecting lines 30 μm wide, corresponding to the desired microfluidic channels in the finished device. The details of the process are as follows:

1. Ensure cleaned slides are dry before processing (see Note 9).
2. Spin coat a layer of AZ 5214E resist onto the glass slide. Spin coating parameters: rotation speed 5,000 rpm, acceleration 5, duration 30 s, resulting in a photoresist thickness of 1.5 μm.
3. Pre-bake: 2 min at 120°C, on the hot-plate.
4. Exposure using UV lithography system: 8 s.
5. Develop for 40 s using AZ726MIF developer.
6. Post-bake: 15 min, at 100°C, in oven.
7. Transfer the pattern from photoresist to the Cr layer (Fig. 1d). For this, immerse the sample in the Cr etching mixture and stir gently. Follow the data from the manufacturer for details on etching conditions, but in general it takes about 1–2 min to remove 150 nm of chromium depending on the etching agent.
8. Wash the slide for a few minutes under running water and dry with compressed nitrogen.

9. Protect the backside of the substrate with sticky-tape (i.e., Scotch tape). This step is optional but extends the lifetime of the glass etching mixture. As only the features on the front side will be etched, the mixture will be saved from premature degradation from the etching by-products originating from the comparatively large backside.
10. Prepare the mixture of buffered oxide etch (BOE), hydrochloric acid and water. In our experiments, two compositions were tested: BOE–HCl–H_2O 1:2:4 or 1:2:2 (v/v ratios). The mixtures differed in etching rates: 0.45 μm/min and 0.75 μm/min at 20°C, respectively (see Note 10). Important: when mixing, always start with the chemical of the lowest specific gravity, i.e., H_2O followed by BOE and HCl. Note that the BOE is based on HF acid, which is toxic and corrosive. Do refer to the MSDS files of both HF and HCl and follow all the safety precautions!
11. Glass etching (Fig. 1e): put the slide into a PTFE or polypropylene beaker (not glass!) and pour in 50 ml of the glass etchant mixture. Use a beaker approximately 8 cm in diameter and 5 cm in height. This will protect your experiment from dangerous spills when stirring (see Note 11).
12. Before proceeding to the etching of the actual device, it is advisable to perform a few test runs for different etching times and measure the respective etched depths with a surface profiler. This series of tests allows establishing the kinetics of etching. Before each measurement, the sample should be plunged into a beaker filled with water, then rinsed thoroughly under running water for approximately 5 min and finally dried with nitrogen. For handling use tweezers made of PTFE or other materials chemically resistant to the etching mixture. All these precautions are aimed at protecting the operator and the equipment against contamination from the mixture.
13. The final etched depth is determined by the process duration (assuming that the process temperature and mixture composition are stable), which is straightforward to calculate. It is recommended, however, to interrupt the etching a few times before reaching the desired depth and check its actual value with a surface profiler. Such intermediate measurements allow verification of whether the process is stable and allow reaching the target depth with a precision of a few tenths of micron.
14. Once the etching has been finished, proceed to the stage "f" (Fig. 1) of the process flow. Remove the Scotch tape from the backside of the slide and remove the photoresist layer with acetone.
15. Rinse the slide with ethanol and water.

16. Remove the Cr layer using the Cr etchant as explained in step 3. If there are any spots of Cr left (most probably due to incomplete removal of photoresist residues), repeat the step.

    Final cleaning is as described in Subheading 3.1 but some of the initial rough cleaning steps can be omitted. Typically, piranha treatment plus rinsing plus drying should suffice.

17. Carry out inspection and characterization of the fabricated devices (see Note 12).

3.3. Pattern Definition and Transfer via Dry Etching Method

1. Clean a glass slide following the protocol described in Subheading 3.1. If the slide is not pre-coated with a Cr masking layer then this Cr layer must be deposited as described at the start of Subheading 3.2; however, the Cr layer is much thicker in this case: 2 μm instead of 150 nm.
2. The photolithography and window opening steps are realized as in the case of the wet etching protocol (see Subheading 3.1, steps 1–8). Obviously, it takes more time to remove the thicker layer of Cr.
3. Having transferred the pattern to the Cr layer, remove the photoresist with acetone. Rinse the sample with ethanol and water, and dry.
4. Place the sample in your RIE etching machine. It is desirable that your set-up is equipped with a laser interferometry end-point-detection system. If this is the case, position the laser spot onto the surface of the Cr.
5. Start etching using the following process conditions:
    (a) SF_6/Ar gas mixture, gas flows: 32.6 sccm/25 sccm respectively (see Note 13)
    (b) Working pressure: 15 mTorr
    (c) R.F. power: 200 W
    (d) Self-bias voltage: −300 V
6. During the process, the Cr masking layer is etched along with the glass substrate (roughly twice as slowly). Our goal is to etch the features as deep as possible, i.e., until the protective layer has been "consumed" completely. This moment is marked by the rapid drop of the intensity of laser light reflected from the sample—stop your etching when it occurs.
7. Retrieve the sample from the RIE machine.
8. Remove remaining traces of Cr mask (if any) as explained in Subheading 3.2, step 7.
9. Rinse the sample with water and dry with compressed nitrogen.
10. Inspect and characterize the fabricated device using (see Note 14).

4. Notes

1. It is assumed that the operator, who will execute the experimental protocols presented here, will receive standard health and safety cleanroom training. Therefore, these issues are not discussed in the text, apart from the exceptionally risky elements at certain stages of the experimental protocol.
2. Mask pattern and resolution are application-specific issues and as such, are beyond the scope of this text. The protocols presented here were elaborated for the pattern consisting of two intersecting channels, each 30 μm wide.
3. When the substrate is washed with acetone only, traces of solid contamination are left behind. In order to remove this contamination it is essential to rinse the substrate with ethanol immediately after the acetone rinsing.
4. The simplest way to assure the free access of water to the entire surface of the slide is to place it in a beaker of the right size, in which the slide shall rest spontaneously in tilted position. For a standard microscope slide a beaker 6.5 cm in height and 9.5 cm in diameter will do just fine.
5. For the "piranha" cleaning step the 3:1 v/v ratio of sulfuric acid and hydrogen peroxide is recommended. However, other ratios, such as 5:1, are also reported.
6. As glass is a hydrophilic material by nature, once the slide is removed from the beaker it should be uniformly covered with a film of water. If any dry spots are noticed, indicating hydrophobic areas, the cleaning procedure should be repeated.
7. It is advisable that any further processing/characterization steps are carried out immediately after the final drying. If it is necessary to prepare and stock a certain number of slides, they may be stored in water for a few hours, after the final rinsing but before drying. Dry substrates can be also stored in a clean vacuum environment.
8. Intuitively, the thicker the Cr layer, the better protection against penetration of the etching agents. On the other hand, however, as the thickness of the Cr layer increases, so does the mechanical stress confined within. This stress may be relaxed upon etching, resulting in formation of defects in the masking layer. The thickness of 150 nm was chosen as a compromise between the described constraints. The reader is encouraged to refer to the papers (21, 23), where the subject is treated in more detail.
9. Prior to the photolithography process, it is recommended to dehydrate the samples by annealing them at 120°C on the hot plate for 5–10 min and afterwards, cool them down to room temperature before spin coating of the photoresist.

10. The etching rates given here are indicative and may vary depending on the composition of your glass. It is possible to etch faster by increasing the temperature of the mixture. However, at elevated temperatures both HF and HCl, (which are actually gases), are released from the mixture at a higher rate. As a result, the composition of the mixture may be perturbed in a rather uncontrollable manner. Nevertheless, the etching rates as high as 5.5 μm/min at 55°C can be achieved, which is useful for etching of rather large (~1 mm in lateral size) and deep (~100 μm) features (36).
11. Stirring is important to remove the by-products of etching from the features. In our experiments a "Thermomixer comfort" apparatus (from Eppendorf) was used at 500–600 rpm. Apart from vigorous stirring, the machine also ensures the equilibration of process temperature (if different from the ambient temperature) at the same time.
12. Regarding channel geometry following wet etching: in theory channel width at the top, W, obeys the simple relation: $W = 2 \times D + X$; where D is the depth of the channel and X is the width of the window in the photoresist layer. For example, in our case: $X = 30$ μm and $D = 30$ μm, hence $W = 90$ μm should be obtained; in practice, however, an additional lateral under-etching of ~5 μm is typically observed. This effect probably resulted from the use of HCl in the etching mixture, which attacks chromium to a certain extent. Nevertheless, for this application a defect of such a size was acceptable. Alternatively, a method where HCl is replaced by HNO_3 has been reported to give less under-etching, (37) but the procedure was carried out at an elevated temperature of 45°C, which is more difficult to control and too risky for inexperienced users. The problem of the under-etching becomes more pronounced when adhesion of the masking layer to the substrate is poor, in which case, the etching agent penetrates between them, resulting in extensive surface damage and jagged channel walls. An example of this highly harmful defect is presented in Fig. 2. Regarding solid precipitates and roughness of the etched features, the composition of etching mixture is devised so as to effectively dissolve and remove all the components of the substrate material. If, however, the concentrations of the constituents in the mixture are not optimized versus the composition of glass, insoluble precipitates are left behind in the etched features. This unwanted phenomenon is illustrated in Fig. 3. For a well-tailored etching agent, it is possible to achieve a surface roughness as low as 5 nm (36). A SEM image of a device fabricated via wet etching is shown in Fig. 4.
13. The SF_6 gas in the process atmosphere is aimed at removing the SiO_2 from the soda lime glass, whereas our solution to

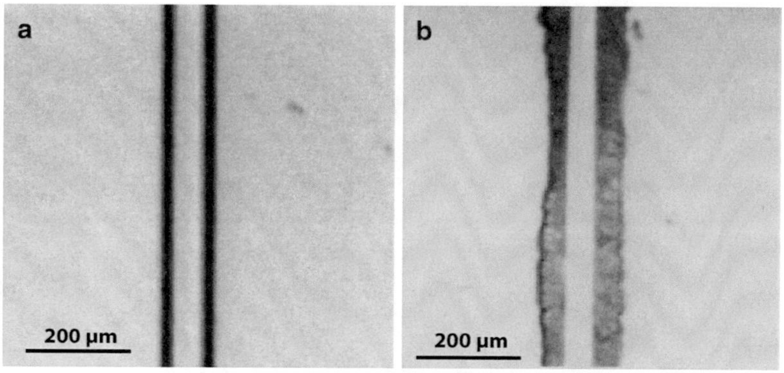

Fig. 2. Problem of under-etching: (**a**) microfluidic channel with well-defined and smooth edges; (**b**) microfluidic channel with irregular edges and extensive surface damage due to the poor mask adhesion.

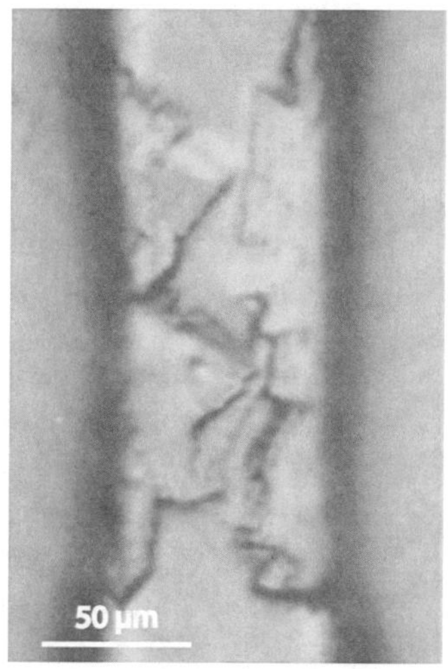

Fig. 3. Insoluble deposits accumulated in the microchannel.

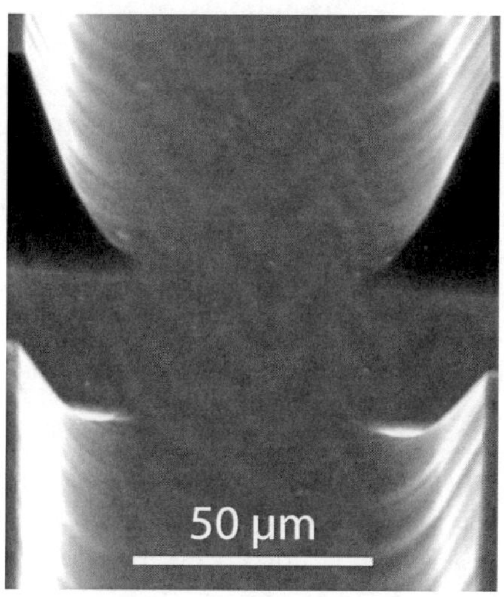

Fig. 4. SEM image of the microchannel fabricated according to the wet etching protocol.

Fig. 5. SEM image of the microchannel fabricated according to the dry etching protocol.

eliminate the nonvolatile compounds is to pulverize them with energetic Ar^{+} ions. Standard $cm^3/min = sccm$.

14. Regarding channel geometry following dry etching: the principal advantage of dry plasma etching over wet processing is the anisotropy of the former—the substrate material is removed more efficiently in the vertical than the horizontal direction. As an example of a finished dry-etched device, the following results were achieved after 40 min of etching: channel depth, 4.5 ± 0.2 μm; channel width at the top, 39 ± 1 μm; channel width at the bottom, 35 ± 1 μm. Taking into account that the initial channel width (the window in the masking layer) was 30 μm, it is clear that the effect of lateral etching is still significant in this protocol. Regarding solid precipitates and roughness of the etched features then in the example above a roughness of 50 nm was found at the bottom of the microchannel, which is considerably higher compared to the results of the wet etching. Most probably a certain amount of nonvolatile by-products are formed in course of the process, perturbing the access of etching species to the substrate. This phenomenon is called a micromasking effect (30). A SEM image of the microchannel is presented in Fig. 5.

Acknowledgments

This research was supported by the NANOLYON technological platform of the Institut des Nanotechnologies de Lyon (INL), which provided us with access to its technological facilities.

References

1. Terry SC, Jerman JH, Angell JB (1979) A gas chromatographic air analyzer fabricated on a silicon wafer. IEEE Trans Electron Dev 26: 1880–1886
2. Oosterbroek RE, Berg van de A (2003) Lab-on-a-chip: miniaturized systems for (bio)chemical analysis and synthesis. Elsevier B.V., Amsterdam
3. Geschke O, Klank H, Tellemann P (2008) Microsystem engineering of lab-on-a-chip devices, 2nd edn. Wiley-VCH, Verlag GmbH & Co. KGaA, Weinheim
4. Huikko K, Kostiainen R, Kotiaho T (2003) Introduction to micro-analytical systems: bio-analytical and pharmaceutical applications. Eur J Pharm Sci 20:149–171
5. Erickson D, Li D (2004) Integrated microfluidic devices. Anal Chim Acta 507:11–26
6. Weigl BH, Bardell RL, Cabrera CR (2003) Lab-on-a-chip for drug development. Adv Drug Deliv Rev 55:349–377
7. Bruin GJM (2000) Recent developments in electrokinetically driven analysis on microfabricated devices. Electrophoresis 21:3931–3951
8. Lee SJ, Lee SY (2004) Micro total analysis system (μ-TAS) in biotechnology. Appl Microbiol Biotechnol 64:289–299
9. Vilkner T, Janasek D, Manz A (2004) Micro total analysis systems. Recent developments. Anal Chem 76:3373–3385
10. Nguyet TT, Ayed I, Pallandre A, Taverna M (2010) Recent innovations in protein separation on microchips by electrophoretic methods: an update. Electrophoresis 31:147–173
11. Franssila S (2004) Introduction to microfabrication. Wiley, Chichester
12. Stjernström M, Roeraade J (1998) Method for fabrication of microfluidic systems in glass. J Micromech Microeng 8:33–38
13. Ronggui S, Righini GC (1991) Characterization of reactive ion etching of glass and its applications in integrated optics. J Vac Sci Technol A 95:2709–2712
14. Li X, Abe T, Esashi M (2001) Deep reactive ion etching of Pyrex glass using SF_6 plasma. Sens Actuators A 87:139–145
15. Metwalli EE, Pantano CG (2003) Reactive ion etching of glasses: composition dependence. Nucl Instrum Methods Phys Res B 207:21–27
16. Leech PW (1999) Reactive ion etching of quartz and silica-based glasses in CF_4/CHF_3 plasmas. Vacuum 55:191–196
17. Simpson PC, Woolley AT, Mathies RA (1998) Microfabrication technology for the production of capillary array electrophoresis chips. Biomed Microdevices 1:7–26
18. Holden MA, Kumar S, Castellana ET, Beskok A, Cremer PS (2003) Generating fixed concentration arrays in a microfluidic device. Sens Actuators B 92:199–207
19. Grosse A, Grewe M, Fouckhardt H (2001) Deep wet etching of fused silica glass for hollow capillary optical leaky waveguides in microfluidic devices. J Micromech Microeng 11:257–262
20. Corman T, Enoksson P, Stemme G (1998) Deep wet etching of borosilicate glass using an anodically bonded silicon substrate as mask. J Micromech Microeng 8:84–87
21. Iliescu C, Miao J, Tay FEH (2005) Stress control in masking layers for deep wet micromachining of Pyrex glass. Sens Actuators A 117: 286–292
22. Mourzina Y, Steffen A, Offenhäusser A (2005) The evaporated metal masks for chemical glass etching for BioMEMS. Microsyst Technol 11: 135–140
23. Iliescu C, Jing J, Tay FEH, Miao J, Sun T (2005) Characterization of masking layers for deep wet etching of glass in an improved HF/HCl solution. Surf CoatTechnol 198:314–318
24. Daykin RNC, Haswell SJ (1995) Development of a micro flow injection manifold for the determination of orthophosphate. Anal Chim Acta 313:155–159
25. Iliescu C, Chen B, Miao J (2008) On the wet etching of Pyrex glass. Sens Actuators A 143: 154–161
26. Ceriotti L, Weible K, Rooij de NF, Verpoorte E (2003) Rectangular channels for lab-on-a-chip applications. Microelectron Eng 67–68: 865–871
27. Rodriguez I, Spicar-Mihalic P, Kuyper CL, Fiorini GS, Chiu DT (2003) Rapid prototyping of glass microchannels. Anal Chim Acta 496:205–215
28. Park JH, Lee NE, Lee J, Park JS, Park HD (2005) Deep dry etching of borosilicate glass using SF_6 and SF_6/Ar inductively coupled plasmas. Microelectron Eng 82:119–128
29. Kolari K (2008) Deep plasma etching of glass with a silicon shadow mask. Sens Actuators A 141:677–684
30. Thiénot E, Domingo F, Cambril E, Gosse C (2006) Reactive ion etching of glass for biochip applications: composition effects and surface damages. Microelectron Eng 83:1155–1158
31. Queste S, Salut R, Clatot S, Rauch J-Y, Khan Malek GC (2010) Manufacture of microfluidic glass chips by deep plasma etching, femtosecond laser ablation, and anodic bonding. Microsyst Technol 16:1485–1493

32. Mazurczyk R, Vieillard J, Bouchard A, Hannes B, Krawczyk S (2006) A novel concept of the integrated fluorescence detection system and its application in a lab-on-a-chip microdevice. Sens Actuators B 118:11–19
33. Bou Chakra E, Hannes B, Vieillard J, Mansfield CD, Mazurczyk R, Bouchard A, Potempa J, Krawczyk S, Cabrera M (2009) Grafting of antibodies inside integrated microfluidic–microoptic devices by means of automated microcontact printing. Sens Actuators B 140:278–286
34. Vieillard J, Mazurczyk R, Morin C, Hannes B, Chevolot Y, Desbène P-L, Krawczyk S (2007) Application of microfluidic chip with integrated optics for electrophoretic separations of proteins. J Chromatogr B 845:218–225
35. Vieillard J, Mazurczyk R, Boum L-L, Bouchard A, Chevolot Y, Cremillieu P, Hannes B, Krawczyk S (2008) Integrated microfluidic–microoptical systems fabricated by dry etching of soda-lime glass. Microelectron Eng 85:465–469
36. Mazurczyk R, El-Khoury G, Dugas V, Hannes B, Laurenceau E, Cabrera M, Souteyrand S, Krawczyka E, Cloarec J-P, Chevolot Y (2008) Low-cost, fast prototyping method of fabrication of the microreactor devices in soda-lime glass. Sens Actuators B 128:552–559
37. Hannes B, Vieillard J, Bou Chakra E, Mazurczyk R, Mansfield CD, Potempa J, Krawczyk S, Cabrera M (2008) The etching of glass patterned by microcontact printing with application to microfluidics and electrophoresis. Sens Actuators B 129:255–262

Chapter 10

Glass Microstructure Capping and Bonding Techniques

Radoslaw Mazurczyk, Colin D. Mansfield, and Marcin Lygan

Abstract

The capping of microfluidic features fabricated in glass substrates is achievable by various technological methods. Of the entire spectrum of possibilities (gluing, glass bonding via intermediate layers, pressure or plasma-assisted glass bonding, etc.) a detailed description of three techniques is presented here. The first is a low temperature PDMS-glass adhesion bonding, the second is medium temperature pressure assisted glass–glass bonding, and finally, high temperature glass–glass fusion bonding. All these protocols allow completion of the manufacturing process for a fully enclosed microfluidic chip. Nevertheless, as they are complementary rather than competing methods, they effectively extend the range of tools available to fabricate lab-on-a-chip microdevices. Each has its own merits and each could feasibly be used at different developmental stages of a given microfluidic device.

Key words: Glass fusion bonding, PDMS bonding, Direct wafer bonding, Microfluidics, Lab-on-a-chip

1. Introduction

With the application of standard microfabrication methods (photolithography and etching), we are able to manufacture a variety of microfluidic components in glass substrates. However, the use of such manufacturing techniques implicates that the resulting features (channels, cavities, etc.) are open from the top; thereby preventing realization of many functions (mixing, separations, filtering) for which entirely enclosed microfluidic channels are a prerequisite. This necessitates the development of appropriate capping methods; in other words, a suitable cover material has to be selected and a bonding process performed to integrate it with a substrate that possesses microfabricated features. Three approaches to the problem can be distinguished.

1.1. Adhesion Bonding

Adhesion bonding refers here to a plastic (polymer) or glass cover, glued onto a glass substrate, which possesses prefabricated

Gareth Jenkins and Colin D. Mansfield (eds.), *Microfluidic Diagnostics: Methods and Protocols*, Methods in Molecular Biology, vol. 949, DOI 10.1007/978-1-62703-134-9_10, © Springer Science+Business Media, LLC 2013

microfluidic features, with the use of intermediate polymer layers (1) or adhesives (1–5). On the positive side, these are low-temperature methods, which eliminate the risk of thermal deformation of microfluidic features. However, channels may easily become clogged if excessive amounts of glue penetrate inside them. As an alternative and a remedy to the problem, self-sustained polydimethylsiloxane (PDMS) polymer layers are a viable solution. They adhere spontaneously to glass substrates, providing leak-tight, glue-free capping (6), even onto non-flat surfaces. PDMS is transparent to visible light and can be fabricated in the form of sheets as thin as 50 μm. These qualities prove particularly valuable for fabrication of integrated microfluidic–microoptic systems as delivery/collection of light can be performed via an optical fiber positioned in very close proximity to the microchannels (6). Unfortunately, in all examples of this capping approach, the final device contains more than just one material in its structure. As a result, liquid–surface interactions in the channels gain in complexity and liquid manipulation becomes more problematic than in the case of purely glass devices. If we take a glass/PDMS microfluidic device as an example, the following problematic phenomena are observed (6, 7):

- Plug flow is distorted at the glass/PDMS interface, deteriorating the performance of electrophoretic separation.
- Proteins are prone to adsorption by the PDMS, hindering the detection of trace proteins.

1.2. Low Temperature, Glass–Glass Fusion Bonding

Low temperature, glass–glass fusion bonding can be realized through various cleaning and/or surface processing steps, via wet chemistry or dry plasma treatment (8–11). Also, the application of external pressure can be used as a factor for promoting bonding (12). Again, these are low (or medium) temperature bonding methods and, usually, no additional material is incorporated into the device structure, with all the advantages this brings. Unfortunately, their application frequently involves manipulation with hydrofluoric acid, which is a highly dangerous chemical (13). Also, specific equipment is often required—a plasma processing setup or a hydraulic press (10, 12). Needless to say that specialized training of the personnel is indispensable for the realization of such processing.

1.3. High Temperature, Glass–Glass Fusion Bonding

High temperature, glass–glass fusion bonding (12, 14–17) provides us with high quality, monolithic integration of glass components and no material other than glass is present in the structure of the final device. Essentially, fusion bonding involves contacting two surfaces together, under slight pressure and at temperatures around the substrate's softening point. Under such circumstances, surface material from both substrates intermixes resulting in the formation of a strong joint once cooled; bond strength is significantly higher than in the case of low temperature methods (12). The processing can be performed by someone possessing basic laboratory skills and experience—even though certain safety precautions must be strictly followed throughout the experiments. However, as noted earlier, high temperature processing has an inherent risk of damaging the chip's microfluidic structures (13, 14, 16).

1.4. Glass Properties and Characteristics

The common element to all these techniques is glass and it is therefore valuable to give a brief overview of glass composition, structure and selected properties, in particular those pertinent to the subject of fusion bonding. Glass is a state of matter, which combines some properties of crystals and some of liquids but is distinctly different from both. Under room temperature, glasses have the mechanical rigidity of crystals, but the random disordered arrangement of molecules characteristic of liquids. The important consequence of such particular properties of glasses is that the term "melting point," defined as the temperature of phase transition between solid and liquid, does not really apply here. Instead, if we consider glass as a liquid of very high viscosity, we can use the notion of "softening" and "softening point" as the temperature (or the range of temperatures), at which glass starts losing its solid character. Evidently, this parameter varies with the type of glass, i.e., its structure and chemical composition. From numerous types of glass, two are most commonly used in microfluidic applications—namely soda lime and borosilicate glass. Their composition (by weight) is as follows:

- Soda lime glass contains: SiO_2 (silica) 60–75%, Na_2O (soda) 12–18%, CaO (lime) 5–12%, Al_2O_3, SO_3, K_2O, MgO, Fe_2O3, TiO_2—less than 1% each.
- Borosilicate glass is any silicate glass having at least 5% of boric oxide, B_2O_3, in its composition. For example, borosilicate glass known under the brand name of Pyrex 7740 contains SiO_2 80.6%, B_2O_3 13%, Na_2O 4%, and Al_2O_3 2.3%.

It is clear that softening temperatures of these two classes of glass must differ, reflecting dissimilarities in their composition. Moreover, in the case of soda lime glass, a significant variation of composition occurs within its own material type. Hence, the softening of soda lime glass may take place over a rather wide range of temperatures, from approximately 570°C to 730°C. For the borosilicate glasses (Pyrex and alike) a much narrower range of 815°C to 820°C is usually quoted. Taking into account the values of softening points for the soda lime and borosilicate glasses, it becomes clear that the former is more favorable for fusion bonding applications. In consequence, the two fusion bonding procedures described later in the text concern this particular type of glass.

Having successfully completed your bonding process, one might want to know how well it is bonded, or whether a consistent bond quality is achieved for a series of chips. The two principal features that define bond quality are leak-tightness and bond strength. The first parameter can be simply assessed via visual inspection under an optical microscope. If possible, it is preferable to fill the channels with dyed liquid as it improves the optical contrast. In a more advanced approach, a fluorescence microscope and fluorescent liquid could be used for detailed inspection of the chip. The quantitative characterization of the bond strength is particularly important for those devices in which pressure is exerted from within the microchannels, imposing the risk of debonding, e.g., microsystems for on-chip high-performance liquid chromatography (HPLC) chromatography. The problem of bond strength measurements is, however, beyond the scope of this text and the reader can refer to review papers treating the subject in detail (17, 18).

This chapter details three bonding/capping protocols, one representative of the three approaches described above:

- Low temperature, glass/PDMS adhesive bonding, which we find particularly useful for fast prototyping and preliminary testing of temperature-sensitive microfluidic chips. This approach also lends itself well to training applications, being cheap, safe, and reversible, as well as facilitating cleaning of the glass substrate's microfluidic features for storage and reuse.
- Medium temperature, pressure assisted glass–glass fusion bonding, which allows the fabrication of a full-glass microdevice at temperatures of 300°C and therefore below the softening-point.
- High temperature, glass–glass fusion bonding, which provides the highest bond quality but is applicable under the condition that the processing at 600°C (approximately) is not detrimental to the chip's prefabricated microstructures.

2. Materials

2.1. Materials Common for All Protocols

1. Soda lime glass slides 75 mm × 25 mm, thickness 0.9–1.1 mm (Corning Glass, Corning, NY USA; product number 2947-75×25).
2. Acetone, ethanol, sulfuric acid 95–98%, hydrogen peroxide 30–35%; purity 95% or better.
3. Deionized (DI) or double-distilled water (henceforth, referred to as "water").
4. Dry compressed nitrogen.
5. Paper towels, lint-free cleanroom tissues and cotton swabs (optional but recommended)—for example, Sontara®

MicroPure SV tissue (DuPont) and Alpha® Swab TX761 (ITW Texwipe).

6. Glassware made from Duran® (Schott) or equivalent material (dimensions given are approximate):

   Crystallization dishes (×2): diameter 9.5 cm, height 5.5 cm.

   Beakers (×2): diameter 6.5 cm, height 9.5 cm.

   Measuring cylinders: 20 and 100 ml.
7. Wash bottles for acetone, ethanol, and water.
8. Tweezers, one set of tweezers made from polytetrafluoroethylene (PTFE), chlorotrifluoroethylene (CTFE), or other fluoropolymer of equivalent chemical resistance for chemical processing use; and another set made of stainless steel for manipulating samples in the furnace.
9. Laboratory stirrer plate.
10. Spinner (optional, see Note 1).

2.2. Materials Specific for Low Temperature, Glass/PDMS Adhesive Bonding

1. Sylgard 184 silicone elastomer kit (Dow Corning, Midland, MI, USA).
2. Xylene.
3. Flat glass support, preferably with raised edges.
4. Cutter, ruler, print-off of device geometry, hole punch.
5. Oven (processing temperatures between 100°C and 200°C).

2.3. Materials Specific for Medium Temperature, Pressure Assisted Glass–Glass Fusion Bonding

1. Hydraulic press with heated plates (up to 300°C), capable of delivering pressures up to 30 MPa (Rondol Technology Limited, Stoke on Trent, UK).

2.4. Materials Specific for High Temperature, Glass–Glass Fusion Bonding

1. Programmable furnace, capable of providing temperatures up to 800°C approximately, with the possibility of defining various temperature profiles (ramps). The processing chamber should be large enough to accommodate the assembly of samples and weights pressing them together.
2. Weight made of stainless steel or brass, of approximately 3 kg (see Note 2). It is assumed that the external pressure in fusion bonding should not exceed 0.1 MPa (13). For the standard microscope slide, this value corresponds to the weight's mass of approximately 19 kg.
3. Separation plates made of vitreous carbon, alumina or other ceramic. The plates have to be smooth and flat, otherwise they will leave imprints in the bonded glass slides.

3. Method

As all the devices described consist of at least one glass substrate into which microfluidic features have been fabricated, the glass substrate cleaning process prior to bonding is common to all three protocols described. Afterwards, the specific processing steps will be described for each method separately.

3.1. Glass Substrate Cleaning: Common to All Protocols

1. The appropriate cleaning procedure is strongly dependent on the initial cleanliness of the slides. First, if the presence of loose particulate matter has been observed on the slides, it should be washed off with either ethanol or under the stream of water and then dried with compressed nitrogen. This will prevent scratching the samples in the subsequent cleaning steps. If surface contamination is still visible to the naked eye, it is advisable to wipe the slides first with a lint-free tissue or a swab soaked with acetone and then again with ethanol. After this rough cleaning, the slides should have a mirror-like appearance, without any visible stains.
2. Place each slide, one by one, into a crystallization dish filled with acetone (make sure that the slide is entirely submerged in the liquid). Place the dish onto the stirrer plate for 5 min.
3. Repeat the above step with ethanol instead of acetone, using the PTFE (CTFE) tweezers to transfer slides between crystallization dishes (see Note 3).
4. Using the wash bottles, rinse the slides with acetone first and then ethanol.
5. Dry the substrates under the nitrogen stream.
6. Place the slides into the beaker filled with either DI or double distilled water. Make sure that the water has free access to both sides of the slides (see Note 4). Rinse the samples under a stream of water running through the beaker for 10 min.
7. Dry the substrates under the stream of nitrogen.
8. Prepare the "piranha" cleaning solution according to the recipe presented below. At this point and throughout the entire processing using this mixture, the operator should wear: acid resistant gloves and apron, safety glasses or, preferably, a face mask (see Note 5).
9. Pour 30 ml of sulfuric acid and 10 ml of hydrogen peroxide into the measuring cylinders (see Note 6).
10. Place the slide into the clean and dry crystallization dish. Preferably, only one slide should be cleaned at a time.
11. Put the dish on the stirrer plate and pour the hydrogen peroxide first and then the sulfuric acid (see Note 7). This will

initialize a very rapid, exothermic chemical reaction with a temperature that may exceed 100°C. The resulting piranha solution is highly efficient in removing organic contaminants. Obviously, it is also very corrosive, so this step has to be performed with extreme care. The reaction extinguishes spontaneously after approximately 10 min. The dish should be stirred gently throughout the cleaning process.

12. Prepare a beaker filled with approximately 200 ml of water.
13. Using the PTFE (CTFE) tweezers, retrieve the slide from the piranha solution and immerse it immediately into water (see Note 8).
14. After this rough removal of the piranha solution, rinse the slide under running water for about 5 min.
15. For the final rinsing, place the slide again into a beaker filled with water and let the water run through the beaker for at least 15 min.
16. Retrieve the slide after the final rinsing. As glass is by nature a hydrophilic material, the slide should be uniformly covered with the layer of water. If this is not the case, the slide should be dried and the cleaning procedure repeated.
17. Dry the slide either under the nitrogen stream or by utilizing a spin-drying method (if a spinner is available). This drying step is critical as it should remove any solid particles floating on the water layer that would otherwise remain on your slide.
18. Dry the slide in the furnace at 120°C for 15 min to remove any traces of moisture, which could negatively affect the bonding procedure (see Notes 9 and 10).

3.2. Bonding

3.2.1. PDMS Layer Preparation and Glass/PDMS Adhesive Bonding

1. Mix the base polymer and curing agent in a ratio of 10:1 (v/v) and stir vigorously for 5 min to obtain a homogenous solution. For a 15 cm × 15 cm glass support, approximately 16.5 and 55 ml total volume produces sheets of thickness 0.75 and 2.5 mm, respectively (see Note 11).
2. Pour this mixture evenly onto the glass support and let it rest for 1 h to degas and flow (see Notes 12 and 13).
3. Place the glass support onto a flat surface in the oven and bake for 1 h at 100°C. (Alternatively, 24 h at room temperature or 30 min at 120°C also work well.)
4. Once cooled, place the glass support over your template and scribe around the chip's perimeter. If access ports are required, use the punch (e.g., 2 mm outer diameter) to cut out holes, taking care not to leave any shreds.
5. Remove the PDMS cover with plastic tweezers and place it over your prepared glass chip, ensuring that the side in contact with the support's surface is facedown. Any trapped air is easily

expelled by applying light pressure to the PDMS and pushing it towards the edge. If you are unhappy with the final alignment or cannot remove some trapped air, simply peel off the PDMS layer and repeat this step.

3.2.2. Medium Temperature, Pressure Assisted Glass–Glass Fusion Bonding

1. Preheat both slides at 150°C for 5 min by placing them on the lower plate of the press.
2. Put the slides in contact and apply slight pressure of 0.5–1 MPa, still at the temperature of 150°C.
3. Increase the temperature up to 250°C.
4. Apply pressure of 30 MPa (see Notes 14 and 15).
5. Maintain the sample under the above conditions for 4 h.
6. Release the pressure and switch the heating off.
7. Retrieve the sample once cooled down to room temperature.

3.2.3. High Temperature, Pressure Assisted Glass–Glass Fusion Bonding

1. Once cleaned and dried, put the slides into contact with one another and place them between two separation plates.
2. Place the entire assembly into the furnace and put the weight on top of it (see Note 2). The weight's footprint should be slightly larger than the sample to ensure that the pressure is applied uniformly.
3. Start the firing sequence, which in our case was as a follows (19):
   (a) Heating from room temperature to 400°C, with a ramp of 280°C/h.
   (b) Incubation at 400°C for 4 h.
   (c) Heating up to 590°C, ramp 280°C/h.
   (d) Incubation at 590°C for 6 h.
   (e) Slow, spontaneous cooling to room temperature—overnight.
4. Having cooled down, the bonded sample can be retrieved from the furnace.

4. Notes

1. For relatively large samples such as microscope slides, the spin drying method usually provides better results than blowing surface liquid off with compressed nitrogen.
2. From our experience, a weight of a few kilograms should suffice. Some authors (19) used a weight as small as 40 g, but in their case the bonding process had to be repeated a number times—each time moving the weight from one area to the next.

3. It is essential to wash the slide with ethanol immediately after acetone rinsing. Acetone itself leaves traces of solid contamination behind, which are otherwise difficult to eliminate.
4. When a standard microscope slide is placed in a 6.5 cm × 9.5 cm beaker, it simply rests in a tilted position, with its corners touching the beaker's bottom and wall, and with both its facets well-exposed to the water flow.
5. Safety issues and waste disposal: it is assumed that all the experiments involving chemical processing will be carried out on a wet bench, under a fume hood. The operator must wear an apron, acid resistant gloves, protective goggles or a face mask—in particular, when dealing with the piranha solution. For details concerning manipulation and disposal of the chemical products used, please refer to their respective material safety data sheet (MSDS) files and consult your local health and safety officer.
6. In the cleaning step, a 3:1 volume/volume ratio of sulfuric acid and hydrogen peroxide was used to prepare the piranha solution. However, other ratios, such as 5:1, can also be found in the literature (12).
7. It is the common knowledge of chemists that acids must be added to water, not the other way round, and this rule also applies to mixing hydrogen peroxide with sulfuric acid. If you add hydrogen peroxide to concentrated sulfuric acid, it can boil and spit. This occurs as hydrogen peroxide is less dense than sulfuric acid, so if you pour hydrogen peroxide on the acid, the reaction occurs on top of the liquid. If you add the acid to the hydrogen peroxide—i.e., in the correct order—it sinks and any reactions take place in the bulk of mixture.
8. Please note that the slide covered with the layer of piranha solution is very slippery and difficult to handle—proceed with caution!
9. The bond quality is extremely sensitive to surface contamination (17, 20). Even though the cleaning recipe described is very effective in removing organic contamination, the slides have to be protected against airborne solid particles. Hence, once the slides are taken out from the liquid environment and until they are put in intimate contact and bonded, they should be manipulated in the low-particle environment, i.e., in the cleanroom laboratory or under the laminar flow hood.
10. It is recommended that the final drying is followed immediately by the actual bonding. If it is necessary to prepare and stock a certain number of slides, they may be stored in water for a few hours, after the final rinsing but before drying.
11. PDMS is messy stuff, so first cover the working area with paper towels to catch any spills.

12. Outgassing of the mixture can be accelerated by placing it into an ultrasonic bath for 5 min before pouring.
13. For cleanup of PDMS pre-polymer liquid, we have found the solvent xylene works well.
14. It is important that the temperature—pressure treatment is carried in the right order, i.e., the temperature is always increased prior to raising the pressure. If the sequence were reversed, the sample would almost certainly fracture.
15. The described cleaning and bonding protocols should be considered as a guideline. The reader is encouraged to modify them if necessary. For example, the cleaning procedure may be altered depending on the initial contamination level of the slides used. Also, as the composition and thermal properties of the soda lime glass may vary between suppliers, the pressures, temperatures, and firing sequence described in these protocols may need to be accordingly adapted.

Acknowledgments

This research was supported by the NANOLYON technological platform of the Institut des Nanotechnologies de Lyon (INL), which provided us with access to its technological facilities.

References

1. Plecis A, Chen Y (2008) Improved glass–PDMS–glass device technology for accurate measurements of electro-osmotic mobilities. Microelectron Eng 85:1334–1336
2. Chang JK, Bang H, Park S-J et al (2003) Fabrication of the PDMS microchip for serially diluting sample with buffer. Microsyst Technol 9:555–558
3. Oh KW, Han A, Bhansali S et al (2002) A low-temperature bonding technique using spin-on fluorocarbon polymers to assemble microsystems. J Micromech Microeng 12:187–191
4. Zhou X, Poenar DP, Liu KY et al (2007) Glass-based BioMEMS devices for optically excited cell impedance measurement. Sensor Actuat A 133:301–310
5. Wilke R, Büttgenbach S (2003) A micromachined capillary electrophoresis chip with fully integrated electrodes for separation and electrochemical detection. Biosens Bioelectron 19:149–153
6. Mazurczyk R, Vieillard J, Bouchard A, Hannes B, Krawczyk S (2006) A novel concept of the integrated fluorescence detection system and its application in a lab-on-a-chip microdevice. Sensor Actuat B 118:11–19
7. Vieillard J, Mazurczyk R, Morin C, Hannes B, Chevolot Y, Desbène P-L, Krawczyk S (2007) Application of microfluidic chip with integrated optics for electrophoretic separations of proteins. J Chromatogr B 845:218–225
8. Chiem N, Lockyear-Shultz L, Andersson P et al (2000) Room temperature bonding of micromachined glass devices for capillary electrophoresis. Sensor Actuat B 63:147–152
9. Jia Z-J, Fang Q, Fang Z-L (2004) Bonding of glass microfluidic chips at room temperatures. Anal Chem 76:5597–5602
10. Bhattacharya S, Gao Y, Korampally V et al (2007) Mechanics of plasma exposed spin-on-glass (SOG) and polydimethyl siloxane (PDMS) surfaces and their impact on bond strength. Appl Surf Sci 253:4220–4225
11. Zhang M, Zhao J, Gao L (2008) Glass wafers bonding via Diels-Alder reaction at mild temperature. Sensor Actuat A 141:213–216

12. Sayah A, Solignac D, Cueni T et al (2000) Development of novel low temperature bonding technologies for microchip chemical analysis applications. Sensor Actuat 84:103–108
13. Iles A, Oki A, Pamme N (2007) Bonding of soda-lime glass microchips at low temperature. Microfluid Nanofluid 3:119–122
14. Stjernström M, Roeraade J (1998) Method for fabrication of microfluidic systems in glass. J Micromech Microeng 8:33–38
15. Fu L-M, Yang R-J, Lee G-B (2003) Electrokinetic focusing injection methods on microfluidic devices. Anal Chem 75: 1905–1910
16. Mao P, Han J (2005) Fabrication and characterisation of 20 nm planar nanofluidic channels by glass–glass and glass–silicon bonding. Lab Chip 5:837–844
17. Plöbl A, Kräuter G (1999) Wafer direct bonding: tailoring adhesion between brittle materials. Mater Sci Eng R25:1–88
18. Vallin Ö, Jonsson K, Lindberg U (2005) Adhesion quantification methods for wafer bonding. Mater Sci Eng R 50:109–165
19. Holden M, Kumar S, Castellana E et al (2003) Generating fixed concentration arrays in a microfluidic device. Sensor Actuat B 92: 199–207
20. Gueguen P, Ventosa C, Di Cioccio L et al (2010) Physics of direct bonding: applications to 3D heterogeneous or monolithic integration. Microelectron Eng 87:477–484

Chapter 11

Rapid Prototyping of PDMS Devices Using SU-8 Lithography

Gareth Jenkins

Abstract

This protocol describes the fabrication of single and multi-layer SU-8 microstructures for generating microfluidic devices via PDMS (polymethyldisiloxane) casting. SU-8 is a negative, thick-film, epoxy based photoresist that has become widespread in the MEMS industry for producing durable, high aspect ratio microstructures for a variety of applications. It has become especially popular with microfluidics researchers to produce molds for PDMS casting since such molds allow for the rapid replication of prototype microfluidic structures made from PDMS. Although SU-8 processing does allow for rapid and straightforward development of devices it is prone to numerous pitfalls which have gained it a reputation of being somewhat of a "black art." This protocol attempts to give as full an account as possible of all the tricks and tips the author has learned over the years for processing SU-8. It also describes the casting of PDMS and plasma bonding for the generation of complete microfluidic devices ready for use in the lab.

Key words: SU-8, PDMS, Polymethyldisiloxane, Soft-lithography, Plasma bonding, Microfluidics

1. Introduction

PDMS (polymethyldisiloxane) has risen in popularity in recent years as a useful material for the rapid production of microfluidic devices. It benefits from being optically transparent (down to ~280 nm) with a low auto-fluorescence and can easily be cast over a mold to replicate features with exceptional fidelity. Its flexible nature gives it unique properties, allowing it to be cut and shaped remarkably easily and also enables it to be used for novel valve and pumping mechanisms. Being gas permeable allows it to be used for cell culturing applications and it is largely biocompatible although nonspecific protein binding can be an issue.

Early work was pioneered in the laboratory of George Whitesides at Harvard University and much of the early techniques developed remain relatively unchanged and used in many laboratories across the globe (1, 2). One of its major applications is its use for "soft-lithography" in which a microstructured PDMS stamp is

Gareth Jenkins and Colin D. Mansfield (eds.), *Microfluidic Diagnostics: Methods and Protocols*, Methods in Molecular Biology, vol. 949, DOI 10.1007/978-1-62703-134-9_11, © Springer Science+Business Media, LLC 2013

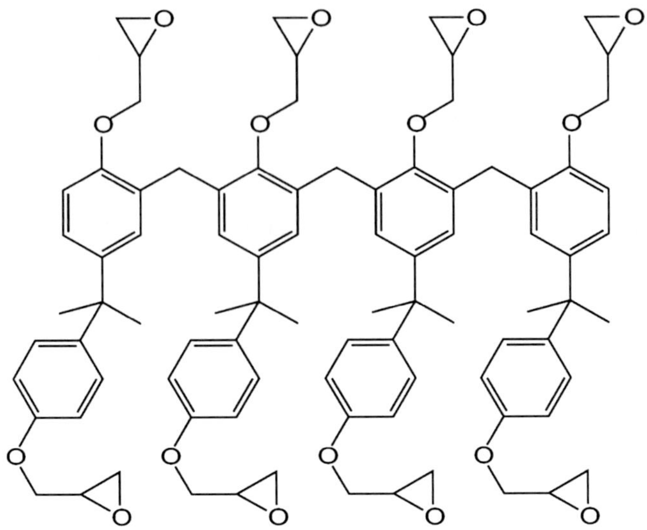

Fig. 1. Chemical structure of the SU-8 molecule.

used to pattern and transfer various materials via micro-contact printing (3). Similar techniques are used to create microfluidic devices by casting PDMS over a microstructure mold and then bonding the resultant, cured PDMS structure to a second, planar substrate to achieve enclosed microchannels.

SU-8 is an epoxy based, negative, thick film photoresist that has become common as a micro-molding material for PDMS microfluidic structures. It is able to generate high aspect ratio structures from a few microns up to several millimeters in height. This ability is due to the low optical absorption at the wavelength to which the resist is photosensitive (365 nm). SU-8 was initially developed by IBM and consists of three main components:

1. EPON SU-8 Epoxy (Shell Chemicals). This contains, on average, eight epoxy groups (hence the "8" in SU-8). Figure 1 shows the chemical structure.
2. Solvent: Gamma-Butyrolactone GBL (used in the standard formulation—the more recent formulations use different solvents—e.g., cyclopentanone for SU-8 2000).
3. Photo initiator: A triarylium–sulfonium salt.

SU-8 is sensitive to broad band UV exposure (350–400 nm, optimized for 365 nm) and upon exposure a Lewis photoacid is generated that catalyzes the cross-linking reaction between the epoxy groups to form a dense, solid network. Unexposed material remains uncross-linked and soluble in the developer solvent (usually PGMEA).

Figure 2 shows an overview of the process for fabricating a SU-8 master mold for generating PDMS microfluidic structures.

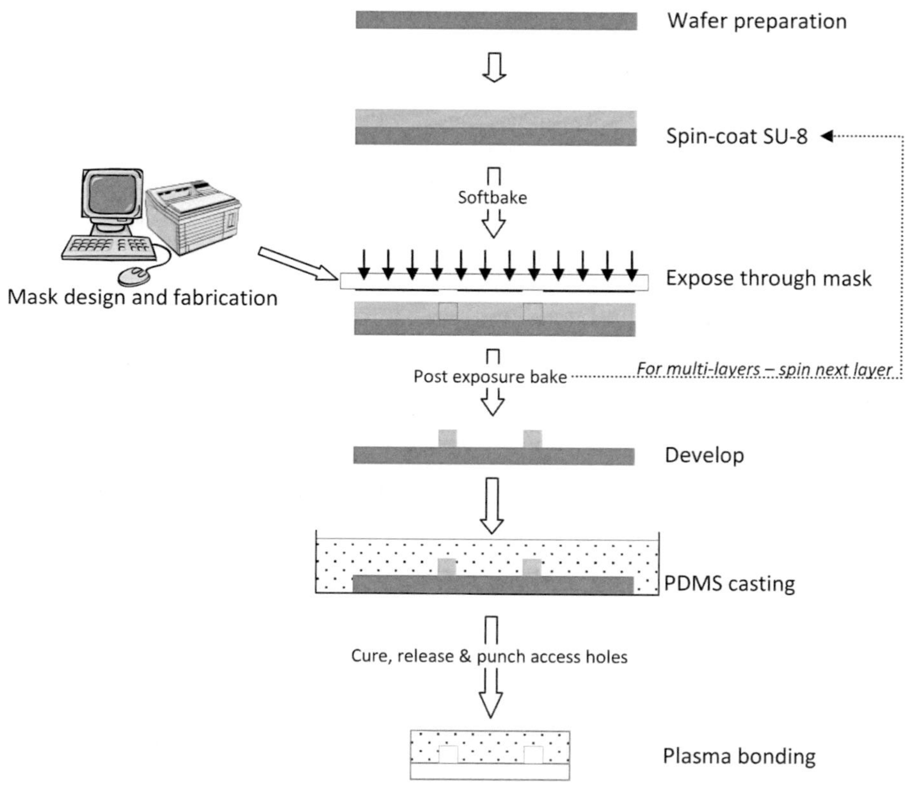

Fig. 2. Overview of the process for making a PDMS microfluidic device using a microfabricated SU-8 mold.

Process guidelines for SU-8 may be found in the relevant MicroChem datasheet (4). The steps needed for a photolithographic fabrication of the SU-8 mold are photomask design and preparation, wafer preparation, spin-coating of SU-8, softbake, UV exposure, post-exposure bake, and development. Using the generated mold, PDMS casting, and plasma bonding is then used to form a completed microfluidic device.

1.1. Photomask Design and Preparation

To lithographically transfer the desired design to the SU-8 photoresist, a photomask is required. For low resolution structures, it is possible to use a desktop printer to print the design onto a transparency foil. However, some issues concerning mask contrast may be encountered for thick layers and good optimization of the print settings are required.

In our experiments we use a commercial, high resolution photoplotting service (JD Photo-Tools, Oldham, Lancs, UK) to plot masks onto transparent film enabling features of 5–10 μm with some optimization. Although the cost is higher, a large number of designs can be included on the same film and the quality is far better than can be achieved with a high resolution desktop printer. Commercial vendors of high resolution chrome masks may be used for even higher resolutions but at increased cost and decreased mask area.

A further advantage of chrome masks is the higher contrast, durability and dimensional stability.

Polarity of the mask is important to achieve the desired results and it should be remembered that SU-8 is a negative photoresist, such that areas exposed (through apertures in the mask) will be cross-linked to form a positive relief structure on the SU-8 mold. When casting the PDMS, this positive relief structure translates into a recess in the final PDMS structure. Hence, to create a simple microfluidic channel in the PDMS, a darkfield mask is required with a clear, thin line through which light can pass. This will generate a positive ridge structure in the SU-8 which then translates to a microchannel in the cast PDMS.

Mask design requires the use of a suitable CAD package. Perhaps the most common file format is DXF (Design eXchange Format) which can be created and edited using AutoCAD or other software packages. Gerber, GDSII and CIF are other common formats. It is important to consider the correct design rules depending on how the mask will be produced. If using a commercial vendor, determine their individual design rules prior to starting the design to avoid lengthy reworking or extra setup fees.

In DXF, polylines are the most commonly used element in designs. Giving a polyline width will enable channel features to be easily defined whereas a zero-width closed polyline can be used to define a more complex shape with all the enclosed area being filled (note: avoid self-intersecting polylines and do not use the Hatch function in AutoCAD for filling—this will not appear in the DXF file). Drawing island features or holes in solid shapes can be achieved in several ways depending on the design rules. Often, shapes can be drawn in different layers such that cut/subtract and even more complex Boolean operations may also be performed. Alternatively the shape may be split up into two or more sections and joined together. Simple text is usually possible, although Truetype fonts may not be supported.

Naturally, for multiple layers, a separate mask will be required for each layer. To ensure accurate registration between different layers it is necessary to include alignment features in each mask. A simple cross structure is fine when using a brightfield mask but for darkfield masks (which are more common for SU-8 molding of microfluidic structures) it can become virtually impossible to find the alignment features since most of the visual field is blocked by the mask. Hence, a large box should be used to define an open brightfield region in which alignment crosses can be placed (see Fig. 3). Several alignment features, preferably in each corner of the design, should be included to ensure good rotational alignment.

Once a film photomask has been printed, it needs to be mounted onto a glass substrate to allow insertion into the mask aligner (unless a collimated UV lamp is being used, in which case the film mask may placed face down, directly onto the wafer being

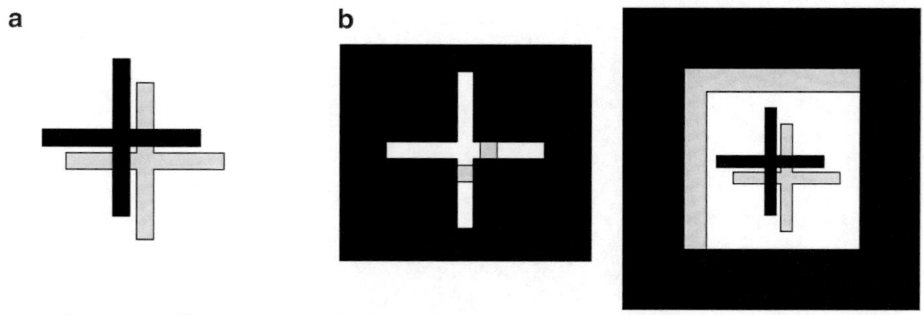

Brightfield mask alignment structure. Black cross is on mask, grey cross is on previous layer).

Darkfield alignment structures. Left example is difficult to align due to mask obscuring view. Right example overcomes this issue using a small bright field box.

Fig. 3. Comparison of alignment structures for use with brightfield and darkfield masks. (**a**) Shows simple brightfield alignment structure, (**b**) Shows good and bad examples for darkfield alignment structures.

exposed—good contact must be ensured during exposure). Film masks may be mounted onto the glass using adhesive tape. It must be ensured that the tape does not interfere with the contact between the mask and the SU-8 wafer otherwise a gap will exist leading to poor results. It should also be ensured that the printed side will contact the wafer, otherwise a gap equal to the film thickness will exist allowing diffraction and poor resolution (5).

1.2. Wafer Preparation

A 4 in. silicon wafer is used as a substrate for depositing the layers of SU-8. Other sizes may be used depending on the size requirements of the design and compatibility with the mask aligner and other tools used. To achieve good adhesion of the SU-8 to the wafer surface requires careful attention to the preparation of the wafer surface. The surface should be hydrophilic after treatment and free from contaminants including adsorbed water and particulate matter. Use of HMDS as an adhesion promoter is not generally recommended. Adhesion can be made worse if not done in a dry environment with a good dehydration bake.

1.3. Spin-Coating of SU-8

Spin-coating the SU-8 onto the silicon wafer allows the definition of the final layer thickness. This performed in a two-step process: (1) a short spin, low speed spin to spread the SU-8 evenly across the wafer; (2) immediately followed by a longer higher speed spin to define the final thickness. Due to the high viscosity of the thicker formulations of SU-8, this step can be problematic particularly with bubble entrapment during resist deposition. This is a common source of problems observed during the softbake step.

1.4. Softbake

The softbake step is necessary to drive the solvent out of the resist and densify the film prior to exposure. Insufficient softbake can lead to problems with the mask sticking to the wafer during exposure, whereas over-baking can lead to brittleness. Hence, careful

Fig. 4. Microscope image showing cross section of PDMS cast on a SU-8 mold exhibiting extreme "T-topping" and negative sidewall slope.

optimization is required. The softbake can also influence the final thickness of the film and it is critical to ensure even temperature uniformity and to avoid airflows over the surface which can lead to poor uniformity of solvent evaporation

1.5. Exposure

SU-8 is optimized for I-line exposure (at 365 nm). Its optical transparency in this region allows for highly vertical sidewalls even for very thick layers. Above 400 nm the resist becomes too transparent for efficient cross-linking to occur and below 350 nm, the resist becomes opaque which can lead to excessive cross-linking in the top of the film compared to the bottom. Hence, for thicker layers, filtering out the deep UV can be critical to avoid an excessive negative sidewall slope and also to avoid the "T-topping" phenomenon in which features in the top layer are much wider than the bottom. This can be achieved by inserting a high-pass filter into the optical path to cut out wavelengths below 350 nm.

Good mask contact is also critical for good results. A gap between the mask and SU-8 layer will cause poor resolution and "T-topping" (5). Figure 4 is a microscope image of a PDMS section from a thick SU-8 mold showing an extreme case of negative sidewall slope and "T"-topping. (See Note 1 for further tips on avoiding T-topping.)

Exposure dosage is dependent on film thickness and must be optimized for best results as a wide variety of factors can influence the final result. In particular, UV can be absorbed by the polymer film used for the mask which can considerably increase the required exposure time.

Over exposure can lead to poor resolution with increased feature dimensions and can also result in a film over the top of masked regions if the mask contrast is too low. Under exposure can result in poor adhesion and delamination of the exposed features. If resolution is not critical, it can be advisable to slightly over expose the resist to improve the adhesion. If features are visible immediately after exposure, this generally indicates a more than adequate dose. In a process optimized for the highest resolution, structures may only become visible during the post exposure bake.

1.6. Post-exposure Bake

After exposure, the photoresist should be baked to further induce cross-linking. In this step it is particularly important to avoid rapid temperature changes which can increase the stress in the SU-8 film and lead to delamination (see Note 2 for further tips to reduce the effects of stress).

If you are looking to generate SU-8 structures with multiple layers, additional layers may be spun and processed according to the datasheet parameters used for each single layer, *after* the post exposure bake of the previous layer but *before* the development of the final structure. Also note it is important to use the same series of SU-8 for each layer. Differences in the solvent system used for different formulations (e.g., SU-8 2000 series and SU-8 3000) will lead to poor results due to different rates of solvent evaporation. Also take note of the previously mentioned tips for the mask design of each layer to allow alignment of additional layers to assist good registration between each layer.

1.7. Development

After processing all of the required SU-8 layers, all of the layers are developed in a single step. Dissolution rates are given in the datasheets, but as with all the parameters, environmental and other factors will influence the actual time required. However, if all layers have been well enough exposed, over developing tends not to produce any problems. Additional development time will also be required for deep trench or well features which limit the rate at which fresh developer can remove the unexposed resist. Such features may also be better developed using an ultrasonic bath. Note also that the hardbake step mentioned in the datasheets is not generally required for PDMS casting and can actually increase the risk of delamination if not well optimized. The hardbake step can, however, be effective in annealing stress cracks which commonly appear on the surface of the SU-8 after development. Such cracks tend to be quite shallow and hence rarely affect fluidic performance of the final PDMS structure

1.8. PDMS Casting and Plasma Bonding

After preparation of the SU-8 mold, it can then be used for PDMS casting. During this step, the two component mixture of base monomer and curing agent are mixed and cast on to the mold. Generally, release from the untreated mold should be OK but in some cases sticking can occur. Hence, it is advisable to treat the mold with trichloro-(1H,1H,2H,2H-perfluorooctyl)silane vapor by placing a small amount in an open container in a vacuum desiccator together with the mold. This will allow smooth release from and increase the usable lifetime of the mold.

Oxygen plasma bonding may be used to seal the PDMS device with a glass (or PDMS) cover lid. This works by creating silanol groups on the glass and PDMS surfaces such that when they are brought into contact, a covalent bond is formed between the silicon and oxygen atoms with water evolved during the reaction ($SiOH + SiOH \rightarrow Si–O–Si + H_2O$). Optimizing the plasma parameters (mainly pressure, power, and time) is key to achieving good bonding. An easy way to optimize such parameters is to perform a series of tests on PDMS to measure the resultant contact angle with water as a function of the plasma power and treatment time (6). Just enough treatment to make the surface hydrophilic is required but over treatment leads to microscopic cracks in the PDMS surface which prevents good bonding (6, 7). Since water is evolved during the bonding reaction, moisture may affect the bonding process; hence, it is important for both surfaces to remain dry.

2. Materials

2.1. SU-8 Photolithography for Mold Fabrication

1. PC with suitable CAD program (e.g., AutoCAD).
2. Photomask: High resolution plotted transparency film (JD Photo-Tools, Oldham, Lancs, UK).
3. Glass plate: For mounting film photomask (e.g., 5 $in.^2$ soda-lime glass, Nanofilm, CA, USA).
4. Adhesive tape.
5. Silicon wafers: 4 in. (100 mm) and approximately 500 μm thick. Reclaim grade is fine and Miller Index is not important.
6. Wafer handling tweezers.
7. Dry nitrogen supply: With gun for drying.
8. Cleaning solvents: Acetone and isopropanol (IPA).
9. Concentrated H_2SO_4.
10. Ultrasonic bath.
11. SU-8: e.g., Microchem SU-8 50. Select grade depending on depth requirements. Should be stored in the dark and at room temperature—ideally in the same room as used for processing.
12. Spin coater.
13. Hotplate (must be level and have uniform temperature profile).
14. Mask aligner: Suss MA6/BA6 (Suss, GmbH, Germany). Or if accurate alignment of multiple layers is not required, a well collimated UV mercury lamp is suitable. I-line filter is advisable for thick layers.

15. Phase contrast microscope/differential interference contrast microscope: for alignment of multiple layers—optional depending on requirements, e.g., Nikon L150 Eclipse.
16. Developer solution: PGMEA (or MicroChem EC Solvent). Ethyl lactate may also be used.

2.2. PDMS Casting and Plasma Bonding

1. Release agent: Trichloro-(1H,1H,2H,2H-perfluorooctyl) silane (CAS no.: 78560-45-9, Sigma-Aldrich, UK).
2. PDMS kit: Sylgard 184 Silicone Elastomer kit (base and curing agent, Dow Corning).
3. Vacuum pump and desiccator.
4. Oven.
5. Plastic/glass Petri dish: For casting—ensure size big enough for wafer and material will withstand curing temperature.
6. Scalpel.
7. Hole punching tool: e.g., biopsy punch or wide bore, syringe needle which has been cut to form a flat end.
8. Oxygen plasma: e.g., Diener Nano UHP 40 kHz/300 W.
9. Microscope slides.

3. Methods

3.1. SU-8 Lithography and Mold Preparation

1. Solvent cleaning of silicon wafer (optional if wafer is new and straight from the box): immerse in acetone and sonicate for 5 min. Rinse in deionized (DI) water and dry under nitrogen. Immerse in IPA and sonicate for 5 min. Rinse in DI water and dry under nitrogen.
2. Wafer surface oxidization: *EITHER* immerse wafer in concentrated H_2SO_4 (see Note 3) for >1 h *OR* perform O_2 plasma *after* dehydration step 3.
3. Dehydration bake: Place wafer on hotplate at 250°C for 30 min. Cool under dry nitrogen stream.
4. Optional instead of step 2: Treat with oxygen plasma for 5 min at~250 W RF power, at 0.8 mbar.
5. Spin coat immediately: Place wafer onto spin-coater and deposit ~4 ml SU-8. Ensure sure wafer has cooled from dehydration bake. Deposit enough SU-8 to cover ~70% of wafer surface. See Note 4.
6. Any bubbles observed on the surface should be removed using a disposable plastic pipette.
7. Spin-coat according to MicroChem datasheets depending upon desired thickness. Typically, a two step process is employed

with the first step ensuring adequate spreading of the resist at 500 rpm for 5 s with a ramp rate of 100 rpm/s. The second step defines the final thickness e.g., 3,000 rpm for 30 s with a 300 rpm/s ramp rate. (See Note 5.)

8. Relaxation step. Allow the wafer to rest on a leveled surface at room temperature (in a clean environment and covered to avoid particle contamination/excessive solvent evaporation). This step is particularly important for thicker layers to reduce the edge-bead effect which can lead to poor mask contact but may be omitted for thinner layers (e.g., <10 μm).
9. Softbake: Place the wafer on a leveled hotplate and ramp up to 65°C. Hold at 65°C for the time specified by the data sheets (or as optimized). Ramp hotplate up to 95°C. Hold at 95°C for the time specified by the data sheets (or as optimized). See Notes 6–8 for common problems observed during the softbake.
10. Allow to cool to <50°C before removing from hotplate.
11. Lithography: Cut out the photoplotted film mask and using Scotch tape attach to a 5 in.^{2}, glass plate (depending on mask aligner). Ensure that the mask has the printed emulsion side facing such that it will contact the wafer and also ensure that the tape used to attach the mask does not contact the wafer. Otherwise a gap will exist between mask and SU-8 which will lead to poor results (see Note 1).
12. Insert the mask in to the aligner and load the silicon coated with SU-8. If aligning to a previous layer, align exposed SU-8 reticule features with mask reticules. Exposed SU-8 features on thin layers (<10 μm) can be hard to see on the aligner. In this case, a phase contrast microscope may be used to find the exposed features and a scalpel blade can be used to carefully scribe a mark in the resist which may then be used for alignment. Take care not to expose critical regions with excessive light from the microscope.
13. Expose in vacuum contact mode. The exposure dose will depend on layer thickness. For long exposures, it may be desirable to break it into multiple exposures. This is especially important if using an unfiltered light source as this may cause heating and sticking of the mask to the resist. However, problems with the mask sticking are usually an indication of insufficient soft bake.
14. Post-exposure bake: Place on a level hotplate at room temperature. Ramp up to 65°C and hold for the time specified in the datasheet (e.g., 1 min). Ramp up to 95°C and hold for the time specified in the datasheet (e.g., 10 min). Allow to cool slowly on the hotplate (do not remove until <40°C). See Notes 2, 9 and 10.

Fig. 5. Microscope image of developed SU-8 mold.

15. If you are looking to generate SU-8 structures with multiple layers, additional layers may be spun and processed according to the datasheet parameters used for each single layer. Return to step 5 and process the additional layers *before* the final development step.
16. Development: Immerse wafer into a dish containing developer solution. (PGMEA, or MicroChem's EC Solvent). Agitation is recommended for faster, more complete development. Using an ultrasonic bath may assist with deep trenches or wells. See Note 11.
17. After the recommended development time, remove from the dish and rinse with IPA straight from a solvent wash bottle (over an empty dish). If white residues are observed, place back in the developer solution and continue developing.
18. Repeat development and IPA rinse steps until no further white deposits are visible. If many repeats are required better results may be obtained by replacing with fresh developer solution.
19. Dry under nitrogen (do not use water to rinse). Check under microscope (taking care not to expose to too much light). If undeveloped resist is still observed, place back in developer. Figure 5 shows a microscope image of a fully developed SU-8 mold.
20. Optional hardbake.

3.2. PDMS Casting and Plasma Bonding

1. Treat the mold with release agent by placing in a vacuum chamber/desiccator with an opened bottle of the release agent. Apply vacuum and leave for approximately 1 h.
2. Mix PDMS monomer base and curing agent in a 10:1 ratio (by weight). For example, if using a 5 in. diameter Petri dish, approximately 30 g of monomer base plus 3 g curing agent is typical. Adjust for desired thickness. Mixing must be very

thorough or curing will not be complete. Inevitably many bubbles will become entrapped giving an opaque appearance.

3. Degas in a vacuum chamber/desiccator. This can take 30 min to 1 h depending on strength of vacuum and volume of PDMS. The mixture may froth up so ensure it does not spill over.
4. When the mixture is clear and bubble free, remove from vacuum and pour carefully over the SU-8 mold. Avoid trapping bubbles. Bubbles are easily removed by popping with a scalpel or by using a pipette. The whole molding dish may be returned to the vacuum if required.
5. Place dish on a level surface in an oven and cure at 60°C for 1 h (see Note 12).
6. Once cured, allow to cool fully (see Note 13) and cut out the device using a scalpel and place face down onto a well cleaned glass surface in order to punch holes without contaminating the surface to be bonded.
7. Punch holes using a biopsy punch of the desired size or other sharpened metal cylindrical tool.
8. Bond immediately (the PDMS surface will quickly attract dust and debris affecting the bonding capability).
9. Plasma bonding: Place the PDMS structure and a clean, dry, glass slide into the plasma chamber (with the surfaces to be bonded face up).
10. Pump down to 0.2 mbar pressure and flush with oxygen at 0.8 mbar.
11. Apply plasma (for a 300 W total power, use 20% RF power for 1 min).
12. Remove from chamber and immediately place the treated side of the PDMS onto the treated surface of the glass slide. Handle both pieces by the edges to avoid touching the activated surfaces and avoid entrapping air by placing the PDMS down in a "rolling" motion.
13. Place the bonded device onto a hotplate at 85°C for 1 h. (This helps to drive out the moisture generated in the bonding reaction and improves bond strength.)
14. The device is now ready for use.

    The notes section below contains a number of troubleshooting tips detailing a few commonly experienced problems with possible causes and solutions. For other issues, there is the excellent MEMS-talk mailing list group which is populated by many helpful experts (8). Another excellent resource is the MEMScyclopia which contains a comprehensive collection of practical tips from many sources in the MEMS community (9).

4. Notes

1. T-topping is where an intended rectangular cross-sectional structure appears more like a "T" shape or mushroom. Three possible causes are as follows:

   Poor filtering of deep UV (<350 nm). Use an I-line optical filter in UV source.

   Poor mask contact. Often due to the edge bead effect. Increase relaxation step before soft-bake. Check mask is being placed with the printed side in contact with wafer. Check tape or other materials are not creating a gap between mask and wafer. A drop of glycerol may also be used between the mask and SU-8 wafer during exposure. This reduces the effects of diffraction by matching the refractive index. Blow the glycerol off with a stream of nitrogen before the post-exposure bake (4).

   Over exposure. If cross section shows exposed features are wider than the mask, even at the base of this structure, this could indicate over-exposure. Reduce dose.

2. During cross-linkage, the SU-8 contracts slightly which results in stress within the film. Stress can be reduced by ensuring that the film does not experience rapid heating or cooling during the post-exposure bake. Wait until the hotplate has cooled to below 40°C prior to removal. Stress becomes more acute for structures with large areas of exposed SU-8. A fully exposed, 100 μm thick SU-8 layer will cause a silicon wafer to visibly bow and highly stressed structures have been known to eject fragments of SU-8 at high speed (goggles must be worn). Avoid unnecessarily large areas in the design. Incorporating "stress relief" structures can help (i.e., surrounding critical features with a border of unexposed resist). Another technique to minimize stress is to perform longer but lower temperature post-exposure bakes. A post exposure bake at 55°C (below the glass transition temperature of uncross-linked SU-8) has shown to decrease stress substantially (7).

3. Piranha solution is also sometimes used for silicon wafer oxidation. This typically consists of 3:1 conc. H_2SO_4:30% H_2O_2. However, care must be taken when preparing piranha due to the heat evolved which can lead to explosion. And it is advisable to wait 30 min after preparing before use to reduce this risk. Also the solution degrades after 12 h rendering it useless for further use. Use of concentrated H_2SO_4 seems to work equally well and can be reused which reduces the associated waste and safety concerns. However, fresh acid should be used if discoloration is observed as impurities can affect resist adhesion.

4. Use of a pipette for dispensing can be problematic due to the viscosity of SU-8. It is often easier to pour direct from the bottle. However, this can lead to a buildup of particulate matter around the neck of the bottle which may then contaminate the SU-8 resulting in "divots" during the softbake (see Note 7). Additionally, repeated pouring leads to bubble entrapment within the bottle which also leads to "divots" (see Note 6). Decanting a smaller quantity first into a smaller bottle is therefore advisable. Wait ~1 h after decanting to allow bubbles to rise to the surface. The use of a small bottle also allows the user to minimize the height above the wafer surface enabling smoother pouring with less bubble entrapment.
5. Achieving the correct thickness often requires some optimization. Although the main parameter which determines film thickness is the final spin speed, other effects also can be major factors. These factors include airflows both over the spin-coater and the hotplate (during softbake) since it will affect the rate of solvent evaporation. Having a cover over the spin-coater can help reduce such airflows.
6. During softbake, large depressions in the SU-8 layer may appear (commonly referred to as "divots" or "dimples"). Three possible causes of such divots are:

   Particulate contamination in the SU-8 or on the wafer (often falling from the neck of the SU-8 bottle). Use fresh SU-8 and take steps to prevent contamination.

   Small microscopic bubbles in the SU-8 (often not visible) which may have become entrained from repeated pouring from the bottle or inversion (e.g., during shipping). Place the *closed* bottle in an oven and heat to 60°C for 1 h and allow to cool. This helps bubbles rise to the surface and come out.

   Contamination of the wafer surface (e.g., from contaminated sulfuric acid used during wafer prep). Replace chemicals and takes steps to reduce contamination.
7. Wrinkles may appear during softbake. This can due to poor hotplate uniformity or uneven evaporation of solvent. Cover wafer with dish (ensuring a gap for solvent to escape). Check temperature uniformity of hotplate.
8. If the SU-8 recedes from the edges of the wafer towards the center during softbake, this is an indication that the wafer has not been properly treated during the preparation stage. The sulfuric acid used may have become neutralized over time or plasma treatment may not have been sufficient.
9. Wrinkles maybe observed across whole wafer during post exposure bake. Such wrinkles may appear around 55°C and then disappear during the higher temperature step. This normally indicates insufficient softbake. In this case, revise the softbake parameters.

10. Wrinkles may also be observed in the UV exposed regions only during post-exposure bake. This is normally due to insufficient UV exposure dose.

11. If delamination occurs after development, possible causes are:

    Insufficient UV exposure dose preventing cross-linking at the wafer interface. This can be checked by looking at the cross-sectional profile (take a PDMS cast and slice a section to view). If features have a very negative sidewall slope and are less wide than expected near the base, exposure dose could be an issue.

    Excessive stress in film and/or thermal shock—see Note 2 for steps to reduce stress.

    Poor adhesion to wafer due to poor wafer preparation or contamination. Avoid contamination and improve wafer preparation.

12. Curing at higher temperatures for shorter times is also possible (a glass Petri dish may be required). Curing at room temperature is also possible but may take 24 h or more. The Young's Modulus of the final cured PDMS is somewhat dependent on the curing parameters. A high temperature cure will give a more rigid result whereas a lower temperature cure results in a more flexible device. Punching holes in room temperature cured PDMS can be more problematic compared to higher curing temperatures as achieving a "clean" punched hole in the softer PDMS is more difficult. Adjusting the ratio of base to curing agent can also be used to vary the final rigidity of the PDMS. Note also that PDMS shrinks slightly on curing and that this shrinkage is also dependent on curing temperature. This can be compensated for in the mask/mold design. A room temperature cure generally results in no observable shrinkage whereas 60°C may yield about 1% shrinkage.

13. A common mistake is to attempt to peel of the PDMS before the PDMS and mold has had sufficient time to cool. This can lead to poor release from the mold. Conversely, lowering the temperature by placing the molding dish in the fridge may also improve release from the mold.

References

1. Duffy DC, McDonald JC, Schueller OJA, Whitesides GM (1998) Rapid prototyping of microfluidic systems in poly(dimethylsiloxane). Anal Chem 70:4974–4984
2. McDonald JC, Duffy DC, Anderson JR, Chiu DT, Wu HK, Schueller OJA, Whitesides GM (2000) Fabrication of microfluidic systems in poly(dimethylsiloxane). Electrophoresis 21: 27–40
3. Whitesides GM, Ostuni E, Takayama S, Jiang XY, Ingber DE (2001) Soft lithography in biology and biochemistry. Annu Rev Biomed Eng 3: 335–373
4. MicroChem (2011) MicroChem Corp. http://www.microchem.com/Prod-SU82000.htm. Accessed 12 Nov 2012
5. Chuang YJ, Tseng FG, Lin WK (2002) Reduction of diffraction effect of UV exposure

on SU-8 negative thick photoresist by air gap elimination. Microsyst Technol 8: 308–313
6. Bhattacharya S, Datta A, Berg JM, Gangopadhyay S (2005) Studies on surface wettability of poly(dimethyl) siloxane (PDMS) and glass under oxygen-plasma treatment and correlation with bond strength. J Microelectromech Syst 14:590–597
7. Owen MJ, Smith PJ (1994) Plasma treatment of polydimethylsiloxane. J Adhes Sci Technol 8:1063–1075
8. MEMS Talk – General MEMS discussion. http://mail.mems-exchange.org/mailman/listinfo/mems-talk. Accessed 12 Nov 2012
9. Chollet F (ed) (2009) SU-8: thick photo-resist for MEMS. http://memscyclopedia.org/su8.html. Accessed 12 Nov 2012

Chapter 12

Microfluidic Interface Technology Based on Stereolithography for Glass-Based Lab-on-a-Chips

Song-I Han and Ki-Ho Han

Abstract

As lab-on-a-chips are developed for on-chip integrated microfluidic systems with multiple functions, the development of microfluidic interface (MFI) technology to enable integration of complex microfluidic systems becomes increasingly important and faces many technical difficulties. Such difficulties include the need for more complex structures, the possibility of biological or chemical cross-contamination between functional compartments, and the possible need for individual compartments fabricated from different substrate materials. This chapter introduces MFI technology, based on rapid stereolithography, for a glass-based miniaturized genetic sample preparation system, as an example of a complex lab-on-a-chip that could include functional elements such as; solid-phase DNA extraction, polymerase chain reaction, and capillary electrophoresis. To enable the integration of a complex lab-on-a-chip system in a single chip, MFI technology based on stereolithography provides a simple method for realizing complex arrangements of one-step plug-in microfluidic interconnects, integrated microvalves for microfluidic control, and optical windows for on-chip optical processes.

Key words: Microfluidic interfaces, Stereolithography, Lab-on-a-chip, Microfluidic interconnects, Microvalves, Optical windows, Genetic sample preparation system

1. Introduction

For more than two decades, analytical microfluidic devices (1–3) have been developed for manipulating and analyzing biological/chemical samples. Microfluidic devices, when compared to macroscale devices, have advantages such as smaller geometrical size, shorter analysis times, less sample/reagent consumption, and disposability. Many researchers have successfully demonstrated singular functional microfluidic devices (4–6) and transducers (7–9) for fluid manipulation. However, the development of lab-on-a-chips (LOCs) has shown relatively modest growth. The reasons for this are at least partially rooted in the fact that integration of sample processing steps and analytical functions into a LOC has numerous technical

Gareth Jenkins and Colin D. Mansfield (eds.), *Microfluidic Diagnostics: Methods and Protocols*, Methods in Molecular Biology, vol. 949, DOI 10.1007/978-1-62703-134-9_12, © Springer Science+Business Media, LLC 2013

difficulties, such as the need for more complex packaging, the possibility of biological/chemical cross-contamination between functional compartments, and the possible need for different substrate materials for the individual compartment functionalities.

One of the major technical difficulties in developing LOCs is realizing complex arrangements of microfluidic interconnects for sample/reagent/buffer introduction, for interconnection between multiple chips, and for interconnection between the microsystem and the macro world. Several microfluidic interconnection technologies have been demonstrated (10), including adhesive bonding of miniaturized connectors (11–14), assembled interconnection blocks (15, 16), silicone rubber O-ring couplers (17, 18), silicon connectors fabricated by deep reactive ion etching (DRIE) (19, 20), flanges (21), injection molded components (14), and socket-type multi-connectors (22). Adhesive bonding of miniature fittings over through-holes in the substrate or coverplate has several inherent problems including large dead volumes, a large footprint, labor intensive assembly, and clogging/contamination by the adhesive.

The microfluidic interface (MFI) technology based on stereolithography (SLA) provides a practical method for realizing complex arrangements of microfluidic interconnects. The primary advantage of the SLA technology is that it allows for the fabrication of arbitrary three-dimensional (3D) structures with very few design constraints. In addition, it can be used to fabricate various 3D structures within minutes to a few hours with 25 μm vertical resolution and ~1 μm horizontal pattern resolution (9). Furthermore, it allows integration of other component functionalities such as the incorporation of electrical, mechanical, and optical components as part of the build process (9). The MFI can easily be aligned onto complex microfluidic systems using tooling holes and/or edge alignment. Additionally, the use of O-rings to implement tight fluidic seals avoids fluid leakage and prevents clogging of the microchannel by the bonding adhesive when the MFI is assembled to the underlying LOC.

One of the advantages of the MFI technology is the ability to integrate elastomeric microvalves into the MFI along with complex arrangements of microfluidic interconnects and optical windows. The integrated microvalves are essential fluid control elements for sequential processing and manipulation of fluids within LOCs. A large number of microvalves have been reported using silicon/glass material (23). While such microvalves have demonstrated good performance, many were realized using hybrid manufacturing approaches, leading to relatively large flow/dead volumes. To overcome this geometrical scaling limitation, elastomeric microvalves have been realized. The microvalves were fabricated using an elastomer substrate of polydimethylsiloxane (PDMS) (8, 24–26). While PDMS based microfluidic devices with integrated microvalves were successfully demonstrated, the hydrophobicity

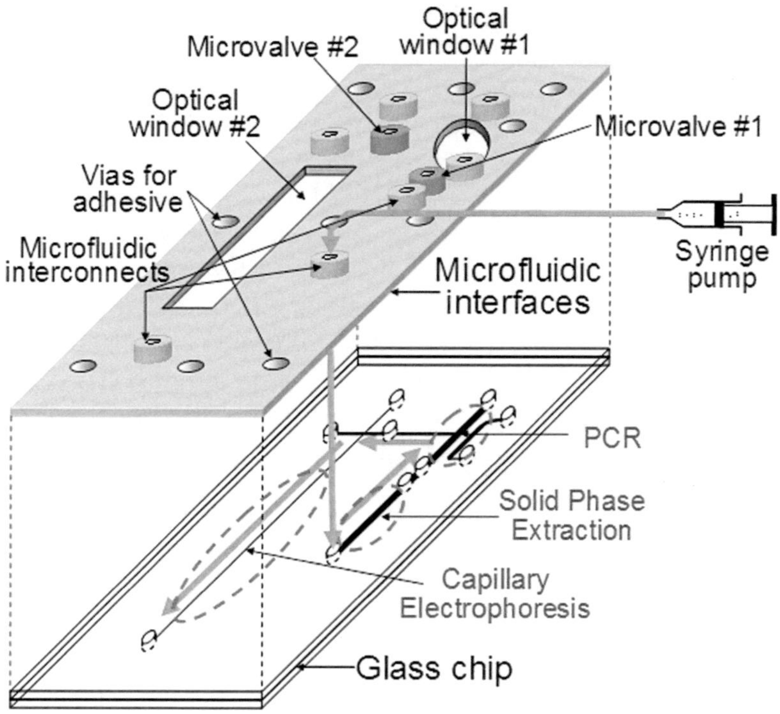

Fig. 1. Schematic of a miniaturized genetic sample preparation system with a MFI.

and porosity of PDMS remains an issue (27, 28). In addition, PDMS has significant auto-fluorescence (8), thus limiting its use in applications requiring high sensitivity fluorescence detection. In contrast, glass-based microfluidic devices have demonstrated advantages in applications for which highly sensitive fluorescence detection was required (29–31). Since LOCs fabricated by glass-to-glass bonding technology result in a permanent mechanical bonding between adjoining layers, the technology lends itself to microfluidic applications requiring mechanical rigidity, high fluid pressures, or relatively high operating temperatures. Furthermore, the development of integrated microvalves for glass-based LOCs is critical as a method for controlling fluid flow within these LOCs.

This chapter introduces the microfluidic interface (MFI) technology based on the stereolithography (SLA), which is a practical interface/packaging methodology for glass-based LOCs requiring complex arrangements of microfluidic interconnects and fluidic control elements such as microvalves and optical interfaces for functions such as thermocycling and fluorescence detection. To demonstrate the usefulness of the MFI for glass-based LOCs, the MFI is applied to a miniaturized genetic sample preparation system (Fig. 1), including genetic functional components, such as a DNA solid phase extraction system, a PCR thermocycler, and a capillary electrophoresis device.

2. Materials

2.1. Fabrication of the Bottom Glass Substrate

1. Borofloat glass slides (0.7 mm thick, Borofloat33 Pyrex, Schott AG, Germany), coated on both sides with chrome (100 Å) and gold (1,000 Å) layers by evaporation (see Note 1).
2. Mask aligner/UV light source (MDA-4000B, MIDAS System Co., Republic of Korea).
3. The photomask for the bottom glass structure with patterns of microfluidic channels and chambers (see Note 2).
4. Spin coater (SPIN-3000D, MIDAS System Co.).
5. Hotplate with stirrer (WiseStir® MSH-20D, DAIHAN Scientific Co., Republic of Korea).
6. Metallurgical microscope (LV 100, Nikon).
7. Convection oven (OV-11, JEIO TECH. Co., Republic of Korea).
8. Surface profiler (Alpha-Step IQ, KLA-Tencor Corp., USA).
9. Magnetic stirrer bar (size: 1/2″×5/16″ Octagon Spinbar® Magnetic Stir Bar, Bel-Art products, USA).
10. Hexamethyldisilazane (HMDS; C40602, J. T. Baker, NJ, USA).
11. Photoresist (AZ1512, AZ Electronic Materials Corp., USA).
12. Developer (AZ400K, J. T. Baker).
13. Gold etchant, which is prepared by mixing 4 g of iodine (I) and 16 g of potassium iodide (KI) in 160 mL of deionized (DI) water.
14. Chrome etchant (CR7, Cyantek Corp., USA).
15. Hydrofluoric acid (HF, Electrical grade).
16. Nitric acid (HNO_3, Electrical grade).
17. Acetone ($(CH3)_2CO$, Electrical grade).
18. Sulfuric acid (SO_4, Electrical grade).
19. Hydrogen peroxide (H_2O_2, Electrical grade).

2.2. Fabrication of the Top Glass Substrate

1. Photomask for the top glass structure with patterns of holes to act as microvalves.
2. Vinyl tape (471 Green, 3M Co., USA) for passivation of the top glass, when the microvalve holes are etched through using HF.
3. A drilling machine (Dremel 220, Robert Bosch Tool Corp., USA) for making reservoirs in the top glass.
4. Diamond drill bits (0.5 mm in diameter, MCDU20, UKAM Industrial Superhard Tools, Valencia, USA).
5. Glycerol.

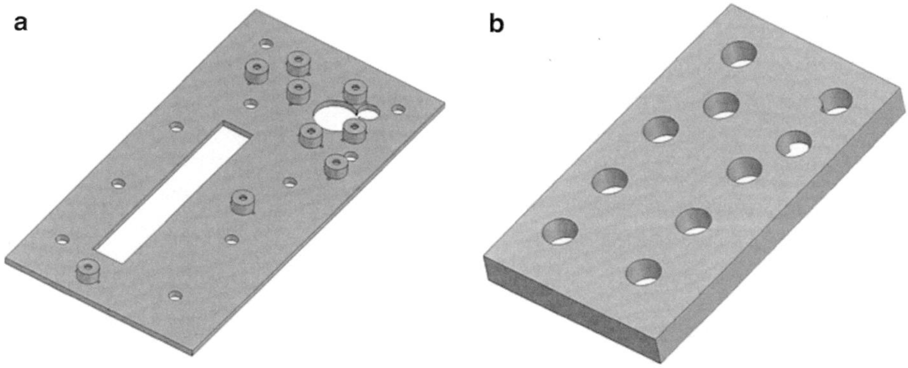

Fig. 2. Perspective views of 3D CAD drawings of: (**a**) the microfluidic interface, and (**b**) clamping jig.

2.3. Glass-to-Glass Thermal Bonding of the Top and Bottom Glasses

1. Two ceramic plates (size: 15 × 15 × 1 cm^3, Macor® Machinable Glass Ceramic, Professional Plastics Inc., USA).
2. Boron nitride (BN) powder (08816LD, Sigma Aldrich).
3. Programmable box furnace (BLF-15, 0–1,050°C, Shin Sung Youl Yeun Ind Co., Republic of Korea).

2.4. Fabrication of MFI and a Clamping Jig Using Rapid Stereolithography

1. Three-dimensional computer-assisted design (3D CAD) software (Solid Edge, Siemens PLM Software Inc., USA) for drawing the MFI and clamping jig (Fig. 2a, b).
2. A rapid stereolithography (SLA) machine (Viper SI2, 3D Systems Corp., CA, USA).
3. Photosensitive polymer (SL5510™, 3D Systems Corp.).
4. Ultrasonic bath (JAC-1505, KODO Technical Research Co., Republic of Korea).
5. Isopropanol (Electrical grade).
6. UV oven (ProCure™ 350 UV Chamber, 3D Systems Corp.).

2.5. Assembly of the Glass Chip and MFI

1. A 120 μm-thick latex sheet (85995K12, McMaster-Carr, GA, USA) for elastomeric membrane for the microvalves.
2. Punching tool (6 mm in diameter, Leather Hole Punch Tool, CVF Supply Company, CA, USA).
3. Instant glue (Perfect-glue™ (PG-00), Akzonobel paints LLC, OH, USA).
4. Nitrile rubber (Buna-N) O-ring (Size-001-1/2, McMaster-Carr).
5. 3 mm-thick glass plate (10 × 15 cm^2).
6. Binder clips (size: 5/4″, Office Depot Corp., FL, USA).
7. UV adhesive (1187-M, DYMAX Co., CT, USA).
8. UV light source (HBO® 350 W/S, OSRAM, Germany).

2.6. Microfluidic and Microvalve Interconnects

1. Nitrile rubber (Buna-N) O-ring (Size-001, McMaster-Carr).
2. Capillary tubing (1/16″ OD, Teflon fluorinated ethylene propylene (FEP) tubing, Upchurch Scientific, WA, USA).
3. Sandpaper (grade: P50).

2.7. High-Pressure Microvalves

1. Tapper (5.5 mm in diameter, M5 × 0.8, Super Tools Co., Republic of Korea).
2. FlushNut (F-354 for 1/16″ OD tubing, Upchurch Scientific).
3. Lock Ring (LT-100 for 1/16″ OD tubing, Upchurch Scientific).
4. A three-way solenoid valve (ST2352, DKC, Republic of Korea).
5. A compressed air source, such as a nitrogen tank.
6. Gas regulator (825, Harris Calorific Co., OH, USA).
7. LabVIEW™ application software (LabVIEW™, National Instrument Co., TX, USA).
8. Relay module (NI9481, National Instrument Co.).

3. Methods

3.1. Fabrication of the Bottom Glass Substrate

1. A Borofloat glass slide with chrome (100 Å) and gold (1,000 Å) layers (Fig. 3a) is spin-coated with HMDS at 2,000 rpm for 30 s (optional).
2. Spin-coat with photoresist (PR; AZ 1512) at 4,000 rpm for 30 s. The PR layer is about 1.2 μm thick (spin coating).
3. Place the bottom glass slide, coated with PR, on a hot plate at 95°C for 1 min (soft bake).

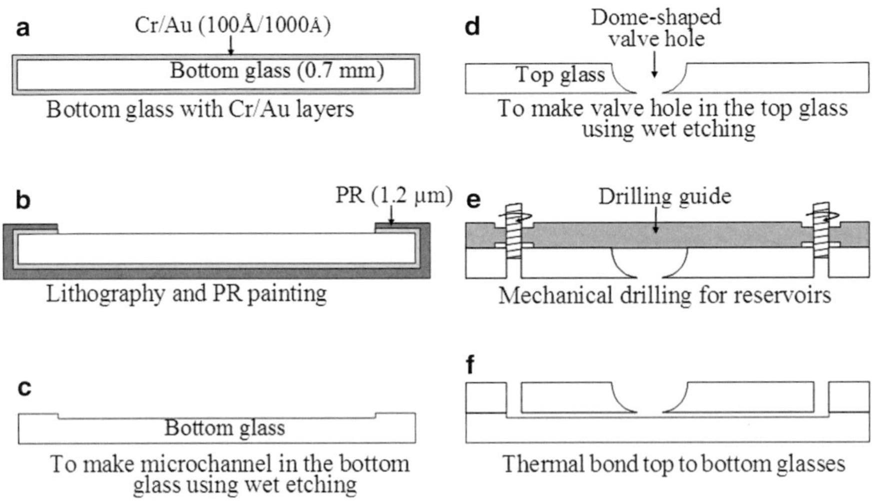

Fig. 3. Fabrication of the bottom and top glass slides.

4. Align the bottom glass slide with the corresponding photomask using the corner patterns with a mask aligner (see Note 3).
5. Expose the bottom glass slide to UV light through the photomask for 2–5 s (exposure).
6. Place the UV-exposed bottom glass slide in developer (AZ400K) for 60 s (develop). After this, inspect the developed patterns using a metallurgical microscope. For accurate inspection, a resolution pattern is recommended to confirm correct size of features.
7. Place the developed bottom glass slide on a hot plate at 105°C for 90 s (hard bake).
8. Paint the backside, edges, and any test or alignment patterns on the front of the glass slide with PR, using sufficient PR to paint the entire area, with no uncovered spots.
9. Place the glass slide in a convection oven at 105°C for 5 min (post bake). To minimize contact between the oven and glass slide covered with excess PR, aluminum foil is used to make a stand for the glass slide (see Note 4).
10. Place the glass slide in gold etchant for 2–5 min. Heating the gold etchant to about 50°C increases the etching rate of the gold layer. Where the gold layer is etched, bright metal can be seen.
11. Rinse the glass slide six times with DI water.
12. Place the glass slide in chrome etchant (CR7) until the pattern is seen as a dark color. Heating the chrome etchant to about 50°C enhances the etching rate of the chrome layer. The pattern where the chrome has been etched shows as dark empty space through the pattern (Fig. 3b).
13. Rinse the glass slide six times with DI water. Following this, inspect it to see whether there are other dark spots that are not part of the pattern. Any unwanted dark spots should be covered with PR.
14. Etch the glass slide to a depth of 60 μm using buffered oxide etchant (BOE), which is made by mixing H_2O:HNO_3:HF in the ratio of 10:3:10 (see Note 5). The etching rate is about 1.9 μm/min using a magnetic stirrer at 200 rpm. The etching rate depends on the patterned area to be etched.
15. The etched thickness is measured using a surface profiler (Alpha-Step IQ). From the measurements, the etching rate can be calculated, then, if necessary, any additional depth required can be etched, based on the calculated etching rate.
16. Rinse the glass slide six times with DI water.
17. Place the glass slide in acetone and gently stir. The acetone solution will turn red.
18. Once gold color becomes visible on the surface of the glass, place in DI water.

19. Place the glass slide in boiling acetone for 10 min.
20. Rinse the glass slide six times with DI water, and inspect the surface of the gold layer.
21. Any residual PR remaining on the gold layer should be removed by placing the glass slide in piranha solution (H_2SO_4:H_2O_2 = 4:1) heated to 100°C for 10 min (see Note 6).
22. Rinse the glass slide six times with DI water.
23. Remove the gold and chrome layers on the glass slide by following steps 10–13 above (Fig. 3c).
24. Rinse the glass slide six times with DI water.
25. Clean with piranha solution at 100°C for 10 min to remove any organic residue on the bottom glass slide.
26. Rinse the glass slide six times with DI water.
27. Place the glass slide in HF:H_2O (1:50) for 30 s to remove the thin oxide film on the surface.
28. Rinse the glass slide six times with DI water.

3.2. Fabrication of the Top Glass Substrate

1. The lithography process for the top glass slide is the same as steps 1–9 in Subheading 3.1.
2. Passivation tape is tailored to cover the entire top glass slide; the valve holes areas are cut out and it is attached to the glass slide. A sheet of paper with the pattern of the top glass photomask drawn on it will help when cutting out the microvalve holes.
3. For etching the gold and chrome layers in the microvalve holes, follow steps 10–13 in Subheading 3.1.
4. Place the glass slide in buffered oxide etchant consisting of H_2O:HNO_3:HF (10:3:10). Etching takes about 12 h with stirring at 200 rpm. To etch the valve holes, the etchant should be flushed every 5–6 h and the glass slide should be rotated by 180° every hour (see Note 7).
5. Rinse the glass slide six times with DI water.
6. Inspect the etched dome-shaped valve holes under a microscope. If the valve holes are not etched sufficiently, 25% HF solution should be dropped into these valve holes, which are then inspected every 5–10 min. The diameter of the valve hole on the backside should be 0.9–1 mm (Fig. 3d).
7. Remove the passivation tape after finishing the etching.
8. Rinse the glass slide six times with DI water.
9. To generate inlets and outlets, mechanical drilling with a diamond bit is used to make holes in the glass (Fig. 3e) (see Note 8). To make fine holes, the drilling speed is set to about 30,000 rpm and glycerol is used as a lubricant.
10. Follow steps 16–27 in Subheading 3.1 to remove the PR, gold and chrome layers.

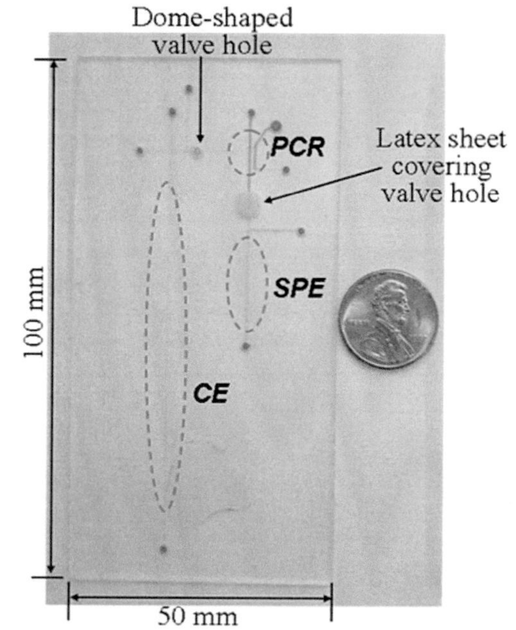

Fig. 4. Photograph of the fabricated glass chip. Red dye was used to highlight the microchannel.

3.3. Glass-to-Glass Thermal Bonding of the Top and Bottom Glasses

1. Place approximately 10 mg boron nitride (BN) powder onto two ceramic plates.
2. Smooth the BN powder over the top surfaces of the two ceramic plates and blow off any excess BN powder residue (see Note 9).
3. Manually align the top glass slide with the bottom glass slide (see Note 10).
4. Place the aligned glass slides on the middle of one ceramic plate and then cover carefully with the second ceramic plate (see Note 11).
5. Load the glass slides, sandwiched between the two ceramic plates, into a furnace for bonded together using thermal bonding at 70°C for 30 min, 550°C for 30 min, and 680°C for 4 h (Fig. 3f) (see Note 12). Figure 4 shows the fabricated glass chip.

3.4. Fabrication of MFI and a Clamping Jig Using Rapid Stereolithography

1. The microfluidic interface (MFI) (Fig. 2a) and clamping jig (Fig. 2b) are drawn using 3D CAD software (Solid Edge).
2. Transfer the CAD files to STL format. For the transfer, "Surface plane" in "Options" should be less than 1°.
3. Import the CAD file in STL format to the stereolithography (SLA) system, at which point multiple copies of a single design can be built in an automated fashion with the photosensitive polymer.

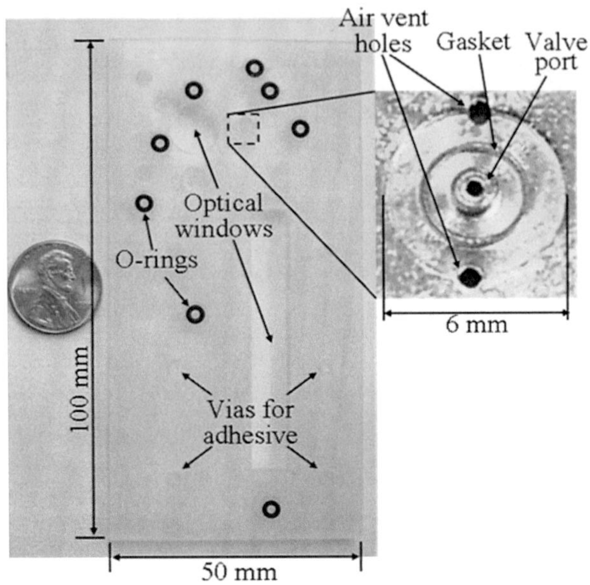

Fig. 5. Photograph of the bottom view of the fabricated microfluidic interface. The *inset* shows the bottom view of the microvalve port with gasket and air vent holes; this is coated with a thin gold film to delineate the image.

4. Build up the MFI and clamping jig parts using the rapid SLA system.
5. Once the parts are formed, remove them from the building platform, trim to remove any supports, and clean to remove the excess uncured photopolymer resin.
6. Clean by immersing in 100% isopropyl alcohol in an ultrasonic bath for 5 min and cure in a UV oven (post-curing apparatus) for 1 h.
7. Place in a convection oven for 15 min at 80°C to increase the mechanical strength of the material by subsequent thermal curing. Figure 5 shows the completed MFI.

3.5. Assembly of the Glass Chip and MFI

1. Punch 6 mm circles in a 120 μm thick latex sheet using the punching tool.
2. Attach the 6 mm latex circles to the valve holes in the glass chip using instant glue (see Note 13).
3. Insert buna-N O-rings (0.015″ OD) beneath the microfluidic interconnects.
4. Using the clamping jig and a 3 mm thick glass plate, the glass chip and MFI are assembled, as shown in Fig. 6. Use binder clips to clamp them to each other tightly.
5. Put UV adhesive into the adhesive vias designed in the MFI, where capillary forces pull the adhesive into the gap between the glass chip and MFI (see Note 14).

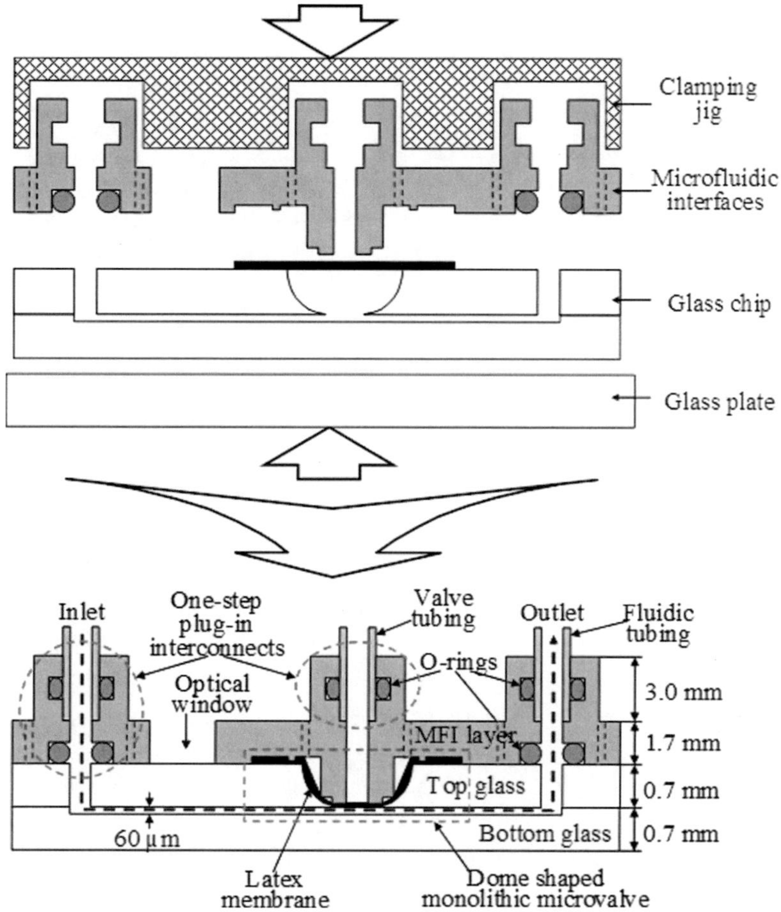

Fig. 6. Schematic of the assembly of the glass chip and MFI.

6. Once the UV adhesive has spread sufficiently, place the assembly under a UV light source for 30 min. Release the microfluidic device, consisting of the assembled glass chip and MFI, from the clamping jig and glass plate, and place it under the UV light source for another 30 min. Figure 7 shows the completed microfluidic device with the MFI.

3.6. Microfluidic and Microvalve Interconnects

1. Insert buna-N O-rings (0.01″ OD) into the microfluidic and microvalve interconnects (see Note 15).
2. Using sandpaper, sand the capillary tubing (1/16″ OD) normal to the direction in which the tubing is to be plugged. Next, plug the tubing into the microfluidic and microvalve interconnects of the MFI in one step (see Note 16).

3.7. High-Pressure Microvalves

High-pressure microvalves (Fig. 8) can be applied for high fluidic pressures (>400 kPa). These are appropriate for PCR applications, because they can withstand high pressure and prevent fluid leaks

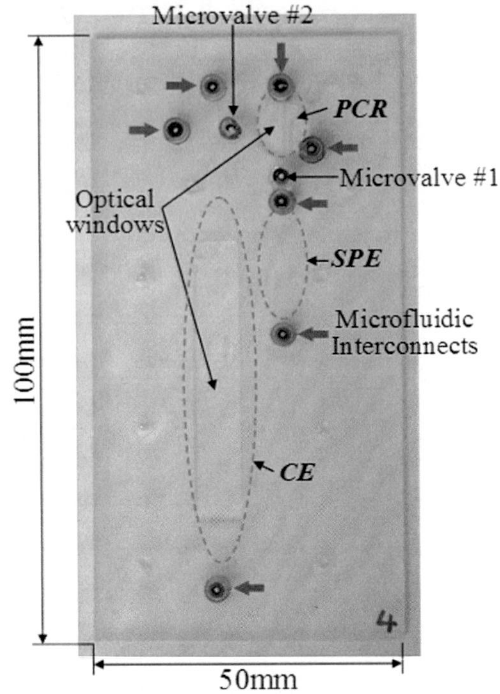

Fig. 7. Photograph of a completed miniaturized genetic sample preparation system, in which the MFI is assembled to realize seven one-step plug-in microfluidic interconnects, two monolithic microvalves, and two optical windows.

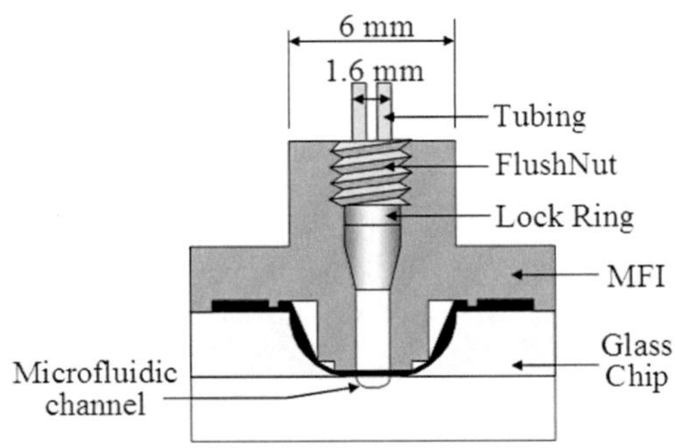

Fig. 8. Schematic of a high-pressure microvalve.

during the PCR processing at temperatures of approximately 95°C. The fabrication protocol is as follows:

1. The high-pressure microvalve interconnects are tapped (using a tapper).
2. Insert the FlushNut (Nut) and Lock Ring in the capillary tubing (1/16″ OD). Next, insert the tubing into the tapped microvalve interconnect and tighten the nut.

3. Connect the other end of the tubing to the outlet of a three-way solenoid valve, in which one inlet is connected to an air pressure source and the other inlet to the atmosphere. The three-way solenoid valve is controlled by a relay module (NI9481, NI) and LabVIEW™ software (see Note 17).

3.8. Optical Windows for PCR Thermocycling and Fluorescence Detection

The optical windows allow analytical operations, such as PCR thermocycling and fluorescence detection for capillary electrophoresis (Fig. 1). They can be built together with the microfluidic interconnects and microvalves using the fabrication steps in Subheading 3.6. Because the maximum PCR temperature of 95°C can generate thermal stress between the glass chip and MFI, the edge of the PCR optical window should be designed to be more than 3 mm from the PCR heating region.

4. Notes

1. The thickness of the glass slides should be based on the application. The thinnest Borofloat glass (0.7 mm thick) can be chosen to reduce the thermal capacity, which allows a reduced PCR processing time.
2. In many cases the microfluidic devices are rectangular. Thus, during the spin-coating process (see step 2 in Subheading 3.1), the photoresist is thicker at the corners compared with that at the center. Therefore, patterns on the mask should be placed about 5 mm away from the corners to develop fine patterns. The distance from the corners depends on the spinning speed and viscosity of the photoresist.
3. The proper contact mode of the mask aligner is "hard contact." If not using a mask aligner, then a UV light source is also suitable. For manual UV exposure, carefully push the photomask over the bottom glass slide using your fingers. Take care to avoid shading the glass slide with your fingers. Suitable gloves and goggles should be worn to avoid skin and eye exposure to UV.
4. During the post bake, the backside should be facing upward. This is because the solvent from the PR painted on the backside and edge is vaporized during the post bake and if the thin PR layer patterned on the front is exposed to the solvent vapor, the thin PR layer will crack.
5. Pour the DI water first, and then add NHO_3 and HF sequentially. Suitable safety precautions must be taken when preparing and using the solution.

6. To prepare the piranha solution, H_2SO_4 solution is poured into a beaker first and then the H_2O_2 solution is carefully added to this. Heat will be evolved naturally and suitable safety precautions must be taken when preparing and using the solution to avoid explosion.
7. If a number of glass slides are etched at the same time, they should be placed at the same distance from the magnetic stirrer bar.
8. For the drilling process, the backside of the top glass slide should be facing upward because the thin part of the dome-shaped valve holes is readily damaged by pressing down and friction. A drilling guide as shown in Fig. 3e may be used to aid alignment of the holes.
9. If there are BN clumps on the surface of the two ceramic plates these will generate pockmarks on the surface of the glass chip.
10. After alignment, a tiny water drop at the corner of the aligned glass slides helps to fix the slides to each other, thereby preventing them from sliding when they are loaded into the furnace.
11. An appropriate load pressure (= loading weight/bonding area) is necessary to prevent air voids forming between the top and bottom glass slides and squeezing the microfluidic channel.
12. Keep at 70°C for 30 min to remove any water between the glass slides. The glass bonding temperature depends on the glass transition temperature (T_g) of the type of glass used; if the bonding temperature is higher than the appropriate bonding temperature, the microchannels can be squashed and the glass slides sometimes adhere to the ceramic plates. In our experience, the bonding temperature for Borofloat glass is 680°C, which is slightly higher than the 660°C required for borosilicate glass.
13. The instant glue should be pasted around the latex sheet circles. If the glue is pasted directly on the surface of the top glass, it may leak into the valve hole and thereby clog the microchannel.
14. Monitor the UV adhesive through the glass plate; if necessary, add more UV adhesive.
15. Curved tweezers are used to help to insert the O-rings into the microfluidic and microvalve interconnects.
16. For a low fluidic pressure, the sanding is optional. Sanding helps prevent the tubing from becoming unplugged by a high fluidic valving pressure.
17. Generally, a low valve pressure <100 kPa is used to control the microfluidic flow, while a valve pressure >300 kPa is used to isolate the PCR chamber during thermocycling.

References

1. Terry SC, Jerman JH, Angell JB (1979) A gas chromatographic air analyzer fabricated on a silicon wafer. IEEE T Electron Dev ED-26:1880–1886
2. Bakajin O, Duke TAJ, Tegenfeldt J, Chou C-F, Chan SS, Austin RH, Cox EC (2001) Separation of 100 kilobase DNA molecules in 10 seconds. Anal Chem 73:6053–6056
3. Huang Y, Joo S, Duhon M, Heller M, Wallace B, Xu X (2002) Dielectrophoretic cell separation and gene expression profiling on microelectronic chip arrays. Anal Chem 74:3362–3371
4. Fu AY, Spence C, Scherer A, Arnold FH, Quake SR (1999) A microfabricated fluorescence-activated cell sorter. Nat Biotechnol 11: 1109–1111
5. Chang KS, Tanaka S, Chang CL, Esashi M, G. Benjamin Hocker (2003). The Institute of Electrical and Electronics Engineers, Inc. (IEEE). Combustor-integrated micro-fuel processor with suspended membrane structure. In: Tech dig 12th int. conf. solid-state sensors and actuators workshop (Transducers'03), Boston, USA, 2003, pp 635–638
6. Han K-H, Frazier AB (2006) Paramagnetic capture mode magnetophoretic microseparator for high efficiency blood cell separations. Lab Chip 6:265–273
7. Man PF, Jones DK, Mastrangelo CH, Kazuo Sato and Shuichi Shoji (1997). The Institute of Electrical and Electronics Engineers, Inc. (IEEE). Microfuidic plastic capillaries on silicon substrates: a new inexpensive technology for bioanalysis chips. In: International workshop on micro electromechanical systems (MEMS 97), Nagoya, Japan, pp 311–316
8. Grover WH, Skelley AM, Lui CN, Lagally ET, Mathies RA (2003) Monolithic membrane valves and diaphragm pumps for practical large-scale integration into glass microfluidic devices. Sensor Actuat B 89:315–323
9. Han A, Graff M, Mohanty SK, Edwards TL, Han K-H, Frazier AB (2003) Multi-layer plastic/glass microfluidic systems containing electrical and mechanical functionality. Lab Chip 3:150–157
10. Fredrickson K, Fan ZH (2004) Macro-to-micro interfaces for microfluidic devices. Lab Chip 4:526–533
11. Stachowiak TB, Rohr T, Hilder EF, Peterson DS, Yi M, Svec F, Fréchet JMJ (2003) Fabrication of porous polymer monoliths covalently attached to the walls of channels in plastic microdevices. Electrophoresis 24:3689–3693
12. Pattekar V, Kothare MV (2003) Novel microfluidic interconnectors for high temperature and pressure applications. J Micromech Microeng 13:337–345
13. Chen H, Acharya D, Gajraj A, Meiners J-C (2003) Robust interconnects and packaging for microfluidic elastomer chips. Anal Chem 75:5287–5291
14. Gray BL, Jaeggi D, Mourlas NJ, van Drieenhuizen BP, Williams KR, Maluf NI, Kovacs GTA (1999) Novel interconnection technologies for integrated microfluidic systems. Sensor Actuat A 77:57–65
15. Nittis V, Fortt R, Legge CH, de Mello AJ (2001) A high-pressure interconnect for chemical microsystem applications. Lab Chip 1:148–152
16. Verlee D, Alcock A, Clark G, Huang TM, Kantor S, Nemcek T, Norlie J, Pan J, Walsworth F, Wong ST (1996) Fluid circuit technology: integrated interconnect technology for miniature fluidic devices. Tech Dig Solid State Sensor Actuat (Hilton Head, USA 1996:9–14
17. Yao TJ, Lee S, Fang W, Tai Y-C, Hiroki Kuwano and Isao Shimoyama (2000). The Institute of Electrical and Electronics Engineers, Inc. (IEEE). A micromachined rubber O-ring microfluidic coupler. In: Proc IEEE micro electro mechanical systems conference (MEMS 2000), Miyazaki, Japan, 2000, pp 624–627
18. Meng E, Wu S, Tai Y-C, A. van den Berg, W. Olthuis and P. Bergveld (2000). Kluwer Academic Publishers. Micromachined fluidic couplers. In: Proceedings of the micro total analysis systems symposium (μTAS), Enschede, Netherlands, 2000, pp 41–44
19. González C, Collins SD, Smith RL (1998) Fluidic interconnects for modular assembly of chemical microsystems. Sensor Actuat B 49: 40–45
20. Gray BL, Collins SD, Smith RL (2004) Interlocking mechanical and fluidic interconnections for microfluidic circuit boards. Sensor Actuat A 112:18–24
21. Puntambekar CA, Ahn CH (2002) Self-aligning microfluidic interconnects for glass- and plastic-based microfluidic systems. J Micromech Microeng 12:35–40
22. Yang Z, Maeda R (2003) Socket with built-in valves for the interconnection of microfluidic chips to macro constituents. J Chromatogr A 1013:29
23. Kovacs GTA (1998) Micromachined transducers sourcebook. McGraw-Hill Co., Boston
24. Hosokawa K, Maeda R (2000) A pneumatically-actuated three-way microvalve fabricated with polydimethylsiloxane using the membrane transfer technique. J Micromech Microeng 10:415–420

25. Wang Y-C, Choi MH, Han J (2004) Two-dimensional protein separation with advanced sample and buffer isolation using microfluidic valves. Anal Chem 76:4426–4431
26. Lee S, Jeong W, Beebe DJ (2003) Microfluidic valve with cored glass microneedle for microinjection. Lab Chip 3:164–167
27. Ren X, Bachman M, Sims C, Li GP, Allbritton N (2001) Electroosmotic properties of microfluidic channels composed of poly (dimethysiloxane). J Chromatogr B 762:117–125
28. Hu S, Ren X, Bachman M, Sims CE, Li GP, Allbritton N (2002) Surface modification of poly(dimethylsiloxane) microfluidic devices by ultraviolet polymer grafting. Anal Chem 74: 4117–4123
29. Wolfe KA, Breadmore MC, Ferrance JP, Power ME, Conroy JF, Norris PM, Landers JP (1997) Toward a microchip-based solid-phase extraction method for isolation of nucleic acids. Electrophoresis 23:727–733
30. Harrison DJ, Fluri K, Seiler K, Fan Z, Effenhauser CS, Manz A (1993) Micromachining a miniaturized capillary electrophoresis based chemical analysis system on a chip. Science 261:895–897
31. Giordano BC, Jin L, Couch AJ, Ferrance JP, Landers JP (2004) Microchip laser-induced fluorescence detection of proteins at submicrogram per milliliter levels mediated by dynamic labeling under pseudonative conditions. Anal Chem 76:4705–4714

Chapter 13

Three-Dimensional, Paper-Based Microfluidic Devices Containing Internal Timers for Running Time-Based Diagnostic Assays

Scott T. Phillips and Nicole K. Thom

Abstract

This chapter describes a method for fabricating three-dimensional (3D), paper-based microfluidic devices that contain internal timers for running quantitative, time-based assays. The method involves patterning microfluidic channels into paper, and cutting double-sided adhesive tape into defined patterns. Patterned paper and tape are assembled layer by layer to create 3D microfluidic devices that are capable of distributing microliter volumes of a sample into multiple regions on a device for conducting multiple assays simultaneously. Paraffin wax is incorporated into defined regions within the device to provide control over the distribution rate of a sample, and food coloring is included in defined regions within the device to provide an unambiguous readout when the sample has reached the bottom of the device (this latter feature is the endpoint of the timer).

Key words: Microfluidic devices, Microfluidic, Paper-based analytical devices (μPADs), Patterned paper, Telemedicine, Fluidic timers, Point-of-care, Diagnostics

1. Introduction

Microfluidic devices that are fabricated out of paper offer new opportunities for creating low-cost diagnostic devices for use in resource-limited environments such as those found in the developing world, emergency settings, home healthcare, and other point-of-care environments (1–5). These types of microfluidic devices possess many of the positive attributes of conventional microfluidic devices that are made out of plastics, elastomers, glass, etc., yet they are as easy to use as paper strip tests and slightly more complicated lateral-flow-type devices (e.g., pregnancy tests) (6). Paper-based microfluidic devices are capable of distributing microliter volumes of fluids by capillary action into multiple regions on a

Gareth Jenkins and Colin D. Mansfield (eds.), *Microfluidic Diagnostics: Methods and Protocols*, Methods in Molecular Biology, vol. 949, DOI 10.1007/978-1-62703-134-9_13,

device without the use of pumps or other external equipment. The ease of sample distribution makes paper-based microfluidic devices a promising platform for developing diagnostic devices that are capable of detecting and measuring multiple analytes simultaneously at exceedingly low cost: only small quantities of reagents are needed, and both paper and tape are inexpensive and readily available (6).

To date, efforts in this burgeoning area of paper-based microfluidics have focused on developing methods for fabricating different types of paper-based microfluidic devices (6) as well as fabricating auxiliary features within the devices to enable control over the flow rate (2, 7, 8), distribution (9), and mixing (10) of a sample with another fluid. A second major focus has been on proof-of-concept demonstrations for different types of assays (e.g., colorimetric (3–6), antibody-based (11), and electrochemical (12, 13)). Several of these assays require external instrumentation both for conducting the assays and for quantifying the results, while others (e.g., colorimetric assays) do not require auxiliary instrumentation (3, 4). Of the assays developed thus far, colorimetric assays on paper-based microfluidic devices have the potential to be lowest cost (6).

Colorimetric assays are attractive for use on paper-based microfluidic devices because many colorimetric solution-phase clinical tests are available already, and because many of these tests are proving to be compatible with paper (6). These colorimetric methods provide quantitative measurements in a time-dependent manner, and therefore, colorimetric assays conducted on paper-based microfluidic devices must be timed precisely (4, 6). Because a timer is an auxiliary instrument that adds to both the cost and complexity of an assay, we have developed a new type of timer that is built within paper-based microfluidic devices (4). These timers are composed of paraffin wax and food coloring, and therefore, do not contribute substantially to the overall cost of the device. These embedded timers enable quantitative, colorimetric assays in paper-based microfluidic devices that are equally—if not more—precise than colorimetric assays run on paper using external electronic timers (4).

This chapter describes a method for fabricating 3D paper-based microfluidic devices that contain internal timers. The general method described herein can be used as a starting point for prototyping a variety of paper-based microfluidic devices. There are a number of methods for preparing two-dimensional (2D), paper-based microfluidic devices for diagnostic applications (6). For example, some methods describe (i) simply cutting paper into defined shapes (14), (ii) patterning hydrophobic and hydrophilic regions into paper using photolithography (15, 16), (iii) plasma etching hydrophobic paper (17), and (iv) printing hydrophobic and hydrophilic regions onto (and into) paper (1, 18–20). The method for patterning paper-based microfluidic devices described herein falls into the latter category. This method is not the least expensive, nor does it provide the smallest channels (and, ultimately, the smallest diagnostic devices), but we believe that it provides a useful balance between cost, throughput, channel size, and production of devices that are easy to use and manipulate. It is also the most convenient of current methods for fabricating 3D microfluidic devices that are made from patterned paper and patterned double-sided adhesive tape (1, 5), which is the focus of this chapter. After an initial outlay for purchasing a wax printer and a laser cutter, the method for fabricating 3D microfluidic devices described herein is inexpensive (the primary consumables are paper and tape), rapid (approximately 100 devices can be fabricated by a single person within ~4 h), and flexible (the design of a 3D device can be changed easily to enable rapid prototyping of new diagnostic devices). An example 3D paper-based microfluidic device that contains an internal timer is shown in Fig. 1.

2. Materials

2.1. Patterning Microfluidic Channels into Paper

1. CleWin Layout Editor (freely accessible on the internet through WieWeb Software) (see Note 1) and Adobe® Illustrator® (Adobe Systems).
2. Whatman Chromatography Paper No. 1 or Whatman Filter Paper No. 1 (20 cm × 20 cm sheets).
3. Wax printer (e.g., Xerox Phaser 8560).
4. Hot plate (e.g., Thermo Scientific Cimarec Hotplate) or another flat surface that can be held at constant temperature (150°C). The flat, hot surface must be larger than 20 cm × 20 cm.

2.2. Patterning Channels into Double-Sided Adhesive Tape

1. Ace Plastic Carpet Tape (Ace Hardware® Item number 50106).
2. Flat sheet of glass or non-vinyl plastic (rectangular in shape, not larger than 25 cm × 30 cm; the thickness should be between 1 and 10 mm).
3. Laser engraver (e.g., Epilog Mini 45 W CO_2 laser cutter; Epilog Laser).
4. Metal tweezers, sharp tip, between 9 and 16 cm long (e.g., World Precision Instruments Economy Tweezers #1 Item number 501974).

2.3. Fabricating Timers in Paper Microfluidic Devices

1. Paraffin wax (mp 53–57°C, Sigma Aldrich) is dissolved in hexanes (≥98.5% purity, VWR) at fixed concentrations that range from 1 to 55 mg of paraffin wax per mL of hexanes (see Note 2). These solutions are made fresh prior to use (see Note 3).

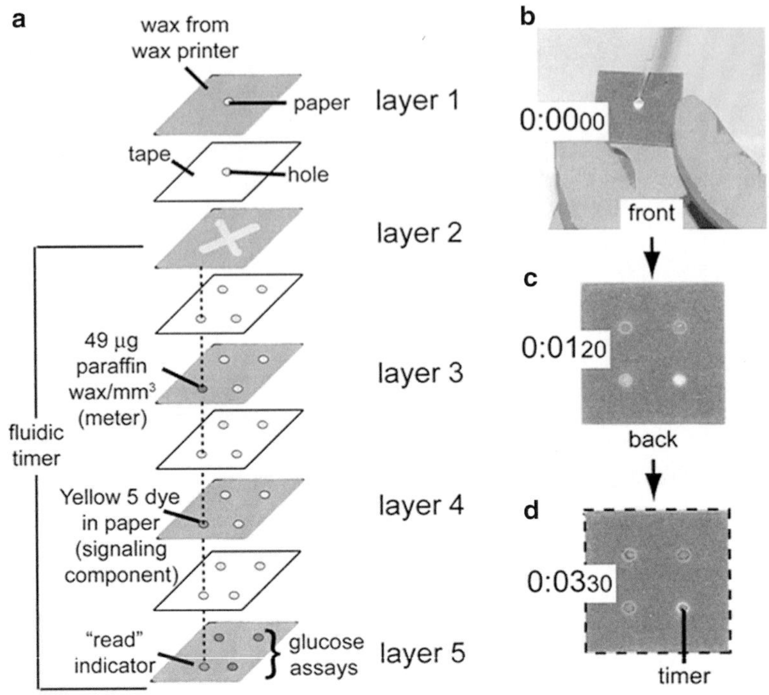

Fig. 1. An example 3D paper-based microfluidic device that contains an internal timer. (**a**) Each layer of the device is 20 mm × 20 mm, the circular hydrophilic regions (*light gray*) are 2.4 mm in diameter, and the hydrophilic channels (*light gray*) are 2.4 mm wide and 4.5 mm long (measured from the center of the X to the first edge of the hydrophilic circle) (layer 2). (**b**) Addition of a sample to the entry point on the device. (**c**) Partial completion of an example assay for measuring the level of glucose in a sample (4). The *light gray* areas represent a blue color that develops during the glucose assay. (**d**) End point of an example time-based assay. The timer region turns orange at 3 min and 30 s to signal the end of the assay.

2. Micropipette (2 µL, VWR Signature Ergonomic High Performance Single-Channel Variable Volume Pipettor, VWR) for depositing 0.5 µL of solutions of paraffin wax in hexanes onto hydrophilic regions within paper-based microfluidic devices (see Note 4).
3. Synthetic food dyes (Assorted Food Colors & Egg Dye; Wal-Mart brand) were used to give colorimetric responses and to track the distribution of fluids within a device (see Note 5). The dyes were used as 1:5 mixtures of dye to distilled water.

2.4. Fabricating Three-Dimensional, Paper-Based Microfluidic Devices

1. Rolling pin. Not larger than 2 ft long, diameter of about 2.5 in. (e.g., OXO Plastic Rolling Pin, available at Target®).
2. Paper discs (2.4 mm diameter) made in bulk using the laser cutter and Whatman Chromatography Paper No. 1 or Whatman Filter Paper No. 1 (20 cm × 20 cm sheets).

3. Methods

3.1. Patterning Microfluidic Channels into Paper

1. Patterns of hydrophobic regions for each layer of paper in the 3D microfluidic device are designed using CleWin Layout Editor (see Note 6) (1), with different layers of the device designed on different layers in CleWin so that the features in each layer are aligned perfectly. The design is saved as a postscript (.ps) file (see Note 7).
2. The postscript file is opened in Adobe® Illustrator® (see Note 8). The size of the page within Adobe® Illustrator® is set to 20.0 cm × 20.0 cm, which is the size of the paper onto which the design will be printed.
3. The patterns of hydrophobic regions designed for each layer of the device are arranged in Adobe® Illustrator® and are separated from one another by 4.5 mm. Multiple copies of these patterns can be arranged in the same Adobe® Illustrator® file. White space (~9 mm) is maintained on all edges of the Adobe® Illustrator® page. The Adobe® Illustrator® file is saved as a portable document format (.pdf).
4. The .pdf file is opened in Adobe® Acrobat® (see Note 9). The page is printed using black wax and a Xerox Phaser 8560 printer (see Note 10) (1). The paper is fed through tray 1, which is set to 200 mm × 200 mm card stock (see Note 11).
5. The printed paper is placed flat onto a 150°C hot plate (printed side up) for 2 min (see Notes 12 and 13). The paper is removed from the hot plate and allowed to cool to room temperature before further use (1). The paper is ready for further manipulation after 10 s (Fig. 2 provides an overview of the fabrication process).

3.2. Patterning Channels into Double-Sided Adhesive Tape

1. Patterns that mimic the desired cut regions in each layer of tape in the 3D microfluidic device are designed using CleWin Layout Editor (see Note 6) (2), with different layers of the device drawn on different layers in CleWin so that alignment of the features is correct. The design is saved as a postscript (.ps) file (see Note 7).
2. The postscript file is opened in Adobe® Illustrator®. The design is highlighted using the "select all" function and the line widths are set to 0.01 point. The patterns for each layer of tape within the 3D device are aligned in rows, maintaining at least 2 mm spacing between rows. The design is moved to the upper left corner of the page, approximately 3 mm from the top and left sides of the Adobe® Illustrator® page (see Note 14). The file is saved as a portable document format (.pdf).

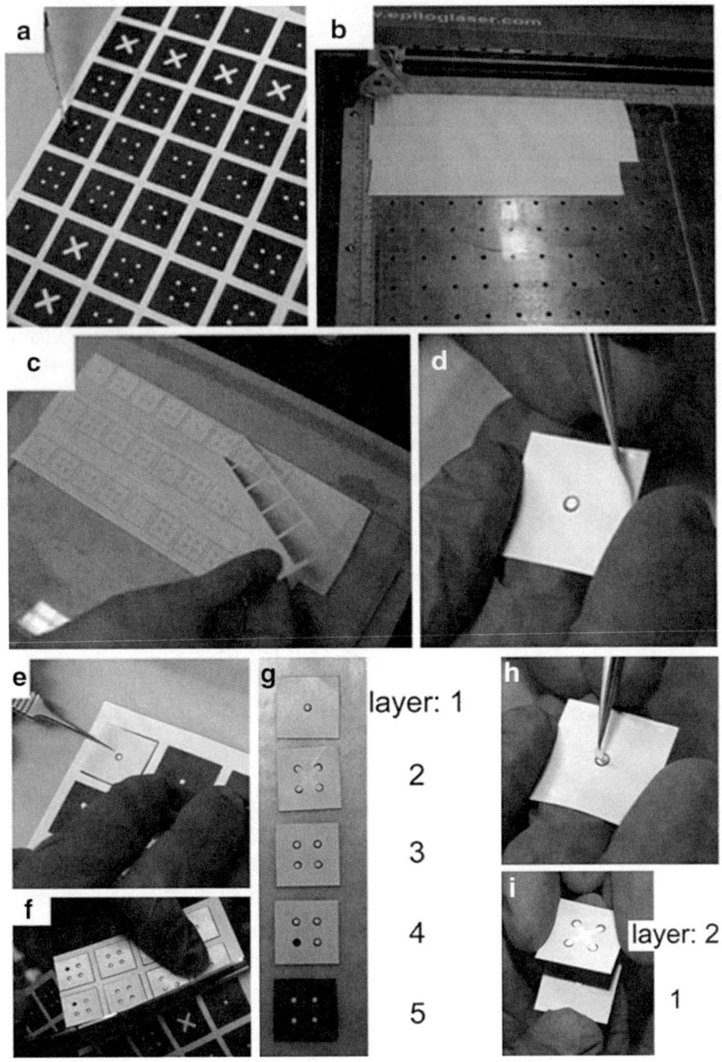

Fig. 2. Photographs of several key steps in the fabrication of an example 3D paper-based microfluidic device that contains an internal timer. (**a**) Paper that has been patterned with multiple components. The photograph shows the deposition of dye to the fourth layer of paper in the example device. (**b**) Double-sided adhesive tape on glass and aligned with the rulers in the laser cutter. (**c**) Patterned double-sided adhesive tape. The excess tape is removed, as shown in the photograph. (**d**) The top layer of protective backing is removed from an individual component of patterned double-sided adhesive tape. (**e**) The sticky side of this individual component is adhered to the appropriate component of patterned paper. (**f**) Each component of patterned tape adhered to patterned paper is cut from the 20 cm × 20 cm piece of Whatman chromatograph No. 1 paper. (**g**) The individual components for the example 3D microfluidic device. (**h**) The second protective backing of the tape is removed and a 2.4 mm disk of Whatman chromatography paper No. 1 is placed in each hole in the tape. (**i**) The component from (**h**) is adhered to layer 2.

3. The roll of double-sided adhesive tape contains a protective backing on one side of the tape. Segments (20 cm) of this tape are adhered (sticky-side down) to the sheet of glass (or plastic). A second 20 cm segment of tape is adhered directly on top of

the first strip of tape (again, sticky-side down). The tape is arranged against the top left corner of the glass in strips across the width of the glass. If the patterns in the .pdf file are designed to cut several strips of tape, then additional strips of tape are attached in rows on the glass, with each row directly below the one before it. This tape/glass unit is placed within the Epilog Engraver laser cutter with the upper left corner of the glass placed against the top/left corner of the perpendicular rulers (Fig. 2b).

4. The .pdf file is opened in Adobe® Acrobat®, and the file is printed using the Epilog Engraver laser cutter using the following settings (see Notes 15 and 16): the job type is set to Vector, the speed is set to 100%, the power is 13%, and the frequency 1,053 Hz. Vector sorting should be selected, the pull down menu should be set to "optimize," and the auto focus box should be selected. The printing options should be set to "no page scaling," and "auto-rotate and center" should be unselected (2).
5. When the laser cutter has finished cutting the tape, the glass and tape are removed from the laser cutter, and the excess tape is removed from the bottom layer of tape using tweezers (Fig. 2c). When each patterned section of tape is removed, it is essential to remove it so that both sides of the patterned tape contain a protective backing (Fig. 2d shows an example of a piece of patterned tape that contains protective backings on both sides). The layers of tape that are attached to the glass are discarded.

3.3. Fabricating Timers in Paper Microfluidic Devices

1. Paraffin wax in hexanes (0.4 μL) is deposited (using a 2 μL micropipette) into the center of a 2.4 mm diameter hydrophilic region of patterned paper that will be designated as the timing layer within the 3D microfluidic device (e.g., layer 3 in the example device shown in Fig. 1a). The hexanes is allowed to evaporate over 10 min, and another 0.4 μL of the wax solution is deposited on the opposite face of the same 2.4 mm diameter hydrophilic region of patterned paper. The hexanes is allowed to evaporate over another 10 min before the layer is incorporated into a 3D microfluidic device (4).
2. The food coloring (1.0 μL) is deposited into the center of a 2.4 mm diameter hydrophilic region of patterned paper that will be designated as the signaling component within the 3D microfluidic device (e.g., layer 4 in the example device shown in Fig. 1a). The layer is allowed to dry for 30 min before it is assembled into a 3D microfluidic device (4).

3.4. Assembling Three-Dimensional, Paper-Based Microfluidic Devices

1. Before the layers of the device are assembled, reagents for assays can be deposited as aqueous solutions into appropriate hydrophilic regions of paper within the device (4, 5). These solutions are allowed to dry for 30 min before the device is assembled.

2. The bottom protective backing on the adhesive is removed from the tape (Fig. 2d). Excess sections of cut tape typically adhere to the protective backing and are removed along with it. The newly exposed adhesive is aligned with, and placed onto the appropriate piece of patterned paper (Fig. 2e). This process is repeated for all pairs of patterned tape and paper. These segments of the final device are cut along the edges of the tape to remove them from the patterned paper. Standard scissors (i.e., ~10–12 cm blade) are used for this operation (see Note 17) (Fig. 2f).
3. Fig. 2g shows the layers for the example 3D paper-based microfluidic device described in Fig. 1. These layers are arranged in the order of assembly. Prior to assembly, each layer is compressed using a rolling pin (see Note 18). The pin is rolled over the layers twice, once each in perpendicular directions.
4. The remaining protective backing on the tape is removed and a 2.4 mm disc of Whatman Chromatography paper No. 1 (see Note 19) is inserted into each of the holes in the tape (4, 5) (Fig. 2h). Layer 1 is arranged so that the sticky side of the tape is facing up, while layer 2 is arranged so that the patterned paper is facing down. These two layers are aligned and pressed together (Fig. 2i). The rolling pin is used to compress the layers (see Note 18). The remaining layers are assembled by repeating these last two steps (4, 5).

3.5. Conducting Quantitative Assays Using Paper-Based Microfluidic Devices with Internal Timers

1. Many different configurations of fluidic channels are possible using the general fabrication procedure described herein. Sample entry points in devices can be designed to allow the sample to wick into the device (4), or they can be designed for delivery of the sample using a micropipette (as shown in Fig. 1b) (see Note 20).
2. The timer in the device functions by slowing the wicking rate of the sample to the bottom of the device (where completion of the assay is indicated) relative to the rate at which the sample reaches the detection zones and the assays develop. This delay is caused by the paraffin wax, which is present in layer 3 in the example device shown in Fig. 1a. Because larger quantities of wax cause decreased flow rates for the sample, the quantity of paraffin wax is used to modulate the flow rate of the sample through this timer region of the device.
3. The exact quantity of wax needed to create a timer with a specific end point depends on the configuration of the 3D paper-based microfluidic device, but reference (4) provides a useful starting point for choosing the appropriate quantity of wax for a desired timer.

4. The timer is read when the "read" indicator in Fig. 1a fills completely with the color of the dye that was deposited into the timer when the device was fabricated.

4. Notes

1. Other vector graphics editor programs may be used instead of CleWin Layout Editor; the program must allow the user to draw and arrange lines (and simple geometric features) with micron-level control over line widths.
2. A range of solutions of paraffin wax is needed to create 1 min up to 2 h timers in 3D paper-based microfluidic devices (e.g., more wax deposited in the "meter" region of Fig. 1a will produce a timer for a longer time period). Depending on the concentration of wax, the solutions may need to be sonicated (for up to 10 min) for the paraffin wax to dissolve entirely.
3. Solutions of paraffin wax dissolved in hexanes tend to change concentration after several hours due to slow evaporation of hexanes, even when a cap is placed on the container holding the solution. Fresh solutions provide the most accurate results for the timers.
4. Glass capillary tubes (1 μL) can be substituted for micropipettes, but we have found that glass capillary tubes give slightly less accurate results than micropipettes.
5. The synthetic food dyes contain the following components: RED 40 (disodium salt of 6-hydroxy-5-[(2-methoxy-5-methyl-4-sulfophenyl)azo]-2-naphthalenesulfonic acid.), BLUE 1 (disodium salt of ethyl [4-[*p*-[ethyl(*m*-sulfobenzyl)amino]-a-(*o*-sulfophenyl) benzylidene]-2,5-cyclohexadien-1-ylidene] (*m*-sulfobenzyl)ammonium hydroxide plus *p*-sulfobenzyl and *o*-sulfobenzyl salts), YELLOW 5 (trisodium salt of 4,5-dihydro-5-oxo-1-(4-sulfophenyl)-4-[4-sulfophenylazo]-1H-pyrazole-3-carboxylic acid), and GREEN (a 1:1 mixture of YELLOW 5 and BLUE 1).
6. The computer designs for creating patterned paper must take into account the following considerations: the hydrophobic regions in patterned paper are the solid lines in the graphics program, and the hydrophilic regions are the regions without shading. For creating patterned tape, the lines on the computer graphics program provide the opposite result: a line corresponds to a cut in the tape.
7. When saving as a .ps file, only one layer of the program can be saved at a time, so all of the device layers need to be moved onto the same layer within the program and arranged laterally

so that there is no overlap between different layers. The paper and tape layers should be put on different layers in CleWin and saved as different files.

8. At this point, each of the layers should contain patterns for regions that will become hydrophobic, as this is where the wax will print.
9. Make sure to check that the size of the page remained as 20 cm × 20 cm square when opened in Adobe® Acrobat®.
10. When printing from the .pdf file, make sure that the page scaling menu is set to "none" and "auto-rotate and center" is not checked.
11. From the main menu on the printer: move the arrow to "Paper Tray Setup" and press OK. This action brings up the "Paper Tray Setup" menu. Move the arrow to "Tray 1 Paper" and press OK. This action will bring up the "Tray 1 Paper" menu that states the current setup for tray 1. Move the arrow to "Change Setup…" and press OK. This action brings up the "Tray 1 Paper Size" menu. Move the arrow to "New Custom Size" and press OK. This action brings up the "Short Edge" menu. Move the arrow to "Change…" and press OK. Move the arrow up or down until you reach 200 mm (the units are set to mm and cannot be changed). Press OK. This action will bring up the "Long Edge" menu. Move the arrow to "Change…" and press OK. Move the arrow up or down until 200 mm is reached, and press OK; the "Tray 1 Paper Type" menu will appear. Move the arrow to "Card Stock" and press OK. This action will return you to the "Paper Tray Setup" menu. Move the arrow to "Exit" and press OK. This action returns to the main menu.
12. A piece of aluminum foil is used to cover the hot plate while heating to provide a clean surface. The printed chromatography paper is placed, wax side up, on the aluminum foil. A large stack (~1.3 kg) of 8.5 in. × 11 in. printer paper is placed on top of the chromatography paper such that it covers completely the chromatography paper to ensure that the printed paper has even contact with the surface of the hot plate. Make sure that the chromatography paper is centered on the hot plate to allow for even diffusion of the wax through the paper.
13. The paper needs to stay on the hot plate for between 1 min 45 s and 2 min. If the paper is heated longer than 2 min, the holes will close, and the sample will not wick through the layer. If the paper is not heated for long enough, the wax will not fully penetrate the paper, and separated holes may become connected.
14. The tape is 36 mm wide, so extra space may need to be added between rows so that the gap between pieces of tape is between the rows in the computer-generated pattern.

15. Changing the properties of the laser cutter: Within Acrobat, select "File" and "Print." Select Epilog Engraver from the drop down menu, and click on "Properties" next to the printer selection. Change the selection to the desired values and click "OK."
16. Choosing to print on the computer: the "Go" button on the printer must be pressed. If after pressing this button, the laser does not print, check to make sure that all lines in the graphics file are set to 0.01 point. If any of the lines are larger than this value, the laser cutter will not be able to use vector cutting. If the lines are smaller than this value, they may be too thin for effective cutting of the tape.
17. The paper can be cut prior to affixing the tape, but this alternative procedure makes the alignment of tape with paper more difficult than when the patterned paper remains as a full 20 cm × 20 cm sheet.
18. The pressure needed to provide uniform contact between layers of paper and tape is similar to the pressure needed to roll dough.
19. The paper discs are made using the laser cutter and the same method described for cutting tape. The discs are made from Whatman Chromatography paper No. 1 and are cut to 2.4 mm diameter.
20. The amount of sample added to the device is dictated by the volume of liquid that is absorbed by the entire device.

Acknowledgements

This work was supported in part by the Bill & Melinda Gates Foundation as a subcontract from Harvard University (subcontract No. 01-270716-00), the Beckman Young Investigators Program, the Camille and Henry Dreyfus New Faculty Award, 3M, Mr. Louis Martarano, and The Pennsylvania State University.

References

1. Carrilho E, Martinez AW, Whitesides GM (2009) Understanding wax printing: a simple micropatterning process for paper-based microfluidics. Anal Chem 81:7091–7095
2. Noh H, Phillips ST (2010) Metering the capillary-driven flow of fluids in paper-based microfluidic devices. Anal Chem 82:4181–4187
3. Martinez AW, Phillips ST, Carrilho E, Thomas SW, Sindi H, Whitesides GM (2008) Simple telemedicine for developing regions: camera phones and paper-based microfluidic devices for real-time, off-site diagnosis. Anal Chem 80:3699–3707
4. Noh H, Phillips ST (2010) Fluidic timers for time-dependent, point-of-care assays on paper. Anal Chem 82:8071–8078
5. Martinez AW, Phillips ST, Whitesides GM (2008) Three-dimensional microfluidic devices fabricated in layered paper and tape. Proc Natl Acad Sci U S A 105:19606–19611

6. Martinez AW, Phillips ST, Whitesides GM, Carrilho E (2010) Diagnostics for the developing world: microfluidic paper-based analytical devices. Anal Chem 82:3–10
7. Fu E, Lutz B, Kauffman P, Yager P (2010) Controlled reagent transport in disposable 2D paper networks. Lab Chip 10:918–920
8. Martinez AW, Phillips ST, Nie Z, Cheng C-M, Carrilho E, Wiley BJ, Whitesides GM (2010) Programmable diagnostic devices made from paper and tape. Lab Chip 10:2499–2504
9. Fu E, Kauffman P, Lutz B, Yager P (2010) Chemical signal amplification in two-dimensional paper networks. Sens Actuators B Chem 149:325–328
10. Osborn JL, Lutz B, Fu E, Kauffman P, Stevens DY, Yager P (2010) Microfluidics without pumps: reinventing the T-sensor and H-filter in paper networks. Lab Chip 10:2659–2665
11. Cheng C-M, Martinez AW, Gong J, Mace CR, Phillips ST, Carrilho E, Mirica KA, Whitesides GM (2010) Paper-based ELISA. Angew Chem 122:4881–4884
12. Nie Z, Nijhuis CA, Gong J, Chen X, Kumachev A, Martinez AW, Narovlyansky M, Whitesides GM (2010) Electrochemical sensing in paper-based microfluidic devices. Lab Chip 10: 477–483
13. Delaney JL, Hogan CF, Tian J, Shen W (2011) Electrogenerated chemiluminescence detection in paper-based microfluidic sensors. Anal Chem 83:1300–1306
14. Fenton EM, Mascarenas MR, López GP, Sibbett SS (2009) Multiplex lateral-flow test strips fabricated by two-dimensional shaping. ACS Appl Mater Interfaces 1:124–129
15. Martinez AW, Phillips ST, Butte MJ, Whitesides GM (2007) Patterned paper as a platform for inexpensive, low-volume, portable bioassays. Angew Chem Int Ed Engl 46:1318–1320
16. Martinez AW, Phillips ST, Wiley BJ, Gupta M, Whitesides GM (2008) FLASH: a rapid method for prototyping paper-based microfluidic devices. Lab Chip 8:2146–2150
17. Li X, Tian J, Nguyen T, Shen W (2008) Paper-based microfluidic devices by plasma treatment. Lab Chip 80:9131–9134
18. Abe K, Suzuki K, Citterio D (2008) Inkjet-printed microfluidic multianalyte chemical sensing paper. Anal Chem 80:6928–6934
19. Lu Y, Shi W, Jiang L, Qin J, Lin B (2009) Rapid prototyping of paper-based microfluidics with wax for low-cost, portable bioassay. Electrophoresis 30:1497–1500
20. Olkkonen J, Lehtinen K, Erho T (2010) Flexographically printed fluidic structures in paper. Anal Chem 82:10246–10250

Chapter 14

Thread Based Devices for Low-Cost Diagnostics

Meital Reches

Abstract

The need for low-cost diagnostic devices, both for developing and industrial countries, has led to the search for inexpensive matrixes that will allow the performance of analytical assays. One approach uses paper to create multiple microfluidic channels which allow analytes in urine or blood to flow to different detection zones the device. The choice of paper arises from its low-cost and its ability to wick biological fluids by capillary forces (i.e., an external power is not required to move fluid in a device). This chapter describes the use of a common material—cotton thread—as an alternative matrix for low-cost diagnostics. Thread-based devices can be fabricated using established techniques that rely on common house-hold tools for manipulating threads (e.g., sewing machines and looms). The fabrication schemes described here could potentially be adapted for large-scale manufacturing of diagnostic devices.

Key words: Thread, Low-cost, Diagnostics, Devices, Microfluidics

1. Introduction

Recent efforts towards the fabrication of low-cost diagnostic devices have led to the development of fabrication schemes of microfluidic channels in paper (1–13). By printing hydrophobic barriers on hydrophilic paper it is possible to create multiple microfluidic channels on a low-cost matrix. An alternative matrix to paper is cotton thread (14–16). Cotton thread shares some advantages with paper: it is inexpensive, flexible and lightweight, hydrophilic and wicks fluids by capillary action, can be chemically modified by various functional groups and can be disposed of simply by burning after contamination with biological fluids. Cotton thread has some additional unique characteristics that make it a useful matrix for the fabrication of biomedical devices: (1) its aspect ratio (length-to- diameter ratio) allow the confinement of the capillary flow to one dimension and therefore the required volume of sample is very small (tens of μL); (2) it can be manipulated easily by sewing, knitting and weaving (14).

Gareth Jenkins and Colin D. Mansfield (eds.), *Microfluidic Diagnostics: Methods and Protocols*, Methods in Molecular Biology, vol. 949, DOI 10.1007/978-1-62703-134-9_14,

This chapter describes the design and preparation of three possible designs of diagnostic devices based on the use of cotton thread (14). All three devices wick fluids by capillary action from a sample application zone/zones (e.g., where the user deposits the sample) to a detection zone (e.g., where the user reads the results of the assay). Two designs, the "woven array" and the "branching design" take advantage of the ease which thread can be woven on a loom. They both rely on similar strategies for loading reagents, defining detection zones, and encapsulating microchannels. The third design—the sewn array—takes advantage of the ease with which thread can be sewn into various substrates.

The choice of cotton as the material of the thread is based on its rapid rate of wicking, its durability, ease of which one can manipulate it, adherence to inexpensive tapes with adhesives that can be used to encapsulate the devices. This chapter demonstrates the ability to use these three designs as diagnostic devices by detecting analytes colorimetrically in artificial urine and in artificial blood plasma samples.

2. Materials

2.1. Device Preparation

2.1.1. Woven Array Design and Branching Design

1. Steel pins (McMaster-Carr, NJ, USA).
2. 1.5 cm thick Delrin block (McMaster-Carr, NJ, USA).
3. 100% mercerized cotton thread with a diameter of 0.3 mm, Cebelia Crochet Cotton Art. G167 (DMC, NJ, USA) (see Note 1).
4. Scotch® MultiTask Tape #25, and 3M™ Vinyl Tape #471 (3M, MN, USA) (see Note 2).
5. Tie-Fast Combo Tool (FishUSA.com, PA, USA).
6. A laser cutter or hole puncher.
7. A laminator (Pro-Etch by Micro-Mark, NJ, USA).

2.1.2. Sewn Array Design

1. Latex free plastic bandages (CVS Pharmacy, Inc., RI, USA).
2. 100% mercerized cotton thread with a diameter of 0.3 mm, Cebelia Crochet Cotton Art. G167 (DMC, NJ, USA).
3. Clear nail polish.

2.2. Colorimetric Assays

2.2.1. Detection of Proteins

1. Solution of 250 mM citric acid (pH 1.8).
2. 3.3 mM tetrabromophenol blue (TBPB) in 95% ethanol.

2.2.2. Detection of Nitrite

1. 2 mg/mL sulfanilamide.
2. 1.7 mg/mL 3-hydroxy-1,2,3,4-tetrahydrobenzo(h)quinoline.
3. 25 mg/mL tartaric acid in methanol.

2.2.3. Detection of Ketones

1. Solution of 20 mg/mL sodium phosphate.
2. 20 mg/mL sodium borate.
3. 10 mg/mL glycine.
4. Solution of 20 mg/mL nitroprusside.
5. 30 mg/mL polyethylenglycol (PEG, molecular weight [MW] = 2,000).
6. 2 mg/mL polyacrylic acid (PAA, MW = 2,000).

2.3. Enzymatic Assays

2.3.1. Detection of Glucose

1. Polyacrylamide spheres (Educational Innovations, CT, USA).
2. 0.6 M aqueous solution of potassium iodide.
3. 1:5 horseradish peroxidase–glucose oxidase aqueous solution (15 units of protein per 1 mL, Sigma G3660).

2.3.2. Detection of Alkaline Phosphatase

1. Polyacrylamide spheres (Educational Innovations, CT, USA).
2. Solution of nitrotetrazolium blue chloride (1.5 mg/mL) and 5-bromo-4-chloro-3-indolyl phosphate (1 mg/mL) dissolved in 100 mM Tris buffer pH 9.5.

3. Methods

3.1. Device Preparation

3.1.1. The Woven Array and Branching Designs

1. Prepare a loom by installing steel pins into a 1.5 cm thick Delrin block (two parallel rows of pins, 6 cm apart, with the pins in each row spaced by 0.5 cm).
2. Wind the thread through the loom and arrange the thread in parallel lines (Figs. 1a and 2b) or with a common branching point (Fig. 1b).
3. To generate the detection zones, immerse a piece of thread in a solution contacting the reagents of the desired assay, and dry for ~1 h. Then, use this thread to create knots on the aligned threads using a Tie-Fast Knot Combo Tool (see Note 3, Fig. 2a, b).
4. Pattern a piece of Scotch tape or Vinyl tape with equally spaced holes using a laser cutter or a hole puncher (see Note 4).
5. Insert the piece of tape patterned with holes underneath the threads aligned on the loom (Fig. 2c).
6. Press the threads against the tape and then press a second piece of tape on top of the thread to encapsulate the threads between the two pieces of tape (Fig. 2d).
7. To disconnect the threads from the loom, cut the thread where it contacts the pins of the looms.
8. Laminate the threads between two pieces of tape using a laminator, repeat this step two additional times.

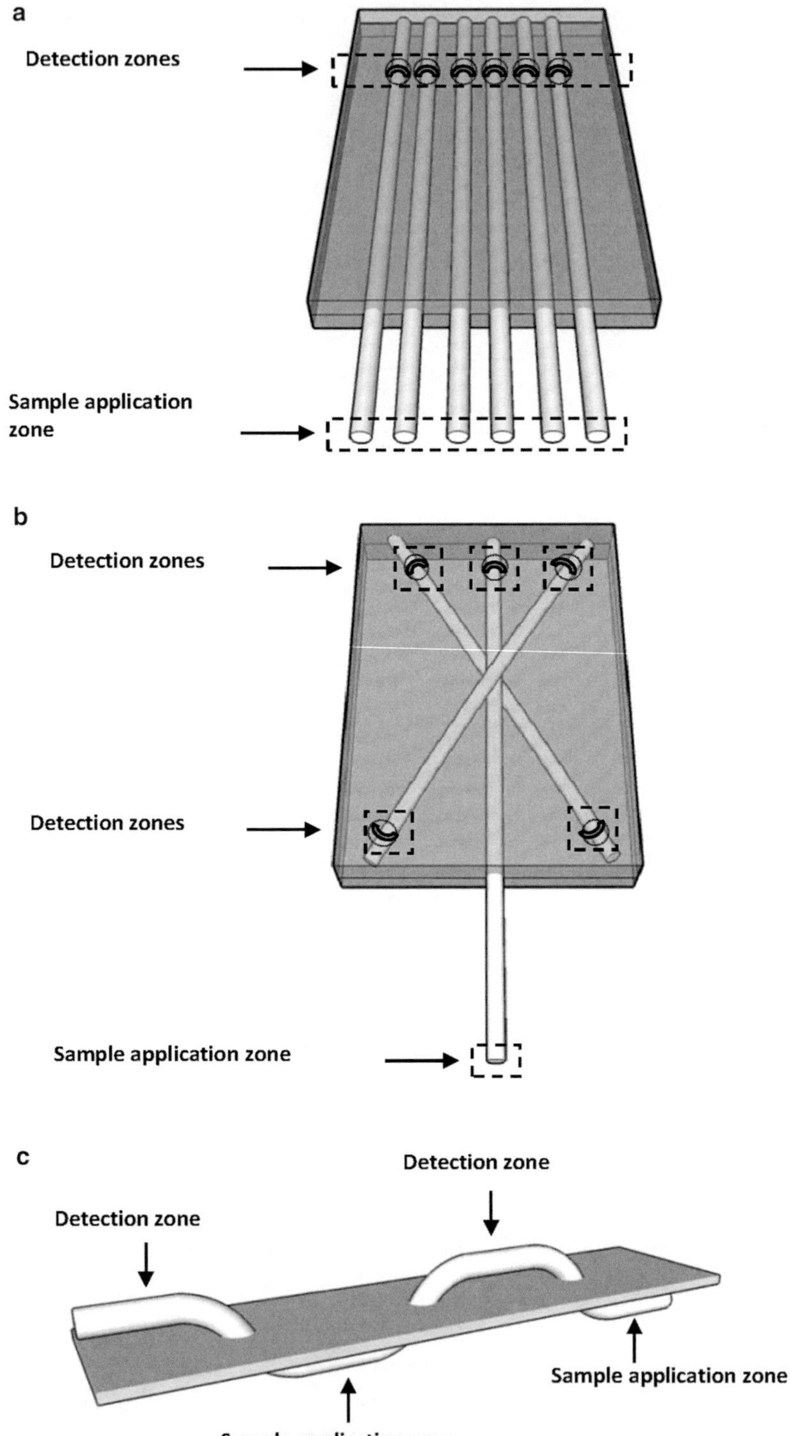

Fig. 1. Schematic illustrations of the thread-based devices. Each device comprises one sample application zone and one detection zone. (**a**) The woven array device. (**b**) The branching design. (**c**) The sewn array design (Reproduced from ref. 14 with permission from American Chemical Society).

9. Cut the threads such that one end of each piece of thread protrudes from the tape portion of the device and serves as the sample application zone (Fig. 2e).

3.1.2. The Sewn Array Design

1. Sew the thread with a needle into all-purpose latex free plastic bandages (or any other desired polymer). Use the small holes in the bandages as guidelines for sewing the threads.
2. One stitch on the upper side of the bandage together with its corresponding lower stitch defines the detection zone and sample application zone, respectively. To physically isolate the assays on two adjacent stitches, block every other hole (formed by the needle) in the bandage with 3 μL droplets of clear nail polish (Fig. 1c).

3.2. Colorimetric Assays

3.2.1. Detection of Proteins

1. Immerse the thread in a solution of 250 mM citric acid (pH 1.8) and 3.3 mM tetrabromophenol blue (TBPB) in 95% ethanol (see Note 5).
2. Dry in air for ~1 h.
3. After the assembly of the device, wick 10 μL of the solution to be analyzed (see Note 6).

3.2.2. Detection of Nitrite

1. Immerse the thread in a solution of 2 mg/mL sulfanilamide, 1.7 mg/mL 3-hydroxy-1,2,3,4-tetrahydrobenzo(h)quinoline and 25 mg/mL tartaric acid in methanol.
2. Dry in air for ~1 h.
3. After the assembly of the device, wick 10 μL of the solution to be analyzed (see Notes 6 and 7).

3.2.3. Detection of Ketones

1. Immerse the thread in a solution of 20 mg/mL sodium phosphate, 20 mg/mL sodium borate, 10 mg/mL glycine, 20 mg/mL nitroprusside, 30 mg/mL polyethylenglycol (PEG, MW=2,000), and 2 mg/mL polyacrylic acid (PAA, MW=2,000).
2. Dry in air for ~1 h.
3. After the assembly of the device, wick 10 μL of the solution to be analyzed (see Notes 6 and 8).

3.3. Enzymatic Colorimetric Assays

3.3.1. Detection of Glucose

1. Incubate dry spherical particles of polyacrylamide in ~50 μL of 0.6 M aqueous solution of potassium iodide *or* 1:5 horseradish peroxidase–glucose oxidase aqueous solution for ~1 h (see Notes 9 and 10).
2. After the beads expand in the solution to a diameter of ~5 mm, thread one sphere containing potassium iodide and one containing horseradish peroxidase–glucose oxidase onto a cotton thread. The distance between the spheres should be ~1 cm (see Note 11).

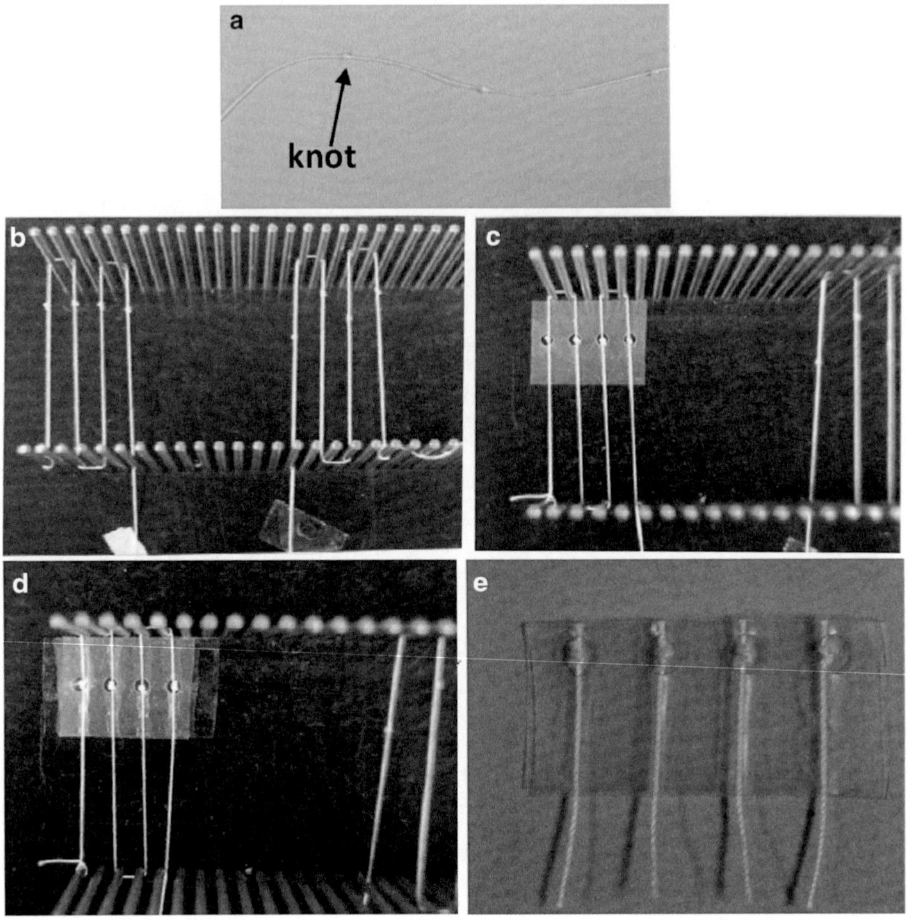

Fig. 2. A sequence of images showing the assembly of the "woven array" device. (**a**) Make knots on a thread. (**b**) Use a loom to arrange the thread as desired. (**c**) Insert a piece of tape (*with holes*) underneath the threads and press the threads against the tape to make intimate contact between the tape and the thread. (**d**) Press a second piece of tape on top of the threads to encapsulate them between the two pieces of tape. (**e**) Disconnect the thread form the loom by cutting the thread where it contacts the pins of the loom (Reproduced from ref. 14 with permission from American Chemical Society).

3. After the assembly of the device, wick 10 μL of the solution to be analyzed (see Note 12).

3.3.2. Detection of Alkaline Phosphatase

1. Suspend spherical particles of polyacrylamide in ~50 μL of nitrotetrazolium blue chloride (1.5 mg/mL) and 5-bromo-4-chloro-3-indolyl phosphate (1 mg/mL) dissolved in 100 mM Tris buffer pH 9.5 for ~1 h (see Note 13).
2. Wick 10 μL of the solution to be analyzed (see Note 14).

4. Notes

1. A thread with a diameter of 0.3 mm is sufficiently wide to be seen by a naked eye and confine the flow of fluid to effectively one dimension.
2. These two types of tape have strong adhesives that allow adherence of the tape to the thread. 3M Vinyl Tape 471 melts upon lamination and makes a conformal seal around the thread.
3. The Tie-Fast Knot Combo Tool, a tool that usually makes tight knots for fishermen, allows fabrication of knots in a reproducible manner.
4. The holes in the tape enhance the evaporation of the solution and therefore allow faster rate of wicking in the desired direction of flow.
5. In the presence of a protein TBPB will change its color from yellow to blue. TBPB is a protein error indicator. Protein error indicators are pH indicators which contain an ionizable group that is displaced in the presence of protein to provide a detectable color change (TBPB changes color from yellow to blue when it ionizes). This is the same color change that the indicator would undergo under the influence of a pH change. The citric acid buffer prevents the change in pH in the absence of the protein (17).
6. It is possible to notice color change on the thread ~10 s after applying the sample. The maximum signal is obtained after ~3 min for protein and nitrite, and after ~6 min for ketones (Fig. 3).
7. In the presence of nitrite the thread changes its color is from faint light pink to red (Fig. 3).
8. In the presence of ketones the colorless knot changes its color to dark purple (Fig. 3).
9. The polyacrylamide gel spheres keep the enzyme active for at least 5 days at 22 ± 2°C and a relative humidity of 20–30%.
10. In the presence of oxygen, glucose oxidase catalyzes the oxidation of glucose to gluconic acid and hydrogen peroxide. The horseradish peroxidase then catalyzes the reaction of hydrogen peroxide with potassium iodide. The colorless iodide is oxidized to brown iodine: this reaction induces a color change from colorless to yellowish brown in the presence of glucose (Fig. 4).
11. A distance of ~1 cm between the spheres allows the transfer of solution along the thread without cross-contamination.

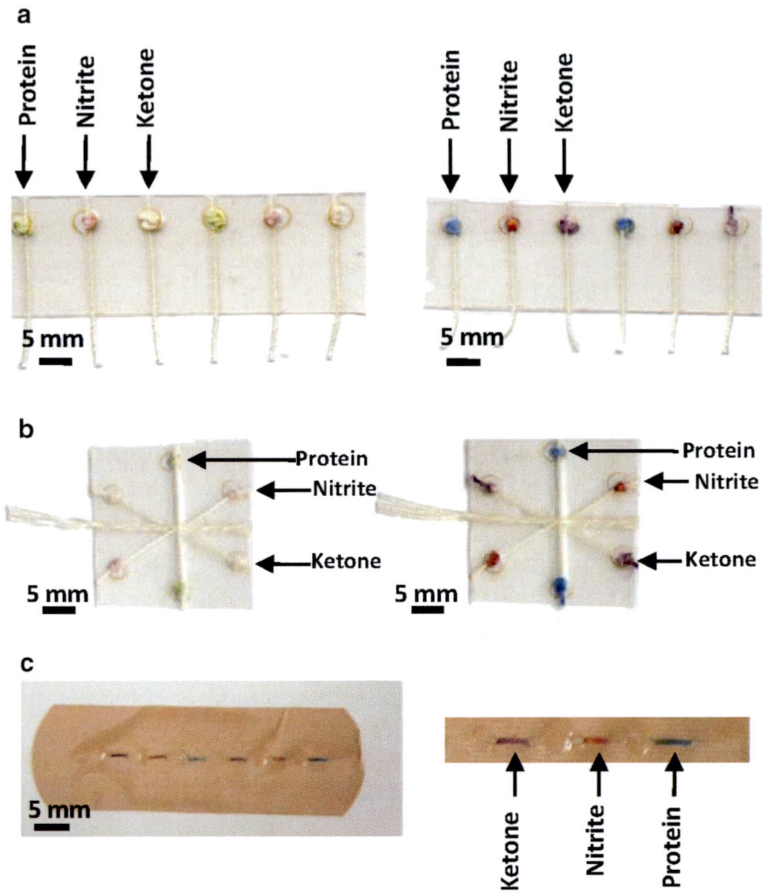

Fig. 3. Colorimetric assays performed using the (**a**) woven array device, (**b**) branching device, (**c**) sewn array design (Reproduced from ref. 14 with permission from American Chemical Society).

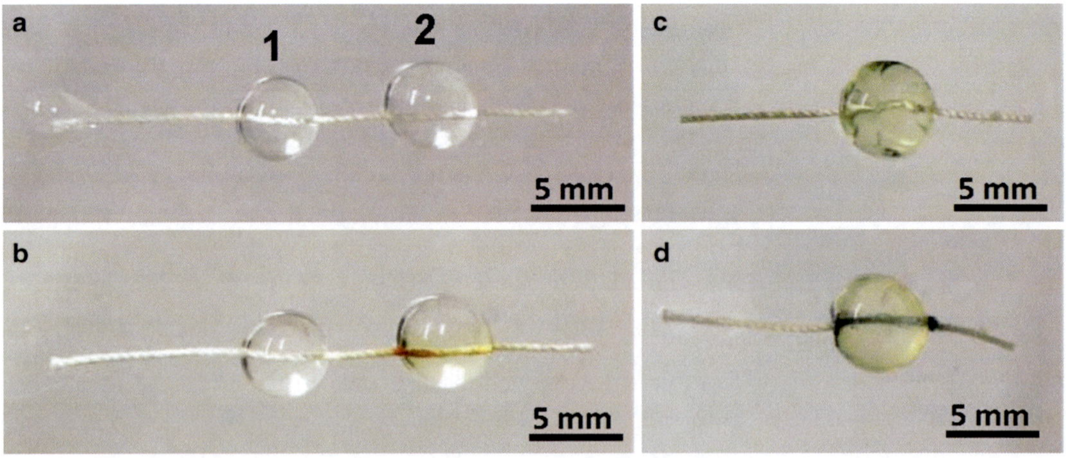

Fig. 4. Enzymatic reactions performed on a cotton thread. (**a**) The sphere labeled 1 contains potassium iodide; the sphere labeled 2 contains horseradish peroxidase–glucose oxidase. (**b**) A change in color is noticeable 15 min after applying artificial urine containing glucose. (**c**) The sphere contains the regents needed to detect alkaline phosphatase. (**d**) The change in color after the application of 140 units/L of alkaline phosphatase (Reproduced from ref. 14 with permission from American Chemical Society).

12. For a solution containing 50 mM glucose, it is possible to detect change in color after ~6 min and full intensity of color after ~15 min (Fig. 4).
13. Alkaline phosphatase catalyzes the cleavage of 5-bromo-4-chloro-indolyl phosphate (BCIP) to metaphosphoric acid and the corresponding indolyl moiety. The indolyl group then rearranges and dimerizes to form either 5,5′-dibromo-4,4′-dichloro-indigo under acidic conditions, or 5,5′-dibromo-4,4′-dichloro-indigo white under alkaline conditions. In the process of dimerization at any pH, the indolyl moiety releases hydrogen ions; these ions reduce the nitro blue tetrazolium, and cause the formation of a blue-colored precipitate (18, 19).
14. For a solution containing 140 units/L of alkaline phosphatase it is possible to detect change in color after ~10 min (Fig. 4d).

References

1. Martinez AW, Phillips ST, Butte MJ, Whitesides GM (2007) Patterned paper as a platform for inexpensive, low-volume, portable bioassays. Angew Chem Int Ed 46:1318–1320
2. Martinez AW, Phillips ST, Carrilho E, Thomas SW, Sindi H, Whitesides GM (2008) Simple telemedicine for developing regions: camera phones and paper-based microfluidic devices for real-time, off-site diagnosis. Anal Chem 80:3699–3707
3. Martinez AW, Phillips ST, Whitesides GM (2008) Three-dimensional microfluidic devices fabricated in layered paper and tape. Proc Natl Acad Sci U S A 105:19606–19611
4. Martinez AW, Phillips ST, Whitesides GM, Carrilho E (2010) Diagnostics for the developing world: microfluidic paper-based analytical devices. Anal Chem 82:3–10
5. Martinez AW, Phillips ST, Wiley BJ, Gupta M, Whitesides GM (2008) FLASH: a rapid method for prototyping paper-based microfluidic devices. Lab Chip 8:2146–2150
6. Bruzewicz DA, Reches M, Whitesides GM (2008) Low-cost printing of poly(dimethylsiloxane) barriers to define microchannels in paper. Anal Chem 80:3387–3392
7. Carrilho E, Phillips ST, Vella SJ, Martinez AW, Whitesides GM (2009) Paper microzone plates. Anal Chem 81:5990–5998
8. Carrilho E, Martinez AW, Whitesides GM (2009) Understanding wax printing: a simple micropatterning process for paper-based microfluidics. Anal Chem 81:7091–7095
9. Nie Z, Nijhuis CA, Gong J, Chen X, Kumachev A, Martinez AW, Narovlyansky M, Whitesides GM (2010) Electrochemical sensing on paper-based microfluidic devices. Lab Chip 10:477–483
10. Abe K, Suzuki K, Citterio D (2008) Inkjet-printed microfluidic multianalyte chemical sensing paper. Anal Chem 80:6928–6934
11. Daar AS, Thorsteinsdottir H, Martin DK, Smith AC, Nast S, Singer PA (2002) Top ten biotechnologies for improving health in developing countries. Nat Genet 32: 229–232
12. Dungchai W, Chailapakul O, Henry CS (2009) Electrochemical detection for paper-based microfluidics. Anal Chem 81:5821–5826
13. Fenton EM, Mascarenas MR, Lopez GP, Sibbett SS (2009) Multiplex lateral-flow test strips fabricated by two-dimensional shaping. ACS Appl Mater Interfaces 1: 124–129
14. Reches M, Mirica KA, Dasgupta R, Dickey MD, Butte MJ, Whitesides GM (2010) Thread as a matrix for biomedical assays. ACS Appl Mater Interfaces 2:1722–1728
15. Reches M, Dickey DM, Butte JM, Whitesides MG (2009) Cotton thread as a low-cost multi-assay diagnostic platform. US 08206992
16. Li X, Tian JF, Shen W (2010) Fabrication of paper-based microfluidic sensors by printing. ACS Appl Mater Interfaces 2:1–6
17. Free HM, Collins GF, Free AH (1960) Triple-test strip for urinary glucose, protein, and pH. Clin Chem 6:352–361
18. McGadey JA (1970) Tetrazolium method for non-specific alkaline phosphatase. Histochemie 23:180–184
19. Blake MS, Johnston KH, Russelljones GJ, Gotschlich EC (1984) A rapid, sensitive, method for detection of alkaline-phosphatase conjugated anti-antibody on western blots. Anal Chem 136:175–179

Chapter 15

Droplet-Based Microfluidics

Sanjiv Sharma, Monpichar Srisa-Art, Steven Scott, Amit Asthana, and Anthony Cass

Abstract

Droplet-based microfluidics or digital microfluidics is a subclass of microfluidic devices, wherein droplets are generated using active or passive methods. The active method for generation of droplets involves the use of an external factor such as an electric field for droplet generation. Two techniques that fall in this category are dielectrophoresis (DEP) and electrowetting on dielectric (EWOD). In passive methods, the droplet generation depends on the geometry and dimensions of the device. T-junction and flow focusing methods are examples of passive methods used for generation of droplets. In this chapter the methods used for droplet generation, mixing of contents of droplets, and the manipulation of droplets are described in brief. A review of the applications of digital microfluidics with emphasis on the last decade is presented.

Key words: Digital microfluidics, Droplet microfluidics, Electrowetting, Dielectrophoresis, Applications

1. Introduction

Microfluidic systems are a consequence of the miniaturization processes that were envisaged in the noted lecture presented by Richard Feynman on December 29th 1959 at the annual meeting of the American Physical Society at the California Institute of Technology (Caltech), and has its roots in the work of Terry in 1975 for a space project involving the miniaturization of a gas-phase chromatographic system onto a silicon wafer (1).

In the early 1990s, miniaturized analysis systems were rejuvenated by the introduction of the concept of Micro Total Analytical Systems (μTAS), or Lab on Chip, by Widmer et al. (2). The basic idea was to miniaturize all functional components of an analytical system such as injection, mixing, separation, and detection into a hand held device that would be portable. As the concept is still at

Gareth Jenkins and Colin D. Mansfield (eds.), *Microfluidic Diagnostics: Methods and Protocols*, Methods in Molecular Biology, vol. 949, DOI 10.1007/978-1-62703-134-9_15,

an early stage and has yet to be fully realized, most of the devices today conform to a "chip in a lab" concept rather than a "lab on a chip" device. Microfluidic systems are based on multidisciplinary concepts borrowed from the fields of physics, chemistry, biology and engineering. The devices used in the early 1990s were mainly made of glass and silicon using fabrication protocols borrowed from the microelectronics industry. However, over the following decades microfluidic devices have not developed as the silicon clones (3) but have rather seen plastics and elastomeric material such as polydimethylsiloxane (PDMS) (4) mainly being used for microfluidic devices. Since their introduction they have been finding applications ranging from simple microreactors to biomedical devices.

An important operational step for a wide range of applications is mixing. With extremely low volumes ranging in the sub-nano to femto-liter volumes the flow of fluids in microfluidic devices is characterized by low Reynolds number and thus laminar flow. One of the major limitations of laminar flows is that it does not facilitate efficient mixing of fluids. Several approaches have been put forth to address this problem. These include use of "herring-bone" structures to introduce turbulence in the flow and the use of micromixer designs integrated into the microfluidic devices (5). As shown in Fig. 1, a micromixer is designed to split a stream of fluid repeatedly into several streams before converging, thus facilitating mixing of fluids (6).

Continuous flow-based microfluidic systems have been popular because of the control they offer over the flow characteristics; however, one of the biggest issues has been scaling up. In droplet-based microfluidics the fluid is broken into tiny droplets and manipulated. Droplet microfluidics is capable of performing a large

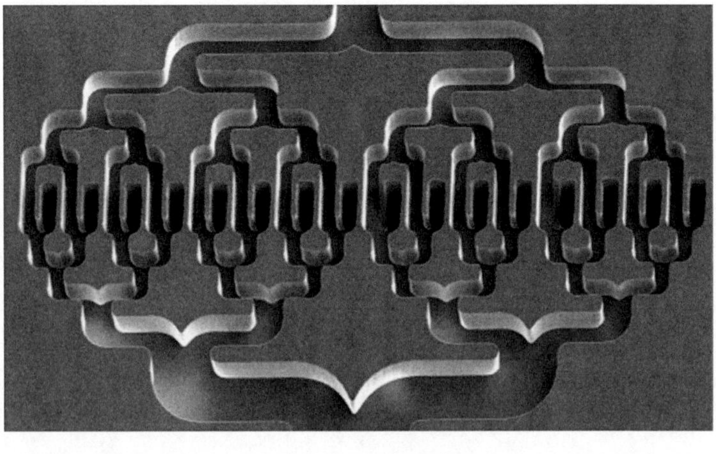

Fig. 1. Photograph of a distributive mixer which splits each solution into 16 separate flows before recombination. Reproduced by permission of The Royal Society of Chemistry http://dx.doi.org/10.1039/A902237F (6).

number of reactions without the need to increase the device dimensions or make the device complex. Since the droplets are amenable to being addressable, synchronized and manipulated, this approach is often referred to as "digital microfluidics." Analogies are being drawn between droplet-based microfluidics and microprocessors (7).

Droplets in microfluidic devices are formed by either active or passive methods. In active methods, an external stimulus such as an electric field is used to generate droplets. Two examples of such techniques are electrowetting on dielectric (EWOD) and dielectrophoresis (DEP). In passive methods multiple streams of aqueous reagents are introduced alongside an immiscible fluid in a spontaneous and chaotic process that is triggered by the flow instabilities of the two immiscible fluids. This is exemplified in droplets generated by T-junction and flow focusing devices.

The basic principles explaining droplet formation have been described elsewhere (8–11). In this chapter, an overview on droplet formation is provided mainly by looking at EWOD, DEP, T-junctions and flow focusing devices. We also look at how these discrete droplets, which represent picoliter volume units offering unique opportunities, are extended to different applications.

2. Droplet Microfluidics

2.1. Mechanism of Droplet Generation

The method of choice for droplet generation is driven mainly by the application and resources available. In general for applications requiring a high throughput (frequency >100 Hz) and large droplets (size in μm range), T-junction and flow focusing techniques are used. Active methods such as EWOD and DEP generate droplets at low frequencies and the droplets are usually small (nL to pL volumes) (10).

The active methods that are used for droplet generation and actuation include surface acoustic wave (12), thermocapillary forces (13), magnetic forces (14), and electrohydrodynamic (EHD) methods. In this chapter we only describe EHD methods used to generate droplets. The droplet formation here depends on the electrical control provided by the electrodes integrated in the microfluidic devices. Dielectrophoresis (DEP) and electrowetting on dielectric (EWOD) are examples of EHD methods used for generation of droplets.

DEP-driven droplet generation is based on the principle that polarizable fluids with high dielectric permittivity are attracted to areas of high electric field intensity. The three main forces that contribute towards droplet generation are: (i) the wetting force on the interfacial line between the droplet, its surrounding medium, and the surface it contacts; (ii) force at the interface between two

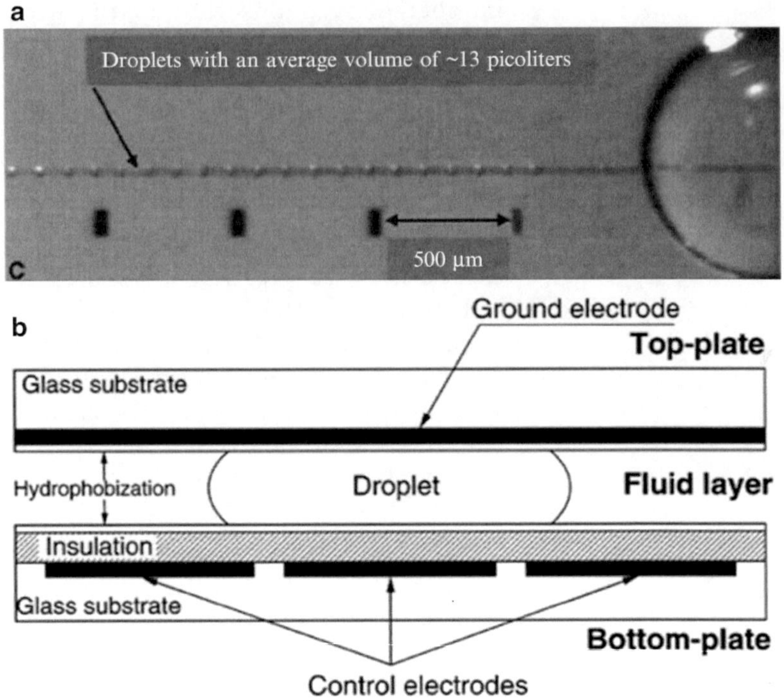

Fig. 2. (**a**) Dielectrophoresis-based generation (16). (**b**) Cross sectional schematic of EWOD-based droplet system. Reproduced by permission of The Royal Society of Chemistry http://dx.doi.org/10.1039/B110474H (26).

fluids; and (iii) body force caused due to pressure gradients in the fluid. The dimension of the droplets and the dispersity depends on the magnitude and frequency of the applied voltage (15–19).

In EWOD method, an electric field is used to change the interfacial energy and henceforth the contact angle between a fluid and the surface it is in contact with. An EWOD-based device may consist of a single or double layer, with the control electrodes on the bottom plate. The hydrophilicity of the area is changed to cause wetting of the surface within tens of microseconds, leading to the formation of liquid fingers between electrodes. When the electrodes are switched off the surface becomes hydrophobic and the liquid finger breaks off from the reservoir forming droplets. The size of the droplets depends on the electric field strength and frequency. At higher frequencies smaller droplets are produced (20–25). DEP-based droplet generation is shown in Fig. 2a and b shows a schematic for an EWOD-based system (26).

Droplets can be passively generated in microfluidic devices using two immiscible phases, usually an aqueous phase and an organic phase which, in most cases, are oil solutions. As described before droplets are examples of passive mixing and depend on the geometries of the device for their formation. The two common

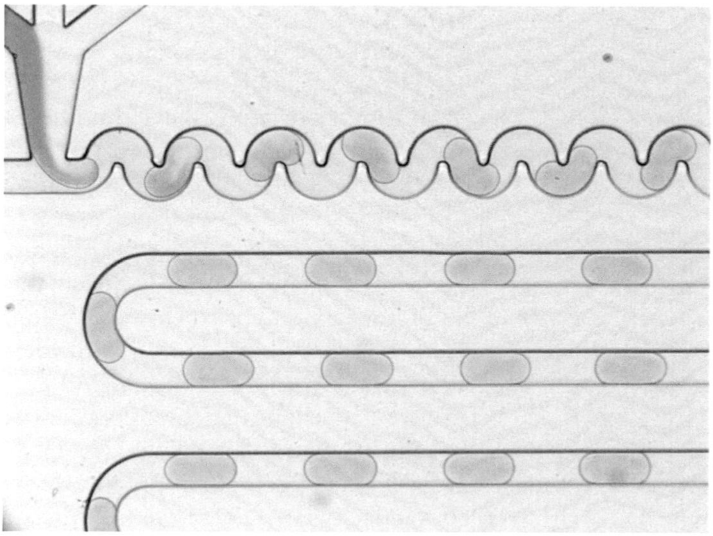

Fig. 3. T-junction droplet device: Water-in-oil droplets are generated from oil and aqueous solutions using a T-junction microfluidic device having three-aqueous inlets, an oil inlet and an outlet. The aqueous flow is pumped into the oil stream and is broken into droplets due to the shear force of the oil phase. The winding channel is used to induce rapid mixing within droplets. The channel dimensions are 50 μm deep, 50 μm wide and 4 cm long, respectively.

geometries used for droplet generation are T-junction (27, 28) and flow focusing designs (11, 29–32).

Because most of the microfluidic-based applications involve analytes soluble in water it is desirable to have aqueous droplets (water in oil or "W/O"). Aqueous droplets are formed when the solution (dispersed phase) comes into contact with an oil stream (carrier fluid). The carrier fluid applies shear force, which is responsible for the droplet formation. The aqueous droplets formed are discrete and separated by the oil phase/carrier fluid. The carrier fluid prevents any dispersion of the aqueous solution during the transportation process. The dimensions of the droplets formed depend on the channel size and the ratio of the oil flow rate to the total aqueous flow rate. Figure 3 shows an example of aqueous droplets generated from oil and aqueous solutions introduced through a T-junction device.

The commonly used organic solutions as carrier fluids for aqueous droplets (W/O) are mainly fluorocarbons. The main advantages offered by fluorocarbons are that they are hydrophobic, immiscible with aqueous solutions, are biocompatible and do not cause swelling of PDMS elastomer used for making devices. To avoid wetting of the channel walls of the PDMS devices a surfactant such as 1H, 1H, 2H, 2H, 2H-perfluorooctanol or Span-80 is added. This reduces the surface tension at the water–oil interface.

These surfactants also stabilize the droplets by preventing adsorption of samples and droplet coalescence (33).

Flow-focusing is another passive method of generating droplets. In this method, the dispersed phase and the continuous phase are forced through a constricted region that causes the shearing of the dispersed phase by the continuous phase, thus leading to a controlled formation of droplets (31, 32, 34, 35).

The research group of Abraham Lee in the University of California, Irvine, has shown that the flow focusing method can be further extended to shear-focusing where a very fine nozzle is used to create a single point with the highest shear leading to the formation of uniform droplets. The size of the droplets can be adjusted by changing the flow rate of the continuous phase (31). In another work this group has shown that the size range of the droplets can be further broadened by the use of pneumatically controlled walls. These walls control the dimensions of the flow focusing region (36).

Gas–liquid dispersion has also been used in microfluidic systems. Microbubbles are generated using designs such as T-junction (37, 38), capillary (39), and flow focusing designs (29, 40). In these cases the size, the frequency and gas fraction of the microbubbles depend on the flow rates of the gas and continuous phase, viscosities of the fluids, and the channel geometry.

The four parameters that characterize droplet formation are Capillary number (C_a), the period, the droplet length and the water fraction (W_f). Capillary number: C_a is a dimensionless quantity defined as: $C_a = v\mu / \gamma$ Here v (m/s) is the flow velocity, μ (kg/m) is the viscosity of fluid, and γ is the surface tension at the aqueous/carrier fluid interface. Droplets are generated at C_a values of less than 0.1 (41). The distance between two adjacent droplets is defined as the period. The droplet length is the length of droplet. Both period and droplet length are independent of the flow rate and capillary number. Water fraction, W_f, is the ratio of the volumetric flow rate of the aqueous stream (V_w) to that of the carrier fluid (V_O). It is defined as:

$$W_f = \frac{V_o}{(V_w + V_o)}$$

The droplet size depends on the water fraction but is independent of the flow rate. The larger the water fraction, the larger the droplet is. Smaller droplets are formed when increasing the ratio between the oil flow rate and the aqueous flow rate. Conversely, decreasing the flow rate ratio between the oil and aqueous solution results in larger droplets. However, changing the total flow rate, whilst maintaining a constant water fraction, will not change droplet size.

Organic droplets (O/W) can be generated when an aqueous solution is used as a carrier fluid and an organic solution is used as a dispersed phase. Double droplets or emulsions can be formed by using two consecutive T-junctions (42, 43). These double emulsions have been extended to various applications in the fields of food science, cosmetics and pharmaceuticals that require storage of ingredients and allow release and exchange to outer layers (43, 44). These applications will be described in detail in later sections.

2.2. Mixing of Contents in Droplet

Recirculation of the contents in the droplet leads to mixing in droplets. This is a slow and inefficient process wherein the two vortices formed within the droplet are mixed only at the left and right half without crossing the boundary (45, 46) in a manner exhibited by steady flows in microchannels.

Mixing in a straight microchannel can be achieved in a rapid and efficient manner if the pure components are introduced in a manner that they form the front and back halves of the droplet (41). This is achieved by a "twirling" process that occurs at the tip of a T-junction. This phenomenon occurs due to the shear forces between the oil and aqueous streams during the process of droplet formation. As the droplets are transported along the channel, a series of recirculation steps ensures efficient mixing.

In general, mixing in large droplets is slow and inefficient while for smaller droplets mixing is rapid and efficient. However, if the droplets are too small then the mixing becomes inefficient because of "over twirling," which leads to the components in the droplets being localized to the left and right halves instead of front and back.

Another means to achieve efficient and rapid mixing in droplets is chaotic advection, which is achieved by the use of curved channels called winding channels. In winding channels, the droplets travel at different velocities relative to the wall, which leads to unsteady flows. The unsteady flow leads to chaotic advection. As shown in Fig. 4, the basic mechanism of chaotic advection involves stretching, folding and reorienting fluids (47). Chaotic advection leads to very fast and efficient mixing. The mixing time is usually in the order of milliseconds (41, 45). Similarly, mixing between viscous and sticky biological samples can be achieved using "bumpy" winding chambers integrated in the device (48–50). In these devices the chaotic advection is induced by the oscillating interfacial shear stress, resulting in rapid mixing of the components.

Active mixing of droplets using an electric field has been demonstrated (51, 52). Mixing in these cases does not depend on the channel dimensions but by moving the droplet back and forth. EWOD-based mixing is influenced by the aspect ratio of the electrode as reported by Pail et al. (53). They demonstrated that a droplet splitting, oscillating and fusing sequence repeated several times over 2 s enabled mixing of the droplet contents.

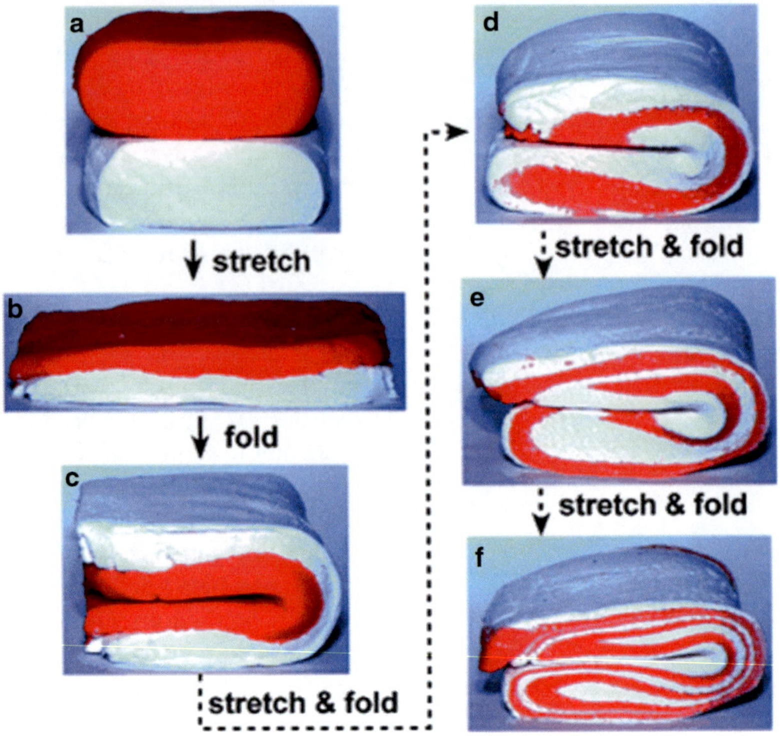

Fig. 4. Modeling clay demonstrates the mixing of two components by chaotic advection within droplets. Stretching and folding are repeated to achieve complete mixing. Reproduced from ref. 47.

2.3. Manipulation of Droplets

Basic processes for droplet manipulation include droplet merging (fusion), droplet splitting (fission), and droplet sorting. Such operations can be exploited for a variety of applications and are expanded upon below.

Droplets can be fused or merged together either actively or passively. In an active mode of droplet fusion, an electric field of the order of 1–3 V is applied to induce instability across the interface of droplets, which results in fusion of the droplets (54–57). The two technologies used for fusion of droplets are EWOD (56, 58, 59) and DEP (60). Cells and liposomes have been merged using electrofusion (61). In addition to electric fields, thermal induction of droplet fusion has also been reported. Thermal induction relies on the change in viscosity of the continuous phase leading to slow movement of droplets consequently leading to fusion of droplets (62). Optical tweezers have also been reported for droplet fusion (63). The passive approach for fusion involves exploitation of channel geometries to achieve fusion of droplets. The device configuration slows down the droplets allowing them to come closer together and fuse (64–66). In a similar approach, pillar-like structures were used to induce passive droplet fusion with precise control of the fusion process (67, 68). Other approaches include fusion based on surface properties (68).

In passive modes of droplet fission, T junctions and other means of obstruction are introduced into microfluidic devices leading to splitting of droplets (37, 69–72). Active fission approaches reported for splitting of droplets include the use of electric fields (54) . The most common technology used for fission of droplets into smaller droplets is EWOD. Splitting is achieved between electrically addressable parallel plates, when the surfaces activated at the opposite ends of the droplets cause the droplet to pinch and divide in the middle (73). Similarly, surface gradients induced by heat have been used for droplet fission in microchannels (74).

Droplet sorting can be achieved based on the size of droplets or content of droplet. Sorting based on size difference can be achieved by the use of the right channel geometry (64, 65, 75). Sorting based on contents involves the use of electric fields in techniques such as DEP (54, 57, 76, 77) and EWOD (78). EWOD, although allowing separation and sorting of droplets, lacks the high throughput speed offered by passive methods such as T-junction and flow focusing. A laser-based localized heating approach facilitates sorting by thermal manipulation of droplets (79).

3. Applications of Droplet Microfluidics

Since its advent droplet-based microfluidic systems, (also referred to as microdroplets), have been extended to several biological and chemical applications. The ability to produce picoliter cells at a frequency of kHz enables high-throughput analysis. Each droplet acts as self-contained unit that prevents the loss of analytes by diffusion, cross talk/contamination (80) between the droplets and surface adsorption of the analytes, thus allowing accurate and controlled screening. In addition, the ease of manipulation of droplets extends the range of applications (10, 47).

3.1. Biological Assays

Droplets can be generated in microfluidic systems at a high frequencies (>1 kHz). The uniform size of droplets and rapid mixing of the components set the stage for high throughput screening and analysis of biological compounds. Compared to conventional screening systems such as 96-well plates, droplet-based assays require samples in the nanoliter range.

Zheng et al. have used a three phase system consisting of liquid–liquid–gas to measure the phosphate activity of enzymes (81). To prevent cross talk between the enzyme sample plug, a buffer plug and a gas plug/bubble was inserted on either side of the enzyme plug. The array of plugs was then merged with a fluorogenic solution specific to the enzyme's phosphate activity. Dittirich et al. have demonstrated in vitro protein expression in microfluidic droplets. As shown in Fig. 5, all components for protein

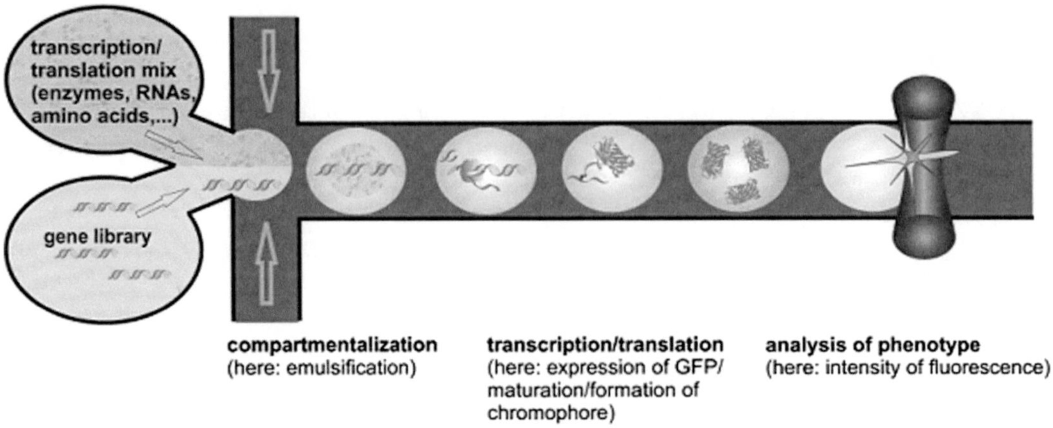

Fig. 5. Schematic of in vitro expression of proteins using droplet microfluidics. Reproduced from ref. 82.

expression were mixed in the droplet followed by off-chip incubation for 50 min. The solution was then reintroduced into the microfluidic device for online detection using an epifluorescence confocal detector. This work has a special significance as it lays the path for in vitro biological assays in droplet units (82). Courtois et al. have successfully fabricated an integrated microfluidic device for droplet formation, a chamber for incubation and analysis of in vitro expression of green fluorescent protein (GFP). The droplets that were generated were incubated in the storage chamber for desired time and the GFP was expressed in 6.5 h (83).

Biocompatible synthetic fluorosurfactants have been reported by Holtze et al. that play a crucial role in stabilization of droplet formation and their incubation and reintroduction back into the device (33). High throughput biological screening of a target from thousands of candidates has been achieved using droplet microfluidics. The common approach is to introduce a single input array of large plugs (~320 nl), followed by breaking them into several output arrays of smaller droplets (20 nl) (70). To prevent coalescence of the plugs due to differential motion, spacers are introduced. These spacers are either bubbles or liquid droplets (84). Using this approach enzymatic activity of thrombin and the coagulation of human blood plasma assays have been successfully demonstrated (85). High-throughput screening of bacteria for antibiotic drug susceptibility in samples without pre-incubation has been demonstrated by Boedicker et al. (86). Lin et al. used droplet-based nuclear magnetic resonance (NMR) as a sample injection/loading method for identification of metabolites in cyanobacterial extract showing antibacterial activity (87).

3.2. Cell-Based Assays

As droplets offer sub-nanoliter compartments they are ideal for cell encapsulation and therefore cell-based assays. The confinement of cells in the droplets ensures that any material released from the cell is not lost but is concentrated to enable detection. In addition, the ease of manipulation of droplets further establishes the applicability of droplet microfluidics for cell-based assays.

The first application of droplets for single cell assays was reported by He et al. In this work, aqueous droplets were used to encapsulate mitochondria (81). Edd et al. have demonstrated that a high aspect ratio microchannel and a high density of cell suspensions enabled passive control of the loading of single cells into droplets (88). The encapsulated cells were subjected to laser-based photolysis to enable assay of enzymatic activity. Cell-based enzymatic assay has been reported by Huebner et al. In this work, the catalytic activity of alkaline phosphatase expressed in *E.coli* cells was investigated as a function of fluorescence intensity (89). Droplets with and without cells, which can be referred to as positive and negative droplets respectively, have been sorted using flow focusing geometries. While the negative droplets under the influence of asymmetric oil flow move towards the right, the positive droplets travel along the center of the focusing channel. This approach has been used to demonstrate encapsulation and sorting of cancerous lymphocytes from whole blood (90). Another important aspect of cell-based assays is incubation of encapsulated cells. Clausell-Tormos et al. used syringes to perform off-chip incubation of the encapsulated cells, which were later reintroduced into the microfluidic devices and analyzed. These cells were found to be fully viable even after several days (80). Figure 6 shows a complete droplet microfluidic chip with functional components for encapsulation, incubation, and analysis. This chip has been used to measure the antibodies released from hybridoma cells (91).

3.3. Double/Multiple Emulsion Systems

Double emulsion systems comprise a smaller droplet trapped in a larger droplet. Since droplet microfluidics allows generation of monodisperse double emulsions at a high frequency (i.e., high throughput) and allows control of the loading volume and the release of encapsulated ingredients, it is suitable for applications that require double emulsions, such as food science, cosmetics, and pharmaceutics (44, 92, 93).

Figure 7 depicts generation of a double emulsion system or W/O/W system (for water-in-oil-in-water). As seen from the picture, an upstream hydrophobic T-junction is used to produce aqueous droplets. These aqueous droplets are moved to a hydrophilic downstream T-junction to form a double emulsion (W/O/W) (42, 94). The number of aqueous droplets (internal droplets) and their size can be controlled by controlling the input flow rates. This approach however has its own limitation; it requires a surface modification step to render different regions of the device hydrophilic and hydrophobic. Utada et al. have introduced an elegant way of generating double emulsions using a coaxial microcapillary device (95). This device produces monodisperse double emulsions

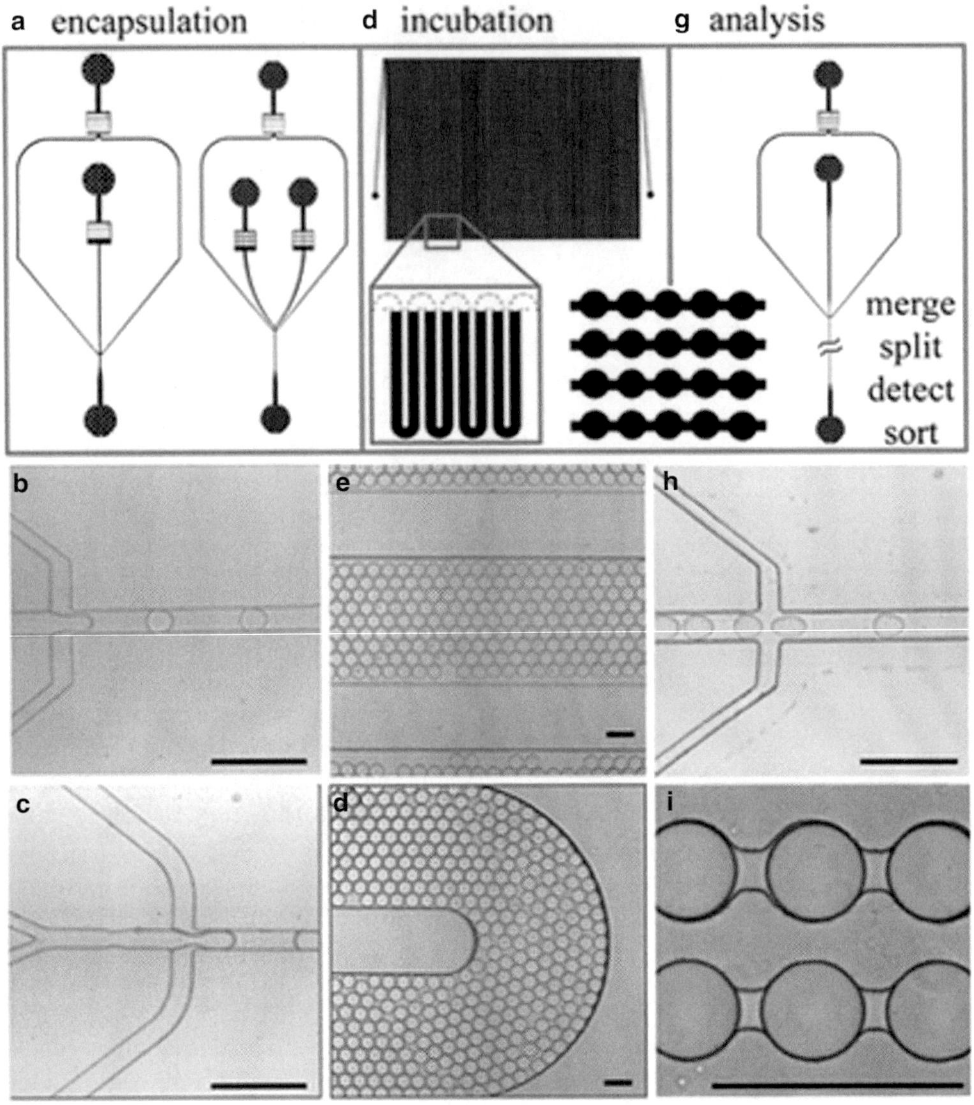

Fig. 6. Microfluidic devices designed for particular purposes. (**a**) Schematic of encapsulation devices having 1 or 2 aqueous inlets. (**b** and **c**) Droplet formation and cell encapsulation. (**d**) Schematic of an incubation chip. (**e** and **f**) Optical micrographs showing droplet encapsulation within microchannels. (**g** and **h**) Devices for analysis (merging, sorting, splitting, or detection). (**i**) An incubation device for time-resolved study. Each device can be flexibly connected using external tubing. All scale bars are 100 μm. Reproduced from ref. 91.

in a one step process precisely controlling the diameters of inner and outer droplets. As seen in Fig. 8, this device consists of two cylindrical microcapillary tubes enclosed in a square tube. A W/O/W double emulsion is produced when the aqueous solution (inner solution) is pumped through the inner tube, an oil phase (middle solution) is introduced through the square tube and a third aqueous solution, (outer fluid), is pumped into the square tube but in a direction opposite to the oil phase. This configuration involving

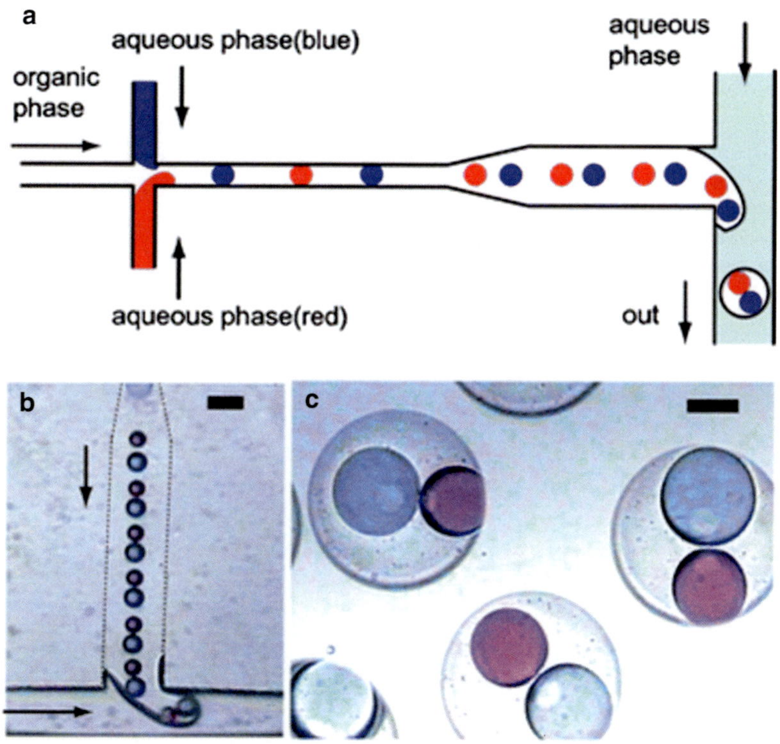

Fig. 7. (**a**) Schematic of a microfluidic device for multiple droplet generation. Aqueous (W/O) droplets are alternatively produced at the upstream junction. (**b**) The formation of organic (W/O/W) droplets by enclosing the *blue* and *red* aqueous droplets with an external aqueous phase. (**c**) Multiple emulsions with different compositions and numbers of internal droplets generated using the device in (**a**). Reproduced by permission of The Royal Society of Chemistry http://dx.doi.org/10.1039/B501972A (93).

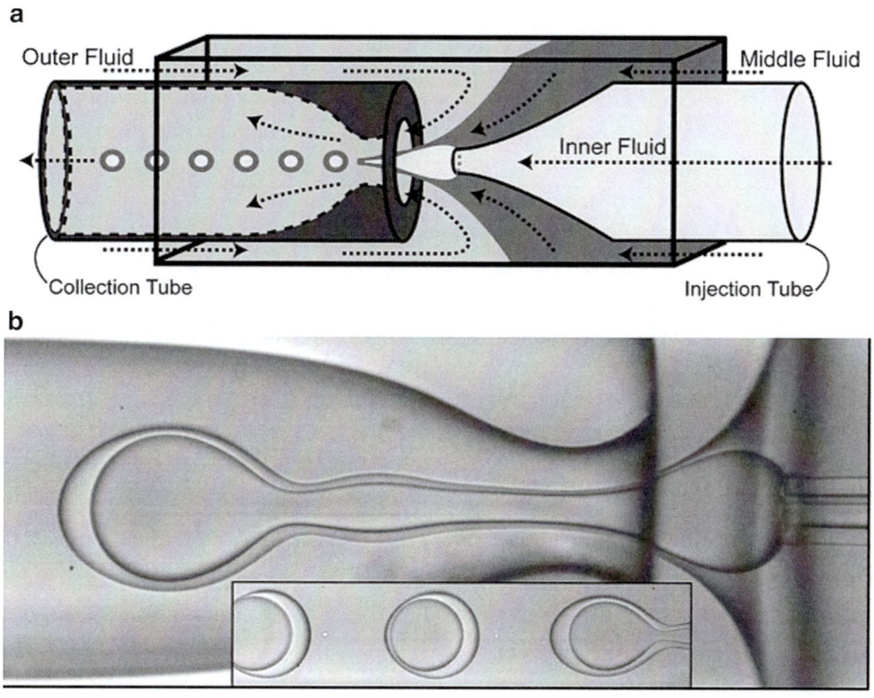

Fig. 8. (**a**) Schematic of a coaxial microcapillary device for the preparation of multiple emulsions. (**b**) Preparation of multiple emulsions with a single internal droplet. Reproduced from ref. 95.

three streams, two in the same direction and one opposite are introduced hydrodynamically. Due to interfacial surface tension the flow breaks at the orifice leading to the formation of double emulsions. Coaxial microcapillary devices have been used to generate double emulsions that have been extended to preparation of well-defined particles and functional vesicles (93), polymer vesicles (96), biodegradable polymerosomes (97), phospholipid vesicles (98), and nanoparticle colloidosomes (99). Multiple emulsions such as W/O/W/O droplets and even higher have been reported using the microcapillary device (92, 93).

3.4. Measurement of Kinetics in Droplets

Rapid mixing of the components/reagents in droplets and the nanoliter volume sample requirement makes droplet microfluidics suitable for measurement of kinetics. Examples of kinetic measurements include binding studies of calcium ions with Fluoro-4 dye (45) and enzymatic reaction involving ribonuclease A (27). The calcium binding to the dye was measured over a distance of 500 μm and exhibited mixing in 2 ms. The rate constant measured for the activity of ribonuclease was in agreement with reported results.

3.5. Protein Crystallization Studies

Due to the low sample requirement (nano-picoliter volumes), droplet-based microfluidic systems are useful for studying protein crystallization. The first demonstration of protein crystallization studies using extremely low volumes of sample (<4 nL) was done by Zheng et al. (100). The nanoliter volume droplets were used for screening thousands of protein crystallization conditions. The conditions were varied by controlling the flow rates of the solutions. Further work done in the research group of Ismagilov on protein crystallization studies, includes indexing of droplets using markers (101), long-term storage of protein crystals inside droplets to assess crystal structure (102, 103), and studying the effect of mixing on nucleation of protein crystals (104). One of the main requirements for protein crystallization studies is to perform nucleation and growth processes separately. This has been achieved on a microfluidic system by using a two stage droplet formation process. In the first stage smaller droplets containing seed precipitants are formed and then transferred to the growth stage, which involves merging with protein and precipitants to form larger droplets that are subsequently transferred to a glass microcapillary for storage and observation. This approach was used to successfully crystallize the SARS nucleocapsid N-terminal domain and oligoendopeptidase F crystals (105, 106). In another reported work, a combined method involving screening and optimization has been reported for studying the crystallization of membrane proteins (107).

3.6. Droplet-Based Microreactors

Droplet-based microfluidic systems, on account of their ability to form well defined independent microreactor units and the ability to manipulate these units, offer a platform for synthetic processes.

Examples of synthesis in droplets include synthesis of nanoparticles, polymeric particles and alginate hydrogels and organic reactions.

Monodispersed nanoparticles such as CdS/CdSe and CdSe crystals have been synthesized using droplet-based microreactors. The reactions were very fast and mainly performed in glass microfluidic devices (30, 108). In another approach, two aqueous channels were allowed to collide in a microchannel with the oil stream running orthogonal to the aqueous streams. Using this method, calcium carbonate particles and silver nanoparticles with a very low degree of agglomeration were produced (109). Droplet-based microfluidic systems have been used to create irregular particles such as non-spherical particles, Janus droplets, and double emulsions (described earlier) that are difficult to create using macroscopic systems, as it allows control over the shape of the droplets and hence the particles. In addition to the spherical particles, disk- and rod-shaped particles have been synthesized in microfluidic systems that facilitate confining of particles and influence the shape of the particles (110–113). In the first step, T-junction or flow focusing is used to generate droplets followed by transporting the droplets to a shape-defining region. For rods, the shape defining region involves microchannels with dimensions (height and width) smaller than the droplet, causing elongation of the droplets to rods. For disk-shaped particles, the height of channel is designed to be smaller than the diameter of the droplet whilst the width is made larger, causing the droplet to flatten. Janus droplet is a term used to describe droplets with two different and distinct surface regions. Generation of Janus droplets in droplet microfluidic systems have been described by Nisisako et al. made from black and white monomeric solutions of carbon black and titanium dioxide with isobornyl acetylate, respectively (114). Other examples of Janus particles reported include spherical particles made up of two hemispheres of two different fluorescent dyes (115).

Polymeric nanoparticles have been synthesized in situ in droplets. Monomer droplets with monomers and photoinitiator have been polymerized in situ using either UV irradiation or thermal initiation (116). Other examples of polymeric nanoparticles synthesized in droplets include polyvinyl alcohol (PVA) (44), biodegradable microgels (117), chitosan microparticles encapsulating ampicillin (118), molecularly imprinted polymer beads (119), copolymer particles (120), and mesoporous silica particles (121). Alginates are extensively used in dentistry and prosthetics as molding material. Alginate hydrogels are becoming popular because of their similarity with natural extracellular matrices, thus leading to tissue engineering applications. Alginates can easily encapsulate biomolecules such as proteins, enzymes, cells and DNA. Alginate hydrogels have been used successfully for encapsulation of gold nanoparticles (122) and polystyrene beads (123). In addition they have also been used for encapsulation of cells such as mammalian cells (109) and yeast cells (124) without loss of their viability. Both off-line and on-line/in situ approaches for synthesis of alginate hydrogels have been reported. In the reported off-line approach alginate droplets are produced in a flow focusing microfluidic device followed by off-line gelations to produce alginate hydrogels in the range of 50–2,000 μm (122). The on-line/in situ approach for producing alginate hydrogels involves calcium initiated polymerization of alginate to give alginate hydrogels. Both T-junction (124) and flow focusing geometries (125) have been used to achieve monodisperse alginate hydrogels. The two streams comprise alginate solution as aqueous phase and calcium iodide in undecanol as the organic phase. The calcium ions in the undecanol diffuse into the alginate droplets leading to the formation of alginate hydrogels (123). The chaotic advection with the droplets leads to rapid mixing of alginate and calcium chloride leading to yield the hydrogels in a fast reaction. In addition to calcium chloride solution, calcium carbonate nanoparticles suspended in oil phase have also been reported (109). One of the major issues with synthesis of alginate hydrogels using microfluidic droplets is coagulation/blocking of microchannels due to formation before the droplets. This issue has been addressed by forming droplets comprising alginate and calcium ions and merging the droplets in a larger/wider chamber (43). Since PDMS molecules swell up when brought into contact with organic solvents, other substrates such as Teflon (126) and thiolene-based resin (127–129) have been reported for use in droplet-based organic reactions. The organic reactions performed in droplet-based microfluidic systems include deacetylation (126) of ouabain hexaacetate and bromination of alkene (127, 128). Gerdts et al. have demonstrated oxidation of Co^{3+} by KH_2PO_4 in a two-stage microfluidic device yielding an amplification of the order of 5,000 times (130).

3.7. Microextraction

Similar to solvent extraction methods, droplet generation also requires two immiscible liquids. Thus, droplet microfluidics provides a platform for rapid microextraction based on liquid–liquid extractions. Fang et al. have demonstrated microextraction of butyl Rhodamine B as organic droplets using 1-hexanol in a glass-based microfluidic device. The droplet composition was determined using laser induced fluorescence (131) and chemiluminescence detection (132). Other examples of microextraction include extraction of aluminum in water using a metal chelate (2,2-dihydroxyazobenzene) and a buffer system. The aluminum complex was compartmentalized in the tributyl phosphate droplets, at a rate 90 times faster than bulk extraction using a separation funnel and an accuracy similar to conventional methods (133). Patrick Tabeling's research group has investigated mass transfer between moving aqueous droplets and a continuous phase to understand the underlying liquid–liquid microextraction in droplets. They used fluorescien in octanol as the continuous phase with droplets to understand extraction of fluorescein into droplets. Purification was studied by monitoring the diffusion of rhodamine from droplets to a continuous phase. The studies on the extraction and purification in droplet microfluidic systems revealed that these processes occur much faster than conventional methods (134). Rhodamine 6G has also been used to study its extraction from aqueous droplets to ionic liquid phase (135). Ionic liquids on account of their superior extraction properties, biocompatibility, hydrophobicity, and hydrophilicity are excellent candidates for microextraction processes.

3.8. Polymerase Chain Reaction

Droplet microfluidic has been applied to polymerase chain reaction (PCR) because of the precise volume control achievable with droplets. Each droplet reactor is a million fold smaller than conventional array-based PCR units. Droplet-based PCR enables issues associated with continuous-flow microfluidic PCR devices, such as contamination and inhibition due to surface adsorption to be overcome (136–139). In addition, droplet-based PCR techniques eliminate the synthesis of short chimeric products and reduce the time required for the thermal cycles. It promises to be a great tool for single molecule and single cell amplification. Droplet-based PCR can be broadly divided into two categories: continuous-flow droplet PCR devices and non-continuous droplet PCR devices. While in the first instance droplets are generated using passive methods (such as T-junction and flow focusing) (140–144) the second category involves the use of active methods such as the EWOD technique (145, 146). There is good evidence to indicate that this technology will emerge as a reliable technique for PCR.

Examples of continuous-flow droplet PCR devices include one reported by Beer et al. that requires only 18 cycles for single copy amplification and is capable of real time detection (140). Besides DNA PCR amplification, the use of droplets has also been reported for single RNA genome copies (141). Novel designs such as the circular radial design configuration shown in Fig. 9 enables alternating temperature zones thereby providing an amplification factor of up to 5×10^6 and thus represents a high throughput platform for PCR in droplets (144). Viovy's group in Institut Curie developed a prototype for an automated device for the sample preparation and analysis of generated droplets for PCR. The PCR products were detected using laser induced fluorescence in a capillary. This device was capable of performing about 3,000 PCR reactions/day showing excellent reproducibility and sensitivity at low concentrations (147). One of the major advantages of droplet-based PCR has been the fact the longer DNA amplicons have been amplified with great efficiency. Researchers from the group of Richard Mathies have demonstrated amplification of DNA fragments with more than 600 base pairs (at 10 Amol concentration) (148).

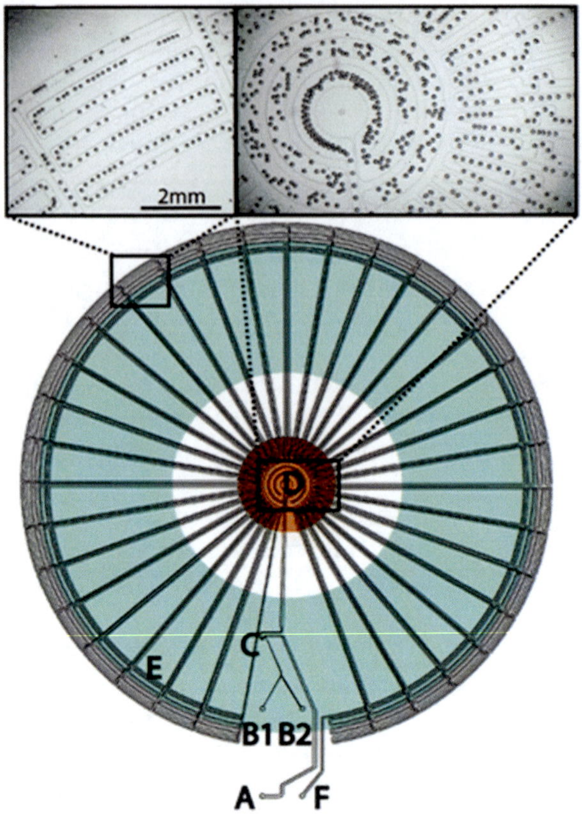

Fig. 9. A radial microfluidic device for continuous-flow PCR in droplets. The device (75 μm in depth) consists of an oil inlet (A), two aqueous inlets (B1 and B2) and an outlet (F). Droplets generated at a T-junction (C) move through a 500 μm-wide inner-circular channel, which contains a hot zone (D) for template denaturation. Droplets then travel on to the periphery (200 μm wide channels) of the device allow annealing and template extension to occur. When the droplets move back to the center, the DNA is denatured and a new cycle begins. The whole process is finished after 34 cycles, in which the droplets move out of the device through the outlet (F). Reproduced from ref. 144.

Hua et al. have described an automated and self-contained multiplex device based on EWOD-droplets wherein the thermocycling is achieved by shuttling a droplet between two fixed zones. Using this device DNA levels of methicillin-resistant bacteria were detected with a remarkably high amplification efficiency of 94.7% (149).

3.9. Titrations

Droplet-based microfluidic systems have been employed for titration analysis. Examples of this include neutralization of formic acid with disodium hydrogen phosphate (150) and titration of anticoagulant against activated partial thromboplastin (151). In both cases, the results were consistent with measurements done in bulk.

3.10. Point of Care (PoC) Diagnostic Devices

Lab on a chip or μTAS devices have become popular because of the benefits offered by them such as low sample requirements and faster analysis times. Efforts are being made to integrate the various components into a hand held device that is portable and can be used as a portable on-site device. Droplets generated in EWOD mode have been used by Srinivasan et al. for measuring glucose concentration in fluids such as saliva, serum, plasma and urine using a colorimetric assay based on Tinder's reaction (152, 153). The optical system comprising an LED as source and a photodiode was used to measure the absorbance change of the quinoneimine.

3.11. Future Applications

Some of the work reported over the last few years has created a lot of enthusiasm towards the realization of droplet-based logic gates and thus a microfluidic computer chip (154–158). Compartmentalized of units and the ability to make multiple emulsions systems indicate that this technique has potential to be used to synthesize artificial cells.

References

1. Terry SC (1975) Gas chromatography system fabricated on silicon wafer using integrated circuit technology. Stanford Electron Lab Tech Rep 4603:1–128
2. Manz A et al (1992) Planar chips technology for miniaturization and integration of separation techniques into monitoring systems: capillary electrophoresis on a chip. J Chromatogr A 593:253–258
3. Whitesides GM (2006) The origins and the future of microfluidics. Nature 442:368–373
4. Whitesides GM et al (2001) Soft lithography in biology and biochemistry. Annu Rev Biomed Eng 3:335–373
5. Nguyen NT, Wu Z (2005) Micromixers—a review. J Micromech Microeng 15:R1
6. Bessoth FG, deMello AJ, Manz A (1999) Microstructure for efficient continuous flowmixing. Anal Commun 36:213–215
7. Gascoyne PRC et al (2004) Dielectrophoresis-based programmable fluidic processors. Lab Chip 4:299–309
8. Gunther A, Jensen KF (2006) Multiphase microfluidics: from flow characteristics to chemical and materials synthesis. Lab Chip 6:1487–1503
9. Malic L et al (2010) Integration and detection of biochemical assays in digital microfluidic LOC devices. Lab Chip 10:418–431
10. Teh SY et al (2008) Droplet microfluidics. Lab Chip 8:198–220
11. Huebner A et al (2008) Microdroplets: a sea of applications? Lab Chip 8:1244–1254
12. Guttenberg Z et al (2005) Planar chip device for PCR and hybridization with surface acoustic wave pump. Lab Chip 5:308–317
13. Chen JZ et al (2005) Effect of contact angle hysteresis on thermocapillary droplet actuation. J Appl Phys 97:014906–014909
14. Lehmann U et al (2006) Two-dimensional magnetic manipulation of microdroplets on a chip as a platform for bioanalytical applications. Sens Act B Chem 117:457–463
15. Jones TB (2001) Liquid dielectrophoresis on the microscale. J Electrostatics 51–52: 290–299
16. Ahmed R, Jones TB (2006) Dispensing picoliter droplets on substrates using dielectrophoresis. J Electrostatics 64:543–549
17. Jones TB et al (2001) Dielectrophoretic liquid actuation and nanodroplet formation. J Appl Phys 89:1441–1448
18. Ahmed R, Jones TB (2007) Optimized liquid DEP droplet dispensing. J Micromech Microeng 17:1052
19. Wang KL et al (2007) Dynamic control of DEP actuation and droplet dispensing. J Micromech Microeng 17:76
20. Lee J et al (2002) Electrowetting and electrowetting-on-dielectric for microscale liquid handling. Sens Act A Phys 95:259–268
21. Sung Kwon C, Hyejin M, Chang-Jin K (2003) Creating, transporting, cutting, and merging liquid droplets by electrowetting-based actuation for digital microfluidic circuits. J Microelectromech Syst 12:70–80

22. Berthier J et al (2006) Computer aided design of an EWOD microdevice. Sens Act A Phys 127:283–294
23. Roux J-M, Fouillet Y, Achard J-L (2007) 3D droplet displacement in microfluidic systems by electrostatic actuation. Sens Act A Phys 134:486–493
24. Zeng J, Korsmeyer T (2004) Principles of droplet electrohydrodynamics for lab-on-a-chip. Lab Chip 4:265–277
25. Jones TB (2002) On the relationship of dielectrophoresis and electrowetting. Langmuir 18:4437–4443
26. Pollack MG, Shenderov AD, Fair RB (2002) Electrowetting-based actuation of droplets for integrated microfluidics. Lab Chip 2:96–101
27. Song H, Ismagilov RF (2003) Millisecond kinetics on a microfluidic chip using nanoliters of reagents. J Am Chem Soc 125:14613–14619
28. Thorsen T et al (2001) Dynamic pattern formation in a vesicle-generating microfluidic device. Phys Rev Lett 86:4163–4166
29. Garstecki P et al (2004) Formation of monodisperse bubbles in a microfluidic flow-focusing device. Appl Phys Lett 85:2649–2651
30. Chan EM, Alivisatos AP, Mathies RA (2005) High-temperature microfluidic synthesis of CdSe nanocrystals in nanoliter droplets. J Am Chem Soc 127:13854–13861
31. Tan YC, Cristini V, Lee AP (2006) Monodispersed microfluidic droplet generation by shear focusing microfluidic device. Sens Act B Chem 114:350–356
32. Yobas L et al (2006) High-performance flow-focusing geometry for spontaneous generation of monodispersed droplets. Lab Chip 6:1073–1079
33. Holtze C et al (2008) Biocompatible surfactants for water-in-fluorocarbon emulsions. Lab Chip 8:1632–1639
34. Anna SL, Bontoux N, Stone HA (2003) Formation of dispersions using "flow focusing" in microchannels. Appl Phys Lett 82: 364–366
35. Woodward A et al (2007) Monodisperse emulsions from a microfluidic device, characterised by diffusion NMR. Soft Matter 3:627–633
36. Chun-Hong L et al (2007) A tunable microflow focusing device utilizing controllable moving walls and its applications for formation of micro-droplets in liquids. J Micromech Microeng 17:1121
37. Garstecki P et al (2006) Formation of droplets and bubbles in a microfluidic T-junction-scaling and mechanism of break-up. Lab Chip 6:437–446
38. Xu JH et al (2006) Formation of monodisperse microbubbles in a microfluidic device. AIchE J 52:2254–2259
39. Gañán-Calvo AM, Gordillo JM (2001) Perfectly monodisperse microbubbling by capillary flow focusing. Phys Rev Lett 87:274501
40. Hettiarachchi K et al (2007) On-chip generation of microbubbles as a practical technology for manufacturing contrast agents for ultrasonic imaging. Lab Chip 7:463–468
41. Tice JD et al (2003) Formation of droplets and mixing in multiphase microfluidics at low values of the Reynolds and the capillary numbers. Langmuir 19:9127–9133
42. Okushima S et al (2004) Controlled production of monodisperse double emulsions by two-step droplet breakup in microfluidic devices. Langmuir 20:9905–9908
43. Nisisako T, Torii T, Higuchi T (2004) Novel microreactors for functional polymer beads. Chem Eng J 101:23–29
44. Nisisako T (2008) Microstructured devices for preparing controlled multiple emulsions. Chem Eng Technol 31:1091–1098
45. Song H, Tice JD, Ismagilov RF (2003) A microfluidic system for controlling reaction networks in time. Angew Chem Int Ed 42:768–772
46. Bringer MR et al (2004) Microfluidic systems for chemical kinetics that rely on chaotic mixing in droplets. Phil Trans Roy Soc Lond Ser A Math Phys Eng Sci 362:1087–1104
47. Song H, Chen DL, Ismagilov RF (2006) Reactions in droplets in microfluidic channels. Angew Chem Int Ed 45:7336–7356
48. Ismagilov RF et al (2000) Experimental and theoretical scaling laws for transverse diffusive broadening in two-phase laminar flows in microchannels. Appl Phys Lett 76:2376–2378
49. Muradoglu M, Stone HA (2005) Mixing in a drop moving through a serpentine channel: a computational study. Phys Fluids 17:073305–073309
50. Liau A et al (2005) Mixing crowded biological solutions in milliseconds. Anal Chem 77:7618–7625
51. Wheeler AR et al (2004) Electrowetting-on-dielectric for analysis of peptides and proteins by matrix assisted laser desorption/ionization mass spectrometry. Am Chem Soc 228:U33
52. Paik P et al (2003) Electrowetting-based droplet mixers for microfluidic systems. Lab Chip 3:28–33
53. Paik P, Pamula VK, Fair RB (2003) Rapid droplet mixers for digital microfluidic systems. Lab Chip 3:253–259

54. Link DR et al (2006) Electric control of droplets in microfluidic devices. Angew Chem Int Ed Engl 45:2556–2560
55. Priest C, Herminghaus S, Seemann R. (2006) Generation of monodisperse gel emulsions in a microfluidic device. Appl Phys Lett 88
56. Ahn K et al (2006) Electrocoalescence of drops synchronized by size-dependent flow in microfluidic channels. Appl Phys Lett 88:264105
57. Ahn K et al (2006) Dielectrophoretic manipulation of drops for high-speed microfluidic sorting devices. Appl Phys Lett 88:024104-1-024104-3
58. Priest C, Herminghaus S, Seemann R (2006) Controlled electrocoalescence in microfluidics: targeting a single lamella. Appl Phys Lett 89:134101
59. Wang J, Lu C (2006) Microfluidic cell fusion under continuous direct current voltage. Appl Phys Lett 89:234102–234103
60. Singh P, Aubry N (2007) Transport and deformation of droplets in a microdevice using dielectrophoresis. Electrophoresis 28:644–657
61. Tresset G, Takeuchi S (2005) Utilization of cell-sized lipid containers for nanostructure and macromolecule handling in microfabricated devices. Anal Chem 77:2795–2801
62. Kohler JM et al (2004) Digital reaction technology by micro segmented flow—components, concepts and applications. Chem Eng J 101:201–216
63. Lorenz RM et al (2006) Microfluidic and optical systems for the on-demand generation and manipulation of single femtoliter-volume aqueous droplets. Anal Chem 78: 6433–6439
64. Tan Y-C et al (2004) Design of microfluidic channel geometries for the control of droplet volume, chemical concentration, and sorting. Lab Chip 4:292–298
65. Tan YC, Ho YL, Lee AP (2008) Microfluidic sorting of droplets by size. Microfluidics Nanofluidics 4:343–348
66. Bremond N et al (2008) Decompressing emulsion droplets favors coalescence. Phys Rev Lett 100:024501
67. Niu X et al (2008) Pillar-induced droplet merging in microfluidic circuits. Lab Chip 8:1837–1841
68. Fidalgo LM, Abell C, Huck WTS (2007) Surface-induced droplet fusion in microfluidic devices. Lab Chip 7:984–986
69. Adamson DN et al (2006) Production of arrays of chemically distinct nanolitre plugs via repeated splitting in microfluidic devices. Lab Chip 6:1178–1186
70. Link DR et al (2004) Geometrically mediated breakup of drops in microfluidic devices. Phys Rev Lett 92:054503
71. De Menech M (2006) Modeling of droplet breakup in a microfluidic T-shaped junction with a phase-field model. Phys Rev E 73:031505
72. Menetrier-Deremble L, Tabeling P (2006) Droplet breakup in microfluidic junctions of arbitrary angles. Phys Rev E 74:035303
73. Cho SK, Moon HJ, Kim CJ (2003) Creating, transporting, cutting, and merging liquid droplets by electrowetting-based actuation for digital microfluidic circuits. J Microelectromech Syst 12:70–80
74. Ting TH et al (2006) Thermally mediated breakup of drops in microchannels. Appl Phys Lett 89:234101–234101-3
75. Tan YC, Lee AP (2005) Microfluidic separation of satellite droplets as the basis of a monodispersed micron and submicron emulsification system. Lab Chip 5:1178–1183
76. Choi S, Park J-K (2005) Microfluidic system for dielectrophoretic separation based on a trapezoidal electrode array. Lab Chip 5:1161–1167
77. Li Y et al (2007) Continuous dielectrophoretic cell separation microfluidic device. Lab Chip 7:239–248
78. Cho SK, Zhao YJ, Kim CJ (2007) Concentration and binary separation of micro particles for droplet-based digital microfluidics. Lab Chip 7:490–498
79. Baroud CN et al (2007) Thermocapillary valve for droplet production and sorting. Phys Rev E 75:046302
80. Clausell-Tormos J et al (2008) Droplet-based microfluidic platforms for the encapsulation and screening of mammalian cells and multicellular organisms. Chem Biol 15:427–437
81. Zheng B, Ismagilov RF (2005) A microfluidic approach for screening submicroliter volumes against multiple reagents by using preformed arrays of nanoliter plugs in a three-phase liquid/liquid/gas flow. Angew Chem Int Ed 44:2520–2523
82. Dittrich PS, Jahnz M, Schwille P (2005) A new embedded process for compartmentalized cell-free protein expression and on-line detection in microfluidic devices. Chembiochem 6:811–814
83. Courtois F et al (2008) An integrated device for monitoring time-dependent in vitro expression from single genes in picolitre droplets. Chembiochem 9:439–446

84. Chen DLL et al (2007) Using three-phase flow of immiscible liquids to prevent coalescence of droplets in microfluidic channels: criteria to identify the third liquid and validation with protein crystallization. Langmuir 23:2255–2260
85. Li L, Boedicker JQ, Ismagilov RF (2007) Using a multijunction microfluidic device to inject substrate into an array of preformed plugs without cross-contamination: comparing theory and experiments. Anal Chem 79:2756–2761
86. Boedicker JQ et al (2008) Detecting bacteria and determining their susceptibility to antibiotics by stochastic confinement in nanoliter droplets using plug-based microfluidics. Lab Chip 8:1265–1272
87. Lin YQ et al (2008) Microscale LC-MS-NMR platform applied to the identification of active cyanobacterial metabolites. Anal Chem 80:8045–8054
88. Edd JF et al (2008) Controlled encapsulation of single-cells into monodisperse picolitre drops. Lab Chip 8:1262–1264
89. Huebner A et al (2008) Development of quantitative cell-based enzyme assays in microdroplets. Anal Chem 80:3890–3896
90. Chabert M, Viovy J-L (2008) Microfluidic high-throughput encapsulation and hydrodynamic self-sorting of single cells. Proc Natl Acad Sci 105:3191–3196
91. Koster S et al (2008) Drop-based microfluidic devices for encapsulation of single cells. Lab Chip 8:1110–1115
92. Chu L-Y et al (2007) Controllable monodisperse multiple emulsions. Angew Chem Int Ed 46:8970–8974
93. Shah RK et al (2008) Designer emulsions using microfluidics. Mater Today 11:18–27
94. Nisisako TT, Okushima S, Torii T (2005) Controlled formulation of monodisperse double emulsions in a multiple-phase microfluidic system. Soft Matter 1:23–27
95. Utada AS et al (2005) Monodisperse double emulsions generated from a microcapillary device. Science 308:537–541
96. Lorenceau E et al (2005) Generation of polymerosomes from double-emulsions. Langmuir 21:9183–9186
97. Shum HC, Kim J-W, Weitz DA (2008) Microfluidic fabrication of monodisperse biocompatible and biodegradable polymersomes with controlled permeability. J Am Chem Soc 130:9543–9549
98. Shum HC et al (2008) Double emulsion templated monodisperse phospholipid vesicles. Langmuir 24:7651–7653
99. Lee D, Weitz DA (2008) Double emulsion-templated nanoparticle colloidosomes with selective permeability. Adv Mater 20: 3498–3503
100. Zheng B, Roach LS, Ismagilov RF (2003) Screening of protein crystallization conditions on a microfluidic chip using nanoliter-size droplets. J Am Chem Soc 125:11170–11171
101. Zheng B, Tice JD, Ismagilov RF (2004) Formation of droplets of alternating composition in microfluidic channels and applications to indexing of concentrations in droplet-based assays. Anal Chem 76:4977–4982
102. Zheng B, Tice JD, Ismagilov RF (2004) Formation of arrayed droplets of soft lithography and two-phase fluid flow, and application in protein crystallization. Adv Mater 16:1365–1368
103. Zheng B et al (2004) A droplet-based, composite PDMS/glass capillary microfluidic system for evaluating protein crystallization conditions by microbatch and vapor-diffusion methods with on-chip X-ray diffraction. Angew Chem Int Ed 43:2508–2511
104. Chen DL, Gerdts CJ, Ismagilov RF (2005) Using microfluidics to observe the effect of mixing on nucleation of protein crystals. J Am Chem Soc 127:9672–9673
105. Gerdts CJ et al (2006) Time-controlled microfluidic seeding in nL-volume droplets to separate nucleation and growth stages of protein crystallization. Angew Chem Int Ed 45:8156–8160
106. Yadav MK et al (2005) In situ data collection and structure refinement from microcapillary protein crystallization. J Appl Crystallogr 38:900–905
107. Li L et al (2006) Nanoliter microfluidic hybrid method for simultaneous screening and optimization validated with crystallization of membrane proteins. P Natl Acad Sci U S A 103:19243–19248
108. Shestopalov I, Tice JD, Ismagilov RF (2004) Multi-step synthesis of nanoparticles performed on millisecond time scale in a microfluidic droplet-based system. Lab Chip 4:316–321
109. Tan WH, Takeuchi S (2007) Monodisperse alginate hydrogel microbeads for cell encapsulation. Adv Mater 19:2696–2701
110. Seo M et al (2005) Continuous microfluidic reactors for polymer particles. Langmuir 21:11614–11622
111. Dendukuri D et al (2005) Controlled synthesis of nonspherical microparticles using microfluidics. Langmuir 21:2113–2116
112. Kobayashi I, Uemura K, Nakajima M (2006) Controlled generation of monodisperse

discoid droplets using microchannel arrays. Langmuir 22:10893–10897

113. Groß GA et al (2007) Formation of polymer and nanoparticle doped polymer minirods by use of the microsegmented flow principle. Chem Eng Technol 30:341–346

114. Nisisako T et al (2006) Synthesis of monodisperse bicolored Janus particles with electrical anisotropy using a microfluidic co-flow system. Adv Mater 18:1152–1156

115. Shepherd RF et al (2006) Microfluidic assembly of homogeneous and Janus colloid-filled hydrogel granules. Langmuir 22: 8618–8622

116. Xu S et al (2005) Generation of monodisperse particles by using microfluidics: control over size, shape, and composition. Angew Chem Int Ed Engl 44:3799

117. De Geest BG et al (2005) Synthesis of monodisperse biodegradable microgels in microfluidic devices. Langmuir 21: 10275–10279

118. Yang C-H, Huang K-S, Chang J-Y (2007) Manufacturing monodisperse chitosan microparticles containing ampicillin using a microchannel chip. Biomed Microdevices 9:253–259

119. Zourob M et al (2006) A micro-reactor for preparing uniform molecularly imprinted polymer beads. Lab Chip 6:296–301

120. Lewis PC et al (2005) Continuous synthesis of copolymer particles in microfluidic reactors. Macromolecules 38:4536–4538

121. Carroll NJ et al (2008) Droplet-based microfluidics for emulsion and solvent evaporation synthesis of monodisperse mesoporous silica microspheres. Langmuir 24:658–661

122. Huang K-S, Lai T-H, Lin Y-C (2006) Manipulating the generation of Ca-alginate microspheres using microfluidic channels as a carrier of gold nanoparticles. Lab Chip 6:954–957

123. Zhang H et al (2006) Microfluidic production of biopolymer microcapsules with controlled morphology. J Am Chem Soc 128:12205–12210

124. Choi CH et al (2007) Generation of monodisperse alginate microbeads and in situ encapsulation of cell in microfluidic device. Biomed Microdevices 9:855–862

125. Tan J et al (2008) Drop dispenser in a cross-junction microfluidic device: scaling and mechanism of break-up. Chem Eng J 136: 306–311

126. Hatakeyama T, Chen DL, Ismagilov RF (2006) Microgram-scale testing of reaction conditions in solution using nanoliter plugs in microfluidics with detection by MALDI-MS. J Am Chem Soc 128:2518–2519

127. Cygan ZT et al (2005) Microfluidic platform for the generation of organic-phase microreactors. Langmuir 21:3629–3634

128. Barnes SE et al (2006) Raman spectroscopic monitoring of droplet polymerization in a microfluidic device. Analyst 131:1027–1033

129. Hung LH, Lin R, Lee AP (2008) Rapid microfabrication of solvent-resistant biocompatible microfluidic devices. Lab Chip 8:983–987

130. Gerdts CJ, Sharoyan DE, Ismagilov RF (2004) A synthetic reaction network: chemical amplification using nonequilibrium autocatalytic reactions coupled in time. J Am Chem Soc 126:6327–6331

131. Chen H et al (2005) Microfluidic chip-based liquid-liquid extraction and preconcentration using a subnanoliter-droplet trapping technique. Lab Chip 5:719–725

132. Shen H, Fang Q, Fang ZL (2006) A microfluidic chip based sequential injection system with trapped droplet liquid-liquid extraction and chemiluminescence detection. Lab Chip 6:1387–1389

133. Kumemura M, Korenaga T (2006) Quantitative extraction using flowing nanoliter droplet in microfluidic system. Anal Chim Acta 558:75–79

134. Mary P, Studer V, Tabeling P (2008) Microfluidic droplet-based liquid-liquid extraction. Anal Chem 80:2680–2687

135. Wang WH et al (2007) Flow-focusing generation of monodisperse water droplets wrapped by ionic liquid on microfluidic chips: from plug to sphere. Langmuir 23:11924–11931

136. Park N, Kim S, Hahn JH (2003) Cylindrical compact thermal-cycling device for continuous-flow polymerase chain reaction. Anal Chem 75:6029–6033

137. Obeid PJ et al (2002) Microfabricated device for DNA and RNA amplification by continuous-flow polymerase chain reaction and reverse transcription-polymerase chain reaction with cycle number selection. Anal Chem 75:288–295

138. Obeid PJ, Christopoulos TK (2003) Continuous-flow DNA and RNA amplification chip combined with laser-induced fluorescence detection. Anal Chim Acta 494:1–9

139. Kolari K et al (2008) Real-time analysis of PCR inhibition on microfluidic materials. Sens Act B Chem 128:442–449

140. Beer NR et al (2007) On-chip, real-time, single-copy polymerase chain reaction in picoliter droplets. Anal Chem 79:8471–8475

141. Beer NR et al (2008) On-chip single-copy real-time reverse-transcription PCR in isolated picoliter droplets. Anal Chem 80:1854–1858
142. Tsuchiya H et al (2008) On-chip polymerase chain reaction microdevice employing a magnetic droplet-manipulation system. Sens Act B Chem 130:583–588
143. Kiss MM et al (2008) High-throughput quantitative polymerase chain reaction in picoliter droplets. Anal Chem 80:8975–8981
144. Schaerli Y et al (2008) Continuous-flow polymerase chain reaction of single-copy DNA in microfluidic microdroplets. Anal Chem 81:302–306
145. Sista R et al (2008) Development of a digital microfluidic platform for point of care testing. Lab Chip 8:2091–2104
146. Chang Y-H et al (2006) Integrated polymerase chain reaction chips utilizing digital microfluidics. Biomed Microdevices 8:215–225
147. Chabert M et al (2006) Automated microdroplet platform for sample manipulation and polymerase chain reaction. Anal Chem 78:7722–7728
148. Kumaresan P et al (2008) High-throughput single copy DNA amplification and cell analysis in engineered nanoliter droplets. Anal Chem 80:3522–3529
149. Hua Z et al (2010) Multiplexed real-time polymerase chain reaction on a digital microfluidic platform. Anal Chem 82:2310–2316
150. Henkel T et al (2004) Chip modules for generation and manipulation of fluid segments for micro serial flow processes. Chem Eng J 101:439–445
151. Song H et al (2006) On-chip titration of an anticoagulant argatroban and determination of the clotting time within whole blood or plasma using a plug-based microfluidic system. Anal Chem 78:4839–4849
152. Srinivasan V, Pamula VK, Fair RB (2004) Droplet-based microfluidic lab-on-a-chip for glucose detection. Anal Chim Acta 507:145–150
153. Srinivasan V, Pamula VK, Fair RB (2004) An integrated digital microfluidic lab-on-a-chip for clinical diagnostics on human physiological fluids. Lab Chip 4:310–315
154. Prakash M, Gershenfeld N (2007) Microfluidic bubble logic. Science 315:832–835
155. Fuerstman MJ, Garstecki P, Whitesides GM (2007) Coding/decoding and reversibility of droplet trains in microfluidic networks. Science 315:828–832
156. Epstein IR (2007) Can droplets and bubbles think? Science 315:775–776
157. Lee W et al (2010) Dynamic self-assembly and control of microfluidic particle crystals. Proc Natl Acad Sci 107:22413–22418
158. Cheow LF, Yobas L, Kwong D-L (2007) Digital microfluidics: droplet based logic gates. Appl Phys Lett 90:054107

Chapter 16

Droplet-Based Microfluidics for Binding Assays and Kinetics Based on FRET

Monpichar Srisa-Art and Sanjiv Sharma

Abstract

Droplet microfluidic systems provide a controlled environment in which to perform rapid mixing, isolation of picoliter size fluid volumes and rapid variation of reaction conditions. Indeed, the ability to controllably form droplets with variable reagent composition at high speed lies at the heart of performance improvements when compared to conventional microfluidic devices operating under laminar flow conditions. Furthermore, it is important to realize that segmented-flow systems can generate droplets at rates in excess of 1 kHz. In theory, this means that millions of individual reactions or assays can be processed in very short times. In addition, since mixing is rapid and reagent transport occurs with no dispersion, microdroplet reactors are superior environments in which to study biological reactions, especially rapid kinetic reactions, when compared to diffusion-limited continuous-flow formats. Accordingly, droplet microfluidics is a promising technology to perform reactions and assays in a high-throughput manner, in which a hugely productive and efficient system for screening a desired component from thousands of samples is necessary.

Key words: Droplet microfluidics, Microdroplet reactors, Binding assays, Binding kinetics

1. Introduction

Herein, droplet microfluidics is demonstrated to perform binding assays and kinetics between streptavidin and biotin via Fluorescence Resonance Energy Transfer (FRET). Typically, determination of binding rate constants is more difficult than determination of dissociation rate constants as most biological molecules bind much more quickly than they dissociate. Indeed, streptavidin–biotin binding demonstrates a high association constant ($K_a \sim 10^{15}$ M^{-1}) (1–4). Accordingly, it is more practical to carry out experiments that measure the dissociation rate constant. However, due to rapid

Gareth Jenkins and Colin D. Mansfield (eds.), *Microfluidic Diagnostics: Methods and Protocols*, Methods in Molecular Biology, vol. 949, DOI 10.1007/978-1-62703-134-9_16, © Springer Science+Business Media, LLC 2013

mixing, droplet systems can be suitably configured to determine binding kinetics (5–7). Exploiting this approach, the binding kinetics of the streptavidin and biotin system can be monitored in real time.

2. Materials

2.1. Chemicals and Oligonucleotides

1. Perfluorodecalin (PFD: 95% mixture of *cis* and *trans*) used as an oil phase, and 1H,1H,2H,2H-perfluoro- octanol (97%) used as a surfactant (Sigma Aldrich).
2. Tris(hydroxymethyl) aminomethane hydrochloride (Tris–HCl) pH 8.0 buffer (Fluka).
3. Binding buffer: pH 8.0, 100 mM Tris–HCl, 10 mM NaCl, and 3 mM $MgCl_2$.
4. Alexa Fluor 488 (AF488) Streptavidin conjugate (Invitrogen, UK). This should be stored at –20°C and protected from light.
5. Oligonucleotides: A biotinylated DNA strand consisting of the followingsequence:**Biotin**-5′-GCGCTAAAATTATTTATTGA TCGATTTTTTTT CGGGCGCGGCGGGC-3′; and Alexa Fluor 647 (AF647) internally labeled on a complementary strand having the sequence: 3′-CG CG[**AF 647**]ATTTTAAA AATAACTAGCTAAAAAAAAGCCCGCGCCGCCCG-5′ (Operon, Germany). These two DNA solutions should be stored at –20°C and protected from light.

2.2. Equipment

1. Genius Thermal Cycler (Techne, UK).
2. Gastight syringes, 250 μL and 2.5 mL (SGE Europe Ltd, UK).
3. Microfluidic device for droplet generation: e.g., a 3-aqueous-inlet T-junction PDMS microfluidic device, having 50 μm wide, 50 μm deep and ~44 cm long channels. The PDMS microdevice is fabricated using standard photolithographic methods. For optical detection, the PDMS device is bonded to a 160 μm thick microscope coverslip.
4. Fused silica capillary (e.g., 100 μm I.D., 360 μm O.D., CM Scientific, UK).
5. Teflon tubing (0.356 mm I.D., 1.55 mm O.D., Upchurch Scientific).
6. PEEK fingertights and unions (1/16″, 10/32 threads, VICI AG International, Switzerland).
7. Precision syringe pump (PHD 2000, Harvard Apparatus, UK).

8. Laser Induced Fluorescence microscope: equipped with 488 nm laser, dichroic mirror (z488rdc, Chroma Technology Corporation, USA) and an emission filter (z488, Chroma Technology Corporation, USA) a second dichroic mirror is incorporated (z630rdc, Chroma Technology Corp., USA), which splits the fluorescence signal such that the AF488 and AF647 dyes can be measured using two APD detectors. The fluorescence signal reflected by the z630rdc dichroic mirror is further filtered by an emission filter (hq540/40m, Chroma Technology Corp., USA). The fluorescence that passes through the z630rdc dichroic mirror is filtered by another emission filter (hq640lp, Chroma Technology Corp., USA).

3. Methods

3.1. DNA Hybridization

1. DNA hybridization is performed at pH 8.0 using the binding buffer (see Note 1).
2. Prepare the mixture of single-stranded DNA (biotin-labeled DNA and AF 647-labeled DNA) by pipetting 50 μl of 200 nM of each DNA solution into 200 μl Eppendorf tubes, resulting in a concentration of 100 nM (see Note 2).
3. Place the Eppendorf tubes into the thermocycler to hybridize the DNA mixture. The hybridization process is carried out by rapidly increasing the temperature to 92°C and holding at this temperature for 2 min The temperature is then slowly decreased to 4°C at a rate of 1.6°C/min.
4. After preparation, hybridized DNA should consist of AF647 on one strand and biotin on the complementary strand.

3.2. Droplet Formation

1. A continuous oil-phase consisting of a 10:1 (v/v) mixture of perfluorodecalin and 1H,1H,2H,2H-perfluorooctanol is filled into a 2.5 mL gas tight syringe, while 250 μL gas tight syringes are used to deliver aqueous solutions.
2. Connect the gastight syringes to the microfluidic device (via fused silica capillaries bonded into the microfluidic device's access holes) using Teflon tubing and PEEK fingertights and unions.
3. Precision syringe pumps are used to deliver reagent solutions into the microfluidic channels.
4. Set the flow rates of both the oil and aqueous solutions to 1.5 μl/min for each phase, resulting in a total flow rate of 3.0 μl/min (20 mm/s) and a water fraction of 0.5.

3.3. Fluorescence Detection

1. Place the entire microfluidic device on a fluorescence microscope stage for droplet measurements (see Note 3).
2. The microscope objective lens brings the laser beam to a tight focus within the microfluidic channel. This objective lens focuses the laser beam in to the sample and collects the resulting fluorescence (see Note 4).
3. For excitation of the AF488, a dichroic mirror and an emission filter are implemented to permit excitation by a 488 nm laser and allow for fluorescence emission of AF488.
4. Fluorescence emission is then split using a second dichroic mirror onto two APD detectors to enable measurement of the fluorescence emission of both AF488 and AF647 dyes (see Note 5).

3.4. Streptavidin–Biotin Binding Assay

1. The assay is carried out using a 3-aqueous-inlet T-junction PDMS microfluidic device, having 50 μm wide, 50 μm deep and ~44 cm long channels.
2. Introduce the acceptor-labeled hybridized DNA and the donor-conjugated streptavidin solutions separately through the aqueous inlets.
3. Pump the AF488-labeled streptavidin at a concentration of 110 nM into the left inlet, while 100 nM AF647 and biotin-labeled DNA is introduced into the right inlet.
4. Pump the binding buffer into the middle inlet in order to prevent the reagents from coming into contact prior to droplet formation and to allow for on-chip dilution. Thus, mixing and binding of streptavidin and DNA will only occur inside a droplet.
5. The initial assay is conducted by fixing the donor-labeled streptavidin concentration and varying the acceptor-labeled DNA concentration. This is achieved by maintaining the flow rate of the 110 nM donor-labeled streptavidin constant at 0.3 μl/min (resulting in a concentration of 22.0 nM), while changing the flow rate of the 100 nM acceptor-labeled DNA between 0.1 and 1.0 μl/min (resulting in a DNA concentration range of 6.67 and 73.33 nM). The flow rate of the binding buffer is changed so as to maintain a total aqueous flow rate of 1.50 μl/min. Consequently, concentration ratios (CR), defined as the ratio of DNA concentration to streptavidin concentration, are varied from 0.3 to 3.33 (see Notes 6 and 7).
6. To study the effect of donor-labeled streptavidin concentration on binding curves, the fixed flow rate of streptavidin is changed from 0.30 to 0.20 μl/min to obtain 22.0 and 14.7 nM streptavidin concentrations, respectively. However, the same CR set

(varied from 0.3 to 3.33) is used to enable comparison of binding results for each streptavidin concentration. Accordingly, the DNA flow rate is changed from 0.07 to 0.73 μl/min for 14.7 nM streptavidin. Conditions for fixed streptavidin concentrations at 0.20 and 0.30 μl/min using the same CR set are shown in Table 1.

7. Online detection is conducted at ~42 cm or ~20 s (the total channel length is ~44 cm) downstream from the droplet-forming region. Both fluorescence signals (green and red) are recorded (see Note 8).
8. For calculation of the binding ratio of this streptavidin and biotin system, a binding curve is plotted using the red fluorescence signal as a function of concentration ratio and the curve is then fitted using an appropriate model, such as the Hill-Waud model. FRET efficiency can be calculated from $I_A/(I_A+I_D)$, where I_A and I_D are fluorescence intensities of the acceptor and donor, respectively.

3.5. Streptavidin and Biotin Binding Kinetics

1. The streptavidin–biotin binding kinetic measurements are performed by fixing the donor-labeled streptavidin concentration and varying the acceptor-labeled DNA concentration.
2. Maintain the flow rate of 110 nM donor-labeled streptavidin at 0.3 μl/min (resulting in a concentration of 22.0 nM), while the flow rates of 100 nM biotin and acceptor-labeled DNA are varied. To obtain uniform droplets, the DNA flow rates are changed from 0.3 to 0.9 μl/min (using steps of 0.1 μl/min), giving a concentration range of 20.0–60.0 nM. Consequently, concentration ratios are changed from 0.91 to 2.73. Accordingly, the flow rate of the binding buffer at the middle inlet is adjusted so as to maintain a total aqueous flow rate of 1.50 μl/min. The experimental conditions are detailed in Table 2.
3. Observe the streptavidin–biotin binding kinetics via FRET by monitoring the acceptor fluorescence intensities at different time points as the detection probe volume is moved downstream in the channels.
4. Ideally, for kinetic measurements, mixing must be complete before acquisition of data. By integrating a winding part into microchannels, complete mixing can be achieved within 60 ms for this device under the applied flow conditions. Additionally, using a microfluidic system allows fine control over the concentration of reagents within the droplets flowing through the microfluidic device. Consequently, it is possible to closely monitor the extent of a reaction and the kinetics with excellent temporal resolution.

Table 1
Conditions for varying the concentration ratio with fixed streptavidin flow rates of 0.20 and 0.30 μl/min. Other parameters: $W_f = 0.5$, an oil flow rate of 1.5 μl/min and a total aqueous flow rate of 1.5 μl/min

F_{St}		F_D		F_B		[St] nM		[DNA] nM		CR	
0.20[a]	0.30[b]	0.20[a]	0.30[b]	0.20[a]	0.30[b]	0.20[a]	0.30[b]	0.20[a]	0.30[b]	0.20[a]	0.30[b]
0.20	0.30	0.07	0.10	1.23	1.10	14.67	22.00	4.40	6.67	0.30	0.30
0.20	0.30	0.13	0.20	1.17	1.00	14.67	22.00	8.95	13.33	0.61	0.61
0.20	0.30	0.20	0.30	1.10	0.90	14.67	22.00	13.35	20.00	0.91	0.91
0.20	0.30	0.27	0.40	1.03	0.80	14.67	22.00	17.75	26.67	1.21	1.21
0.20	0.30	0.33	0.50	0.97	0.70	14.67	22.00	22.29	33.33	1.52	1.52
0.20	0.30	0.40	0.60	0.90	0.60	14.67	22.00	26.69	40.00	1.82	1.82
0.20	0.30	0.47	0.70	0.83	0.50	14.67	22.00	31.09	46.67	2.21	2.21
0.20	0.30	0.53	0.80	0.77	0.40	14.67	22.00	35.49	53.33	2.42	2.42
0.20	0.30	0.60	0.90	0.70	0.30	14.67	22.00	40.04	60.00	2.73	2.73
0.20	0.30	0.67	1.00	0.63	0.20	14.67	22.00	44.44	66.67	3.03	3.03
0.20	0.30	0.73	1.10	0.57	0.10	14.67	22.00	48.84	73.33	3.33	3.33

F flow rate (μl/min), *St* streptavidin, *D* DNA, *B* binding buffer
[a]Fixed F_{st} at 0.2 μl/min
[b]Fixed F_{st} at 0.3 μl/min

Table 2
Experimental conditions for the FRET kinetic measurements

Flow rate (μl/min)			Concentration (nM)		
Streptavidin	**DNA**	**Buffer**	**Streptavidin**	**DNA**	**CR**
0.3	0.3	0.9	22.0	20.0	0.91
0.3	0.4	0.8	22.0	26.7	1.21
0.3	0.5	0.7	22.0	33.3	1.52
0.3	0.6	0.6	22.0	40.0	1.82
0.3	0.7	0.5	22.0	46.7	2.12
0.3	0.8	0.4	22.0	53.3	2.42
0.3	0.9	0.3	22.0	60.0	2.73

5. To investigate the binding kinetics, the FRET efficiency is plotted as a function of time. Generally, binding interaction can be described by a mono-exponential growth equation or a pseudo-first order integrated rate equation (8), i.e.

$$\mathrm{SL}_t = \mathrm{SL}_{\max}\left(1 - \mathrm{e}^{-k_{\mathrm{obs}}t}\right). \quad (1)$$

In this equation, SL_t defines the amount of product at time t, $\mathrm{SL}_{\max}$ is the maximum amount of product (the plateau value), and k_{obs} is an observed rate constant for each condition. k_{obs} is concentration dependent and is related to the binding rate constants (k_1) by the following expression (8):

$$k_{\mathrm{obs}} = k_1[L] + k_{-1}. \quad (2)$$

To extract k_1, the values of k_{obs} are plotted against the concentration of the biomolecule L. According to Eq. 2, this plot should yield a linear relationship, with a slope equal to k_1 and a y-intercept equal to k_{-1} (the dissociation rate constant).

6. In many cases, binding cannot be described by a mono-exponential expression (Eq. 1), but rather is more accurately described by a bi-exponential model (9, 10);

$$\mathrm{SL}_t = A_1[1 - \mathrm{e}^{(-k_{\mathrm{obs}1}t)}] + A_2[1 - \mathrm{e}^{(-k_{\mathrm{obs}2}t)}]. \quad (3)$$

Here, k_{obs1} and k_{obs2} are the observed binding rate constants of each process, with A_1 and A_2 representing the component amplitudes. This model, expressing a biphasic interaction, indicates an unusual bimolecular interaction that involves two distinguishable processes. This biphasic dissociation or association can occur due to heterogeneity of analytes or multiple binding sites causing different affinities (11).

7. For the streptavidin and biotin interaction, the reaction is categorized as having biphasic association and dissociation kinetics due to multiple biotin-binding sites on the streptavidin molecule (12, 13). Hence, to extract the binding rate constant of the streptavidin and biotin system, all kinetic data are analyzed using a bi-exponential model (Eq. 3), in which SL is the FRET efficiency in this case. Observed rate constants (k_{obs1} and k_{obs2}) for each condition are extracted from the fitting. To extract the binding rate constants (k_1 and k_2), the k_{obs1} and k_{obs2} values are separately plotted against the concentration of DNA using Eq. 2 and these plots show linear trends. The slopes obtained from these plots define k_1 and k_2. It should be noted that k_2, representing the binding kinetics of the second process, should be lower than k_1. This is likely to be due to steric hindrance from the first biotin binding.

4. Notes

1. All solutions should be filtered using 0.2 μm syringe filters before use.
2. Care must be taken during sample preparation and handling of fluorescence dyes to avoid excessive exposure to light.
3. For fluorescence detection of droplets, using a confocal detection system reduces background light due to PDMS, buffer and oil phase.
4. The laser beam is aligned directly to the central portion of the microfluidic channel width. Expanding the laser beam further to probe the entirety of the droplet can in principle decrease the spread of fluorescence intensities; however, this is not performed to ensure a sufficiently high signal-to-noise (S/N) ratio. In practice, decreasing the width of the channel should allow for a large percentage of the droplet to be probed while maintaining sufficiently high contrast in terms of S/N.
5. AF488 and AF647 are selected as a FRET donor and a FRET acceptor in order to minimize cross talk and avoid direct excitation of the acceptor as a result of spectral overlap. Due to the high sensitivity of the APD detectors, fluorescence from FRET can be detected accurately and precisely even at low FRET efficiencies. We have found that cross talk of this FRET pair was only ~1.2%. This investigation can be performed on a cover slip by measuring the red fluorescence signals of the

donor with and without the acceptor. The subtraction of these two values was used to calculate the cross talk percentage.

6. For droplet formation, flow rates faster that 0.3 μl/min (2 mm/s) for individual inlet are recommended to obtain uniform droplets and stable fluorescence signals. In addition, syringe sizes can affect the uniformity of droplet formation. Small syringes have less variation of the flow than large syringes.
7. To study binding assays via FRET, fixing the donor concentration and varying the acceptor concentration is a suitable approach. This is because laser intensity can be adjusted to maximize the fluorescence intensity from the acceptor prior to starting measurements. In contrast, when fixing acceptor-labeled DNA concentration, fluorescence intensities from the acceptor are relatively low when compared with the donor fluorescence. Technically, fluorescence intensity can be enhanced by adjusting laser intensity (assuming negligible photodegradation). However, in this case, the laser intensity could not be increased significantly due to extremely high photon counts originating from the donor when using a high streptavidin concentration.
8. Due to the streptavidin and biotin linkage, the donor and acceptor molecules are brought into close proximity and some energy from emission of AF488 can be transferred to excite AF647, resulting in emission of red fluorescence at wavelengths between 650 and 710 nm. In effect, two fluorescence signals, from AF488 and AF647, are detected simultaneously using separate avalanche photodiode (APD) detectors (green and red channels) and recorded for periods of 60 s using a 50 μs resolution.

Acknowledgments

M. Srisa-Art would like to thank Prof Andrew J. deMello and Dr Joshua B. Edel for their supervision, advice, and encouragement and also acknowledge the Royal Thai Government for provision of a research scholarship. This work was supported by the EPSRC and the RCUK Basic Technology Programme.

References

1. Piran U, Riordan WJ (1990) Dissociation rate-constant of the biotin-streptavidin complex. J Immunol Methods 133:141–143
2. Bayer EA, Wilchek M (1990) Biotin-binding proteins—overview and prospects. Methods Enzymol 184:49–51

3. Chilkoti A, Stayton PS (1995) Molecular-origins of the slow streptavidin-biotin dissociation kinetics. J Am Chem Soc 117:10622–10628
4. Livnah O et al (1993) 3-Dimensional structures of avidin and the avidin-biotin complex. Proc Natl Acad Sci U S A 90:5076–5080
5. Griffiths AD, Tawfik DS (2006) Miniaturising the laboratory in emulsion droplets. Trends Biotechnol 24:395–402
6. Kelly BT et al (2007) Miniaturizing chemistry and biology in microdroplets. Chem Commun 18:1773–1788
7. Taly V, Kelly BT, Griffiths AD (2007) Droplets as microreactors for high-throughput biology. Chembiochem 8:263–272
8. Goodrich JA, Kugel JF (2007) Binding and kinetics for molecular biologists. Cold Spring Harbor, New York
9. Oshannessy DJ et al (1993) Determination of rate and equilibrium binding constants for macromolecular interactions using surface-plasmon resonance—use of nonlinear least-squares analysis-methods. Anal Biochem 212: 457–468
10. Edwards PR, Leatherbarrow RJ (1997) Determination of association rate constants by an optical biosensor using initial rate analysis. Anal Biochem 246:1–6
11. Edwards PR et al (1995) Kinetics of protein-protein interactions at the surface of an optical biosensor. Anal Biochem 231: 210–217
12. Gonzalez M et al (1997) Interaction of biotin with streptavidin—thermostability and conformational changes upon binding. J Biol Chem 272:11288–11294
13. Buranda T et al (1999) Ligand receptor dynamics at streptavidin-coated particle surfaces: a flow cytometric and spectrofluorimetric study. J Phys Chem B 103:3399–3410

Chapter 17

Surface Treatments for Microfluidic Biocompatibility

N.J. Shirtcliffe, R. Toon, and P. Roach

Abstract

Microfluidic systems allow small volumes of liquids to be manipulated, either by being passed through channels or moved around as liquid droplets. Such systems have been developed to separate, purify, analyze, and deliver molecules to reaction zones. Although volumes are small, reaction rates, catalysis, mixing, and heat transfer can be high, enabling the accurate sensing of tiny quantities of agents and the synthesis of novel products. The incorporation of multiple components, such as pumps, valves, mixers, and heaters, onto a single microfluidic platform has brought about the field of lab-on-a-chip devices or micro total analysis systems (μTAS). Although used in the research laboratory for numerous years, few of these devices have made it into the commercial market, due to their complexity of fabrication and limited choice of material. As the dimensions of these systems become smaller, interfacial interactions begin to dominate in terms of device performance. Appropriate selection of bulk materials, or the application of surface coatings, can allow control over surface properties, such as the adsorption of (bio)molecules. Here we review current microfluidic technology in terms of biocompatibility issues, examining the use of modification strategies to improve device longevity and performance.

Key words: Lab-on-a-chip, Micro total analysis systems, Biocompatibility, Antifouling surfaces, Protein absorption, Cell adherence

1. Overview of Lab on a Chip Devices

The advancement of microfabrication technology has driven the growth of microfluidic systems, adding to the arsenal of micro total analysis systems (μTAS) used in analytical chemistry, genomics, tissue engineering, and related disciplines (1). Microfabricated cell sorters (2), cell-based assays (3), blood type analysis (4), purification and analysis, and DNA (5) and protein (6, 7) separations have benefited from such systems. Microfluidic devices are attractive for biological applications due to their versatility, including the ability to control the physical and chemical characteristics at the micro- (and submicro-) meter scale. Miniaturization of bioassays is advantageous

Gareth Jenkins and Colin D. Mansfield (eds.), *Microfluidic Diagnostics: Methods and Protocols*, Methods in Molecular Biology, vol. 949, DOI 10.1007/978-1-62703-134-9_17,

with regard to reduction in reagent consumption and therefore cost, and in the ability to parallel-process samples. This is particularly important when considering high-throughput screening strategies, such as drug discovery or medical diagnostics.

Lab-on-a-chip systems offer considerable advantages over current medical diagnostic methodologies, such as the ability to incorporate separation and sorting of analytes on chip immediately prior to analysis. Patients can benefit from these devices due to their ease of use; for example, in making multiple analyte screening possible, either in a clinical environment or in the development of home test kits. Healthcare services would benefit as a result of lower costs associated with simultaneous sampling and lower staffing costs per sample due to ease of use. High-throughput screening, made possible by such assay technology, may also give rise to earlier diagnosis of illness, impacting on treatment costs and also leading to better patient care. Development of this technology is, therefore, both consumer and industry driven, although there are obstacles to be overcome before these devices emerge from the research labs into common use within healthcare (8).

The commercialization of any medical technology is problematic, due to the nature of the products, but the further development of microfluidic systems is also hindered by specific difficulties, including the necessity for interaction between the developer and end-users throughout the development process of the device (e.g., lab-on-a-chip and diagnostic assay), a lack of standardization of development and fabrication technologies, and a lack of standardization of the materials used for assays.

Advances in diagnostic capability presented by microfluidic systems are well documented and have not gone unnoticed in the medical arena. Biopharma companies have been investigating the development of microfluidic assays for some years, not least of which is one of the basic, but well-known, lateral flow "dip-stick" tests for early term pregnancy. However, more complex systems require much more advanced input in terms of the selection and development of the assay and materials used for the "lab-on-a-chip" design. Lab-on-a-chip devices have significantly higher surface area-to-volume ratios compared with their macroscopic counterparts, resulting from the miniaturization of flow channels. This gives rise to increased difficulties of surface interaction, both of the liquid and of any analytes of interest. These devices are therefore, heavily reliant on surface characteristics, such as wettability, chemical functionality, charge, topography, and stiffness.

Biocompatibility issues within microfluidic devices can be of the utmost importance when dealing with protein-, platelet-, or cell-based samples, such as blood or cells being cultured within the confines of a device. A variety of surface coatings have been investigated in order to better control the environment presented within microfluidic devices. Oxygen plasma treatment to oxidize the surface, silylation, polymer grafting, plasma polymerized coatings, and preadsorption of proteins are just some of the methods investigated to reduce device variability. Here, we discuss various approaches to obtain improved surface characteristics for biomicrofluidic applications, with particular attention towards biomedical diagnostic systems.

2. Fabrication of Microfluidic Devices

Microfluidic devices range in complexity from those based on lateral flow with the sample being wicked across the surface using a paper- or silica-based material, to those with multiple input/output flows for more complex applications. Both require careful consideration of the material used to account for any interaction with the sample. Most commonly, glass and silicon formed the base for early μTAS (9) although advances in molding methodologies have led to the use of elastomers for multifaceted devices, with poly(dimethylsiloxane) (PDMS) now widely used.

Complex microfluidic systems can be difficult to fabricate, due to the size of the channels and reproducibility of the devices being produced. A common method is to use photolithography to pattern a metallized masking layer onto a glass substrate, which is then etched to form the channels. Such lithographic processing can be extremely reproducible, as the pattern is transferred from a master template, the depth of channels being controlled by the etching parameters. However, lithographic equipment can often be expensive and difficult to use. If, for example, glass is used as a base, hydrofluoric acid is required as the etchant making facilities and safety procedures expensive. Simple milling of materials has been used for the fabrication of microchannels, which is a relatively inexpensive method although largely limits the substrate material to polymers. The resolution of the patterned channels is restricted by the tool size and so miniaturization of the fluidic components is difficult (10, 11). Photolithography offers the ability to produce devices with (sub)micron features, utilizing light-sensitive polymers that cross-link or degrade on the action of UV light. One such epoxy-based polymer, commonly used for microfabrication, is SU-8, providing biocompatibility (12) mechanical stability and chemical inertness (13) against most biological substances. SU-8 facilitates low-cost batch processing and is suitable to form smooth, high aspect ratio channels in a range of dimensions, allowing micro- and nano-fluidics. The possibility of integrating functional components (e.g., electrodes, waveguides) in intermediate layers, and low temperature bonding capability, make SU-8 a suitable material for the manufacture of microfluidic devices.

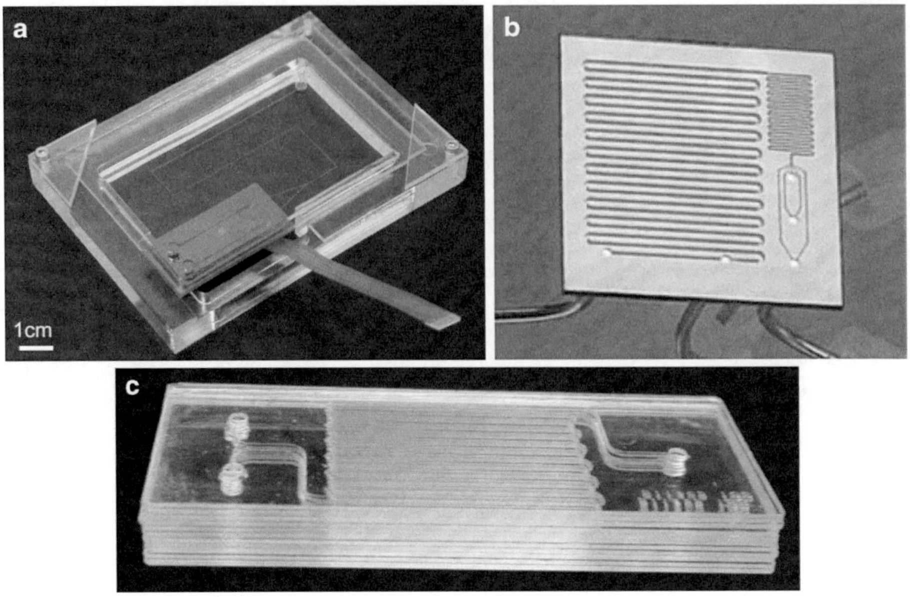

Fig. 1. Microfluidic devices fabricated from (**a**) PDMS (18) (**b**) silicon (20) and (**c**) glass (19). Reproduced with permission from Royal Society of Chemistry, 2002, 2003, and 2005.

In many ways, glass is an optimal material for microfluidic devices being extremely chemically robust, corrosion resistant, and non-swelling and is compatible with a wide range of chemicals, including organic solvents. Glass devices are more resistant to pressure and heat than polymer devices; however, as previously mentioned, they are difficult and costly to fabricate. Nonetheless, for the advantages outlined above they are likely to remain important for the foreseeable future. Polymeric devices are, however, much more rapidly produced, being cast over a predetermined inverted-master. Here the patterning, width, and depth of the channels are all copied from the master in a single step, making the process very simple and devices extremely reproducible. PDMS devices can be easily, rapidly, and inexpensively fabricated with superb resolution (Fig. 1a). The only major cost is the production of the original master, which is usually performed using standard lithographic processing (10, 11). PDMS can be used to either produce a complete device or, more simply, as open channels to be sealed against another material, often silicon or glass. PDMS is considered to have many favorable properties offering optical transparency, good flexibility, and elasticity (useful in terms of valve incorporation) and allows some of the highest copying resolutions. Whilst glass and silicon are inherently hydrophilic, PDMS is hydrophobic, which can steer the adsorption and accumulation of proteinaceous material onto the surface affecting the ability of the device to function properly, especially if the device is to be used for blood or cell sampling. PDMS also suffers from numerous problems hindering its use in commercial device fabrication. Changes in mechanical properties can be observed as a result of aging or variability in curing parameters (14), and flow channels have been found to be unstable at high pressures, leading to inconsistent delivery of reagent volumes and flow rates. Cured PDMS also remains permeable to liquids and gases, which limits control and can interfere with reactions in the channels. Small molecules can diffuse into PDMS walls, fouling channel surfaces and altering device behavior (15). Organic solvents, such as toluene and chloroform, are necessary for many applications including the formation of vesicles and the syntheses of drugs. However, these chemicals swell PDMS, collapsing microfluidic channels and significantly degrading device performance (16, 17). The poor chemical compatibility of PDMS is, therefore, a major challenge that limits its applicability to lab-on-a-chip microfluidics (17).

Glass offers robustness, although microfluidic devices fabricated from glasses or ceramics can be difficult to seal and are often bonded using a form of welding. Adhesives can be used, but are prone to dissolution or swelling in organic solvents and have limited pressure resistance, limiting their applicability. The design of microfluidic devices, therefore, centers on the materials chosen for their production, with appropriate bulk material and coatings/surface modification impacting on device functionality and practicality. Whereas silica glass, silicon, and PDMS have been predominantly used for μTAS systems to date (Fig. 1) (18–20), other bulk materials are being investigated to arm the "designer" with a range of tools from which advanced devices may be fabricated. One of the main areas of interest in this field, however, is the investigation of surface modification strategies.

3. General Considerations for Surface Modification

Despite the variety of potential applications, a major impediment to the evolution of microfluidic devices into mainstream technological use has been the limited materials available, restricting both the nature of structures that can be realized and their applications. Modification of a material surface allows the chemical and mechanical properties of the bulk to be unchanged whilst other properties can be presented at the surface. This can be achieved either by altering the bulk material itself within a small depth below the surface, or by the addition of a coating covering the underlying bulk material. For microfluidic applications, surface modification to alter the wettability of the material can be used to minimize fluid pressures. Water is a common solvent used in microfluidic applications, particularly those handling biological samples and therefore, hydrophilic surfaces are mostly required. A problem arising in these cases is the ability of the surfaces to become fouled by species adsorbing from the sample, such as proteins or blood platelets. For lab-on-a-chip applications involving live cells the surface properties are again of major importance. In both these situations, the term "biocompatibility" can be used to describe desired surface properties in relation to an interaction with biological species. In the case of protein fouling the adsorption of species can cause a blockage within the microchannels and therefore restrict the flow and/or increase fluid pressure. Protein adsorption is however (mostly) a prerequisite for cell adhesion; hence the term biocompatibility often being used here is to describe the ability of the surface to allow cells to adhere.

Wettability is a very general term used to describe the interaction of a liquid on a solid, being underpinned by many physicochemical properties. Many of these characteristics can be influenced by surface modification including surface charge, chemical functionality, chemical composition, degradability, topography, and electrical/thermal conductivity. A very common method of bio-functionalization is to coat a surface with amine-containing units, either by the use of polymers, wherein the functionality is randomly distributed, or via ordered self-assembled monolayer formation. This presents a suitable attachment for a variety of biological molecules. Chen et al. used APTES (3-aminopropyltriethylsilane) to form self-assembled monolayers (SAMs) of alkane chains, presenting amines away from the surface (21). Amines are commonly used as a biologically relevant functionality within microchannels, which can be further modified by attachment of antibodies through an amide bond.

Surfaces for the selective recognition of proteins may also be formed by making use of biotin–avidin interactions (22). Linking chains attaching the head group to the surface can also infer desirable characteristics, such as incorporating ethylene glycol units to prevent nonspecific protein binding (Fig. 2). Diluent molecules can also be used to space out specific functionalities on the surface, reducing the density of active head groups presented. These types of surface in lab-on-a-chip devices allow the direction of very specific bio-affinity between the surface antibody and an antigen being delivered in the solution, forming the basis of many sensor or assay recognition elements. Other functionalities commonly presented using SAMs include, carboxyl, hydroxyl and alkanes, allowing a range of surface wettabilities and charges to be formed. SAM coatings are attractive for microfabricated structures because of their facile formation and nanometre scale thickness. However, the permeation of ions and other small molecules into the substrate is difficult to inhibit because of inherent defects and the chemical instabilities associated with monolayer films. This can lead to material fouling and degradation as well as a limited shelf life (23, 24).

Polymer coatings are normally at least an order of magnitude thicker than SAMs and can inhibit the permeation of smaller species.

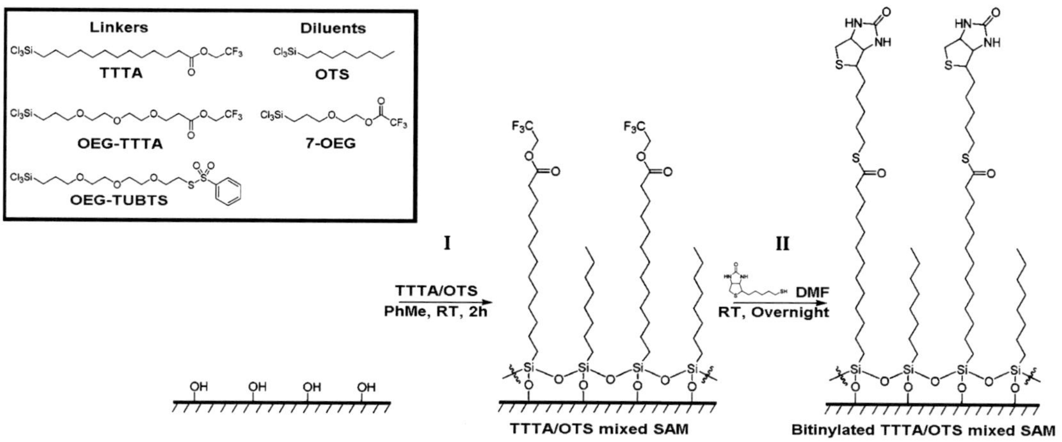

Fig. 2. Formation of a mixed SAM (*step I*) and the subsequent site-specific covalent immobilization of biotinthiol (*step II*). Adapted from Thompson et al. (22). Reproduced with permission from The Royal Society of Chemistry 2010.

The application of polymers into microfluidic structures as thin films and the resultant biocompatibility of these surfaces are areas of interest that are still being addressed (25). PDMS has many desirable qualities, in terms of microfluidic device fabrication and also offers excellent biocompatibility. PDMS normally forms the basis of the microfluidic device by casting around a suitable template and using a cross-linker to harden the material. PDMS can also be spin-coated as an additional coating onto a substrate to form a conformal, pin-hole free layer. Biopolymers such as collagen, chitosan, and silk are promising candidates for coating microfluidic devices, because they can be doped with bioreactive reagents that would otherwise not survive the harsh processing environments used in other microfabrication processes (26). These polymers are also inherently biocompatible, allowing biological interaction without any adverse effects as they are derived from biological systems. However, much work in terms of material characterization and quality control is required in order to be able to use these in large quantities. Therefore, these materials have been mostly limited to research, rather than commercial μTAS.

Forming a coating within preformed channels is often difficult and depends on the method used to seal the device. Lee et al. demonstrated excellent separation efficiencies of amino acids (Fig. 3), proteins and peptides in devices prepared via in-channel atom transfer radical polymerization (ATRP) (27). Here the channels were formed from a poly(glycidyl methacrylate)-co-(methyl methacrylate) (PGMAMMA) allowing an atom transfer radical polymerization (ATRP) initiator to be attached to the surface and subsequent poly(ethylene glycol) (PEG) modification.

SU-8 is a commonly utilized material for microfluidic fabrication, although it is often considered to be difficult to chemically

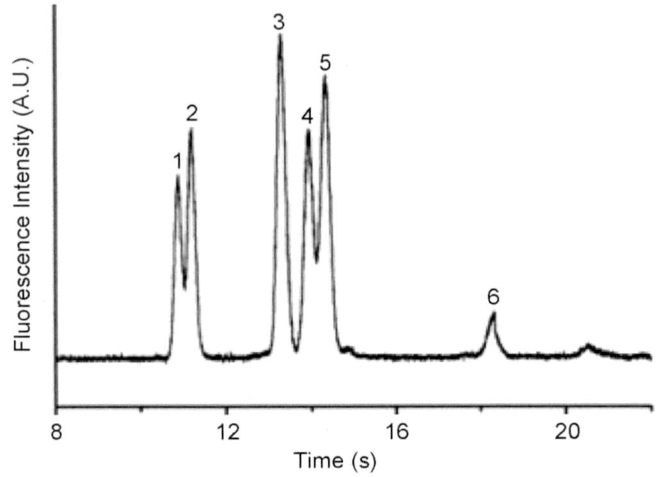

Fig. 3. Capillary electrophoresis separation of FITC-labeled amino acids in a PGMAMMA modified microTAS. Peak identifications: (1) aspartic acid, (2) glutamic acid, (3) glycine, (4) asparagine, (5) phenylalanine, and (6) FITC (27). Reproduced with permission from Wiley-VCH Verlag GmbH & Co, 2008.

modify. It has been shown that, depending on the manufacturing process, excess epoxy groups on the surface of SU8 can be changed to hydroxyl groups by treatment with strong acids (28). These hydroxyl groups can then be used for covalent binding of suitable molecules (e.g., 3-aminopropyltrimethoxysilane), allowing further functionalization with biomolecules, as described earlier.

PDMS devices are very difficult to functionalize, which significantly limits many applications (29). Attempts have been made to increase the utility of PDMS by modifying its surface properties, although some strategies lead only to temporary surface modifications (30, 31). For example, the diffusion of small molecules can be reduced by infusing PDMS surfaces with metal-oxide precursors (15) and swelling by organic solvents can be retarded by coating PDMS with poly(urethaneacrylate) (32). PDMS can be grafted onto by oxidizing its surface to form silica (usually using plasma process) and then using silane coupling agents. The major drawback of this is that the soft polymer tends to re-order over time and surface groups can migrate below the surface, depending on the storage conditions.

In many biochemical applications, proteins or enzymes are used as reactants or catalysts. In microfluidic environments, these species can adsorb onto the channel walls reducing reaction efficiency. One strategy to avoid this is to pre-coat the surface with proteins, blocking binding sites in order to hinder further adsorption. Modification to present an anti-fouling surface is another method used and various coatings having been investigated with varying degrees of success. In digital microfluidics, where reagents are held in separate droplets resting on a surface and are brought together to initiate reactions, the interface must be functionalized to control wettability. Droplets must not spread much on the surface in order to ensure that the drops stay mobile on the surface. In other applications, the channel properties must be controlled spatially so that different regions of the device have different properties. This is particularly important for μTAS where sequential reactions or separations take place.

Plasma treatment to oxidize or deposit thin polymer films onto the surface can be used to provide varying modifications. Typically, an electric field will be used to generate ions and excited neutral species in a gas mixture. The ions have relatively high energy so can etch or react with otherwise un-reactive surfaces. In many cases air is used as the plasma gas, grafting oxygen and nitrogen into polymer surfaces and oxidatively etching it. The end effect of air plasma treatment is a roughening of the surface and an increase in surface energy. These combined effects are very effective at generating biologically compatible surfaces and have been used for modifying polymer surfaces for some time (33). Additionally, plasma treatment is also suitable for pretreating polymers to accept other coatings. This method is often used in industry to produce tissue culture plastics. The polystyrene is treated in air plasma, oxidizing the surface and rendering it more wettable. Further treatment with adhesive molecules such as collagen, produce surfaces very compatible for cell culture.

For microfluidic devices, open channel structures are often treated in a plasma chamber, prior to device assembly; the gas feed and/or monomer is chosen depending on the type of surface required. Oxygen plasma (or UV irradiation in air) causes oxidation of surface species, giving rise to hydrophilic surfaces and it is often used for removing organic contaminants. PDMS is relatively hydrophobic and can be rendered hydrophilic by exposure to oxygen plasma or superhydrophobic by roughening using a CO_2 laser (31, 34). Atmospheric pressure plasmas formed by dielectric barrier discharge across a sealed channel can be used to modify the inner surface of devices after they are closed. This is a rapidly expanding area described in some depth in a review by Thomas et al. (35) Many methods of device bonding for enclosing microchannels, require the two surfaces of the device to be incredibly clean and plasma treatment is the most common method for achieving this. Plasmas can also be used to modify surfaces with specific functionalities, by incorporating additional species into the gas phase. Being excited in the plasma, these species can react with each other forming the monomer units of a deposited polymer. For instance, plasma polymerized allylamine is often used to present amines on surfaces (36).

Chemical modification can typically be carried out either prior to or after complete assembly of a complex microfluidic device.

In the latter case, channels can be filled with the reactant species allowing modification of all the channel walls. For glass microfluidics silanes are often the modification of choice, due to their versatility and ease of use. Silanes can be used to form SAMs, covalently attaching to the substrate via a siloxane bond, whilst presenting their terminal chemistry away from the surface. Although very reproducible and cost effective, this method has limitations based on the construction of the device.

Silanes will only form SAMs on substrates with sufficient free hydroxyl groups, such as silica or some ceramics, so microfluidic base substrates must be of a suitable material. SAMs can also be formed using thiols, which assemble on metallized surfaces. Gold is commonly used as the metal layer, deposited onto substrates by thermal evaporation or sputtering. Thiols, similar to silanes, have limited applications due to the use of solvents and the metallized portion of the surface becoming opaque due to the metal layer. Gold layers can, however, be coated onto any material, whereas silanes are limited to hydroxyl-functionalized surfaces, such as those presented by glass. This also allows regions to be locally treated by patterning the metal on the microfluidic device. In this way, surface functionality can be used to alter processes in the fluid through selective adsorption, entrapment, release of pre-attached reactants, etc., as the flow progresses along the channel.

In order to generate advanced, functional lab-on-a-chip systems for biological and chemical applications, devices must be fully integrated. Such complex devices are usually composed of many interconnected components with coupled supporting structures. One approach, used to reduce the complexity of these devices, is to take advantage of the material properties of hydrogels. Hydrogels translate stimuli from the aqueous environment into mechanical responses (swelling and contraction). They can act as sensors and actuators at the same time and are driven by molecular processes. The use of hydrogels for biomimetic microfluidic applications has led to self-regulating valves, micro-lens arrays, and drug release systems. However, a challenge remains due to the slow response time to the environmental stimuli, which is in the order of hours. Matrigel™ is one such hydrogel commonly being used to support cell growth in vitro, although its brittle nature does not allow it to be used as a device material and therefore it can only be used as a coating or gel within a microfluidic chamber (37). Biomaterials that are used to provide the structure of a μTAS must fulfill a range of requirements. The material should allow microfabrication of channels and, where necessary, sealing the channels with different materials, they should allow good fluid flow without degradation or contamination, and permit diffusion of large and small molecules (38). Calcium alginate hydrogel (4% w/v) is important because it fulfills the above criteria. It also has low cytotoxicity, biodegradability, the ability to be molded under mild conditions and mechanical stability at low solid fractions (39).

4. Antifouling Surfaces

The accumulation of biomatter on surfaces, or "biofouling," is a major problem in many applications, not least in bioreactor or medical technologies. Flow systems are particularly vulnerable due to surface contamination causing heterogeneities in the flow profile. This is important in microfluidics where the surface area-to-volume ratio is very high. One commercially important example of this is a biocavity laser instrument, which is used to rapidly screen cells for disease. Based on a microfluidic structure, the operation of this device is severely hindered by the adhesion of cells or accumulation of protein within and surrounding the microchannels (40). Various coating strategies have been developed over the years to attenuate protein and cell adhesion, each having their own particular strengths and weaknesses. Protein adsorption occurs almost immediately on contact between a material and a protein-containing solution. This is driven by removal of water from hydrophobic protein regions or localized electrostatic interaction. This is even more thermodynamically favorable when the surface is also hydrophobic. The interaction of platelets, cells, and single cell organisms with surfaces is mediated by this pre-adsorbed proteinacious layer (41). Most cell types must adhere to surfaces in order to proliferate, although some cell adhesion is driven by other processes, such as foreign body response where tissue encapsulation ultimately results in the foreign materials being ejected from the biological system. The development of advanced biomaterials and surface coatings is a major part of regenerative medicine and tissue engineering, with the understanding and control of biological-material surface interactions being central to the progression of these fields.

Microfluidic silica structures can be modified using silanes to prevent biofouling, as demonstrated by Sasaki et al., using alkane and poly(ethylene oxide) SAMs to compare hydrophobic and hydrophilic anti-fouling abilities (42). Surface adhesion studies on neuronal cells showed that hydrophilic surfaces were more resistant to protein adsorption and therefore, cell adhesion, with hydrophobic surfaces promoting fouling. Hydrophobic interaction is understood to drive protein adsorption, although this study shows time-dependent behavior exhibited by the surface, with hydrophilic surfaces becoming more able to sustain cell attachment over time, compared to hydrophobic surfaces. This may result from orientational, conformational, and/or compositional changes of the adsorbed protein layer on the surface (43).

Monolayer silane coatings are thermally and chemically stable (44), forming via reaction with surface hydroxyl groups and also to neighboring attached silanes (45). Silanes are often used in surface modification for this reason, as well as for their commercial availability and variety that allows further chemical linkages to them. PEG is well known for its antifouling capabilities. This is thought to be due, in part, to its hydrophilicity and also because of the very high water content residing between PEG chains, which allows reversible binding of protein molecules (46). There have been several attempts to modify surfaces, including silicon and PDMS, with PEG. These modifications include passive adsorption (47) and chemical vapor deposition (CVD) (48, 49). These efforts have proved problematic, due to their fabrication complexity or poor long-term stability; for example, some PEG coatings may eventually be removed from microchannel surfaces by rinsing (50). PEG-SAMs can be formed via reaction with surface hydroxyl groups, which are abundant on silica materials but less so on polymeric surfaces (42, 51). For this reason, pretreatment with strong acids, oxygen plasma, or ozone is required prior to SAM attachment. Sui et al. used one such method wherein piranha etch solution (H_2O_2/H_2SO_4) was flushed through a microfluidic channel to oxidize the PDMS, and PEG-silane then used to form an antifouling surface (30). Chemical grafting of PEG onto silicon or PDMS has also been attempted, which provided a better quality final coating, having superior mechanical and chemical robustness (5). However, this method required complex processing steps in order to achieve high-quality modification, and the organic solvents required could also swell the PDMS device. Kim et al. have recently fabricated a microchannel comprised entirely of PEG by cross-linking it through exposure to UV radiation (52). Their device was successful in resisting protein fouling and subsequent cell adhesion (Fig. 4),

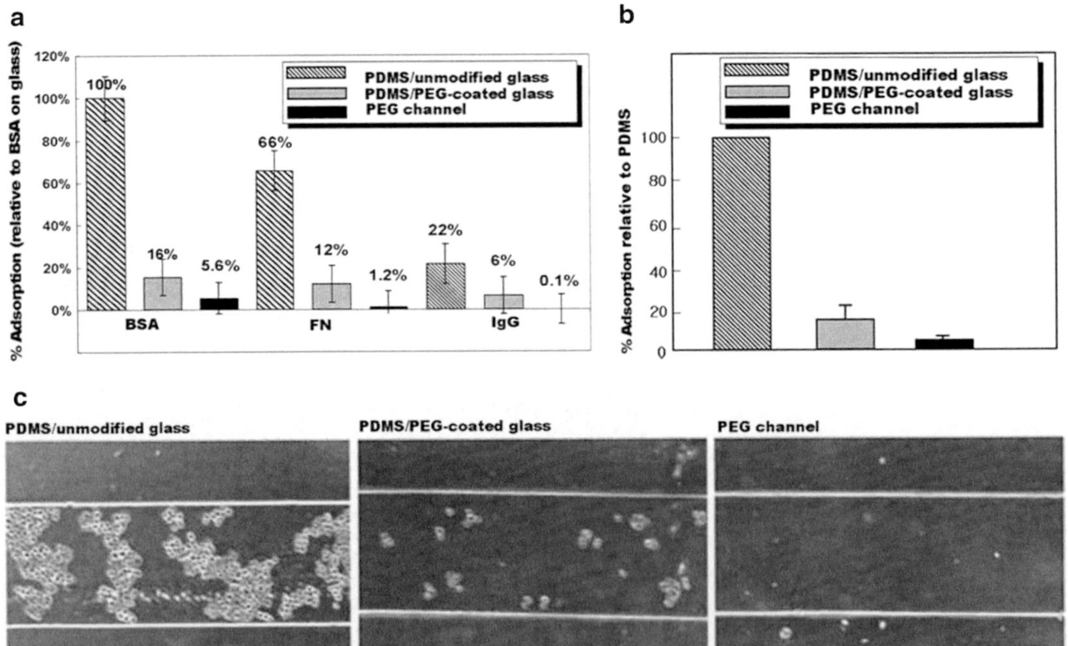

Fig. 4. Comparison of (**a**) BSA, FN, and IgG adsorption and (**b**) NIH-3T3 fibroblasts adhesion onto PDMS/glass, PDMS/PEG-coated glass and PEG channels/unmodified glass microchannels. (**c**) Optical micrographs of NIH-3T3 fibroblasts in microchannels (52). Reproduced with permission from The Royal Society of Chemistry, 2006.

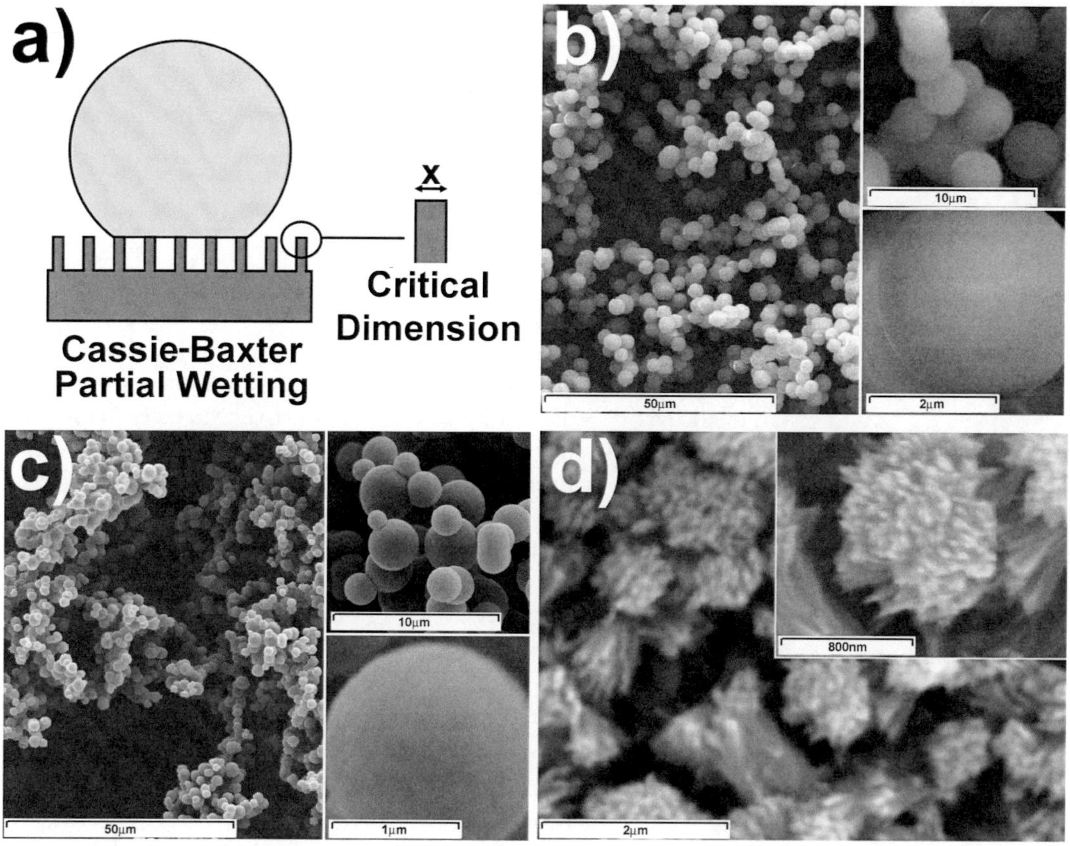

Fig. 5. (**a**) Schematic showing Cassie-Baxter superhydrophobicity and the critical dimension of feature size. Electron micrographs of (**b**) and (**c**) show silica sol–gel materials and (**d**) copper oxide nano-pillars (54). Reproduced with permission from The Royal Society of Chemistry, 2008.

with approximately 5% of protein and 2% of cell adhesion, as compared to unmodified PDMS microchannels.

Hydrophobic surfaces strongly promote protein adsorption. However, by roughening such a hydrophobic material, a highly non-wettable or "superhydrophobic" surface can be formed. These coatings have been shown to have excellent anti-fouling behavior, under flow conditions. Superhydrophobic surfaces are extremely rough and hydrophobic; water cannot penetrate into the roughness and so becomes suspended, bridging the topographic peaks in what is known as a Cassie-Baxter wetting state (53). The liquid-volume contact area is reduced compared to that of a flat surface. Therefore, the surface area available for protein adsorption is also lower. Shirtcliffe et al. coated the base of a microfluidic channel with different types of superhydrophobic surface to investigate the effects of surface chemistry and roughness feature size on protein adsorption (Fig. 5) (54). Micro-roughness of different sizes was produced via sol–gel casting to form porous organo-silica coatings and copper nano-hairs were used to produce much smaller features.

All coatings were chemically modified to present either hydrocarbon or more hydrophobic fluorocarbon surfaces. Smaller fluorocarbon features were found best at resisting protein adsorption under static conditions. Furthermore, under flow conditions, almost no protein was found remaining on these surfaces.

5. Protein Adsorption and Cell Adhesion

Many biotechnological devices could benefit if effective control over molecular and cellular interactions at material surfaces could be achieved. Major goals include the control over surface adsorption of biomolecules, such as DNA and proteins, and also the control over subsequent interfacial interactions between materials and cellular entities, such as mammalian cells and bacteria (55, 56). Proteins adsorbed onto synthetic material surfaces are of particular interest as cell attachment to them activates various cell-signaling pathways and cell functions. Changes in morphology, proliferation, and phenotypic changes have been shown to be linked to the concentration, composition, and conformation of protein molecules in an adsorbed protein layer at the material interface (57, 58). Protein molecules adsorb reversibly to surfaces, establishing equilibrium between the adsorbed and desorbed state. Once adsorbed however, structural changes can occur within the protein maximizing interaction with the surface, and therefore desorption is usually less favored (58). Adsorption of proteins at interfaces is undesirable in many cases, as it can lead to biofouling (as discussed earlier), thrombosis, reduced dynamic range of biosensors and other adverse events (59). Cell-adhesive proteins are, however, essential for cell and tissue colonization of material surfaces.

Over the past decade there has been considerable progress in the design of materials and interfaces that can be described as having both adsorbent and adsorption resistant properties. This has been driven by a number of applications where it may be desirable to attract a protein, cell, or DNA layer temporarily, followed by release from the surface at a specific time (60, 61). Such control over bio-interfacial interactions requires material surfaces that can be switched between protein adsorbing and protein-repellent states. Advances in the control of protein adsorption and desorption at synthetic interfaces are expected to benefit many applications including diagnostic microarrays, biomaterials and regenerative medicine, biosensors (59), chromatography (62), and drug delivery. A wide range of stimuli that may be applied to switchable systems are available. Of these triggers, light- (63), temperature- (64), pH- (65), ionic strength-, and electric field- (66) based methods are the most common. Other triggers include solvents, pressure, ultrasound (67), and magnetic fields (68).

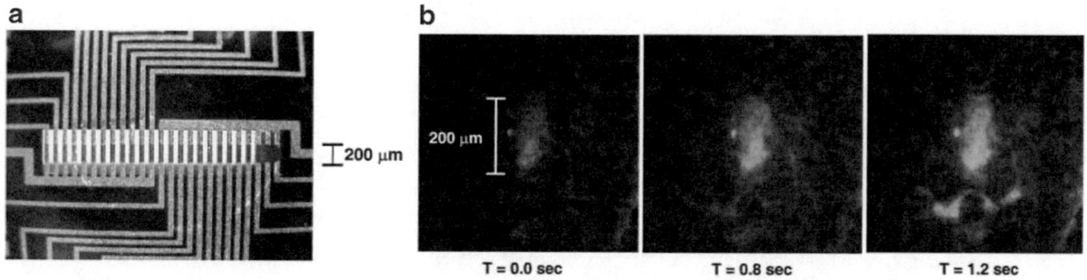

Fig. 6. (**a**) Photograph of microfluidic device with a central 200 µm-wide hot plate zone (**b**) Fluorescein-labeled myoglobin interacting with a single heated line. Image obtained on heating a line above the PNIPAM transition temperature after exposure to a 0.5 mg/mL myoglobin solution and rinse (*left*). Images were obtained 0.8 and 1.2 s after the heating stopped (*center* and *right*) (62). Reproduced with permission from American Association for the Advancement of Science (AAAS), 2003.

Thermoresponsive surfaces have been investigated for the growth and controlled release of cells. An example of this is the use of thermo-responsive poly(*n*-isopropylacrylamide) (PNIPAM), which alters from being hydrophilic to hydrophobic on collapse of its structure. PNIPAM is soluble in water at room temperature, presenting a hydrophilic surface onto which protein molecules can adsorb. Above approximately 35°C the polymer becomes insoluble and hydrophobic, contracting and repelling water from within its network. Bunker et al. demonstrated the use of this switchable characteristic for the controlled adsorption and release of proteins from solution in a microfluidic platform (Fig. 6) (62). The collapsed state of the polymer provides similar surface properties as tissue culture plastic, allowing adsorption of serum proteins and subsequent cell attachment and spreading required for the proliferation of cells. A reduction in the temperature causes the PNIPAM to change configuration, becoming more hydrophillic, absorbing water and releasing living cells (69, 70). The incorporation of such coatings into microfluidic systems will allow for cell culture and passage without the need for lengthy cell culture techniques, which use enzymes to release cells from the surface.

Switchable surfaces used for cellular release are restricted to a range of conditions that do not harm or cause adverse effects to the cells in culture. However, pH, ionic strength, and temperature may be less restrictive for other applications, such as bio-diagnostics where cells do not necessarily need to be viable. The pH of an aqueous solution affects the properties of many surfaces, due to protonation/deprotonation of ionizable groups and the consequent interactions between these charged surface groups, ions, and biomolecules (71, 72). The main difficulty faced by the use of solution-based switching mechanisms is that the solution environment must be changed in order to elicit a response. As the volumes used in microfluidic systems are low and channels are sealed with inlet/outlets along their length, this commonly means that the solvent used must be exchanged by flushing with another solvent. This makes it a more complicated approach, compared to remote switching modalities such as light (and heat), where changes can be imposed rapidly and even locally. The use of light to study and control chemical and biological systems is a well-established area of research. Photosensitive molecules, such as photo-cleavable protecting groups, have been applied extensively to conventional and solid phase synthesis. Although less common in biochip and biomedical devices, photo-isomerization has received some attention. In one study, cell adhesive RGD peptides were attached to the azobenzene derivative and their availability to sustain cell culture was controlled via photo-switching (73).

6. On-Chip Electroosmosis

Capillary electrophoresis (CE) is a powerful technique for the separation of similar ions or even neutral species in solution and so is well suited for on-chip separation of analytes, such as blood components, RNA, DNA, proteins, etc. The separation efficiency is better than that of high performance liquid chromatography (HPLC) because the mechanism of movement reduces band broadening, which is often the ultimate limitation of chromatographic methods. As the flow driver is an electric field this technique is ideal for adaptation to microfluidics, being useful in terms of on-chip rapid separation of species. Commercial CE systems using microfluidic chips are common and are the main route to market of this technology because fluid handling is simplified in μTAS and the path length and detection windows can be prepared reproducibly.

The method involves the analyte being injected into a capillary filled with buffer solution and a high D.C. voltage being applied along it (Fig. 7). Ions are affected by the field gradient and tend to

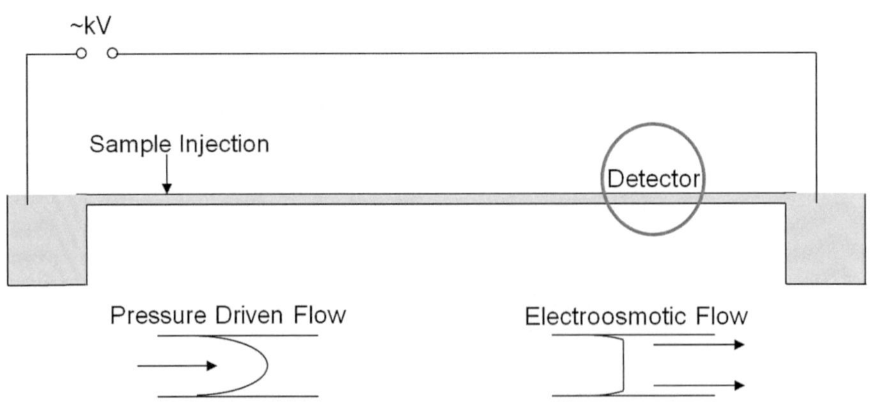

Fig. 7. A schematic of capillary electroosmosis.

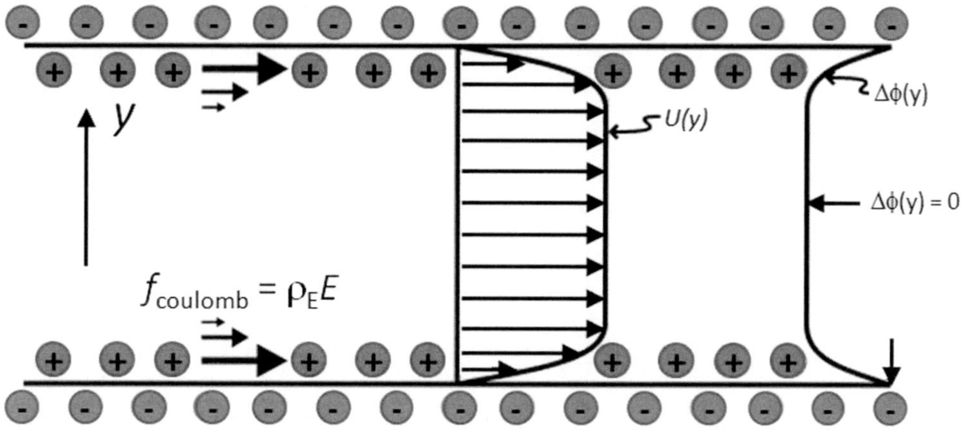

Fig. 8. Schematic of electric double layer and electroosmotic flow conditions. The electrical potential imposed on the liquid is denoted as ϕ.

move towards one of the electrodes. The strength of this effect depends upon the size of the ions and their charge, with separation depending on the ion flow rate. Neutral species are obviously not affected directly by the field. The whole of the solution moves towards one of the electrodes (and hence the detector) by electroosmosis. The microchannel walls are treated to present a defined charge, generating a double layer of accumulated oppositely charged ions at the surface (Fig. 8). Part of this double layer (called the mobile layer) can move in the field, which drags the layer of liquid close to the surface with them via shear forces. This generates a net bulk flow in the microchannel because of the fact that the charge density of the mobile layer is much higher than the oppositely charged species within the center of the capillary.

In most cases the microchannels are functionalized to present a negative charge, giving rise to bulk flow towards the negative electrode as the ion layer is positively charged and ions then arrive at the detector in the order, positive, neutral then negative. The negative ions are swept along in the bulk flow but are retarded by their tendency to move towards the positive electrode. Electroosmotic flow has a much flatter profile than pressure driven flow because the driving force is near the surface and all along the microchannel, instead of in the center and only at one end. The flat flow profile affects the spread of residence times and therefore influences the separation of species, giving rise to sharp peak shapes. Neutral species show poor separation as they are effectively swept along with the bulk flow at the same rate. To improve their separation, micelles of a surfactant such as sodium dodecyl sulfate (SDS), can be added to the solution to generate charged micelles over which partitioning of the target analytes can occur leading to separation much like in other types of liquid chromatography. Here, a greater band broadening of neutral species would be expected than for ions, but the flat flow profile across the capillary still improves separation and chromatographic peak shape.

Electroosmosis requires the surface of the capillary to have a reproducible charge density. This can be achieved on glass by using a strong base to generate extra hydroxyl groups that take on a negative charge at most pH values. Negatively charged glass has been used for some time, but causes certain limitations. In order to suppress binding of proteins or nucleotides to the surface, a very high pH is necessary so that both the surface and the analytes have the same charge. Whilst this prevents adsorption it limits the materials that can be used for the rest of the device and prevents the use of silane coupling agents, which are removed by hydrolysis at high pH. As electroosmosis was first developed in silica capillaries, many of the methods are not optimized for use in microfluidic systems although considerable development is currently underway (74). To improve or vary the technique the surface of the channel can be modified in a more targeted fashion. Many methods of coating are possible, with varied advantages and disadvantages. Polymers can be deposited, although the degree of hydrophobicity must be controlled because the target analytes are usually biomolecules and would show nonspecific adsorption. Different polymers allow the charge sign and density to be varied, but often have to be re-applied after sometime, due to contamination, degradation and/or removal from the surface. Epoxy coatings are amphiphilic, so can be switched from positive through zero to negative by varying solution pH and will have the same charge as typical amphiphilic proteins. Usually a small residual electroosmotic effect is desired so that all species travel towards the detector, even when surface charge is deliberately reduced to improve separation of charged species. Early measurements used PTFE (Teflon®) capillaries or coatings, but these are limited by hydrophobic adsorption of proteins (75). Fragile layers of surfactant molecules can be used as dynamic coatings, where the monomer is present in the buffer to form a dynamic equilibrium at the walls. This is very successful, but usually requires conditioning of the surfaces after each run for relatively long periods (a measurement typically takes a few minutes, whilst conditioning takes of the order of an hour). An intermediate between these ephemeral coatings and permanent ones are layer by layer coatings. This technique, developed by Decher, among others, involves the sequential deposition of negative and positive polyelectrolytes, reversing surface charge at each step and building up to a threshold value (76–78). Commercial products are available, terminating with a negative or positive surface and allow repair of the surface after each measurement cycle. These are complex enough that they can be marketed and are very reliable.

7. Commercial Lateral Flow Systems

Enzyme-linked immunosorbent assays (ELISA) and related techniques are commonplace within medical diagnostics for the qualitative and quantitative measurement of protein or nucleic acid fragments, forming the backbone of current hospital pathology practice. Briefly, antibodies specific to the target analyte of interest are bound to the surfaces within multiwell plates. The sample, which can be derived from bodily fluids such as blood or urine, is then incubated allowing the analyte to bind. A second antibody can then be used to introduce an enzyme, which either reacts with a dye molecule causing a color change or generates another molecule that can be measured. A stronger color change reflects a higher concentration of the enzyme, and therefore, of the analyte.

Many of these enzyme-based tests are quite laborious to perform, although a huge variety of versions exist. Such assays are one of the mainstays of modern diagnostics, with a significant proportion of the medical budgets of most nations being spent carrying out vast numbers of tests. Much of the cost is incurred by employing skilled staff to run the tests with a small proportion being spent on the tests themselves. This means that any new technology capable of reducing the complexity of these tests is very valuable and explains the low take-up of microfluidic technologies, which as an infant technology, tends to increase rather than decrease the complexity of the procedure.

Lateral flow (or "dipstick") tests are a more familiar version of this antibody binding diagnostic technology, being used extensively at the point of care for analysis of urine samples, and also within clinics for blood samples. One of the most recognized of these tests is the pregnancy test, with many other lateral flow tests now in common clinical use for a range of conditions. Pregnancy tests measure levels of the hormone human chorionic gonadotropin (hCG), which is produced at higher levels during pregnancy. A sandwich ELISA method is usually used wherein two antibodies specific to hCG are employed, one to bind the hCG to a certain place along the length of a test strip, and the other generating a visible change; for example, a line on the test strip. A schematic of such a lateral flow system is shown in Fig. 9. The urine is drawn along a wicking material dissolving the first antibody. If hCG is present in the sample it binds to the antibody in the moving phase. The liquid continues along the strip, becoming attached to a test area, via interaction of hCG with a second antibody, which is immobilized to the surface. If hCG is not present, then no attachment will occur. Excess of the first antibody can also attach to a second strip where another antibody is immobilized, giving rise to a positive control. Dyes incorporated into the test strip allow a

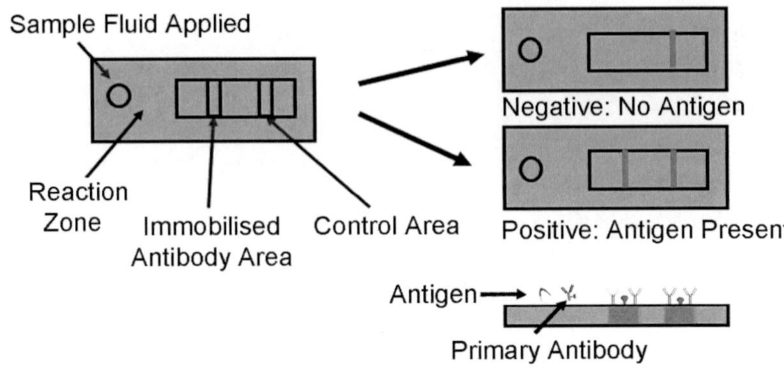

Fig. 9. Schematic of a lateral flow "dipstick" test.

visual readout. Dye molecules embedded in the reaction zones along the length of the strip can be activated by enzymes linked to antibodies. Other tests make use of dye particles such as latex (giving a blue color) or gold particles (red color), with more complex readouts also making use of magnetic particles. Modern liquid crystal readout, lateral flow tests use the same principle of sample detection. Embedded electronics compare the coloration of the test line, for comparison with the control line, before the result is indicated on the display. In this way a calibration of color density can also be incorporated (deeper color compared with the reference, inferring a higher concentration of hCG is present) and so a measure of how long the pregnancy has evolved can also be displayed.

Lateral flow tests can be viewed as simple microfluidic devices with the general principle being suitable for many substrates. Anything that can cause a specific immune response can be a target. Lateral flow devices have been prepared for many analytes of interest, including drugs of abuse such as cannabis, toxins such as aflatoxin and common food proteins (allowing the determination of meat product source or the presence of peanut proteins). When designing these devices, it is important that the surfaces exposed to the sample do not display nonspecific protein adsorption and that the detection zone antibody/protein stays bound (and active) in storage and during use. The chemistry presented in these devices varies, but usually a polymer is grafted to the surface of the attachment zones. This can be achieved in a variety of ways, making use of silane coupling groups or reactive groups within a solid polymer.

Plasma or UV treatment of polymer substrates can be used to promote reactive groups at the surface, which can then be used to attach polymers, via free radical addition, epoxide-amine addition, or disulfide bridging. The next part of the molecule is usually a flexible linker—a short polymer chain. This serves to allow the active part of the surface to sit above the substrate in order for subsequently immobilized protein/enzyme molecules to reside as close to their normal environment as possible. Without this, much of

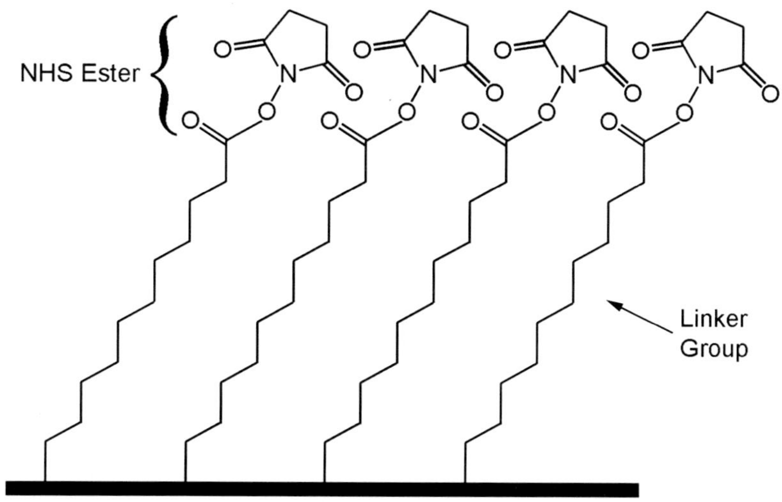

Fig. 10. *N*-hydroxysuccinimide linker attached to surface via a flexible linker group.

the tethered protein would become denatured due to interactions with the surface, or become bound to neighboring molecules. Highly hydrophilic polymers are used here to form a hydrogel on interaction with water, which is a suitable environment for stable protein molecules. Reactive groups are presented at the end of the polymer, allowing the protein to be attached.

Immobilization strategies vary, often driven by the relative costs of the antibodies, the surface treatments, and the required accuracy of the test. Nonspecific attachment using *N*-hydroxysuccinimide to bind to amine groups within protein molecules offers the advantage of ease of reaction (Fig. 10). However, many of the active sites within the bound protein will not be available as there is no way to control the orientation of the molecule on the surface. The same group can be used more selectively on DNA if an amine is added to it at a desired position as no other groups on the DNA will react. Epoxy groups are even more effective, as they bind to a variety of groups present on proteins and DNA, but again their efficiency at collecting molecules from solution comes at the cost of deactivation of much of the protein. *N*-hydroxysuccinimide can also be used to attach an avidin link, which can then be used to specifically bind proteins in a predetermined position by linking a biotin group to them. Specific attachment is more complex (and expensive) but does offer a higher proportion of active bound protein, which is important if the protein is expensive. Surmodics Inc. among others, provides a range of such linkers under the trademark Tridia™ for developers and users of ELISA and lateral flow devices.

The simplicity of lateral flow tests for the end user hides a large amount of technology required to make them reliable. As the devices have a large surface area-to-volume ratio, the inactive areas of the device have to be treated with a blocking material to prevent adhesion of analyte or antibodies carried in the liquid. This can be a serious problem; proteins bind nonspecifically to many surfaces and the high surface-to-volume ratio means that a large proportion of the analyte can be lost even if the binding efficiency is low. As protein molecules contain both positive and negative regions over a wide pH range (urine and blood pH ~7.25) any charged groups on a surface presented to a protein solution will promote adsorption. Hydrophobic areas, Lewis acids, and Lewis bases can also induce protein adsorption. For this reason, many strategies have been investigated to hinder or reduce fouling, as described earlier. Apart from commercial preparations, one of the most commonly used methods is to pre-block the surface so that no sites are available for further adsorption. Bovine serum albumin is a commonly used blocking protein. It is used because of its high abundance (and therefore low cost) and its "sticky" nature resulting from its hydrophobic core and variously charged end groups. Other blocking agents are commercially available for ELISA-based methods, being composed either of modified albumin and/or surfactants with varying characteristics. Mild surfactants can block surfaces and also protect the bound target proteins from decomposition during storage.

Preventing nonspecific binding and promoting deliberate interaction between desired species are key to the generation of reliable assays. Advances in the incorporation of on-chip sample processing, such as separation and concentration of analytes, are pushing the development of lateral flow tests forward. Detection of multiple analytes, along with information on concentrations present, is sometimes necessary and is often diagnostically useful. This is typically achieved by using an attached enzyme to generate a molecule that alters the potential of an electrochemical cell so the concentration can be read off directly as a voltage. The design of devices allowing such sample monitoring require very selective and sensitive "capture" in defined regions along the length of the flow. Surface functionality is therefore of the utmost importance in the development of these devices.

8. Summary and Future Directions

Microfluidic technology is being used as a tool in many research fields, the simplest lateral flow devices being used for home diagnostic testing whilst other, more complicated, fluidic technology is advancing towards commercial viability. Lateral flow microfluidics already has huge commercial value, whilst advances in material and device fabrication methods are driving the application of more complex systems in a variety of areas. With the substantial increase in surface-to-volume ratio when moving into microscale fluidics, consideration of surface chemistry presented is vitally important in order to optimize devices for practical use. One of the major issues encountered when using biological fluids is the nonspecific adsorption of protein onto the inner walls of the device due to interaction with the surface. This is particularly problematic if the species of interest is lost from solution, due to adsorption, rendering it unavailable for use. Examples of this could be the loss of hCG onto the wicking material within a lateral flow device, or loss of an enzyme required for an on-chip microfluidic polymerase chain reaction.

Modification of materials used for the fabrication of microfluidic devices can be achieved in many ways, rendering their surface anti-fouling simply by altering hydrophobic or electrostatic characteristics. For polymers, this can be readily accomplished via plasma treatment, either before or after final construction of the device. SAMs can be applied if more specific chemical modification is required, with the SAM functionality being used as the final layer itself or acting as an anchor layer for subsequent attachment of polymer chains or biomolecules. Structured surfaces presenting superhydrophobic properties have also been used for their antifouling behavior. Hydrophobic surfaces generally promote fouling, due to increased hydrophobic–hydrophobic interaction. Superhydrophobic surfaces present a decreased liquid–solid contact area and where protein molecules do adhere an increase in shear stress (due to increase in interfacial slip) causes protein removal under flow conditions.

Advances in tissue engineering and medical technology may make it possible to implant microfluidic devices for various applications. However, such implantable devices must overcome immune rejection responses and contamination from adsorbing molecules/tissues that may cause adverse effects to the patient or impairment of the device function. PDMS and silicon/silica materials, commonly used for the fabrication of microfluidic devices, have limited biocompatibility and are therefore not suitable for implantation without modification. Anti-thrombogenic coatings have been investigated, including heparin (79), although these commonly have limited lifespan due to deactivation by adsorbing species and are more suited to short-term use. A further limitation of PDMS/silicon bulk materials is that they are not biodegradable and so require extraction after their usefulness is exceeded. Biodegradable microfluidics based on natural polymers have been investigated, such as poly(L-lactic-co-glycolic acid) (PLGA) (80). These have shown promise but are largely held back by their undesirable mechanical properties; brittleness, hardness, and plasticity. Another biodegradable polymer, poly(glycerol sebacate) (PGS) (81), has received attention due to its superior properties compared to PLGA, both mechanically and by invoking a reduced immune response

(82). One of the other limitations for implantable microfluidics is the possible requirement to measure the output, for instance, in diagnostics. Langer et al. have designed an implantable device to monitor the concentrations of multiple cancer biomarkers and chemotherapy agents (83). This system is based on a PDMS chip design, housing magnetic nanoparticles, coated with antibodies selective for varying analytes. In the presence of the analyte, the particles aggregate, with this effect being monitored using magnetic resonance imaging (MRI). Using this device the group have shown real time and noninvasive repeat simultaneous monitoring of several biomarkers and evaluation of chemotherapy agent concentrations.

The plethora of applications for microfluidic devices, from the simplest lateral flow test to more sophisticated and complex μTAS, are not restricted at this stage by technological development or requirement of purpose, but largely by the proper selection of bulk materials and surface coatings/modifications. The driver for biomedical microfluidics, particularly for clinical diagnostics, is the enhancement of healthcare through rapid, high-throughput, and point-of-care testing. Lateral flow tests have now been commercially available for decades with little major advancement in their design. The focus of current research, both for clinical and tissue engineering applications (in vitro and in vivo), is the development of strategies to encompass desirable properties of the bulk material and fine tuning of surface chemical functionality in order to further develop more complex systems. One of the long-term goals for diagnostic microfluidic technology is the production of devices that can rapidly detect multiple indicative state markers without the need for costly laboratory equipment and experienced technical staff. Such devices may encompass multiple technologies, such as separation, purification, concentration, and detection of species. Tests which make available advanced preliminary information from a patient during consultation, or indeed via self-test at home, will continue to change the face of modern healthcare. Furthermore, development of the materials used for microfluidic technology is permitting such devices to be implanted into patients to monitor state and administer drugs when required. Commercial exploration of microfluidics manufacture will inevitably be highly advantageous, with the evolution of devices being possible though the understanding and application of materials research.

Acknowledgements

The authors acknowledge financial support from the National Endowment for Science, Technology and the Arts (NESTA).

References

1. Berg A, Olthius W, Bergveld P (2000) Micro total analysis systems. Kluwer Academics, Dordrecht
2. Fu AY, Spence C, Scherer FH, Arnold FH, Quake SR (1999) A microfabricated fluorescence-activated cell sorter. Nat Biotechnol 17:1109–1111
3. Grafton MM, Zordan MD, Chuang HS, Rajdev P, Reece LM, Irazoqui PP, Wereley ST, Byrnes R, Todd P, Leary JF (2010) Portable microfluidic cytometer for whole blood cell analysis (microfluidics, biomems and medical microsystems VIII). Proc SPIE Int Soc Opt Eng 7593:75930M–75930M-8
4. Khan MS, Thouas G, Shen W, Whyte G, Garnier G (2010) Paper diagnostic for instantaneous blood typing. Anal Chem 82(10):4158–4164
5. Effenhauser CS, Bruin JM, Paulus A, Ehrat M (1997) Integrated capillary electrophoresis on flexible silicone microdevices: analysis of DNA restriction fragments and detection of single DNA molecules on microchips. Anal Chem 69:3451–3457
6. Mao H, Yang T, Cremer PS (2002) Design and characterization of immobilized enzymes in microfluidic systems. Anal Chem 74:379–385
7. Chen SH, Sung WC, Lee GB, Lin ZY, Chen PW, Liao PC (2001) A disposable poly(methylmethacrylate)-based microfluidic module for protein identification by nanoelectrospray ionization-tandem mass spectrometry. Electrophoresis 22:3972–3977
8. Blow N (2007) Microfluidics: in search of a killer application. Nat Methods 4:655–670
9. Quake SR, Scherer A (2000) From micro- to nanofabrication with soft materials. Science 290:1536–1540
10. Duffy DC, McDonald JC, Schueller OJA, Whitesides GM (1998) Rapid prototyping of microfluidic systems in poly(dimethylsiloxane). Anal Chem 70:4974–4984
11. Whitesides GM (2006) The origins and the future of microfluidics. Nature 442:368–373
12. Voskercian G, Shive MS, Shawgo RS, von Recum H, Anderson JM, Cima MJ, Langer R (2003) Biocompatibility and biofouling of MEMS drug delivery devices. Biomaterials 24:1959–1967
13. Christensen TB, Pedersen CM, Goenahl KG, Jensen TG, Sekulovic A, Bang DD, Wolff A (2007) PCR biocompatibility of lab-on-a-chip MEMS materials. J Micromech Microeng 17:1527–1532
14. Khanfer K, Duprey A, Schlicht M, Berguer R (2009) Effects of strain rate, mixing ratio, and stress–strain definition on the mechanical on the mechanical behaviour of the polydimethylsiloxane (PDMS) material as related to its biological applications. Biomed Microdevices 11(2):503–508
15. Roman GT, Hlaus T, Bass KJ, Seelhammer TG, Culbertson CT (2005) Sol–gel modified poly(dimethylsiloxane) microfluidic devices with high electroosmotic mobilities and hydrophobic channel wall characteristics. Anal Chem 77:1414–1422
16. Lee JN, Park C, Whitesides GM (2003) Solvent compatibility of poly(dimethylsiloxane)-based microfluidic devices. Anal Chem 75:6544–6554
17. Rolland JP, Van Dam RM, Schorzman DA, Quake SR, DeSimone JM (2004) Solvent-resistant photocurable "Liquid Teflon" for microfluidic device fabrication. J Am Chem Soc 126:2322–2323
18. Fujii T, Sando Y, Higashino K, Fujii Y (2003) A plug and play microfluidic device. Lab Chip 3:193–197
19. Ratner DM, Murphy ER, Jhunjhunwala M, Snyder DA, Jensen KF, Seeberger PH (2005) Microreactor-based reaction optimization in organic chemistry—glycosylation as a challenge. Chem Comm 10:578–580
20. Kikutani Y, Hibara A, Uchiyama K, Hisamoto H, Tokeshia M, Kitamori T (2002) Pile-up glass microreactor. Lab Chip 2:193–196
21. Zhang ZL, Crozatier C, Le Berre M, Chen Y (2005) In situ bio-functionalization and cell adhesion in microfluidic devices. Microelectron Eng 78–79:556–562
22. Sheikh S, Chih-Chieh SJ, Blaszykowski C, Thompson M (2010) New oligoethylene glycol linkers for the surface modification of an ultra-high frequency acoustic wave biosensor. Chem Sci 1:271–275
23. Aramaki K, Shimura T (2004) Self-assembled monolayers of carboxylate ions on passivated iron for preventing passive film breakdown. Corrosion Sci 46:313–328
24. Zamborini FP, Crooks RM (1998) Corrosion passivation of gold by N-alkanethiol self assembled monolayers: effect of chain length and End group. Langmuir 14:32793286
25. Angelova N, Hunkeler D (1999) Rationalizing the design of polymeric biomaterials. Trends Biotechnol 17:409–421
26. Sofia S, McCarthy MB, Gronowicz G, Kaplan DL (2001) Functionalized silk-based biomaterials for bone formation. J Biomed Mater Res 54:139–148

27. Sun X, Liu J, Lee ML (2008) Surface modification of polymer microfluidic devices using in-channel atom transfer radical polymerization. Electrophoresis 29(13):2760–2767
28. Wang Y, Pai J-H, Lai H-H, Sims CE, Bachmann M, Li GP, Allbritton NL (2007) Surface graft polymerization of SU-8 for Bio-MEMS applications. J Micromech Microeng 17:1371–1380
29. Anna SL, Bontoux N, Stone HA (2003) Formation of dispersions using "flow focusing" in microchannels. Appl Phys Lett 82:364–366
30. Sui G, Wang J, Lee CC, Lu W, Lee SP, Leyton JV, Wu AM, Tseng HR (2006) Solution-phase surface modification in intact poly(dimethylsiloxane) microfluidic channels. Anal Chem 78(15):5543–5551
31. Ren XQ, Bachman M, Sims C, Li GP, Allbritton N (2001) Electroosmotic properties of microfluidic channels composed of poly(dimethylsiloxane). J Chromatogr B 762:117–125
32. Lee J, Kim MJ, Lee HH (2006) Surface modification of poly(dimethylsiloxane) for retarding swelling in organic solvents. Langmuir 22:6544–6554
33. Rhodes NP, Wilson DJ, Williams RL (2007) The effect of gas plasma modification on platelet and contact phase activation processes. Biomaterials 28(31):4561–4570
34. Kyriakides TR, Leach KJ, Hoffman AS, Ratner BD, Bornstein P (1999) Mice that lack the angiogenesis inhibitor, thrombospondin 2, mount an altered foreign body reaction characterized by increased vascularity. Proc Natl Acad Sci 96:4449–4454
35. Klages C-P, Berger C, Eichler M, Thomas M (2007) Microplasma-based treatment of inner surfaces in microfluidic devices. Contrib Plasma Phys 47(1–2):49–56
36. Zelzer M, Majani R, Bradley JW, Rose FRAJ, Davies MC, Alexander MR (2008) Investigation of cell-surface interactions using chemical gradients formed from plasma polymers. Biomaterials 29(2):172–184
37. Lii J, Hsu W-J, Parsa H, Das A, Rouse R, Sia SK (2008) Real-time microfluidic system for studying mammalian cells in 3D microenvironments. Anal Chem 80:3640–3647
38. Cabodi M, Choi N, Gleghorn J, Lee C, Bonassar L, Strook AD (2005) A microfluidic biomaterial. J Am Chem Soc 127: 13788–13789
39. Li R, Altreuter D, Gentile F (1996) Transport characterization of hydrogel matrices for cell encapsulation. Biotechnol Bioeng 50:365–373
40. Gourley PL, Hendricks JK, McDonald AE, Copeland RG, Yaffe MP, Naviaux RK (2007) Reactive biomolecular divergence in genetically altered yeast cells and isolated mitochondria as measured by biocavity laser spectroscopy: rapid diagnostic method for studying cellular responses to stress and disease. J Biomed Optics 12(5) (Article Number: 054003)
41. Roach P, Eglin D, Rhode K, Farrar D, Perry CC (2006) Modern biomaterials: a review: bulk properties and implications of surface modifications. J Mater Sci Mater Med 18(7):1263–1277
42. Cox JD, Curry MS, Skirboll SK, Gourley PL, Sasaki DY (2002) Surface passivation of a microfluidic device to glial cell adhesion: a comparison of hydrophobic and hydrophilic SAM coatings. Biomaterials 23(3):929–935
43. Roach P, Farrar D, Perry CC (2006) Surface tailoring for controlled protein adsorption: effect of topography at the nanometer scale and chemistry. J Am Chem Soc 128(12):3939–3945
44. Geerken MJ, van Zanten TS, Lammertink RGH, Bornemann Z, Nijdam W, van Rijn CJM, Wessling M (2004) Chemical and thermal stability of alkylsilane based coatings for membrane emulsification. Adv Eng Mater 6:749–754
45. Allara DL, Parikh AN, Rondelez F (1995) Evidence for a unique chain organisation in long chain silane monolayers deposited on two widely different solid substrates. Langmuir 11:2357–2360
46. Harris JM (1992) Poly(ethylene glycol) chemistry: biotechnical and biomedical applications. Plenum Press, New York
47. Gingelll D, Owens N, Hodge P, Nicholas CV, Odell R (1994) Adsorption of a novel fluorescent derivative of a poly(ethylene oxide)/poly(butylenes oxide) block-co-polymer on octadecyl glass studied by total internal reflection fluorescence and interferometry. J Biomed Mater Res 28:505–513
48. Popat KC, Johnson RW, Desai TA (2003) Characterisation of vapour deposited poly(ethylene glycol) films on silicon surfaces for surface modification of microfluidic systems. J Vac Sci Technol B 21:645–654
49. Lahann J, Balcells M, Lu H, Rodon T, Jensen KF, Langer R (2003) Reactive polymer coatings: a first step toward surface engineering of microfluidic devices. Anal Chem 75:2117–2122
50. Harbers GM, Emoto K, Greef C, Metzger SW, Woodward HN, Mascali JJ, Grainger DW, Lochhead MJ (2007) Functionalised poly(ethylene glycol)-based bioassay surface chemistry that facilitates bioimmobilization and inhibits non-specific protein, bacterial and mammalian cell adhesion. Chem Mater 19:4405–4414

51. Ostuni E, Chapman RG, Liang MN, Meluleni G, Pier G, Ingber DE, Whitesides GM (2001) Self-assembled monolayers that resist the adsorption of proteins and the adhesion of bacterial and mammalian cells. Langmuir 17:6336–6343
52. Kim P, Jeong HE, Khademhosseini A, Suh KY (2006) Fabrication of non-biofouling polyethylene glycol nicro and nanochannels by ultraviolet-assisted irreversible sealing. Lab Chip 6(11):1432–1437
53. Cassie A, Baxter S (1944) Wettability of porous surfaces. Trans Faraday Soc 40:546–551
54. Koc Y, de Mello AJ, McHale G, Newton MI, Roach P, Shirtcliffe NJ (2008) Nanoscale superhydrohobicity: suppression of protein adsorption and promotion of flow induced detachment. Lab Chip 8:582–586
55. Alexander C, Shakesheff KM (2006) Responsive polymers at the biology/materials science interface. Adv Mater 18(24):3321–3328
56. Kikuchi A, Okano T (2005) Nanostructured designs of biomedical materials: applications of cell sheet engineering to functional regnerative tissues and organs. J Control Release 101(1–3):69–84
57. Allen LT, Tosetto M, Miller IS, O'Connor DP, Penney SC, Lynch I, Keenan AK, Pennington SR, Dawson KA, Gallagher WM (2006) Surface-induced changes in protein adsorption and implications for cellular phenotypic responses to surface interaction. Biomaterials 27(16):3096–3108
58. Nath N, Hyun J, Ma H, Chilkoti A (2004) Surface engineering strategies for control of protein and cell interactions. Surf Sci 570(1–2):98–110
59. Sung WJ, Bae YH (2003) A glucose oxidase electrode based on polypyrrole with polyanion/PEG/enzyme conjugate dopant. Biosens Bioelectron 18(10):1231–1239
60. Fuentes M, Pessala BCC, Maquiese JV, Ortiz C, Segura RL, Palomo JM, Abian O, Torres R, Mateo C, Fernández-Lafuente R, Guisán JM (2004) Reversible and strong immobilization of proteins by ionic exchange on supports coated with sulfate-dextran. Biotechnol Prog 20(4):1134–1139
61. Kikuchi A, Okano T (2002) Intelligent thermoresponsive polymeric stationary phases for aqueous chromatography of biological compounds. Prog Polym Sci 27(6):1165–1193
62. Huber DL, Manginell RP, Samara MA, Kim B-I, Bunker BC (2003) Programmed adsorption and release of proteins in a microfluidic device. Science 301:352–354
63. Edahiro J, Sumaru K, Tada Y, Ohi K, Takagi T, Kameda M, Shinbo T, Kanamori T, Yoshimi Y (2005) In situ control of cell adhesion using photoresponsive culture surface. Biomacromolecules 6(2):970–974
64. Yamada N, Okano T, Sakai H, Karikusa F, Sawasaki Y, Sakurai Y (1990) Thermoresponsive polymeric surfaces: control of attachment and detachment of cultured cells. Makromol Chem Rapid Commun 11(11):571–576
65. Zhu X, De Graaf J, Winnik FM, Leckband D (2004) Tuning the interfacial properties of grafted chains with a pH switch. Langmuir 20(4):1459–1465
66. Wang J, Jiang M, Mukherjee B (2000) On demand electrochemical release of DNA from gold surfaces. Bioelectrochemistry 52(1): 111–114
67. Kwok CS, Mourad PD, Crum LA, Ratner BD (2001) Self-assembled molecular structures as ultrasonically-responsive barrier membranes for pulsatile drug delivery. J Biomed Mater Res 57(2):151–164
68. Ainslie KM, Sharma G, Dyer MA, Grims CA, Pishko MV (2005) Attenuation of protein adsorption on static and oscillating magnetostrictive nanowires. Nano Lett 5(9): 1852–1856
69. Kushida A, Yamoto M, Konno C, Kikuchi A, Sakurai Y, Okano T (1999) Decrease in culture temperature releases monolayer endothelial cell sheets together with deposited fibronectin matrix from temperature-responsive culture surfaces. J Biomed Mater Res 45(4):355–362
70. Canavan HE, Cheng XH, Graham DJ, Ratner BD, Castner DG (2005) Surface characterisation of the extracellular matrix remaining after cell detachment from a thermoresponsive polymer. Langmuir 21(5):1949–1955
71. Mart RJ, Osborne RD, Stevens MM, Ulijn RV (2006) Peptide-based stimuli-responsive biomaterials. Soft Matter 2(10):822–835
72. Murthy N, Campbell J, Fausto N, Hoffmann AS, Stayton PS (2003) Bioinspired pH-responsive polymers for the intracellular delivery of biomolecular drugs. Bioconjug Chem 14(2):412–419
73. Aurenheimer J, Dahmen C, Hersel U, Bausch A, Kessler H (2005) Photoswitched cell adhesion on surfaces with RGD peptides. J Am Chem Soc 127(46):16107–16110
74. Schrott W, Pribyl M, Stepanek J, Snita D (2008) Electro-osmotic characteristics of polystyrene microchips—experiments and modelling. Microelectron Eng 85(5–6):1100–1103
75. Horvarth J, Dolnik V (2001) Wall coatings in capillary electrophoresis. Electrophoresis 22:644–645

76. Decher G, Hong JD, Schmitt J (1992) Buildup of ultrathin multilayer films by a self-assembly process: III. Consecutively alternating adsorption of anionic and cationic polyelectrolytes on charged surfaces. Thin Solid Films 210–211:831–835
77. Lvov Y, Decher G, Mohwald H (1993) Assembly, structural characterization, and thermal behavior of layer-by-layer deposited ultrathin films of poly(vinyl sulfate) and poly(allylamine). Langmuir 9:481–486
78. Hammond PT, Whitesides GM (1995) Formation of polymer microstructures by selective deposition of polyion multilayers using patterned self-assembled monolayers as a template. Macromolecules 1995(28):7569–7571
79. Lin QK, Van JJ, Qiu FY, Song XX, Fu GS, Ji JA (2011) Heparin/collagen multilayer as a thromboresistant and endothelial favorable coating for intravascular stent. J Biomed Mater Res A 96A(1):132–141
80. King KR, Wang CCJ, Kaasempur-Mofrad MR, Vacanti JP, Borenstein JT (2004) Biodegradable microfluidics. Adv Mater 16:2007–2012
81. Wang Y, Ameer GA, Sheppard BJ, Langer R (2002) A tough biodegradable elastomer. Nat Biotechnol 20:602–606
82. Sundback CA, Shyu JY, Wang Y, Faquin WC, Langer RS, Vacanti JP, Hadlock TA (2005) Biocompatibility analysis of poly(glycerol sebacate) as a nerve guide material. Biomaterials 26:5454–5464
83. Daniel KD, Kim GY, Vassiliou CC, Jalali-Yazdi F, Langer R, Cima MJ (2007) Multi-reservoir device for detecting a soluble cancer biomarker. Lab Chip 7(10):1288–1293

Chapter 18

Superhydrophobicity for Antifouling Microfluidic Surfaces

N.J. Shirtcliffe and P. Roach

Abstract

Fouling of surfaces is often problematic in microfluidic devices, particularly when using protein or enzymatic solutions. Various coating methods have been investigated to reduce the tendency for protein molecules to adsorb, mostly relying on hydrophobic surface chemistry or the antifouling ability of polyethylene glycol. Here we present the potential use of superhydrophobic surfaces to not only reduce the amount of surface contamination but also to induce self-cleaning under flow conditions. The methodology is presented in order to prepare superhydrophobic surface coatings having micro- and nanoscale feature dimensions, as well as a step-by-step guide to quantify adsorbed protein down to nanogram levels. The fabrication of these surfaces as coatings via silica sol–gel and copper nano-hair growth is presented, which can be applied within microfluidic devices manufactured from various materials.

Key words: Superhydrophobicity, Non-wettable, Antifouling, Protein adsorption

1. Introduction

Microfluidic technology has become increasingly of interest in the field of medical diagnostics due to the low volumes of liquids that can be manipulated. The use of microfluidics for blood analysis has been reported (1, 2), with the main driver being that the volume of sample needed for each test is drastically reduced compared to conventional analysis methods. This means that either much less sample volume needs to be taken from the patient or that more tests can be run on the same aliquot—with microfluidic devices also providing a route for simultaneous testing. Of somewhat keener interest is the extension of such tests to other, less plentiful biological samples such as saliva, sweat, and tears (2, 3). Other industrially relevant uses for biological microfluidics include production, purification, and high throughput evaluation of proteins, antibodies, RNA, and DNA (4).

Gareth Jenkins and Colin D. Mansfield (eds.), *Microfluidic Diagnostics: Methods and Protocols*, Methods in Molecular Biology, vol. 949, DOI 10.1007/978-1-62703-134-9_18,

Upon presentation of a solution containing biological components to a material of any kind, water and solvated ions are the first to interact due to their abundance and mobility. The surface becomes solvated allowing larger species to interact, usually resulting in the adsorption of proteinatious material. The accumulation of protein molecules then proceeds, later mediating the attachment of other biological species, such as blood platelets and cells. For microfluidic systems such surface fouling can cause major problems, not least in causing blockages and reducing fluid flow but also by the reduction in concentration of species held in solution. Consider a polymerize chain reaction (PCR) wherein an enzyme is required to catalyze a reaction; the loss of the active enzyme species or polynucleotides onto the walls of the reaction vessel through adsorption reduces the amount available in solution and therefore results in a reduction of reaction rate and yield. This problem is particularly evident in microfluidic systems as the surface area to volume ratio is extremely high.

Antifouling surfaces have been of interest for some time, with various strategies being used to reduce adsorption including chemical functionlization with "protein resistant" coatings such as poly(ethylene glycol) (PEG) (5, 6), prefilling surface sites with other molecules (7), and attaching proteases to surfaces that enzymatically cleave proteins upon interaction (8). Removal of surface contaminants by flow shear has been restricted to a small number of applications due to the high shear rates required and the fragility of surfaces used (9).

Biomolecule adsorption onto hydrophobic surfaces is thermodynamically driven, reducing unfavorable interaction of water with nonpolar regions within the biomolecule and the surface. On binding the molecules can become distorted to maximize interactions, leading to orientational and conformational changes that can impair molecular activity in the case of enzymes. Since hydrophobic surfaces are well known to promote fouling and highly solvated PEG chains can hinder protein adsorption, the use of superhydrophobic surfaces as antifouling surfaces has only recently been considered (10).

Highly non-wettable "superhydrophobic" surfaces present a hydrophobic surface chemistry along with a roughness that in combination reduces the contact area of the water on the solid surface due to the droplet suspension on the peaks of the roughness. In some cases these extremely water repellent surfaces allow water to roll off, taking particulate contamination along with the droplets (11). The basic principles of surface roughness and water repellency were originally defined by Wenzel (12) and Cassie and Baxter (13). The Wenzel wetting state is the simplest model, wherein the liquid conforms to the contours of the surface roughness. In contrast, the Cassie–Baxter wetting state involves the liquid being suspended on the peaks of the

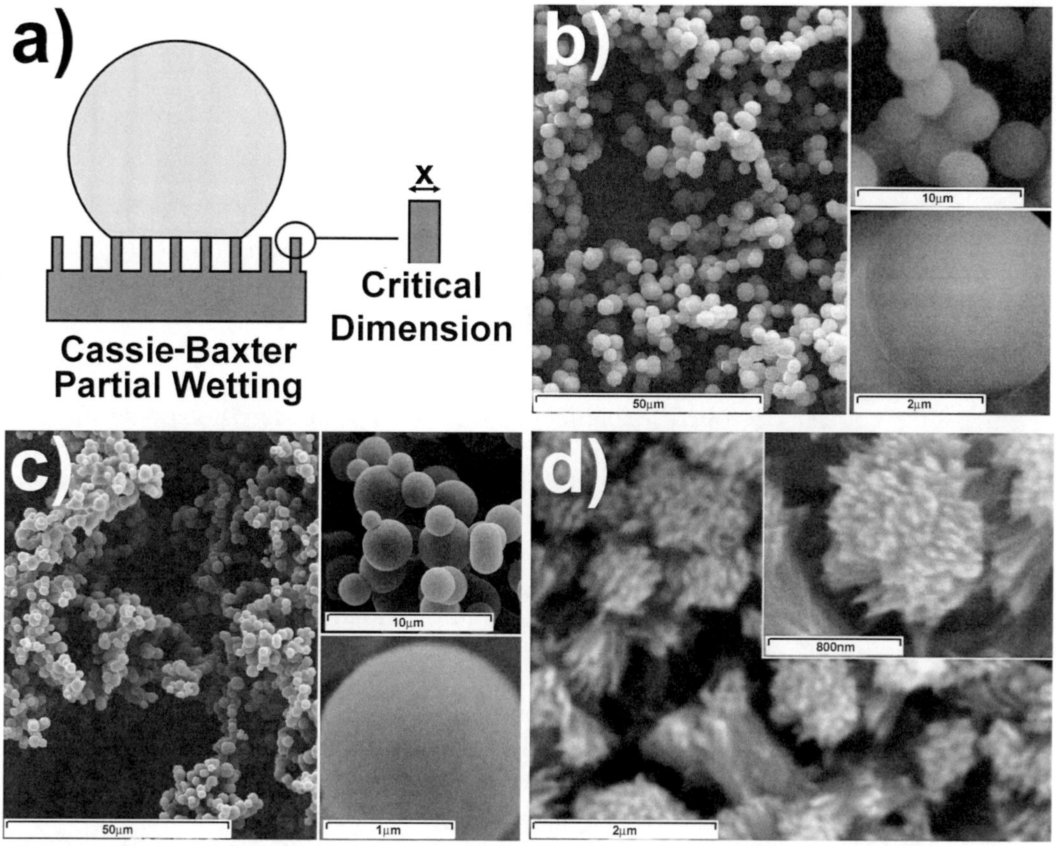

Fig. 1. Illustration of surfaces prepared via sol–gel and copper oxide growth showing, (**a**) critical dimension for Cassie–Baxter wetting, (**b**, **c**) sol–gel prepared coatings, and (**d**) copper oxide nano-needles. Reproduced with permission from the Royal Society of Chemistry Publishing 2008 (20).

roughness with air filling the volume under the droplet (Fig. 1a). Surfaces allowing water to bridge between roughness features provide a lower solid–liquid contact area permitting water to slide or roll off easily. Superhydrophobicity has been explored for numerous applications including self-cleaning and anti-mist surfaces and controllable liquid optics, with recent reviews detailing the range of applications and varied methods of fabrication (14).

Biomolecule adsorption onto superhydrophobic surfaces does occur, although adsorption kinetics and the total amount adsorbed is much less than compared to that on comparatively flat surfaces, possibly due to the reduced solid–liquid contact area in the Cassie–Baxter state (15–18). A review of the developments and uses of superhydrophobicity highlight how little work has been carried out in the area of biofouling (19). Here we describe methodology to prepare superhydrophobic surfaces, presenting data on protein adsorption onto these coatings under both static and flow conditions (20). Particular attention is made in describing the effects of surface feature dimension and chemistry, with results of protein quantification presented on micro- to nanoscale topography overlaid with hydrocarbon and fluorocarbon chemistry. Surface coating methodology is described that can be applied to various materials and geometries useful in the development of superhydrophobic microfluidic devices.

The surface coating procedure presented here is designed to minimize biofouling of surfaces under static conditions (incubation of the surface in protein containing media) and to drastically reduce the amount of protein adsorbed via removal of adsorbed species under flow conditions. The use of a range of superhydrophobic surfaces, varying in feature dimension and surface chemistry, is described. Experiments were conducted using bovine serum albumin as a model biomolecule as this is known to nonspecifically bind to surfaces (15), although the method can be applied to the use of other similar molecular species. Preparation of two different types of structured surface coatings are described, each allowing subsequent modification by hydrocarbon and fluorocarbon chemistries, although this method can be extended to other surface feature sizes or morphologies and a range of chemistries. Sol–gel methodology for the preparation of porous materials is well understood and has been previously reported for thin film coatings (21, 22).

Flat surfaces of comparable chemistry were used as controls. Scanning electron microscopy (SEM) was used to evaluate surface topography, taking an average of 100 surface features to ascertain average dimensions. Water contact angle measurements (WCA) using distilled deionized water, were used to measure wettability, taking an average of six repeats on each sample type. A fluorimetric assay using NanoOrange (Molecular Probes) is described for the quantification of adsorbed protein. This method has been previously optimized for the investigation of protein adsorption onto similar surfaces and the method compared to an amido black colorimetric assay and acoustic microbalance measurements (15). An average of three separate repeats was taken of all samples used for adsorbed protein quantification, using the same stock protein solutions, incubation, and flow conditions. All surfaces of the same type were produced and batch analyzed. Chemical modification and evaluation was also carried out in sample batches.

Following the protocol described here, this study showed that albumin was adsorbed in similar amounts under static conditions to flat and nano-structured surfaces presenting either hydrocarbon or fluorinated functionality, although notably lower amounts were observed on fluorinated surfaces as compared with hydrocarbon surfaces (Fig. 2). With increasing feature size larger amounts were found to adsorb, possibly due to the increasing solid–liquid contact area. Under flow conditions similar trends were observed. As expected the shear force gave rise to a mechanical removal of protein, resulting in lower levels of protein remaining adsorbed

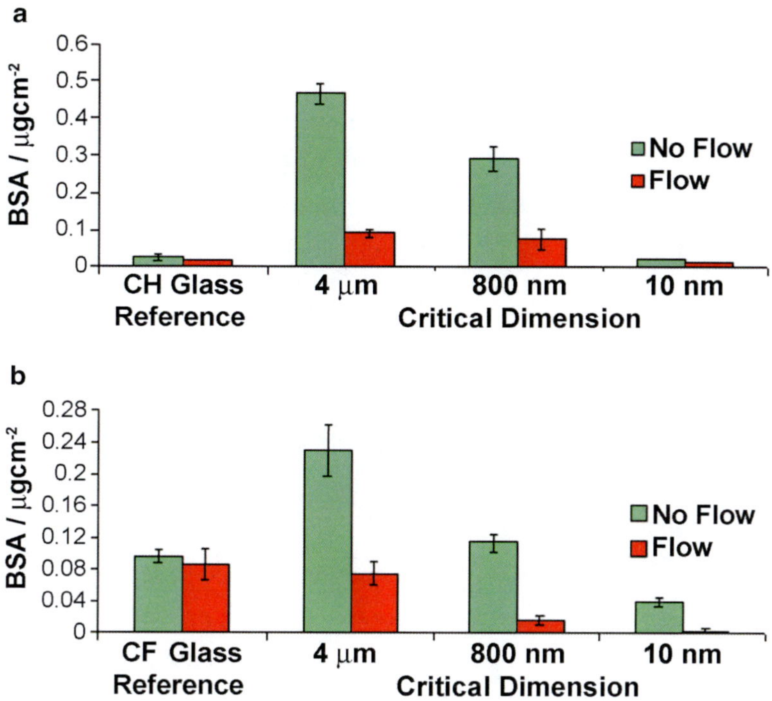

Fig. 2. Quantification of adsorbed albumin onto (**a**) hydrocarbon and (**b**) fluocarbon surfaces under static and flow conditions.

compared to the experiments conducted under static adsorption. For fluorinated nano-hair surfaces almost all protein was completely removed under flow conditions, demonstrating excellent antifouling capabilities.

Protein molecules will adsorb onto superhydrophobic surfaces, but several factors may contribute to their effective removal under flow, particularly if micrometer scale roughness is replaced with nanometer scale roughness. The change in feature size to nanoscale dimensions will result in an increase in liquid flow rate near the surface due to interfacial slip (23–25). Protein molecules adsorbed onto these features would then experience greater shear forces. Nanoscale roughness also presents features on the same scale as protein molecular dimensions, reducing the contact area unless the protein molecules deform (26, 27). The silica sol–gel samples used here are porous, although it is important to recognize that only the upper surface of the material is in contact with liquid due to Cassie–Baxter wetting. Under increased pressure the solution may penetrate the structure, with nanoscale superhydrophobic surfaces remaining in the bridging state under much higher hydrostatic pressure compared to microstructured surfaces. For this reason nano-structured coatings may be of more use within micro/nano-fluidic systems.

2. Materials

2.1. Sol–Gel Coatings

1. Standard microscope slides.
2. Deionized water, approximately 18 MΩ cm^{-2}.
3. Toluene (low sulfur).
4. Ethanol (absolute).
5. Methyltriethoxysilane (>98%).
6. Aqueous HCl (diluted to 0.37%).
7. Aqueous ammonia solution (35% diluted with deionized water).
8. Dimethylformamide.
9. Ethylene glycol (>99%).

2.2. Copper: Flat and Nano-Hair Coatings

1. Copper foil (>99.95%).
2. Standard microscope slides.
3. Deionized water, approximately 18 MΩ cm^{-2}.
4. Aqueous ammonia solution (35% diluted with deionized water).
5. Ethanol (absolute).

2.3. Electron Microscopy and Water Contact Angle

1. Deionized water, approximately 18 MΩ cm^{-2}.
2. Titanium (>99.99%).

2.4. Surface Chemical Modification

1. Octyltriethoxysilane (>95%).
2. “Grangers Wash In” solution (Grangers, UK).

2.5. Protein Adsorption and Analysis

1. Phosphate-buffered saline pellets, 200 mM phosphate, 100 mM sodium chloride pH = 7.4 at 25°C (Gibco, UK).
2. Deionized water, approximately 18 MΩ cm^{-2}.
3. Bovine serum albumin (BSA) (>98%, HPCE).
4. NanoOrange protein quantification kit (Molecular probes, Invitrogen, UK).

2.6. Equipment

1. Imaging system for water contact angle measurements: KSV CAM200 instrument.
2. Syringe pump (Pump 11, Harvard Apparatus, Holliston, MA, USA).
3. Black 96-well plates (Nunc S/A, Denmark).
4. Flow chamber to model microfluidic fluid shear. Here the coatings to be analyzed were prepared either on microscope slides or on copper foil and then assembled into a flow system

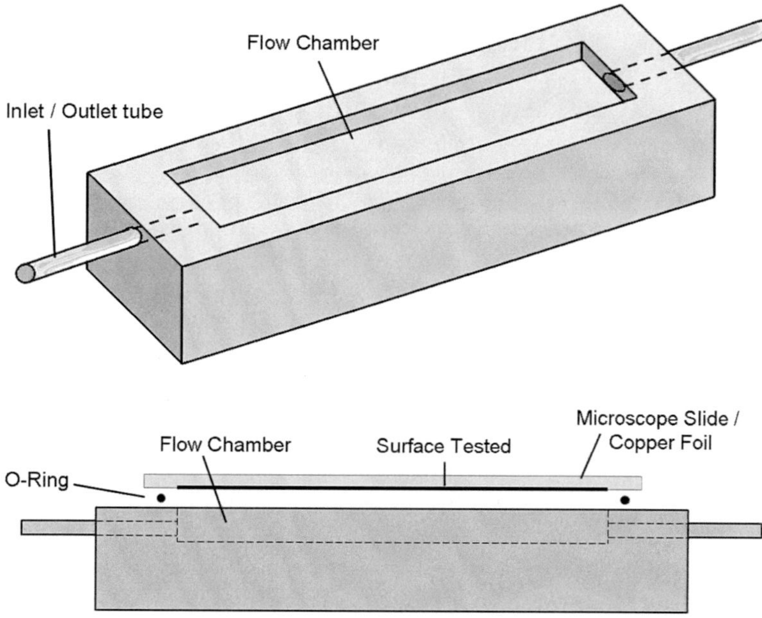

Fig. 3. Schematic of the flow chamber used in this study.

as shown schematically in Fig. 3. The body of the chamber was fabricated by milling acrylic, using stainless steel tubes for the inlet/outlet. Samples were secured onto an o-ring to maintain a liquid seal. The following methodology can be used with any flow chamber setup.

3. Methods

3.1. Sol–Gel Coatings

Sol–gel films can be prepared using various solvents to produce a range of surface feature sizes (Fig. 1b, c). Included here is a generalized preparation using dimethylformamide as an example solvent (see Note 1).

1. Mix methyltriethoxysilane (MTEOS), aqueous hydrochloric acid (HCl), and a solvent in the ratio 2:3:2. The mixture should be continuously stirred for 1 h.
2. Add and mix ammonia solution in a ratio 1:4 with the original total mix volume prepared in step 1 above. Varying concentrations of ammonia can be used to prepare differing surface features (see Note 2).
3. Aliquot the gel solution immediately onto the surface to be coated. The solution should be cast between the surface required and an upper plate to ensure uniformity. In this study two glass slides were used as the upper and lower surfaces.

Cover slips were used as spacers to define the thickness of the sol–gel film. To prevent the upper surface from sticking to the prepared coating it should be hydrophobic. Here we hydrophobized the upper glass slide by immersion into "Grangers Wash In solution" for this step, see Note 3.

4. Incubate the glass slide in "Grangers Wash In" solution for 10 min, rinse in distilled water and heat above 70°C for 10 min. Cool before use.
5. After casting leave the films at least 6 h to solidify before removing the upper plate. If using solvents with low volatility these should be exchanged for solvents such as ethanol or ether. Immerse films for 10 min, changing the solvent at least three times to ensure complete exchange.
6. Heat films to 500°C at a rate of 2°C min^{-1}, holding the upper temperature for 1 h before cooling slowly.
7. Chemically modify the surfaces as required.

3.2. Copper Oxide Nano-Needle Preparation

Copper oxide nano-needle surfaces can be prepared by the following route, giving rise to surface features as shown in Fig. 1d.

1. Cut copper foil to the size required (76×25 mm in this study).
2. Sonicate the copper in ethanol for approximately 20 min (depending on cleanliness of sample) to remove surface contaminants.
3. Immerse the copper sample in 36 mM ammonia solution at 4°C until they become uniformly black (see Note 4).
4. Remove samples and rinse in distilled deionized water.
5. Heat to 180°C at a rate of 1°C min^{-1} holding the upper temperature for 3 h.
6. Cool and chemically modify the surfaces as required.

3.3. Surface Chemical Modification

Materials can be chemically modified to present various functionalities at their surfaces. Here we describe the modification using "Grangers Wash In" solution (fluorocarbon) and octyltriethoxysilane (hydrocarbon).

3.3.1. Fluorocarbon Coatings

1. Fully immerse samples (glass or copper oxide as used in this study) in "Grangers Wash In" solution diluted to 10% concentration in distilled water.
2. Incubate at room temperature for approximately 10 min.
3. Remove and rinse in distilled water.
4. Heat either on a hotplate or in an oven: glass/silica sol–gel samples up to 100°C for 1 h, copper samples 40°C for 24 h.
5. Allow to cool before use.

3.3.2. Hydrocarbon Coatings

1. Fully immerse samples (glass or copper oxide as used in this study) in a 2% octyltriethoxysilane solution in dry toluene.
2. Incubate at room temperature for 24 h.
3. Remove and rinse in toluene.
4. Air dry before use.

3.4. Scanning Electron Microscopy (SEM)

1. Sputter coat (or evaporate) samples with approximately 5–20 nm conductive material. Here we used titanium although gold or carbon is also commonly used.
2. SEM images should be taken at appropriate acceleration voltages in order to minimize risk of sample degradation whilst maximizing image resolution.

3.5. Water Contact Angle Measurement

Measurement of water contact angles allows the wettability of surfaces to be assessed. A KSV CAM200 instrument was used in this study. An example image of water droplets resting on a hydrophobized copper nano-needle surface is presented in Fig. 4.

1. Dispense approximately 1 μL droplet of distilled deionized water (18 MΩ cm^{-2}) onto the sample surface taking an image immediately as the droplet rests and the injection needle is removed.
2. The drop shape can be used along with tangents to the drop at the solid/liquid contact line to estimate the water contact angle through the droplet. A Young-Laplace function can be fit to the droplet profile and the angle measured between this and a straight line placed through the intersection of the droplet and the surface.
3. At least five repeating static contact angle measurements should be carried out to give an average value. Surface contamination or irregularities in the surface due to fault or damage can give rise to variable measurements.
4. As another indication of hydrophobization the driving edge of the liquid through a treated microfluidic channel can be observed. As shown in Fig. 5 the water becomes unable to contact with the surfaces of the hydrophobized channels giving rise to a change in flow profile.

3.6. Protein Adsorption, Flow Removal, and Quantification

1. Decide how many samples you will require for experimentation, including repeats and if you will be conducting static adsorption and/or flow removal experiments.
2. Incubate all samples in protein solution (in this case bovine serum albumin) in 200 mM phosphate buffer at pH 7.4 for 1 h (see Note 5).
3. Using the flow chamber (or other microfluidic chamber allowing the investigation of these coatings) flow buffer solution over

Fig. 4. Water droplets resting on a fluorinated copper oxide nano-needle surface.

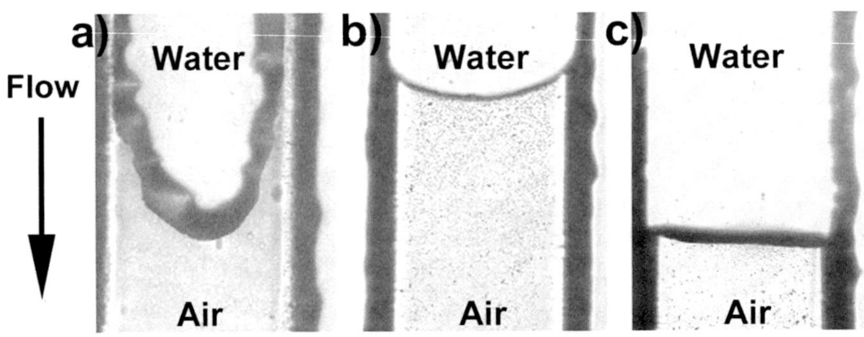

Fig. 5. Flow profiles in microfluidic channels having varying surface modifications, (**a**) fluorinated copper nano-needles, (**b**) fluorinated flat glass, and (**c**) non-modified glass

samples requiring flow removal of protein at the desired rate. The chamber used in this study allowed large surface areas to be examined easily, having dimensions 1,500 × 650 μm cross section, 65 mm in length, Fig. 3. At 20 mL min^{-1} similar shear forces as those expected in much smaller microfluidic systems were achieved; equivalent to a flow rate of 3 μl min^{-1} in a channel having cross-sectional dimensions 500 × 50 μm.

4. Samples not requiring flow removal of protein should be carefully rinsed in three sequential cycles using PBS to remove any loosely bound species.

5. Surfaces should then be rinsed thoroughly in three sequential wash cycles of ethanol and distilled water to detach any adsorbed protein. All washings should be collected separately and reduced to dryness by vacuum centrifugation (see Note 6).

6. Prepare the NanoOrange protein quantification working solution as per manufactures instructions (see Note 7.)Prepare 1×

protein quantitation solution by diluting to 10% in distilled water. For each sample 250 μL will be required.

(a) Prepare 1× NanoOrange working solution by diluting the NanoOrange protein quantitation reagent (Component A) to 0.2% in the 1× protein quantitation solution prepared in the step above.

(b) Prepare calibration plot by making up stocks of protein solution in the concentration range 0–10 μg/mL by serial dilution in 1× NanoOrange working solution.

7. Redissolve collected desorbed protein in 250 μL fluorescent probe working solution in Eppendorf tubes. Protect from light to minimize bleaching effects.
8. Prepare protein standards for a calibration plot also in the fluorescent probe working solution. These should be treated in exactly the same way as the sample solutions.
9. Heat all solutions to 95°C for 5 min and then cool to room temperature (see Note 8).
10. Aliquot 200 μL of each protein solution into wells of a black 96-well plate, along with the calibration standards.
11. Using a plate reader set the excitation and emission filters to that appropriate for the fluorescent probe used (NanoOrange Excitation wavelength: 485 nm, Emission wavelength: 590 nm) and take readings.

4. Notes

1. With reference to previous work the materials formed by the sol–gel process can be adjusted to form coatings having ranging feature sizes, feature morphology, and pore connectivity (21, 22). For repeatability it is critical that the mixing time prior to the addition of ammonia be constant so that the degree of hydrolysis is kept the same from sample to sample. The temperature at which the films are prepared has also shown to impact on the final materials formed. For this reason note that the temperature at which the films were prepared in this study was 24°C.
2. Large grained sol–gel materials can be prepared using ethylene glycol as the solvent and 0.9 M ammonia solution; small grained features can be prepared using dimethylformamide as the solvent and 3.6 M ammonia solution (22).
3. To hinder the sol–gel network attaching to the mold surface functional groups any surface reactive groups should be blocked. In this study we used a fluorinating agent "Grangers,"

although a hydrophobic silane could also be used such as hexadimethyldisilazane (HMDS).

4. The time required for this step will vary greatly depending on the number of samples prepared and the volume of ammonia solution used. As an example six samples of this size incubated in 500 mL solution were ready in about 3 days.
5. Incubation in protein solution should be conducted for all samples simultaneously to avoid differences in amounts of protein adsorbed. In this study we examined the adsorption of bovine serum albumin due its nonspecific binding characteristics. However, substitution of this protein for another biomolecule would be possible due to the quantification assay used being largely independent of protein type. Here we have chosen to use very high concentration of protein solution (3 mg mL^{-1}) to create a uniform layer of protein on all samples. Varying this concentration will allow for adsorption isotherms to be established and/or investigation of energies of adsorption.
6. Here we used vacuum centrifugation to minimize time of the experimentation although freeze drying can also be used.
7. The molecular probe used in this study was NanoOrange due to its highly sensitive and nonspecific detection of protein. However, another assay can be substituted here as others have now been marketed giving even higher sensitivity, or indeed if the reader wishes to choose selectivity they can do so. NanoOrange becomes highly fluorescent when bound to protein allowing quantification of very small amounts of protein. Reference samples of each type should be used as background.
8. After heating some solution may have condensed on the inside of the Eppendorf tube cap. Centrifuge at approximately $50 \times g$ for 1 min to ensure consistency.

Acknowledgments

The authors acknowledge financial support from EPSRC (grant EP/D500826/1).

References

1. Henderson JC, Yacopucci M, Chun CJ, Lenghaus K, Sommerhage F, Hickman JJ (2010) Investigation of the behaviour of serum and plasma in a microfluidics system. J Vac Sci Technol B 28(5):1066–1070
2. Srinivasan V, Pamula VK, Fair RB (2004) An integrated digital microfluidic Lab-on-a-chip for clinical diagnostics on human physiological fluids. Lab Chip 4(4):310–315

3. Saitoh T, Suzuki N, Furuse T, Hiraide M (2009) Heat induced solution mixing in thermo-responsive polymer-coated microchannels for the fluorometric determination of polyamines in saliva. Talanta 80(2): 1012–1015
4. Kim J, Johnson M, Hill P, Gale BK (2009) Microfluidic sample preparation: cell lysis and nucleaic acid purification. Integr Biol 1(10):574–586
5. Bearinger J, Terretaz S, Michel R, Tirelli N, Vogel H, Textor M, Hubbell J (2003) Chemisorbed poly(propylene sulphide)-based copolymers resist biomolecular interactions. Nat Mater 2:259–264
6. Huber D, Manginell D, Samara M, Kim B, Bunker B (2003) Programmed adsorption and release of proteins in a microfluidic device. Science 301:352–354
7. Taylor S, Smith S, Windle B, Guiseppi-Elie A (2003) Impact of surface chemistry and blocking strategies on DNA microarrays. Nucleic Acids Res 31(16):e87
8. Asuri P, Karajanagi S, Kane R, Dordick J (2007) Polymer-nanotube-enzyme composites as active antifouling films. Small 3(1):50–53
9. Callow M, Fletcher R (1994) The influence of Low surface energy materials on bioadhesion—a review. Int Biodeterior Biodegrad 34: 333–348
10. Scardino AJ, Zhang H, Cookson DJ, Lamb RN, De Nys R (2009) The role of nano-roughness in antifouling. Biofouling 25(8):757–767
11. Furstner R, Neinhuis C, Barthlott W (2000) The lotus effect: self-purification of microstructured surfaces, nachr. Chim 48(1):24–28
12. Wenzel R (1936) Resistance of solid surfaces to wetting by water. Ind Eng Chem 28:988–994
13. Cassie A, Baxter S (1944) Wettability of porous surfaces. Trans Faraday Soc 40:546–551
14. Roach P, Shirtcliffe N, Newton M (2008) Progress in superhydrophobic surface development. Soft Matter 4:224–240
15. Roach P, Shirtcliffe NJ, Farrar D, Perry CC (2006) Quantification of surface-bound proteins by fluorometric assay: comparison with quartz crystal microbalance and amido black assay. J Phys Chem B 110(41):20572–20579
16. Sun T, Tan H, Han D, Fu Q, Jiang L (2005) No platelet Can adhere—largely improved blood compatibility on nanostructured superhydrophobic surfaces. Small 1(10):959–963
17. Zhang H, Lamb R, Lewis J (2005) Engineering nanoscale roughness on hydrophobic surface—preliminary assessment of fouling behaviour. Sci Technol Adv Mater 6(3–4):236–239
18. Toes G, van Muiswinkel K, van Oeveren W, Suurmeijer A, Timens W, Stokroos I, van den Dungen J (2002) Superhydrophobic modification fails to improve the performance of small diameter expanded polytetrafluoroethylene vascular grafts. Biomaterials 23(1):255–262
19. Genzer J, Efimenko K (2006) Recent developments in superhydrophobic surfaces and their relevance to marine fouling: a review. Biofouling 22(5):339–360
20. Koc Y, de Mello AJ, McHale G, Newton MI, Roach P, Shirtcliffe NJ (2008) Nano-scale superhydrophobicity: suppression of protein adsorption and promotion of flow-induced detachment. Lab Chip 8:582–586
21. Shirtcliffe N, Mchale G, Newton M, Perry C, Roach P (2005) Porous materials show superhydrophobic to superhydrophilic switching. Chem Commun 25:3135–3137
22. Shirtcliffe N, McHale G, Newton M, Perry C, Roach P (2007) Superhydrophobic to superhydrophilic transitions of Sol–gel films for temperature. Alcohol or surfactant measurement. Mater Chem Phys 103(1):112–117
23. Choi CH, Kim CJ (2006) Large slip of aqueous liquid flow over a nanoengineered superhydrophobic surface. Phys Rev Lett 96(6): 066001
24. Truesdell R, Mammoli A, Vorobieff P, van Swol F, Brinker C (2006) Drag reduction on a patterned superhydrophobic surface. Phys Rev Lett 97(4):044504
25. Ou J, Rothstein J (2005) Direct velocity measurements of the flow past drag-reducing ultrahydrophobic surfaces. Phys Fluids 17(10):103606
26. de Vasconcelos C, Bezerril P, Dantas T, Pereira M, Fonseca J (2007) Adsorption of bovine serum albumin on template-polymerized chitosan/poly(methacrylic acid) complexes. Langmuir 23(14):7687–7694
27. Roach P, Farrar D, Perry C (2006) Surface tailoring for controlled protein adsorption: effect of topography at the nanometer scale and chemistry. J Am Chem Soc 128(12): 3939–3945

Section III

Microfluidic Diagnostics: Application Protocols

Chapter 19

The Application of Microfluidic Devices for Viral Diagnosis in Developing Countries

Samantha M. Hattersley, John Greenman, and Stephen J. Haswell

Abstract

Whilst diseases such as diabetes and cardiovascular disorders are increasing in the developed world, the main threat to global health remains viral-based infectious disease. Such diseases are notably prevalent in developing countries, where they represent a major cause of mortality; however, their detection and prevention is typically hampered by poor infrastructure and a lack of resources to support the sophisticated diagnostic tools commonly found in modern laboratories. Microfluidic-based diagnostics has the potential to close the gap between developed and developing world medical needs due to the robustness and reduced operating costs such technology offers. The most recent developments in microfluidic diagnostics for viral infections have explored the separation and enumeration of immune cells, the capture and identification of viral particles, and antiviral drug evaluation within microchannels and chambers. Advances in solid-phase separation, isothermal amplification, real-time detection of nucleotide products, and improved efficiency of detection systems in microfluidic platforms have also opened up opportunities for diagnostic innovation. This chapter reviews the potential capability microfluidic technology can offer in addressing the practical challenges of providing diagnostic technology for developing countries, illustrated by research on key viral diseases.

Key words: Microfluidic, Diagnostic, Viral, Developing world, HIV, Influenza, Dengue fever

1. Introduction

Viruses were first identified as infectious agents in the 1800s and today we know that these are characterized by a small genome, either DNA or RNA, enclosed in a protein capsid or membrane envelope. The viral genome occurs as either double- or single-stranded DNA/RNA, which can only reproduce within a host cell. Viruses exploit the host cell's internal replication machinery to synthesize progeny viruses. Each virus particle has its own individual host range, which is determined by specific cell receptors on the target cell's membrane. Whilst animal viruses are diverse in their

Gareth Jenkins and Colin D. Mansfield (eds.), *Microfluidic Diagnostics: Methods and Protocols*, Methods in Molecular Biology, vol. 949, DOI 10.1007/978-1-62703-134-9_19, © Springer Science+Business Media, LLC 2013

modes of infection and replication, they invariably use an envelope obtained from a host cell membrane to enable them to enter and exit from the host cell minimizing an immunological response.

Due to their ability to replicate rapidly, viruses are known to be able to mutate into new strains and disease vectors over relatively short periods almost anywhere in the world. The vast majority of human viral infections were first recognized in the areas of the most deprivation where medical care is at its most primitive or even nonexistent. Human immunodeficiency virus (HIV), for example, dramatically arrived in the West in the early 1980s (1), whereas the Ebola virus has emerged sporadically in central Africa since its initial identification in 1976 (2). More recently a newly discovered set of viruses, which cause encephalitis, have now been identified following the 1999 outbreak of Nipah virus leading to the deaths of 105 Malaysians and decimation of the country's pig industry (3). Furthermore, the public is all too familiar with almost annual reports of new serotypes of Influenza A virus causing disease across the world. Accordingly infectious diseases caused by viruses represent one of the major causes of morbidity in the developed world and mortality in the developing world (4).

To prevent the spread of infectious diseases, clinicians require accurate analytical tools to identify the relevant pathogen rapidly in order to assess the nature and progress of a patient's illness and where possible provide appropriate treatment. Recent epidemics, which devastated under-developed regions of the world, have however highlighted the issue that appropriate diagnostic tests are still not available. In well-equipped laboratories the "gold standard" techniques for diagnosis of infectious diseases are microscopy, tissue culture, lateral flow immunoassays, and enzyme-linked immunosorbant assays (ELISA) (5). These techniques are however expensive, offer limited ability to discriminate between multiple and/or diverse pathogens samples, are cumbersome, relatively slow, and have an inadequate detection threshold when the test sample is not stored or prepared adequately (6). Recent techniques using the polymerase chain reaction (PCR) to detect viral infection have been developed and these do offer orders of magnitude improvements in detection threshold and can selectively distinguish between different viral strains, such as seasonal Influenza A and Influenza A/2009 H1N1 (7, 8). However, this technique is problematic as it still requires expert technicians, high cost equipment and reagents coupled with major contamination control issues (9).

Whilst the difficulties associated with PCR can be overcome in developed countries, they are generally insurmountable in developing countries where multiple viral and bacterial infections are highly prevalent. This is typically due to the insidious problems including a lack of regional clinical and laboratory facilities, environmental conditions such as poor sanitation and/or insufficient water, unreliable electricity, high humidity, and temperatures (10).

Table 1
World Health Organization criteria for diagnostic tools (12)

Assured criteria

A	Affordable
S	Sensitive
S	Specific
U	User-friendly (minimal expertise necessary)
R	Robust and rapid (results accessible in under 30 min)
E	Equipment-free
D	Deliverable to remote areas

Furthermore, in developing countries diagnostics represent only a small percentage of the research investment into poverty-related diseases (PRDs). The recent 2010 report by Policy Cures: Global Funding of Innovation for Neglected Diseases (G-FINDER) showed that in 2009 HIV and Acquired Immunodeficiency Syndrome (AIDS) research received $1,139 M in global R&D funding, of which only $42.8 M (3.8%) was committed for diagnostics (11). A similar trend in the proportion of R&D investment being dedicated to diagnostics can be seen for malaria and Dengue fever, which received a $9.1 M (1.5%) and $4.0 M (2.4%) share of the global R&D funding, respectively. Clearly new and appropriate diagnostic technology would have a major impact on the development of novel drug regimens, as the ability to determine if a patient is infected, will be instrumental in the effective treatment and monitoring of diseases as well as minimizing the onset of drug resistant strains. However, the required investment is not yet being made.

A key question therefore would seem to be "What is the ideal diagnostic device for poverty stricken countries"? A guide to this answer is offered by the World Health Organization (WHO, Table 1, (12)), which identifies the following design criteria (the ASSURED criteria) for diagnostic devices for use in developing countries. In this chapter we examine how the introduction of microfluidic-based technology may go some way to mitigating the apparent impasse in introducing diagnostics methodology into the developing world and meet many, if not all, the ASSURED criteria. This would be achieved through exploiting the inherent advantages of microfluidic technology such as portability, reusability, or disposability of equipment, reduced costs through mass manufacture of plastic and paper devices, together with the integration of sample pretreatment and detection modules combine to make microfluidic devices more robust than conventional diagnostic assays.

There have been a limited number of studies for the identification of virus-mediated infections and many of the current microfluidic approaches for virus detection have utilized molecular biology, genomics, and proteomic methodology. The challenge that now needs to be addressed, in order to engineer appropriate diagnostic microfluidic devices, is the integration of all the sample handling processes on a single device: these include a "real world" sample interface, preparation, assay, separation, detection, and collection of waste products. There is a requirement to characterize analytes in a range of biological samples, such as blood, saliva and urine, each of which pose a number of distinct challenges related to factors such as viscosity and potential interferences. The processing of blood represents a number of difficulties due to its complex composition and very low concentration(s) of the analyte of interest. For example, the median JC virus, formally known as papovavirus, load concentrations in infected blood can be in the region of 1.9–5.7 $\log_{10}$ copies ml^{-1} (13). In contrast, the hepatitis B viral load in saliva is in the order of 10^5–10^7; however the presence of mucin molecules cause significant difficulties as they readily aggregate and can lead to blockages in microfluidic channels when extracting viral components (14, 15). Furthermore, the need for additional processing of the biological sample to remove inhibitors increases also with the specimen size. In practical terms processing procedures involve dilution, filtration, reagent mixing, and buffer exchange that in turn leads to further dilution of the target analyte and associated detection issues. DNA amplification using PCR and immunoassay detection within microfluidic channels and chambers offers many advantages over traditional methodology including, faster processing speeds, smaller volumes of samples and reagents, limited sample pretreatment and purification, and for PCR the ability of integration and automation of preparation, amplification, and detection steps (16–18). The challenge therefore facing microfluidic technology in many cases is to extract and pre-concentrate the target viral nucleic acids selectively from relatively large (>1 ml) sample volumes. Such processing can include cell-sorting, separation of plasma/serum from whole blood or cell lysis for immunoassays or nucleic acid extraction and amplification using solid phase and gel methodology where appropriate.

Cell separation in microfluidic-based devices offers the advantages of fully automated processing, portable instrumentation for field and point-of-care use, on small input sample volumes (μl or less of blood) along with corresponding reagent volumes. Separating target cells from contaminating cells or fluids prior to characterization represents the first step in many bioanalytical procedures. Cells range in sizes from 130 to 8 μm, and exist in complex populations, therefore, sophisticated separation methods are needed as shown in Fig. 1 (19).

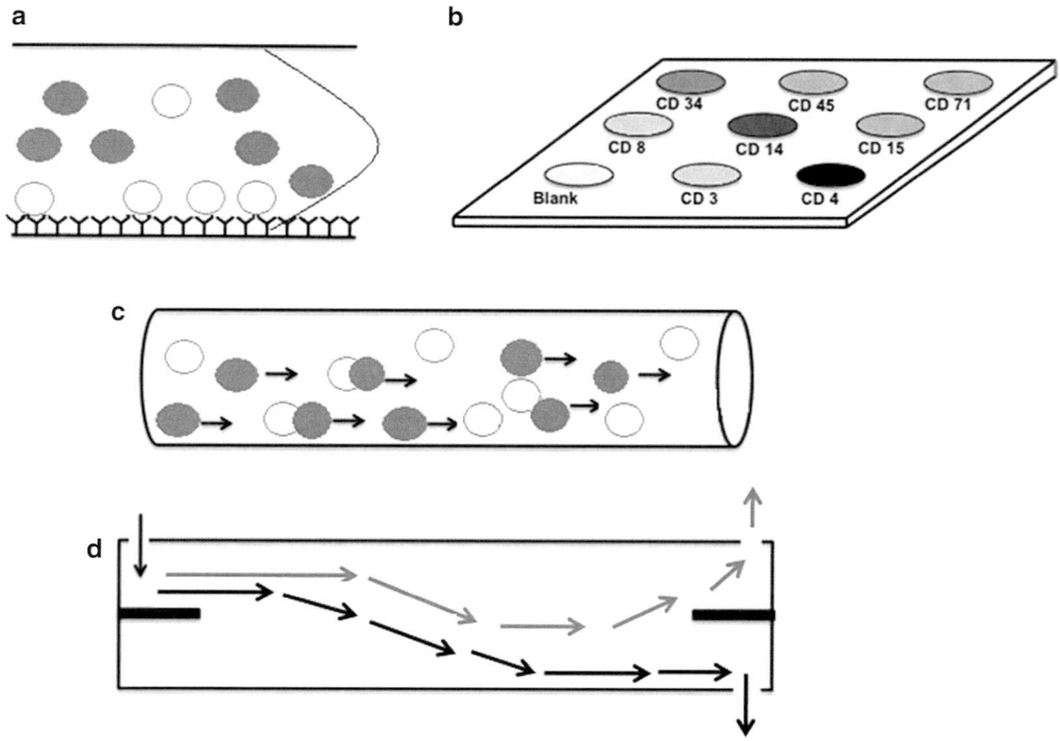

Fig. 1. Overview of common cell separation strategies (19). Affinity capture separations rely on the adhesion of cells to an immobilized, specific antibody (**a**). Unretained cells are washed to waste. The affinity format can be applied to array formats (**b**), where more than one capture molecule is attached to a surface. Here, antibodies corresponding to different cell surface antigens are used to separate blood cells. Affinity separations can occur in a column (**c**), facilitating elution and collection. Cells can also be separated by mechanical methods as shown by the different *colored arrows* (**d**). Copyright (2007), with permission from Elsevier.

Another robust development in microfluidic fabrication has also occurred in the last few years with the arrival of paper-based devices (20, 21). This advance in fabrication technique has the ability to produce diagnostic devices that are greatly reduced in cost with the move away from glass and plastics, simplified fabrication from the removal of need for etching, vacuum pumps and clean rooms, elimination of pumps or power, and the removal of bubbles which currently cause experimental failure in many a microfluidic laboratory. The arrival of paper-based microfluidics could have immense impact on the development of diagnostic devices for healthcare in developing countries.

2. HIV Diagnostic Tools

Globally, HIV/AIDS is most prevalent in Eastern and Southern Africa. Obstacles to HIV diagnosis and treatment in developing countries include funding shortages, lack of political will, limited human resources, and weak procurement and supply management systems (22). The diagnostic criteria for HIV infection are typically made on the basis of an antibody response to HIV (seroconversion). Serological tests for the detection of HIV antibodies are generally classified as screening or first-line assays, which provide the presumptive identification of antibody-positive patients, and supplemental or confirmatory assays, which confirm whether samples are reactive to discrete HIV strains and/or HIV antigen in specific screening assays (23). In recent years, the diagnosis of HIV infection and status of the infection in response to therapy has included assays that detect virus particles (HIV p24 antigen) or viral nucleic acid (24). In clinical HIV/AIDS management, the absolute number of CD4+ T cells is an indicator of the extent of damage to the patient's immune system. Regular examination of CD4 counts (two to four times per year) is recommended after infection has been identified. The most commonly used methods for the differential separation of leukocyte subpopulations are magnetic-activated cell sorting (MACS) and fluorescence-activated cell sorting (FACS). These methods rely on monoclonal antibodies tagged with magnetic or fluorescent compounds, which are used to separate cells based on their specific antigenic determinants; however the process is expensive and requires laboratory facilities. A microfluidic-based system to minimize monocyte contamination and isolate CD4-positive T lymphocytes (CD4+ T cells) has been reported based on a polydimethylsiloxane (PDMS) and glass microfluidic device (25).

The device consisted of two interconnected chambers in series, the first for monocyte depletion and the second for CD4+ T cell capture. The different chambers were functionalized with separate antibodies, the first with a combination of anti-CD36 and anti-CD14; these two surface antigens are widely expressed on non-CD4+ leucocytes. The second chamber was coated with anti-CD4 antibodies. Venous whole blood (10 μl) from a panel of 11 HIV positive patients was injected at 5 μl min^{-1} into the first chamber, followed by phosphate-buffered saline (PBS) with 5% dextran at the same flow rate to propel the unattached cells into the second chamber. Cells in the second chamber were enumerated using "off chip" flow cytometry and compared to the CD4 count achieved by flow cytometry alone. Using the microfluidic device it was found that cell capture of CD4 +ve cells was >80% and results were comparable with conventional flow cytometry in the clinically relevant range of 0–500 cells/μl.

The separation and counting of both CD4 and CD8 T lymphocytes has also been reported using an integrated single device to carry out sample incubation and microflow cytometry (26). The design of the microdevice involves a pneumatically driven mixer, consisting of three flange-like PDMS arms, used to combine the CD4+/CD8+ T cells with differentially labeled detection reagents, i.e., anti-CD4 (fluorescein), anti-CD8 (recombinant phycoerythin), and anti-CD3 (CyQ®) linked to a micro-flow cytometer by micro-pumps generating the different streams for sample and sheath flows.

Pretreated human blood to remove erythrocytes (400 μl) mixed with an excess (8 μl) of anti-human CD4/CD8/CD3 antibodies were placed into the incubation chamber for 30 min with the micro-mixer operating at 1 Hz under 20 psi (138 kPa) of pressure. The cell–antibody complexes were then shuttled and hydrodynamically focused by the two parallel sheath flows. The sample and the sheath flows were 75 and 150 μl min^{-1}, respectively. The sample was illuminated using a 473 nm laser diode for optical detection. Differences in light emissions were used to count and differentiate the labeled T cells (CD3+/CD4+; CD3+/CD8+ cells). The results from the microdevice were compared with those obtained from a 400 μl pretreated sample which was incubated using a bench-top shaker and standard flow cytometer. Based on seven blood samples it was shown that sample processing and analysis by both the microdevice and traditional methodology gave comparable results. However, the microfluidic procedure offered the benefits of lower reagent volumes, faster assay times (1 h for the traditional, 35 min for the microdevice) and a single, simpler, experimental platform.

Additionally, as HIV/AIDS testing relies not only on CD4+ counts but also on the viral load, a microfluidic device has also been reported that will separate viral particles from the plasma based on superparamagnetic nanoparticles as shown in Fig. 2 (27).

The basic steps used to separate the target analyte magnetically from the medium are (1) the mixing of sample with functionalized magnetic particles; (2) retention of the analyte bound particles; (3) removal of any unbound sample; and (4) elution of the target analyte for subsequent detection and analysis.

The design of the device exploited the magnetic field gradient to intensify the force acting on the nanoparticles while reducing the separation distance between the individual nanoparticles and the trapping surface. This allowed the nanoparticles to be more difficult to separate and decreases the time needed to separate them later. The PDMS device used a passive micro-mixer of various designs (10–37 cm length, 200–300 μm wide, 100 μm depth) and a magnetic separator made of close packed polydisperse iron particles (25–75 μm in diameter) with an external neodymium (NdFeB) permanent magnet. Infected plasma (10^6 virons ml^{-1}) was premixed with anti-CD44-coated nanoparticles at a ratio 5:1 for 60 min before being injected on to the device at varying flow rates (30–100 μl min^{-1}). After the addition of wash then lysis buffer, the processed sample (200 μl) was analyzed using an HIV p24 protein ELISA. This was achieved after 20 min for a 0.5 ml sample (after 60 min pretreatment). The magnetic separator demonstrated

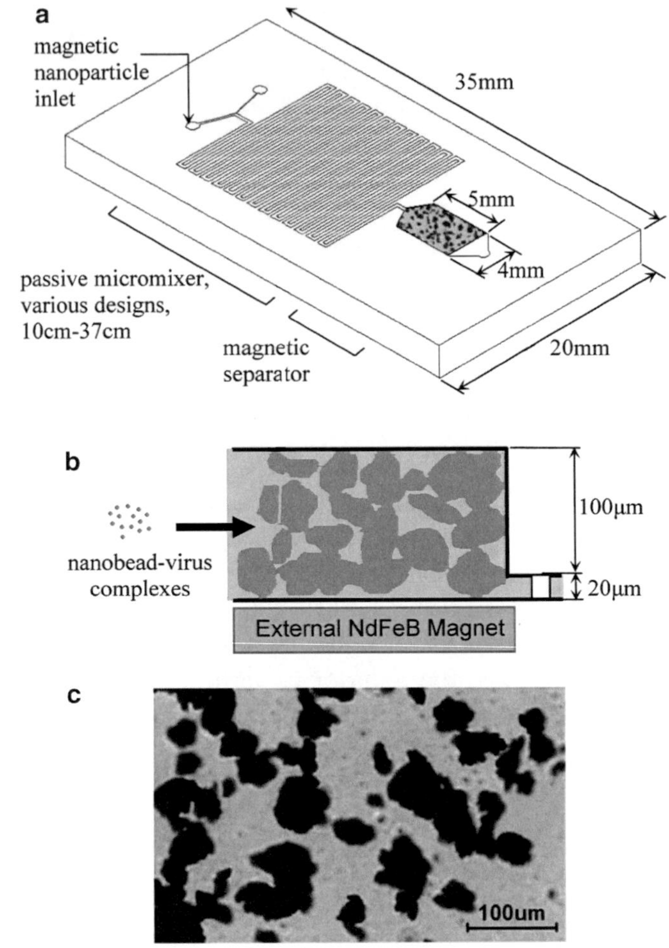

Fig. 2. (**a**) Schematic of the device used for the concentration of HIV-1 virions by superparamagnetic nanoparticles (27). An infected plasma sample is mixed with functionalized magnetic nanoparticles, and then the mixture is trapped in the magnetic separation chamber. (**b**) Schematic of the magnetic separator. The microfluidic trapping chamber is 5 mm long × 4 mm wide × 120 μm high. Iron particles (25–75 μm) are physically trapped before the 20 μm high outlet channel and are compacted into random close packing by the force of the flow. (**c**) Suspension of polydisperse iron particles. Copyright (2010) American Chemical Society.

78% efficiency for viral extraction for plasma in this study, which is comparable with traditional procedures.

Many microfluidic protocols published in the last 5 years have focused on nucleic acid analysis, DNA or RNA templates, as a diagnostic tool, with a large proportion of these exploiting PCR-methodologies for identifying common viruses found in developed countries. Reverse transcriptase-PCR (RT-PCR) uses an additional step before the standard PCR process to convert RNA into complementary DNA (cDNA). This is required for analysis of retroviruses, such as HIV, where the genetic code is stored in a form of messenger (m) RNA rather than DNA. A diagnostic disposable

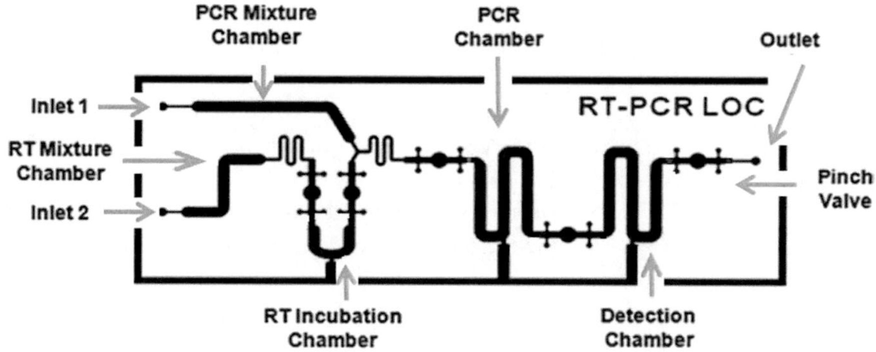

Fig. 3. Design for the polymer RT-PCR lab-on-a-chip for HIV diagnostics (28). Reproduced by permission of The Royal Society of Chemistry.

cyclic olefin copolymer (COC) device for RT-PCR has been described that uses five embedded pinch valves to control the different elements of the procedure as shown in Fig. 3 (28).

The microdevice used embedded cylindrical silicone tubing (6 mm long, 1 mm width, 0.2 mm depth) with external mechanical plungers to isolate different solutions and procedures in specific areas of the device, e.g., RT incubation, PCR amplification and detection. Cycling was achieved using three external infra-red 80 W halogen lamps and two DC cooling fans. Thermocouples with direct contact to the reaction solutions were situated in the RT incubation, PCR and detection chambers, with the temperature being controlled by a LabVIEW program. Optical detection of the chemiluminescence assay was achieved with 0.05 nA photodiode and high-gain signal amplifier with a conversion gain of 1.0×10^8 V/A. To demonstrate the ability to detect HIV *gag* and *envelope* genes, RNA was obtained from HEK 293 cells previously transfected with these viral sequences. The RNA was injected into the embedded tubing, mixed with 2 μl RT enzyme mixture, and incubated at 40°C for 2 min. This was then injected into the PCR chamber with 3 μl of PCR regents at unstipulated flow rates and thermally cycled (94°C for 10 s; 56°C for 10 s; 72°C for 10 s) for 30 cycles after an initial 60 s heating of 95°C and then maintained at 72°C for 60 s as the final extension. The success of the RT-PCR was first assessed by gel electrophoresis "off-chip" and then repeated on-chip using chemiluminescence detection. Gene-specific probe sequences were immobilized in the detection chamber and 5 μl of the cDNA sample was transferred with the chamber heated to 95°C for 60 s to denature the DNA. After binding of the now single-stranded cDNA to the probe DNA, the chamber was washed with PBS to remove any unbound sample then injected with 1 μg ml^{-1} streptavidin-labeled horseradish peroxidase and incubated for 5 min. After a final rinse, 5 μl Supersignal® femto chemiluminescent substrate solution (Rockford, USA) was added with the resulting emitted light measured. Comparisons between the on-chip RT-PCR and conventional procedure were made using gel electrophoresis, and the portable optical analyzer was compared with a conventional scanner. Quantitative comparison data were not shown; however, both the on-chip procedures of RT-PCR and detection were successful.

The methodology described above was used on blood and cell samples however, for different biomedical samples such as salvia, an immunoassay microdevice has been described that uses finger-actuated pouches to move samples and reagents for anti-HIV antibody detection (29). This disposable luminescence-based immunoassay microdevice consists of a top plate containing reagent storage chambers, air-filled pouches, and metering chamber combined with a lower, processing, base plate with connection needles, mixing chamber and detection chamber. Subsequent to sample induction, the two component plates are joined with the needles (internal diameter 500 μm, 900 μm long) forming connections or closing valves. Samples were clarified saliva samples that were spiked with anti-HIV antibodies. The pouches are depressed forcing the reagents around the microdevice. Passive "zigzag" mixing was employed with the contents of two (100 mM HEPES–0.5% (v/v) Tween-1% bovine serum albumin) filled buffer chambers (20 μl), one preceding and one following the sample, and sample metering chamber (10 μl) being forced into the 50 μl mixing chamber. The contents of the mixing chamber were then forced onto an HIV lateral flow strip functionalized with HIV-specific antigens, after a 2 min interval 20 μl of buffer was flushed through to remove any unbound analytes. After a further 2 min an immunolabeling buffer containing 100 ng protein-A-coated with up-converting phosphor (UCP) reporter molecules was run through to interact with a control line functionalized with anti-human IgG to verify assay completion. General applicability of this approach was further demonstrated through its ability to identify interleukin 8 with an integrated microbead array. The ability to identify anti-HIV without instrumentation makes it ideal for developing-world diagnostics, although as mentioned earlier, a problem still to be resolved with this methodology is the issue of real-world samples with mucin molecules in saliva.

3. Influenza A Diagnostic Tools

WHO have identified the importance of influenza research seeking to limit and minimize the spread and impact of influenza, optimizing the treatment of patients, and promoting the development and application of modern public health tools (30). This need was clearly highlighted by the 2009 global pandemic of H1N1 influenza

A serotype, which caused symptoms ranging from mild upper-respiratory tract illness to severe or fatal pneumonia. In Mexico alone, there were 3,734 laboratory-confirmed cases and 74 fatalities. It has been recognized that the virological course of influenza A (H1N1) infection needs to be defined based on the analysis of a range of sample types including extrapulmonary secretion, feces, or vomitus (30). A recent methodology utilizing PCR to differentiate between influenza A serotypes in a silicon-based microfluidic system was reported (31). The PCR chips comprised a micro-PCR chamber (300 μm depth, 300 μm width, 10 μl volume), micro-heaters and temperature sensors at the edges of the device with access ports for the addition and removal of sample and reagents.

A PCR protocol replicated specific sequences of the two different serotypes, FluA and H1N1 using templates and primers taken from conventional PCR methodology. Single-stranded DNA (ssDNA) templates of both serotypes (20.8 fg μl^{-1} FluA and 29.62 fg μl^{-1} H1N1) were mixed with the PCR assay reagents before being added to the chamber for amplification. The process was compared with an assay carried out in a conventional 96-well thermal cycler using the same schedule as in the traditional methodology. The products were flushed from the chip using mineral oil and then separated "off-chip" using gel electrophoresis and quantified after removal from the gel using Picogreen assay in a standard spectrofluorometer. Using multiplexed amplification, viral DNA was increased 10^4–10^5 times in magnitude. This was 50% lower than the sample amplified in the PCR cycling machine. Subsequently, 1 μl of the chip amplified double stranded was denatured and injected into a microchannel (10 mm length, 3 mm width, 200 μm depth) at 10 μl min^{-1} for 15–20 min. The channel was separated into three sections using silicon nanowires (SiNW) functionalized with (1) H1N1 peptide nucleic acid (PNA) probes, (2) FluA PNA probes, and (3) control with no PNA probes. The resistance measurements of the nanowires were recorded and compared with a baseline control. The differential (I_d) exhibited an exponential-shaped hybridization curve on the functionalized SiNW suggesting a binding of the ssDNA in the sample. The average I_d were recorded (H1N1 SiNW cluster was 45.7 ± 158 pA, FluA SiNW cluster was 400 ± 134 pA) after the sample was introduced and compared to the control. As expected, it was shown that the FluA sample measurements were highest as the DNA sequence is common to all FluA serotypes including H1N1.

A similar study reported the extraction of viral RNA using a disposable plastic, Zeonex® medical grade cyclic polyolefin, with channels fabricated measuring 400 μm width, 100 μm depth, and 15 mm in length (32). Circular 1.5 mm wells were drilled at each end of the channel for sample introduction and collection. Within the channel a porous poly (butyl methacrylate co-ethylene) monolith was prepared. Cultured Madin Darby Canine Kidney (MDCK) epithelial cells were inoculated with influenza A for 5 days. Both the cell pellet and supernatant were diluted 1:1 with lysis buffer and poly A carrier RNA and 15 μl aliquots was flowed over the solid phase extraction (SPE) columns. The samples were trapped, washed (70 and 100% ethanol), and eluted (RNase water) using a flow rate of 300 μl min^{-1} for 3 min (sample), 2 min (washes), and 3 min (elution). RT-PCR was then achieved "off-chip" using a PCR cycling machine against a commercially available RNA extraction kit (QIAamp®, Qiagen, Inc.). The threshold cycle value for the μSPE was 29.58 ± 2.28 compared to Qiagen 26.65 ± 1.1. This corresponded to a 7.62 fold drop in extraction from the microfluidic method. However, the μSPE demonstrated that flu-A viral RNA could be extracted and eluted; however for clinical use further refinement and optimization of procedure will be required.

Finally, methodology for the detection of influenza A (H1N1) based on the use of electrophoretic immunoassay has been proposed (33). Two different configurations of microchip immunoassay were examined. The first microdevice used open channel electrophoresis and laser-induced fluorescence detection, a later design integrated a polymerized 6% polyacrylamide membrane into part of the microchannel (150 μm long, 100 μm wide, and 25 μm deep) to remove sample contaminants and to concentrate and separate the viral particles. The pore sizes within the gel were approximately 8–12 nm, allowing the flow of antibody molecules but not influenza virus that has an approximate diameter of 160 nm. Thus, the viral particles were concentrated at the boundary of the gel. The immunoassay was performed at a pH 8.9 with a polyacrylamide coating added to the microchannels to control bulk electroosmotic flow. Platinum electrodes were placed in fluid reservoir wells (~100 μl) and 400 V was applied at the sample waste (SW) well, while the sample (S) well was grounded for 120 s. This caused the negatively charged sample to move to the polyacrylamide membrane as shown in Fig. 4. A voltage (93 V) was applied at the buffer waste (B) to stop the sample entering the separation channel. Separation took 120 s with the sample moving down the channel using 800 V at the waste (W) while grounding the B well, with voltages of 325 V at SW and 275 V at S to pull the sample into the side channels.

To remove the viral sample from the interface of the gel, 800 V was applied to W for 4 min and SW was grounded. Epi-fluorescence detection and imaging was conducted using Alexa Fluor 488-conjugated murine anti-influenza A antibodies. The open channel configuration exhibited several drawbacks during testing with the most important being the limit of detection (LOD) of 2750 $TCID_{50}$ ml^{-1}, which is extremely low sensitivity compared to the negative control sample. Using the polyacrylamide membrane, a viral concentration detection of 7.5×10^2 $TCID_{50}$ ml^{-1} was achieved, which is comparable to traditional LOD, with the advantages

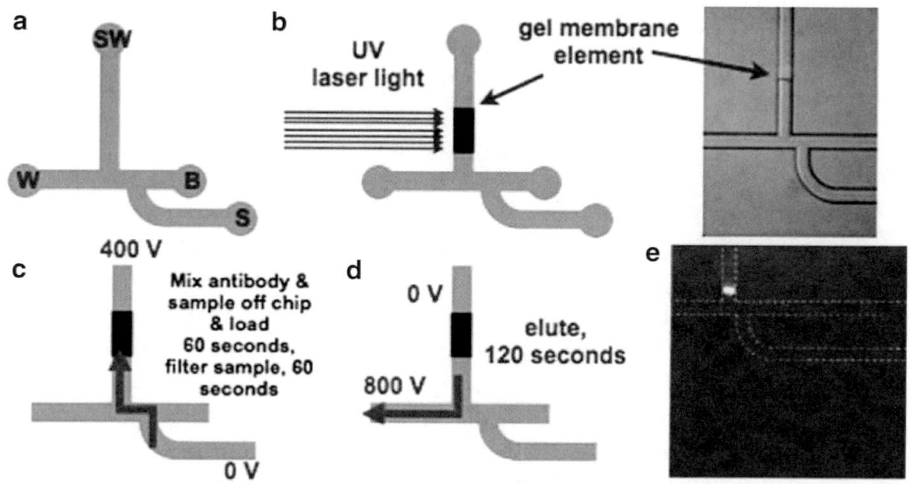

Fig. 4. (**a**) Schematic of microfluidic chip used for electrophoretic immunoassays (33). Reservoir designations: *S* sample, *SW* sample waste, *B* buffer, *W* waste. (**b**) A 6% polyacrylamide plug formed in a glass microchip by projecting a shaped UV laser beam onto the microchannel. (**c**) Virus and antibody are moved by electrophoresis. (**d**) Fluorescence can be measured at the membrane or the retained analyte are eluted towards the waste. (**e**) Labeled antibody complexed with viral particles is detected by epi-fluorescence microscopy with the virus–antibody complex concentrated at the gel interface. Reproduced by permission of The Royal Society of Chemistry.

that the microfluidic device uses smaller sample volumes, allowing for serial/repetitive analysis, and the process is completed within a 6 min cycle time.

4. Dengue Virus Diagnostic Tools

Dengue fever occurs in tropical and subtropical regions of the world and the virus is transmitted by the bite of an *Aedes* mosquito (34). Symptoms are a severe flu-like illness exhibited by all ages and occasionally a potentially fatal complication called dengue hemorrhagic fever (DHF), which may occur 3–14 days after receiving an infectious bite. Whilst there is currently no specific treatment, medical care can reduce the prevalence of the DHF, which if left untreated can lead to a mortality rate of up to 50%. A one-step RT-PCR diagnosis system that integrates the sample purification and enrichment steps using superparamagnetic beads on a single device has been reported (35). The integrated microsystem includes a PDMS microfluidic device, magnetic microcoils, microheaters, and temperature sensors. Dengue virus serotype 2 and enterovirus (EV) 71 virus RNA templates were used in this study. Cavities for wash buffer, sample, and RT-PCR reagents were fabricated for volumes of 50 μl with microchannels connecting the various chambers. The solutions were pumped by three PDMS valve micropumps driven by an electromagnetic valve; the volumes of the reaction chambers were not given. PBS buffer (25 µl) was mixed with 10^7 Dynabeads® 4.5 µm sized M-450 coated with antibodies before loading into the device's pretreatment chamber. Dengue viral fragments (511 bp ssDNA) were then injected onto the device for capture on the beads followed by a PBS wash solution (volumes not described). Aliquots of the viral-bound nanoparticles (5 µl) were removed to the RT-PCR chamber and mixed with a pre-loaded lysis buffer at 95°C for 95 min. The RT-PCR regents (DEPC solution and Mastermix, Promega) were pumped onto the chip and the RT-PCR procedure was performed for 40 cycles at 94°C for 35 s, 55°C for 30 s, and 75°C for 35 s followed by a supplementary 5 min at 72°C to allow for completion of elongation. The PCR products were finally removed from the reaction chamber and separated using gel electrophoresis. The microfluidic procedure was compared with a commercial RNA extraction kit (QIAGEN, Germany) and shown to have the same detection efficiency as the traditional method.

A second microfluidic method has also been reported for the determination of the Dengue virus using a CD-shaped PDMS microdevice with 12 DNA hybridization chambers (15 mm length, 400 mm width, 46 µm depth) temporally sealed onto a glass substrate immobilized with a Dengue serotype DNA array (36). Hydrophobic valves allowing geometric constriction of a hydrophobic channel from a cross-section of 400–50 µm, coupled with centrifugal force were used to generate reciprocating flow in the PCR chamber. Fluorescein-labeled synthetic Dengue virus oligonucelotides (350 nl), in high concentrations ranging from 25 to 200 nM, and wash buffer (1 µl) to remove unbound target DNA, were added to the device and loaded at the sample entry port and the wash reservoir, respectively. The CD-shaped microdevice was then spun between 8.5 and 25 Hz for 3 s and stopped for a further 3 s. Repeated rotation-stop processes continued for 90 s to allow hybridization to occur. The device was then spun for 10 s at 22 Hz to drain the remaining sample into the waste chamber, and subsequently spun for 30 s at 38 Hz to drive the wash buffer over the hybridization zones to remove any unbound target molecules. To detect and analyze the effectiveness of the device, epi-fluorescence microscopy was used.

Characterization and comparisons of different rotation speeds (8.5–25 Hz) and target DNA concentrations on the effects of hybridization rates were investigated. Increased rotational frequencies amplified the amount of DNA hybridization; however rates above 22 Hz had little additional effect on the amount of fluorescent signal detected. This is due to the centrifugal pressure being directly proportional to the square of rotational frequency. An increase in rotation speed results in higher oscillation amplitude of the sample. At maximum amplitude of the sample, increased rotational frequencies have minimal effect on the hybridization signal. As expected, an increase in target DNA resulted in a higher amount of fluorescent signal being detected.

5. Other Viral Diagnostics

It has been shown through the analysis of various genomes that some of the nucleotide sequences contain retroviral genes, presumably due to infection in the distant past (37, 38). One virus, human papilloma (HPV), has become prominent in the last few decades as it has been demonstrated that the presence of the virus is strongly associated with a number of malignancies, most notably cervical and oropharyngeal carcinoma. Screening of cervical cells for viral incidence has been hampered by methodology that has limited sensitivity with poor reproducibility and specificity in the developed world. A microfluidic system has been proposed that could potentially improve current screening procedures (39), as shown in Fig. 5. The COC polymer-based microdevice has ten parallel (80 nl) reaction chambers. A SiHa cell model, which has been previously successful in conventional tests, was transfected in the device with one to two copies of HPV 16 per cell, before lysis and extraction of the nucleic acids for analysis.

To define the lower detection limit of the methodology, cell lines were tested 2×10^{-2} cells μl^{-1} to 2×10^{-3} cells μl^{-1}. A single abbreviated cycle of RT-PCR using PreTect® HPV-Proofer kit (NorChip, Norway) was achieved off-chip before being distributed to the channels then left to run for 2.5 h at 41°C. Detection of the PCR products was achieved by the use of light emitting diodes (LED). The detection capabilities of the microdevice were compared with conventional methods using 20 μl sample volumes and demonstrated the same sigmoid curvature profile.

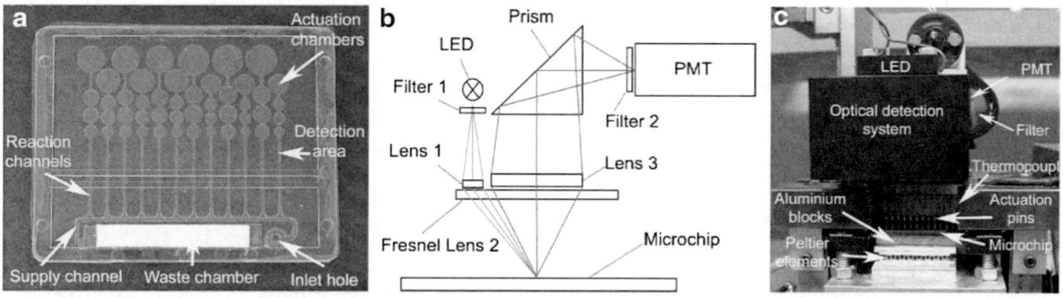

Fig. 5. (**a**) An image of the COC microchip used to detect human papilloma (HPV) with dimensions of 50 × 40 mm. (**b**) Sketch of the optical geometry. (**c**) An image of the major components in the instruments (39). Reproduced by permission of The Royal Society of Chemistry.

A microfluidic-PCR method has also been demonstrated for the detection of BK virus particles in unprocessed whole blood samples (40). The BK virus was first identified in 1971, although the majority of infected people are asymptomatic unless the patients are suffering from immunosuppression where clinical presentation can include renal dysfunction and abnormal urinalysis (41). The microdevice consisted of a 300 μm PDMS layer sandwiched in between two layers of glass and the PCR chamber volume was 600 nl. A 1 μl aliquot of blood was contaminated with 10^5 copies of purified BK virus before the addition of the 20 μl BKV-Alexa labeled primer PCR mix. A rapid two-step PCR was performed with an initial denaturizing period (94°C for 120 s) was followed by 35 cycles of 94°C for 10 s, 64°C for 20 s, 72°C for 120 s using a Peltier element controlled by a calibrated thermocouple situated in the reaction chamber. Detection was achieved by micro-capillary electrophoresis (μCE) using laser-induced fluorescence (excitation wavelength 635 nm; emission wavelength 670 nm) optical detection. The results were compared to those achieved using a traditional thermocycler. Electropherograms show the successful amplification of on-chip BKV PCR with a slight shift of the fluorescent signal due to the overloading of primers. The signal was more condensed than those of the thermocycler control; however the magnitude of intensity was comparable.

The plaque assay is the "gold-standard" to measure virus infection and test for antiviral drug efficacy. To perform a plaque assay, serial dilutions of a virus stock solution are prepared and inoculated onto susceptible cell monolayers. After an incubation period, permitting the virus to attach to the cells, the monolayers are covered with a nutrient medium agar. After a further period of incubation, the original infected cells release the virus progeny. The presence of the gel restricts the spread to neighboring cells and the viruses cause a zone of infected cells called a plaque, which eventually becomes apparent to the naked eye. Antiviral drugs inhibit the spread of the virus progeny providing the means to assess a drug's potency. Although it is one of the most important procedures in virology, it is associated with many drawbacks, e.g., the methodology is very labor-intensive, it lacks sensitivity, requires a relatively large volume of reagents, and is difficult to automate. In a novel approach, an infection assay was performed within microchannels (32 mm × 2 mm × 250 μm) with an inoculation port 8 mm from the inlet port (42). Baby hamster kidney (BHK) cells were injected into the microchannels at 4×10^6 ml^{-1} and cultured for 12 h before inoculation with vesicular stomatitis virus (VSV), which has been used extensively in the study of virus properties and viral evolution. A recombinant form of VSV carrying the gene for green fluorescent protein (GFP) was used in the study and is expressed in infected cells early in the mitosis cycle. After the cell surface density in the microchannels was measured, 6 μl VSV-GFP medium was dropped into the inoculation port and 20 μl of medium was set at the inlet port. The inoculum size, measured in plaque-forming units (PFU), had a serial dilution range from 3×10 to 3×10^6. Passive pumping was employed by the generation of a pressure drop between the inoculation and outlet ports allowing the virus-laden droplet to enter the channel without the need for additional apparatus. After 24 h, having added tissue culture medium after 12 h, the microchannels were imaged with the cells subsequently fixed and stained with crystal violet. The average GFP intensity was quantified using a MATLAB program, which calculated the average pixel value of every channel. These results were compared to those obtained by use of the traditional plaque assay. The BHK cells remained viable for 4 days but had slower replication rates than those grown on standard plates. The spatial patterns of infection were analogous for 3×10^2 to 3×10^6 PFU with a steep decline in the fluorescent signal at the lowest dilution as expected with the low volume of inoculum. The 3×10^2 plaque assays were subsequently subjected to an antiviral drug (5-fluorouracil, 5-FU) at a concentration range of 0 to 64 mg ml^{-1} and regions of cell death visualized with crystal violet stain. As expected the higher the concentration of 5-FU, the lower the intensities of infection-mediated cell death in the channels. This methodology has the benefits of simplicity, and smaller drug reagent volumes, however, the major advantage was that the assay sensitivity was twofold more sensitive than the standard plaque-reduction assay used routinely in virology labs.

6. Conclusions

Infectious disease remains the major threat to global health with 95% of infections occurring in developing countries, which are the least able to cope. Innovations in microfluidic technology have the ability to offer simple, accurate, mass produced, and low-cost diagnostics that are essential and unavailable in those areas that need them the most. Developments in all aspects of the separation, amplification, and real-time detection methodologies have stimulated diagnostic innovations in the past few years. The ability to combine complicated procedures with sophisticated reagents and detection systems into a single easy to use platform that delivers point of care information is the aspiration of many microfluidic groups. One of the main obstacles in supplying appropriate technology to patients with infectious disease is the lack of investment in the engineering and packaging of these diagnostic tools. The patients with the greatest need, in tropical and subtropical climates, are usually some of the poorest in the world and therefore present a low return on the high investment costs to develop, assess, and commercialize new assays. The only path to success, which seems available in the present financial climate, is the amalgamation of the private and public sectors to provide the essential investment coupled with the need for innovative solutions driven by medical necessity.

References

1. Siefkes D (1993) The origin of HIV-1, the AIDS virus. Med Hypotheses 41(4):289–299
2. WHO (1978) Ebola haemorrhagic fever in Sudan, 1976. Bull World Health Organ 56(2):247–270
3. Ng CW, Choo WY, Chong HT, Dahlui M, Goh KJ, Tan CT (2009) Long-term socioeconomic impact of the Nipah Virus encephalitis outbreak in Bukit Pelanduk, Negeri Sembilan, Malaysia: A mixed methods approach. Neurology Asia 14(2):101–107
4. WHO (2009) Global Health Risks—Mortality and burden of disease attributable to selected major risks. Available from: http://www.who.int/healthinfo/global_burden_disease/global_health_risks/en/index.html Accessed 12 Nov 2012.
5. Welch RJA, Anderson BL, Litwin CM (2008) Evaluation of a new commercial enzyme immunoassay for the detection of IgM antibodies to West Nile virus using a ratio method to eliminate nonspecific reactivity. J Clin Lab Anal 22(5):362–266
6. Verweij PE, Erjavec Z, Sluiters W, Goessens W, Rozenberg-Arska M, Debets-Ossenkopp YJ et al (1998) Detection of antigen in sera of patients with invasive aspergillosis: Intra- and interlaboratory reproducibility. J Clin Microbiol 36(6):1612–1616
7. Cass T, Toumazou, C (2006) State of Science Review: Biosensors and Biomarkers. In: Foresight Infectious Diseases: Preparing for the Future: Office of Science and Innovation, London.http://www.bis.gov.uk/assets/foresight/docs/infectious-diseases/s7.pdf. Accessed 12 Nov 2012
8. Tian Y, Madanahally K, Rao H, Mackwan R, Chen L (2010) Sample-to-result nucleic acid test enables accurate detection of Influenza A/2009 H1N1 in 26 min in near-patient settings. Europ Infect Dis 4(4):26–30
9. Hauck TS, Giri S, Gao Y, Chan WCW (2010) Nanotechnology diagnostics for infectious diseases prevalent in developing countries. Adv Drug Deliv Rev 62(4–5):438–448
10. Yager P, Edwards T, Fu E, Helton K, Nelson K, Tam MR et al (2006) Microfluidic diagnostic technologies for global public health. Nature 442(7101):412–418
11. Moran MM, Guzman J, Henderson K, Abela-Oversteegan L, Wu L, Omune B, Gouglas, D, Chapman N, Zmundzki F (2011) Neglected disease research and development: Is the global financial crisis changing R&D? Global Funding of Innovation for Neglected Diseases.
12. Peeling RW, Holmes KK, Mabey D, Ronald A et al (2006) Rapid tests for sexually transmitted infections (STIs): the way forward. Sex Transm Infect 82(Supplement 5):v1–v6
13. Mengelle C, Kamar N, Mansuy JM, Sandres-Saune K, Legrand-Abeavanel F, Miedouge M, Rostaing L, Izopet J (2011) JC virus DNA in the peripheral blood of renal transplant patients: a 1-year prospective follow-up in France. J Med Virol 83(1):132–136
14. Liu C-J, Chen B-F, Chen P-J, Lai M-Y, Huang W-L, Horng K, Chen D-S (2006) Role of Hepatitis B Viral Load and Basal Core Promoter Mutation in Hepatocellular Carcinoma in Hepatitis B Carriers. J Infect Dis 193:1258–1265
15. Lai SK, Wang Y-Y, Wirtz D, Hanes J (2009) Micro- and macrorheology of mucus. Adv Drug Deliv Rev 61:86–100
16. Mairhofer J, Ropper K, Ertl P (2009) Microfluidic systems for pathogen sensing: a review. Sensors 9:4804–4823
17. Whitesides GM (2006) The origins and the future of microfluidics. Nature 442(7101):368–373
18. El-Ali J, Sorger PK, Jensen KF (2006) Cells on chips. Nature 442(7101):403–411
19. Pappas D, Wang K (2007) Cellular separations: A review of new challenges in analytical chemistry. Anal Chim Acta 601(1):26–35
20. Lu R, Shi W, Jiang L, Qin J, Lin B (2009) Rapid prototyping of paper-based microfluidics with wax for low-cost, portable bioassay. Electrophoresis 30(9):1497–1500
21. Osborn JL, Lutz B, Fu E, Kauffman P, Stevens DY, Yager P (2010) Microfluidics without pumps: Reinventing the T-sensor and H-filter in paper networks. Lab Chip 10(20):2659–2665
22. Namjilsuren T (2010) More developing countries show universal access to HIV/AIDS services is possible. World Health Organization [10 December 2010]; Available from: http://www.who.int/mediacentre/news/releases/2010/

hiv_universal_access_20100928/en/index.html. Accessed 12 Nov 2012

23. Organization WH (1999) Operational characteristics of commerically available assays to determine antibodies to HIV-1 and/or HIV-2 in human sera. World Health Organization
24. Feardon M (2005) The laboratory diagnosis of HIV infections. Can J Infect Dis Med Microbiol 16(1):26–30
25. Cheng X, Gupta A, Chen C, Tompkins RG, Rodriguez W, Toner M (2009) Enchancing the performance of a point-of-care CD4+ T-cell counting microchip through monocyte depletion for HIV/AODS diagnostics. Lab Chip 9:1357–1364
26. Wang J-H, Wang C-H, Lin C-C, Lei H-Y, Lee G-B (2010) An integrated microfluidic system for counting of CD4+/CD8+ T lymphocytes. Microfluidics and Nanofluidics.
27. Chen GD, Alberts CJ, Rodriguez W, Toner M (2010) Concentration and Purification of Human Immunodeficiency Virus Type 1 Virions by Microfluidic Separation of Superparamagnetic Nanoparticles. Anal Chem 82(2):723–728
28. Lee SH, Kim S-W, Kang J-Y, Ahn CH (2008) A polymer lab-on-a-chip for reverse transcip-tion (RT)-PCR based point-of-care clinical diagnostics. Lab Chip 8:2121–2127
29. Qiu X, Thompson JA, Chen Z, Liu C, Chen D, Ramprasad S, Mauk MG, Ongagna S, Barber C, Abrams WR, Malamud D, Corstjens PLAM, Bau HH (2009) Finger-actuated, self-contained immunoassay cassettes. Biomed Microdevices 11:1175–1186
30. Public health research agenda for influenza A(H1N1) (2009) pandemic, World Health Organisation Technical Consultation Report, 2011, Tunisia, World Health Organisation http://whqlibdoc.who.int/hq/2011/WHO_HSE_GIP_ITP_2011.3_eng.pdf [Accessed 12 Nov 2012].
31. Kao LT-H, Shankar L, Kang TG, Zhang G, Tay GKI, Rafei SRM, Lee CWH (2011) Multiplexed detection and diffferentiation of the DNA strains for influenza A (H1N1 2009) using silicon-based microfluidic system. Biosens Bioelectron 26(5):2006–2011
32. Bhattacharyya A, Klapperich CM (2008) Microfluidics-based extraction of viral RNA from infected mammalian cells for disposable molecular diagnostics. Sensors Actuators B-Chemical 129:693–698
33. Reichmuth DS, Wang SK, Barrett LM, Throckmorton DJ, Einfeld W, Singh AK (2008) Rapid microchip-based electrophoretic immunoassays for the detection of swine influenza virus. Lab Chip 8:1319–1324
34. Dengue and dengue haemorrhagic fever (2009) World Health Organization [December 2010]; Available from: http://www.who.int/mediacentre/factsheets/fs117/en/index.html. Accessed 12 Nov 2012
35. Lien K-Y, Lee W-C, Lei H-Y, Lee G-B (2007) Integrated reverse transcription polymerase chain reaction systems for virus detection. Biosens Bioelectron 22:1739–1748
36. Li C, Dong X, Qin J, Lin B (2009) Rapid nanoliter DNA hybridization based on reciprocating flow on a compact disk microfluidic device. Anal Chim Acta 640:93–99
37. Marco A, Marin I (2009) CGIN1: A retroviral contribution to mammalian genomes. Mol Biol Evol 26(10):2167–2170
38. Horie M, Honda T, Suzuki Y, Kobayashi Y, Daito T, Oshida T et al (2010) Endogenous non-retroviral RNA virus elements in mammalian genomes. Nature 463(7277): 84–87
39. Gulliksen A, Solli LA, Drese KS, Sorensen O, Karlsen F, Rogne H, Hovig E, Sirevag R (2005) Parallel nanoliter detection of cancer markers using polymer microchips. Lab Chip 5:416–420
40. Manage DP, Morrissey, Y.C., Stickel, A.J., Lauzon, J., Atrazhev, A., Acker, J.P., Pilarski, L.M. (2011) On-chip PCR amplification of genomic and viral templates in unprocessed whole blood Microfluidics and Nanofluidics 10(3):697–702
41. Gardner SD, Field AM, Coleman DV, Hulme B (1971) New human papovavirus (BK) isolated from urine after renal transplantation. Lancet 7712:1253–1257
42. Zhu Y, Warrick JW, Haubert K, Beebe DJ, Yin J (2009) Infection on a chip: a microscle platform for simple and sensitive cell-based virus assays. Biomed Microdevices 11:565–570

Chapter 20

Applications of Microfluidics for Molecular Diagnostics

Harikrishnan Jayamohan, Himanshu J. Sant, and Bruce K. Gale

Abstract

Diagnostic assays implemented in microfluidic devices have developed rapidly over the past decade and are expected to become commonplace in the next few years. Hundreds of microfluidics-based approaches towards clinical diagnostics and pathogen detection have been reported with a general theme of rapid and customizable assays that are potentially cost-effective. This chapter reviews microfluidics in molecular diagnostics based on application areas with a concise review of microfluidics in general. Basic principles of microfabrication are briefly reviewed and the transition to polymer fabricated devices is discussed. Most current microfluidic diagnostic devices are designed to target a single disease, such as a given cancer or a variety of pathogens, and there will likely be a large market for these focused devices; however, the future of molecular diagnostics lies in highly multiplexed microfluidic devices that can screen for potentially hundreds of diseases simultaneously.

Key words: Microfluidics, Micro-total-analysis-systems, Lab-on-a-chip, Point-of-care devices, Sample preparation, MEMS, Rapid prototyping, Biomarker detection, Personalized medicine, Global health care

1. Introduction

The role of molecular diagnostics is critical in today's global health care environment. In the developing world, 95% of deaths are due to a lack of proper diagnostics and the associated follow-on treatment of infectious diseases; i.e., acute respiratory infections (ARIs), malaria, HIV, and tuberculosis (TB) (1). Recent pandemics like the 2009 H1N1 Influenza A pandemic, have accentuated the need for tools to effectively detect and control infectious diseases. Factors like "rapid pathogen mutation rates, transformation of nonhuman pathogens into human pathogens, and recombination of nonhuman pathogen with human pathogens" have added to the challenge of managing novel infectious diseases (2). Increased global

Gareth Jenkins and Colin D. Mansfield (eds.), *Microfluidic Diagnostics: Methods and Protocols*, Methods in Molecular Biology, vol. 949, DOI 10.1007/978-1-62703-134-9_20,

mobility has aided the rapid spread of infectious diseases from region of origin to other parts of the world as seen during the 2009 H1N1 pandemic. This mobility has highlighted the need for rapid, portable diagnostic (point-of-care [POC]) devices at ports of entry to prevent global spread of infections. Current laboratory culture methods for pathogens take a day or more to provide results (2). Clearly, methods need to be developed to aid rapid and site-relevant diagnosis of disease.

For certain other types of infections, in both the developed and developing worlds, the diagnostic tests need to be repeated periodically to measure response to therapy and monitor the disease condition. One such case is monitoring the viral load (number of viral particles per milliliter of blood) for infections like HIV (Human immunodeficiency virus) and hepatitis C. Sub-Saharan Africa is a region heavily affected by the AIDS pandemic. The lack of standard laboratory facilities and trained laboratory technicians in these regions is a serious bottleneck (3). Similar problems exist in medically underserved areas of the USA. A simple POC platform could enable increased access to treatment for patients in such low-resource settings.

In the developed world, the strategy to deal with major disease burdens such as cancer is shifting from a therapeutic to diagnostic mode (4), as the cost of treating disease falls dramatically if it is found early. Ischemic heart diseases and cerebrovascular diseases, which are the major causes of mortality in the developed world, can be targeted by effective diagnostics (1). With projected US healthcare costs of $4.4 trillion by 2018, expanding conventional expensive diagnostic solutions is not a viable option (5). Rapid, low-cost diagnostic tools that can be dispersed throughout a community for easy access, possibly even in the home, would provide substantial benefit by allowing more rapid diagnosis and monitoring of disease and infection.

Homeland security is another key sector where portable molecular biology tools are needed to detect a variety of biological agents (6). The US Departments of Health and Human Services (HHS) and Agriculture (USDA) maintains a list of biological agents and toxins defined as select agents "that have the potential to pose a severe threat to public, animal or plant health, or to animal or plant products" (7). Again, there is a need for rapid, inexpensive detection, identification, and quantification of pathogens to help reduce this threat.

Hence, there is an unmet need for simple, low-cost/cost-effective, accurate, portable/point-of-care diagnostic tools for rapid identification of disease markers and pathogens in a variety of settings. The FDA (Food & Drug Administration), definition of a "simple test" provides a benchmark for features for an ideal diagnostic tool (Table 1, (1, 8)).

Table 1
Features of the ideal diagnostic tool based on FDA's definition of a "simple test"

Is a fully automated instrument or a unitized or self-contained test
Uses direct unprocessed specimens, such as capillary blood (fingerstick), venous whole blood, nasal swabs, throat swabs, or urine
Needs no operator intervention during the analysis steps
Needs no electronic or mechanical maintenance beyond simple tasks, e.g., changing a battery or power cord
Produces results that require no operator calibration, interpretation, or calculations
Produces results that are easy to determine, such as "positive" or "negative," a direct readout of numerical values, the clear presence or absence of a line, or obvious color gradations
Has test performance comparable to a traceable reference method, as demonstrated by studies in which intended operators perform the test? (Intended operator refers to a test operator with limited or no training or hands-on experience in conducting laboratory testing)
Contains a quick reference instruction sheet that is written at no higher than a seventh grade reading level

There is consensus that for such an ideal diagnostic tool, microfluidics will certainly be required and will likely make up the critical components of the device (9). Microfluidics can be defined as "science and technology of systems that process or manipulate small (10^{-9} to 10^{-18} liters) amounts of fluids, using channels with dimensions of tens to hundreds of micrometers" (10). Lab-on-a-chip (LOC) refers to the application of microfluidics in chemical, biological analysis and diagnostics. The ultimate objective of LOC devices is to integrate the entire gamut of laboratory capabilities on a microfluidic chip (11–13).

Some of the features of microfluidics that make the technology attractive for lab-on-a-chip point-of-care applications are:

- The availability of fabrication methods to manufacture small hand-held devices on a large scale at a lower cost.
- The ability to manipulate small volumes of sample, requiring lower amounts of reagents.
- The ability to analyze small volumes for applications like single-cell analysis, multiplexed analysis, or forensic trace analysis (14).
- Smaller length scales result in faster analyses and higher separation efficiencies, reducing response times. The high speed analysis also makes microfluidics a suitable candidate for high-throughput applications.

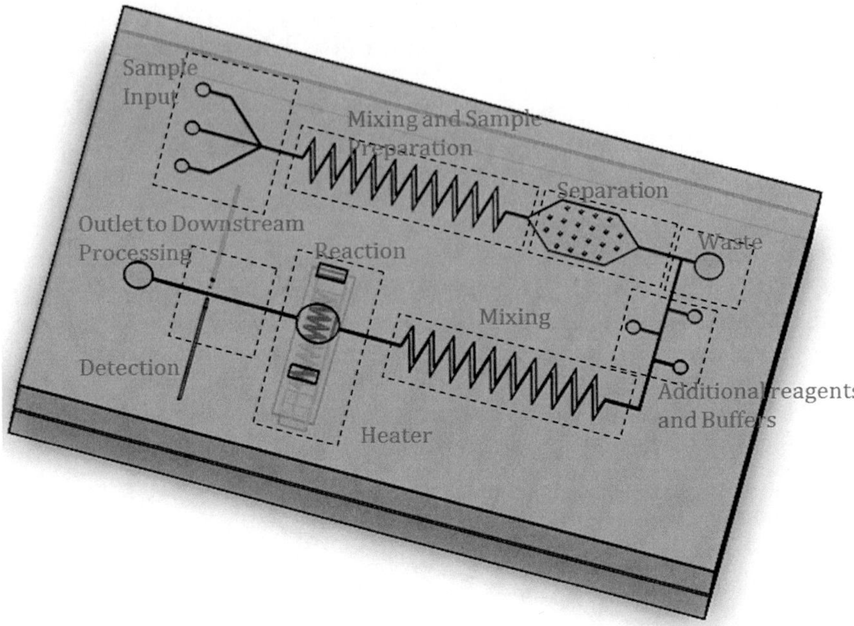

Fig. 1. A schematic diagram of a conceptual lab-on-a-chip device designed to perform a variety of unit operations and unit processing steps including: sample preparation (e.g., fluid handling, , derivatization, lysis of cells, concentration, extraction, and amplification), sample separation (e.g., electrophoresis, liquid chromatography, molecular exclusion, field-flow fractionation), and detection (e.g., fluorescence, UV/Vis absorption, amperometric, conductivity, Raman, electrochemical).

- Straightforward integration of multiple components/functionalities (sample preparation, detection, data processing) on a single device.
- Potentially fully automated and simple to use, enabling use by laypeople.
- Portability and a small footprint should allow field and clinic use, as well as possibly allowing more widespread diagnostics. Pervasive diagnostics should greatly increase the likelihood of personalized medicine having a significant impact on society.
- Highly parallel analyses will allow multiple tests to be run simultaneously, either on the same sample or multiple samples. Microfluidic devices can in principle be used to obtain parameters like proteomic, metabolomic, and genetic data of each individual for personalized care (15).

Figure 1 provides a basic generalized schematic of a microfluidic LOC device with sample-in/readout capabilities. The figure shows some of the various technologies that might be involved in sample preparation, analysis or separation, and detection. Figure 2 is an example of a real nucleotide analysis system developed at the State of Utah Center of Excellence for Biomedical Microfluidics.

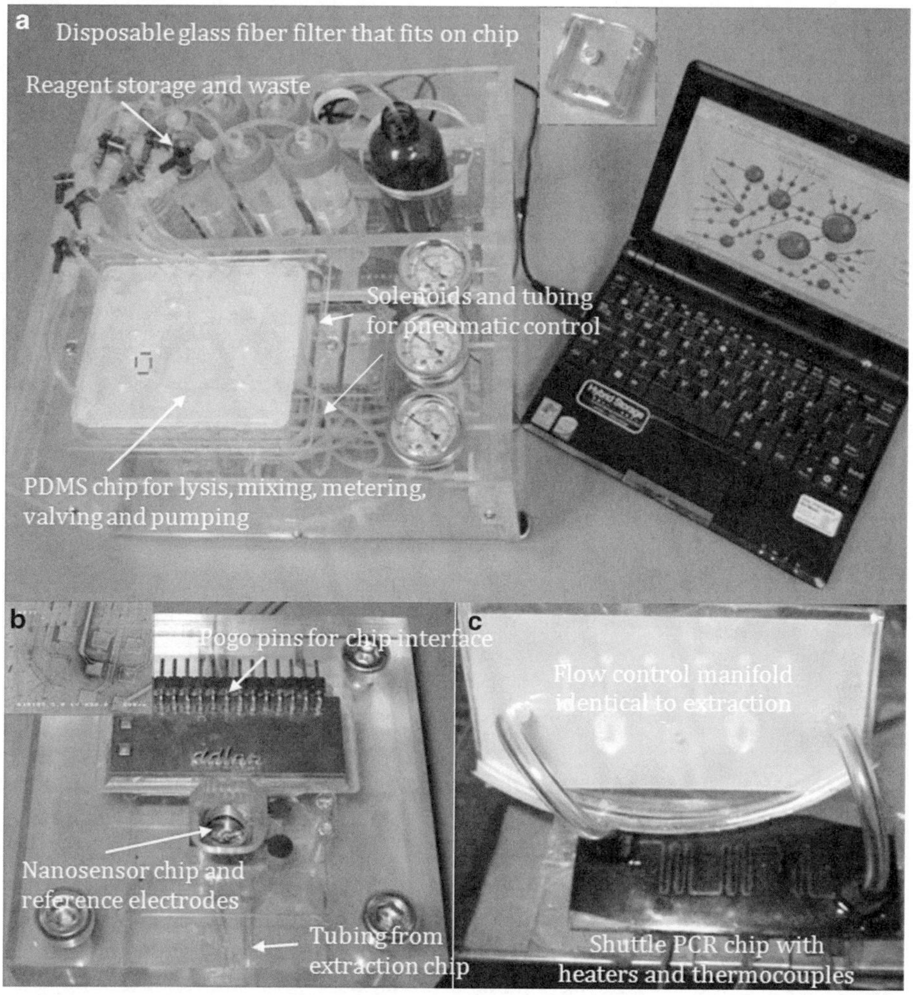

Fig. 2. (**a**) Prototype of an automated nucleotide extraction platform. The microfluidic system consists of five different components: (i) a disposable microfluidic cartridge containing a glass fiber filter (inset figure); (ii) a PDMS-microfluidic chip for flow control; (iii) microfluidic chambers for mixing, metering, pumping, and reactions; (iv) a pneumatic micropump to deliver the eluted sample to downstream assays; and (v) a vacuum pump to control the on-chip valves. The extraction chip also has provision for thermal lysis and reverse transcription (not shown). (**b**) Prototype of a test socket for characterization of a carbon nanotube-based electrochemical nanosensor array. The test socket provides both fluidic and electrical interface to the nanosensor chip (inset figure) that detects nucleotide hybridization. (**c**) Prototype of a shuttle PCR chip with three temperature zones and which is fabricated using polycarbonate lamination. The heaters and thermocouples are shown with a manifold for on-chip fluidic control. The fluidic interface for the extraction system is designed so that it can be readily connected to the downstream assays such as hybridization and PCR.

2. Early Development of Microfluidics

Microfluidic devices have been steadily developing over the past 30 years, but most of the progress related to diagnostic applications has been made in about the past 15 years (10, 16). A major driver for microfluidic development was the focus on genomics and molecular biology in the 1980s especially on microanalysis techniques like high-throughput DNA sequencing. The initial microfluidic devices were inspired by the microelectronics industry and relied on photolithography and MEMS fabrication techniques. Hence, most of the earliest microfluidic devices were fabricated in silicon and glass.

The origins of microfluidics as used in diagnostic and molecular biology applications can be traced to microanalytical tools like gas-phase chromatography (GPC), high-pressure liquid chromatography (HPLC), and capillary electrophoresis (CE) developed in the mid-1990s (16). Rapid progress was made on these tools at this time and many of the developed concepts are still in use today. A summary of some of the best examples follow: Jacobson et al. reported separation of complexed metal ions in polyacrylamide-modified channels (17) using electrophoresis. Micellar electrokinetic capillary chromatography (MECC) separation of biological samples (18) and neutral dyes (19) were reported. Wooley et al. reported ultra-high-speed DNA sequencing and separation using microfabricated capillary electrophoresis chips (20). Surface passivation of silicon-based PCR chips to obtain amplifications comparable to conventional PCR systems was accomplished (21). Koutny et al. reported a competitive immunoassay for separation and quantification of free and bound labeled antigens by capillary electrophoresis (CE) (22). Hadd et al. presented an automated enzyme assay in which nanoliter volumes of substrate, enzyme, and inhibitor were mixed using electrokinetic flow (23). Microchip-based capillary electrophoresis (CE) for separation and relative quantitation of human serum proteins was achieved (24). Some of the other separation methods like free-flow electrophoresis (FFE) (25), capillary gel electrophoresis (26) and capillary array electrophoresis (CAE) (27) were reported. These devices were primarily fabricated in silicon and glass and lead to the work on related components like micropumps, microvalves and sensors.

There are a few examples of plastic devices before 2000. Delamarche et al. used elastomeric microfluidic networks to pattern immunoglobulin with high resolution on a variety of substrates (gold, glass, polystyrene) (28). Freaney et al. developed a prototype miniaturized chemical analysis system comprising biosensors and a microdialysis interface for on-line monitoring of glucose and lactate in blood (29).

In the 1990s, to counter the threat of biological and chemical weapons, the US Defense Advanced Research Projects Agency (DARPA) supported development of "field-deployable microfluidic" devices and was a driver for academic research in microfluidics (10). The first lab-on-a-chip emerged with the concept of a "miniaturized total analysis system" or μTAS, involving a silicon chip analyzer with sampling, sample pretreatment, separation, and detection functionalities embedded on an integrated system (30).

Electroosmotic pumping was the primary actuation mode used in these early μTAS systems especially since separation was one of the objectives and pumping could be controlled using simple electronics and no moving parts (31). Seiler et al. reported amino-acid separation on chip and their detection using laser-induced fluorescence (32). Other applications involving biomolecules and cells emerged during the period. These include flow cytometry (33), DNA amplification (PCR) (34) and cellular metabolism studies (35) on a microfabricated chip.

A host of innovations in microfluidic devices came forth in the period from 1994 to 1997. These include, "reactor chambers for continuous precolumn and postcolumn labeling reactions" (36, 37), high speed efficient separations (38), on-chip static mixing (39), separation of oligonucleotides (40), DNA (41), and amino acids (42), and cell manipulation by electrical fields (43). There was also work on separation modes like synchronized cyclic capillary electrophoresis (44) and free flow electrophoresis (FFE) (45). Verpoorte et al. devised a 3D microflow manifold system incorporating electrochemical and optical detection systems (46). Seiler et al. studied current and electroosmotic fluid flows in microchannels using Kirchhoff's law (47). Jacobson et al. developed glass microchips with octadecylsilane surface modified channels as a stationary phase for open channel chromatography (48). Feustel et al. came up with a miniaturized mass spectrometer incorporating an integrated plasma chamber for electron generation, an ionization chamber, and an array of electrodes acting as the mass separator (49). All of these systems would find their way into later diagnostic microfluidic devices.

The introduction of polymer-based soft lithography offered a cheaper alternative to silicon and glass in microfluidic device fabrication (50). Most of the exploratory research in microfluidics is currently performed on polymer-based devices primarily made of poly(dimethylsiloxane) (PDMS), a soft elastomer (10). Soft lithography techniques for microfluidic devices have been reviewed multiple times along with many of the structures and devices than can be produced (51, 52). Related polymer-based methods like microcontact printing and microtransfer molding enabled rapid fabrication of micrometer scale structures (53). Three dimensional structures were reported using a layer-by-layer structuring using microtransfer molding (54).

Other plastics, hybrid materials, and packaging techniques were soon developed, including a variety of low cost plastic prototyping and manufacturing methods for microfluidics (55, 56). Martynova et al. reported microfluidic devices fabricated in Poly-(methyl methacrylate) (PMMA) by imprinting them with an inverse three-dimensional image of the device micromachined on silicon (57). Wang et al. developed a low temperature bonding process using a sodium silicate layer as an adhesive for glass microfluidic devices. Microfluidic interconnects for connecting vertically stacked micromachined channels and to external tubing on the same plane was demonstrated by Gonzalez and co-workers (58).

Additional landmark work included Johnson et al. fabricating nanometer wide channels on silicon, SiO_2, and gold substrates by exposing them to a metastable argon atom beam in the presence of dilute vapors of trimethylpentaphenyltrisiloxane (59). Lorenz et al. reported the characterization of SU-8 negative photoresist for the fabrication of high aspect-ratio structures (60). Larsson et al. fabricated 3D microstructures by conventional CD-injection molding against a silicon master produced by wet and deep reactive ion etching (DRIE) (61). Silicon micromachining methods based on DRIE, silicon fusion bonding (SFB) (62), and electron cyclotron resonance (ECR) source were reported (63). Dozens of other techniques have also been reported, but cannot all be reviewed here.

3. Modern Microfluidics Fabrication

Microfluidics was inspired by the microelectronics industry and hence most of the initial devices were fabricated in silicon using photolithography and related technologies. The success of the microelectronics and MEMS industries in manufacturing thousands of miniaturized components in parallel at very low costs was thought to be applicable to microfluidics. While this may eventually prove to be true, low cost microfluidic devices made using photolithographic techniques have proven to be the exception rather than the rule, since the numbers of identical microfluidic chips manufactured for current and foreseeable markets tend to be more in the 10,000 s –100,000 s, where batch processing does not provide sufficient cost savings. Packaging and other post processing steps like reagent introduction have also proven challenging and expensive, and consequently other manufacturing methods currently appear to be more in favor. Thus, while most of the earliest work in microfluidics was in silicon, the majority of current devices are now made in glass or a variety of plastics. Nonetheless, silicon and glass manufacturing technique are important in microfluidics, because molds for rapid and inexpensive manufacturing of plastic devices are still often made of silicon or glass.

Standard silicon and glass manufacturing techniques are based on microlithography, subtractive techniques (etching), and additive techniques (64). Microlithography involves the use of an energy beam to transfer a geometric pattern to a substrate. Depending on the type of energy beam used, these can be divided into: photolithography, electron beam lithography, X-ray lithography, and ion lithography. Photolithography involves using light to transfer a geometric pattern from a photo mask to a light sensitive chemical called photoresist. This is followed by a development process using a developer solution to create a positive or negative image of pattern onto the photoresist. Other techniques like X-ray lithography, extreme ultraviolet (UV) lithography, ion particle lithography, scanning probe lithography, and nanoimprint lithography are being increasingly used due to their capability in producing sub-100 nm structures. Of these, nanoimprint lithography, a type of embossing, is a low cost, high throughput and high resolution method that has the potential to be used for low-cost mass manufacture of micro and nanofluidic devices in a variety of materials, but especially for direct embossing of plastics (65).

Subtractive techniques involve dry and wet etching, which are primarily used with glass and silicon devices. Wet etching involves chemical removal of layers from a material and is typically used to etch silicon, silicon dioxide, silicon nitride, metals, and glass. Dry etching refers to the removal of material by bombarding it with ions. Sputtering, ion beam milling and plasma etching (reactive ion etching and deep reactive ion etching) are some of other methods used in silicon etching.

Additive technologies involve techniques to deposit films. Methods to deposit thin films include: thermal oxidation of silicon, chemical vapor deposition (CVD), and physical vapor deposition (PVD). Methods to deposit thick films usually involve a spinning or electroplating technique. These thick films are often patterned using photolithography and then used as molds for microfluidic devices. The reader can refer to the text by Madou (64) for an in-depth description of the techniques described above, as well as a description of other micromanufacturing techniques.

In the past decade, silicon and glass have been largely displaced by plastics as the ideal substrate for microfluidic devices (10). Six primary considerations have been behind this transition. First, silicon is relatively expensive compared to plastics because microfluidics tends to take up larger areas than microelectronic chips (and silicon costs are measured by area). Second, the electronic advantages of silicon are not typically required in microfluidic devices. Third, silicon is not transparent, so troubleshooting microfluidic devices during development can be difficult and optical detection techniques cannot be employed. Fourth, silicon processing typically requires processes found in expensive cleanrooms that are also relatively slow. The development of rapid and inexpensive polymer processing methods has proven compelling. Fifth, silicon is relatively brittle and is not ideal for devices that experience significant "handling." Sixth, silicon is incompatible with the strong potentials used in electrokinetic pumping and capillary electrophoresis (CE). Silicon does have some advantages, such as well controlled surface properties, but these have not proven sufficient to drive microfluidic development.

Polymers, due to their lower cost, ease of fabrication and physical properties, are now the primary materials used in microfluidic research. Many microfluidic components, such as pumps and valves, work better when fabricated in the less rigid polymer medium as compared with silicon. The permeability of polymer to gases make it suitable for work with living mammalian cells. PDMS, an optically transparent, soft elastomer has been used for various microfluidic devices since its introduction (10). Most polymer devices are made using a molding, embossing, or casting techniques, although direct processing means, such as laser-based or knife-based manufacturing is increasing (56). Soft lithography is the technique of replicating structures from a master mold or stamp onto an elastomeric (PDMS) substrate. Fabrication using PDMS is simple and does not require expensive facilities, and prototyping can often be done in less than a day. The reader can refer to in-depth reviews of soft lithography for detailed insight into the method (51, 52). Interestingly, not many commercial products use devices fabricated in PDMS due to a gap between academic and industrial practices, although this is starting to change (66). In addition, PDMS has limited application due to its hydrophobic surface and tendency to swell in organic solvents (67). Although polymers are the preferred material for most microfluidic applications today, silicon and glass are still relied upon for building specialized devices that need chemical and thermal stability (68). In the nascent field of nanofluidics, silicon and glass are used due to their mechanical stability (10). Some of other methods that are used in microfluidic fabrication are xerography (69), laser micromachining (70) and polymer stereolithography (71).

An innovative method for creating low cost disposable microfluidic diagnostic devices (paper-based analytical devices [μPADs]) was introduced by Martinez et al. (72, 73). The fluid movement is controlled primarily by evaporation and capillary forces. Although the technology is very promising, more work needs to be done to bring forth real world μPAD applications. Recently, microfluidic devices fabricated on engineered plastics, such as cyclo-olefin copolymers (COC) (74), and photocurable perfluoropolyether (PFPE) (65) have been reported.

4. Microfluidic Diagnostics in the Past Decade

A major boom in microfluidics research has occurred in the last 10–12 years as is reflected by the number of published journal papers using the term: 26 papers were published before 1990, 341 in the 1990s, 15773 in the 2000s and 3,322 in the first 2 years of this decade. While the number of papers each year appears to be leveling off, the impact of microfluidics is likely to continue to grow. Another consequence of this large body of literature is that it becomes infeasible to cover all the important papers and developments in a chapter such as this. As most reviews on microfluidics for diagnostic applications have focused on the physical methods behind the device operation and not as much on the applications, this work will focus on some specific diagnostic developments and application areas. We examine the microfluidic applications in diagnostics for diabetes, cardiac related conditions, and infections related to bacteria, virus, and HIV. We also review the applications in pharmacogenomics and devices for low resource settings. We discuss some of the methods used in fabricating these microfluidic devices and the challenges in mass production. Included is a section on some of the commercial diagnostic products using microfluidic technology. The reader may refer to supplemental reviews for the theory behind microfluidics (10, 75–78) and methods used in microfluidic LOC detection (14, 67, 79).

5. A Global Health Perspective

There is an increasing need in the developing and developed world for new cost-effective diagnostic technologies, albeit for different reasons. In developed countries, health care costs are rising rapidly, and containment is an issue. In developing countries, delivery of medical services to remote and resource poor areas is difficult and the needs are enormous, as infectious disease is a critical barrier to economic and social development. Interestingly, the two problems tend to converge towards one solution: microfluidic diagnostic devices. The Grand Challenges in Global Health (GCGH) initiative, a major effort to achieve scientific breakthroughs against infectious diseases that cause significant problems in the developing world, has identified seven long-term goals in global health (80), most of which revolve around eliminating infectious disease.

Infectious diseases constitute a huge burden in developing countries (32.1%) using disability-adjusted life year (DALY) metrics compared to developed countries (3.7%) and account for 50% of infant deaths (1, 81). The major concerns in terms of DALY are infections due to viruses (HIV/AIDS, measles, hepatitis B, hepatitis C, and viral gastroenteritis [rotavirus]); bacteria (cholera, tuberculosis, pertussis, tetanus, and meningitis); and parasites (malaria, Lymphatic filariasis, leishmaniasis, and trypanosomiasis). The three most devastating diseases are malaria, tuberculosis, and HIV. In 2009, there was an estimated 169–294 million cases of malaria worldwide, resulting in about 781,000 deaths. Of these 85% of deaths were in children under 5 years of age [82). There was an estimated 14 million people infected with TB and about 1.7 million related deaths in 2009. TB is a major cause of deaths in

HIV infected patients with about 380,000 of the 1.7 million deaths being reported in people with HIV (83). An estimated 33.3 million people are living with HIV worldwide of which about 67.5% live in sub-Saharan Africa (84). There has been an estimated 1.8 million AIDS related deaths, 73% of those being in sub-Saharan Africa. Thus, early infectious disease detection and management is a high priority in low-resource settings and a major driver of microfluidic diagnostic devices. Infectious diseases are not limited to developing countries. Recent outbreaks like H1N1 influenza A demonstrate the rapid spread of infectious diseases from a country of origin to the rest of the world. In April 2009, USA and Mexico reported 38 cases of H1N1 influenza. By June 2009 when World Health Organization declared a pandemic, there were a reported 28,774 cases and 144 deaths in 74 countries. The H1N1 influenza pandemic had a total of 43,677 reported cases in the USA as of July 2010 (85). Estimates of unreported cases are a much higher figure at 1.8–5.7 million cases (86).

In contrast, chronic diseases that require consistent monitoring are the major disease burden for high-income countries. These diseases include: ischemic heart disease, cerebrovascular disease, cancers, and diabetes mellitus. Global mortality and disease burden projections suggest that these chronic conditions common to high-income countries will also become a priority for low-income countries by the year 2030 (1). Thus, the driver for microfluidic diagnostics in developed countries is the need for consistent, accurate, and affordable diagnostics for chronic disease.

6. Microfluidics in Diagnostics

6.1. Bacterial Detection

Bacterial detection is a key need in areas including: clinical diagnostics, monitoring of food-borne pathogens, and detection of biological threat agents. Harmful bacteria are the source of diseases like gastroenteritis and cholera. From a bioterrorism perspective, pathogenic bacteria pose serious risk. Under favorable temperature and in the presence of moisture and nutrition, bacteria spread rapidly. For a list of bacterial diseases and corresponding causative agents the reader can refer to a review by Ivnitski et al. (87).

Conventional methods to detect and identify bacteria require growing a small number of bacteria into colonies of higher numbers. Hence conventional methods take 18–24 h at a minimum (87). Also, conventional methods require complex equipment, highly trained technicians, and cannot be field deployable or used in point-of-care settings.

There are primarily two modes of pathogen detection: immunosensing and nucleic acid-based detection. In immunosensing, a binding interaction between probing antibodies and antigens of target cells (analyte) is detected. A variety of mechanisms can be used to detect this interaction, such as: optical, electrical or electrochemical impedance, cantilever, quartz crystalline microbalance, surface plasmon resonance (SPR), and magnetoresistivity. Nucleic acid-based sensors detect DNA or RNA targets from the analyte organisms (88, 89). The polymerase chain reaction (PCR) or reverse transcription PCR (RT-PCR) is used to amplify the nucleic acids to enhance the detection signal (90).

Optical detection methods are often preferred due to their high selectivity and sensitivity (91, 92). A variety of microfluidic devices have been developed for bacteria using optical means. A microfluidic system for detection of Escherichia coli using laser-optical fiber fluorescence detection was reported by Xiang et al. They reported detection limits an order of magnitude higher than that achieved for conventional fluorescence microscope (93). Gao et al. developed a multiplexed microfluidic device for the fluorescence detection of bacterial antibodies in human serum. TRITC-labeled detection antibodies were captured by antigens pre-patterned on the microchannels (94). An integrated microfluidic platform for fluorescence-based detection of Shiga toxin I (*Shigella dysenteriae*) and Staphylococcal enterotoxin B (*Staphylococcus aureus*) was developed by Meagher et al. (95).

Electrical and electrochemical modes of bacterial detection have also been widely reported. The primary advantage of the method is the ease of fabricating microelectrodes in the microchannel by lithography and the absence of labeling steps (96). A microfluidic sensor based on impedance measurement of *E. coli* was constructed by Boehm et al. (97). The selectivity of the sensor to different bacterial strains was demonstrated by positive identification of E. coli in a suspension of E. *coli* and M. *catarrhalis*. A microfabricated electrochemical sensor array for detection of bacterial pathogens in human clinical fluid samples was demonstrated. The device consisting of a set of 16 sensors was able to detect relevant bacterial urinary pathogens (*E. coli*, *Proteus mirabilis*, *Pseudomonas aeruginosa*, *Enterocococcus spp.*, and Klebsiella–Enterobacter group) and could in principle be used as a point-of-care device for rapid diagnosis of urinary tract infections (98). Table 2 lists a number of detection methods for bacterial diagnostics and, for a comprehensive list of electrical and electrochemical bacterial detection, the reader can refer to a review by Jinseok et al. (96).

Microfluidic devices have also been applied to the detection of parasites. A "microfluidic flow-through membrane immunoassay with on-card dry reagent storage" was developed by Stevens et al. for the detection of *Plasmodium falciparum* (99).

6.2. Viral Detection and Disease Management

As noted earlier in this chapter, effective virus detection and disease management is critical in public health, the biotechnology industry, and biodefense. Some of the most deadly historical epidemics

Table 2
Detection methods of POC devices for bacterial diagnostics

Analyte	Detection method	Material	Limit of detection	Reference
Escherichia coli O157:H7	Fluorescence	PDMS	0.3 ng/μL	(76)
Escherichia coli O157:H7	Fluorescence	PDMS	0.02 μg/mL	(77)
Helicobacter pylori	Fluorescence	PDMS	0.1 μg/mL	(77)
Shiga toxin I	Fluorescence	Glass	500 pM	(78)
Staphylococcal enterotoxin B	Fluorescence	Glass	300 pM	(78)
Escherichia coli	Electrical (impedance)	Silicon	10^4 CFU/mL	(80)
Escherichia coli, *Proteus mirabilis*, *Pseudomonas aeruginosa*, Enterocococcus spp., and the Klebsiella–Enterobacter group	Electrochemical (amperometric)	Au on plastic	Not specified	(81)
Plasmodium falciparum	Optical	Mylar/PMMA	10 ng/mL	(98)

like smallpox, yellow fever, and Spanish flu were due to viral agents. In the twenty-first century, HIV, rotavirus, and measles are found to be among the leading contributors to global disease burden (100). Many deadly viruses such as Variola virus (small pox), Rift Valley fever virus, and Venezuelan Equine Encephalomyelitis virus have been known to be developed as potential biological agents (101). POC devices to detect these bio-agents are extremely critical for global biosecurity. The small size, simple biology, and lack of reproductive ability outside the host cell add to the complexity in detecting viruses.

The primary methods for virus detection are serology, viral antigen detection, and nucleic acid detection. Serologic tests detect the presence of antibodies that the immune system produces in response to viral infection. Viral antigen detection typically relies on immunoassays as described previously. Nucleic acid detection involves amplification of the viral genome using PCR and the subsequent detection of the amplified genome.

6.3. HIV Detection and Monitoring

HIV is one of the primary targets of microfluidic diagnostic research efforts. Conventional HIV diagnostic assays are based on an enzyme assay (EIA/ELISA) followed by western blot and requires trained laboratory personnel. Universal access to HIV diagnostics is stymied by the lack of trained technicians, patient motivation, and laboratory access especially in rural areas and the developing world. For instance, about 83% of HIV patients remain undiagnosed in Kenya (102). Thus, a simple, inexpensive diagnostic tool for HIV would be readily welcomed.

The number of $CD4^+$ T-lymphocytes per microliter of HIV-infected blood is a critical monitor of disease state and this measurement is needed to make informed antiretroviral therapy (ART) treatment decisions. Therefore, the primary mechanisms for HIV detection in POC microfluidic devices are enumeration of $CD4^+$ T-lymphocytes and HIV viral load quantification. To be successful, the POC device needs to detect around 200 $CD4^+$ cells/μL and 400 copies/mL of HIV from whole blood (103). Towards this goal, Sia et al. reported a microfluidic immunoassay, "POCKET (portable and cost-effective)" for quantifying anti-HIV-1 antibodies in the sera of HIV-1 infected patients. The device consisted of a PDMS slab with microchannels placed orthogonally to a polystyrene stripe patterned with HIV-enveloped antigen. The HIV-1 infected patient serum sample is flowed through the microchannels to quantify anti-HIV-1 antibodies. Although the device was able to identify the sera of HIV-1-infected patients from those of non-infected patients, it could not make a correlation of the output data with HIV disease states (103). Lee et al. developed a RT-PCR-based POC diagnostic chip for HIV. The chip relies on HIV markers p24 (a major core protein encoded by the HIV gag gene) and gp120 (an external envelope protein encoded by envelope gene) for diagnostic purposes (104). Cheng et al. reported a POC microfluidic $CD4^+$ T-cell counting device. The device works in two stages, initial depletion of monocytes from whole blood and subsequent $CD4^+$ T cell capture. The strategy of contaminant (monocytes) depletion prior to $CD4^+$ T cell isolation enhances the performance in low CD4 count (200 cells/μL) scenarios (105). Other label-free $CD4^+$ T-lymphocyte capture techniques have been reported. Although the microfluidic devices themselves are disposable and usually cheap, they still require expensive optical microscopes to count the captured $CD4^+$ T-cells (106, 107). A lensless portable CCD-based microfluidic platform developed by Demirci et al. overcomes this limitation. The captured label-free $CD4^+$ T-lymphocytes are detected by a charge coupled device (CCD) sensor using lensless shadow imaging techniques and counted using automatic cell counting software in a few seconds (108). Cheng et al. developed a non-optical method of counting CD^+ T-cells. The cell count is enumerated by measuring the changes in conductivity of the surrounding medium due to ions released from the surface-immobilized cells within a microfluidic channel (109). Gohring et al. demonstrated the detection of $CD4^+$ and $CD8^+$ T-Lymphocyte whole cells and $CD4^+$ T-Lymphocyte cell lysis using an optofluidic ring resonator (OFRR) sensor. This sensor measures the presence of T cells based on a change in refractive index in the microfluidic channel due to the presence of immobilized T cells (110). Wang et al. reported a microfluidic chip with an integrated micromixer for fluorescent labeling of $CD4^+$/$CD8^+$ T-cells and their subsequent counting using a microflow cytometer (111).

6.4. Diagnostics for Other Viral Agents

Microfluidic diagnostics have been designed for other viral agent infections like influenza, severe acute respiratory syndrome (SARS) and dengue fever. These diseases have been of serious concern to global public health organizations especially in the last few years.

The influenza virus causes respiratory tract infection and is found to be severely morbid in children and the elderly (112). The challenge with diseases like influenza is that there is a large variety of the viruses and they are constantly changing. For example, the influenza A virus can subdivided into H1N1 and H1N3 based on the glycoproteins (hemagglutinin and neuraminidase) present in the viral envelope. The 2009 influenza pandemic was caused by a novel H1N1 strain with genes from five different flu viruses (113). Thus, the diagnosis of influenza alone is not sufficient; discovery of the type of influenza is also critical.

Some of the conventional diagnostic methods for influenza virus are enzyme-linked immunosorbent assays (ELISA), immunofluorescence assays, serological hemagglutination inhibition assays, real-time polymerase chain reaction (PCR) assays, and complement fixation tests. Most of these methods are complicated, relatively costly and require a lengthy process and expensive apparatus (114).

Several microfluidic systems have been shown for influenza detection. An immunomagnetic bead-based microfluidic system for detection of influenza A virus has been demonstrated recently. Influenza A viral particles are initially bound to monoclonal antibody (mAb)-conjugated immunomagnetic beads using a suction type micro-mixer. Subsequently the virus-bound magnetic complexes are fluorescently labeled by developing mAb with R-phycoerythrin. An external optical detection module is used to analyze the optical intensity of the magnetic complex. The system displayed better performance than conventional flow cytometry systems in terms of limit of detection (114). However, the expensive external optical detection module could restrict its use in POC low-resource settings. Yamanaka et al. reported a microfluidic RT-PCR chip for rapid detection of influenza (AH1pdm) virus of swine-origin. A disposable electrical printed chip was used for electrochemical detection of the PCR amplicon (112). The electrochemical method is better for use in low-resource settings compared to the optical methods reported above due to the absence of expensive external detection units. A Magnetic Integrated Microfluidic Electrochemical Detector (MIMED) for detection of H1N1 influenza virus from throat swab samples has recently been developed (115).

Microfluidics detection methods for other types of viral agents have been reported. Weiss et al. reported a microfluidic chip-based electrophoretic analysis and laser-induced fluorescence detection of human rhinovirus serotype 2 (116). Zhu et al. developed an optofluidic micro-ring resonator-based system for detection of bacteriophage M13 (117). They reported a detection limit of 2.3×10^3 PFU/mL. An on-chip surface enhanced Raman spectroscopy (SERS)-based biomolecular device for detection of Dengue virus sequences was developed by Huh et al. (118). The fluid is actuated using electrokinetic methods and the limit of detection was reported to be 30 pM. Weng et al. developed a suction-type, pneumatically driven microfluidic device for the detection of dengue infection (119). A detection limit of 10 PFU/ml was reported for the device. A "lab-on-a-disc" centrifugal microfluidics-based portable ELISA system was developed for detection of the antigen and the antibody of Hepatitis B virus (120). The limit of detection of antigen and antibody were reported as 0.51 ng/mL and 8.6 mIU/mL, respectively. Heinze et al. developed a microfluidic immunosensor for detection of bovine viral diarrhea virus (121). An integrated microfluidic assay for targeted ribonucleic acid (RNA) extraction and a one-step reverse transcription loop-mediated-isothermal-amplification (RT-LAMP) process for the detection of nervous necrosis viruses was reported by Wang et al. (122).

6.5. Cancer Biomarker Detection

In 2010, there were an estimated 1,500 cancer related deaths per day in the USA and about 1.4 million new cases of cancer were reported. By 2020, cancer related deaths are estimated to be 10.3 million globally. The cancer mortality rate per 100,000 Americans has dropped from 194 to 190 since 1950, an insignificant drop compared to drop in mortality rates for other diseases. Most of the improvements in cancer survival rates are due to improvements in early diagnosis rather than treatment. For instance, for cancers of the breast, colon, rectum, and cervix, early detection has proved to reduce mortality significantly. Hence, the National Cancer Institute has emphasized a shift from therapeutic to preventive mode in its 2010 vision document.

Existing methods of cancer diagnostics rely on invasive techniques like taking a biopsy and then examining the cell morphology. Further, conventional methods could be inconclusive in disease detection in its early stages (123). Other techniques like immunoassays (ELISA) have been used to detect cancer biomarkers. Although ELISAs are very sensitive, they can be time consuming, expensive and are mostly carried out in a laboratory requiring skilled personnel. In most cases, immunoassays look for only one biomarker and are not sensitive enough to detect very low biomarker levels especially at early stages of the disease. POC devices which are accurate, fast and economic are needed. This would enable improved diagnosis, monitoring of the progress of the disease, and response to therapy.

Advances in oncology have led to identification of biomarkers associated with different kinds of cancers. For a comprehensive list of cancer biomarkers, the reader can refer to reviews in literature (124–129). There are multiple factors responsible for carcinogenesis.

This along with the "heterogeneity in oncogenic pathways" makes it imperative that a range of biomarkers need to be analyzed for cancer diagnostics (123, 130). Hence POC devices with multiplexed capability to detect multiple biomarkers are needed. Although research into cancer diagnostic devices is moving forward, commercialization of the technology still remains a challenge (123).

Here, we review some of the recent research in microfluidics POC devices for cancer diagnostics. Legendre et al. reported work into the design and development of a microfluidic device for diagnosis of T-cell lymphoma. The system accepts a whole blood sample as the input, extracts the DNA, amplifies target sequences of the T-cell receptor-gene, and eletrophoretically resolves the products for detection of a signature consistent with monoclonality (131). Diercks et al. demonstrated a microfluidic device that measured multiple proteins (tumor necrosis factor, CXC chemokine ligand 2, interleukin 6 and interleukin 1b) at pg/mL concentrations in nanoliter volumes. Antibody-coupled polystyrene microspheres labeled with embedded fluorophores were used to detect the analyte (proteins). Optical detection of captured analyte was performed off-chip using a confocal microscope, which proved to be a disadvantage in terms of lack of device portability (132). A similar fluorescence approach has been used to detect vascular endothelial growth factors in human plasma (133). An on-chip nuclear magnetic resonance (NMR)-based biosensor was developed for the multiplexed identification of cancer markers (epidermal growth factor receptors EGFR and Her2/neu). The design consists of a microcoil array for NMR measurements, microfluidic channels for sample handling and a permanent magnet to generate a polarizing magnetic field, all integrated into a handheld device (134). Mass spectroscopy-based microfluidic detection of cancer-specific biomarkers (proliferating cell nuclear antigen, cathepsin D, and keratins K8, K18, and K19) was demonstrated by Lazar (135).

Other mass-based methods like quartz crystal microbalance (QCM) have been used in cancer biomarker detection. For instance, Zhang et al. demonstrated detection of human lung carcinoma cells using a microfluidic surface modified piezoelectric sensor (136). Recently, Von Muhlen et al. have reported a microcantilever-based "suspended microchannel resonator" sensing device for detection of activated leukocyte cell adhesion molecules (137). Zani et al. demonstrated an electrochemical method for detection of prostate specific antigen (PSA) cancer markers. The method works based on the differential pulse voltammetry-based electrochemical detection of protein coated paramagnetic microparticles that selectively capture the analyte (PSA) (138). Similar electrochemical detection methods for breast cancer markers have been reported (139). A microfluidic-based amperometric electrochemical detection system for carcinoembryonic antigen (CEA) and cancer antigen 15–3 (CA15-3) was developed by Kellner et al.

The on-chip fluid function is handled by computer controlled syringe pumps and reports enhanced performance due to fully automated fluidic operations (140). But the external computer control system and syringe pumps prove to be a bottleneck in their use for POC applications. Hence miniaturization and integration of the fluid handling functions within the microfluidic chip is necessary for POC use.

6.6. Cardiac Biomarker Detection

Cardiovascular diseases (CVD) are responsible for nearly half of the deaths in the western world. Studies suggest the acute and long term financial burden of cardiac disease to be substantial (141). It is reported that 5% of myocardial infarction (MI) patients are incorrectly discharged from emergency departments (ED). Hence for timely and effective intervention against cardiovascular diseases, there is a need for rapid and accurate diagnostic tools (142).

For the accurate "diagnosis, prognosis, monitoring and risk stratification of patients with acute coronary syndromes" (ACS), biochemical markers play a fundamental role (142). In clinical settings, in 50–70% of patients with ACS related cases, ECGs give ambiguous results. In such cases, cardiac marker levels could provide critical information for informed decision on the suitable treatment. As a definite indicator of disease condition a combination of cardiac markers need to be explored (143). For a review of cardiac biomarkers, the reader can refer to McDonnell et al. (142).

There is a difference of opinion with regard to the use of POC technologies for cardiac biomarker diagnostic, with some suggesting it to be an alternative to conventional lab analyzers (144, 145) and others questioning the accuracy of the technologies (146, 147). The following section provides a review of microfluidic devices used in cardiac biomarker detection.

Most of the diagnostic mechanisms for biomarkers involve two steps, an initial immunoassay to capture the analyte (biomarker) and subsequent detection of the captured analyte. Using optical methods, Jönsson et al. demonstrated a lateral flow polymer chip for detection of C Reactive Protein (CRP) (148). Gervais et al. demonstrated a microfluidic device for one step detection of CRP in serum. The device works based on capillary action for fluid actuation and does not need any external power requirements (149), which makes it extremely useful in a POC, low-cost setting. A multiplexed cardiac biomarker detection prototype device was developed by Hong et al. The MEMS-based device detected four different cardiac markers viz. myoglobin, CRP, cTnI and BNP using Au nanoparticle-based fluorescence detection (150). Bhattacharyya & Klapperich developed a disposable microfluidic chip for detection of CRP including an on-board detection module (151). A microfluidic chemiluminescence-based immunoassay system for detection of cardiac troponin I (cTnI) was reported by Cho et al. (152). Huang et al. demonstrated a microfluidic chemiluminescence-based detection sensor for alpha-fetoprotein (AFP). Super-paramagnetic microbeads were used to capture the biomarker (153). Use of magnetic microbeads results in higher surface to volume ratio for efficient analyte capture and enables on-chip actuation using an integrated electromagnet. A digital microfluidic platform detection device for cTnI was developed by Sista et al. (154). The fluidic actuation is performed by electrowetting, obviating the need for any off-chip fluid handling apparatus. SPR-based microfluidic detection of cardiac marker B-type natriuretic peptide (BNP) was reported by Kurita et al. (155).

Electrochemical methods have been applied to detection of cardiac markers. Unlike optical methods, these do not need an often expensive, off-chip optical detection device and could be suitable for POC applications. Tweedie et al. presented a microfluidic-based impedimetric sensing device for cardiac enzymes (156). The i-STAT system (Abbott Point of Care Inc., USA) is a commercial test cartridge for electrochemical detection of cTnI (157). The device can detect cTnI in the range of 0–50 ng/ml and has gained good acceptance as a diagnostic tool for MI (143). Other electrochemical-based detection methods for detection of myoglobin (158), cTnI (159) and CRP (160) have been reported. Recently, Mitsakakis and Gizeli have developed an integrated microfluidic surface acoustic wave (SAW) platform for detection of cardiac markers creatine kinase MB (CK-MB), CRP, and D-dimer (161).

Many commercial systems for cardiac marker detection are currently available, which could possibly limit the impact of microfluidic devices in this area. These include Triage® [Biosite Diagnostics Inc., USA] (myoglobin, CK-MB, and cTnI), Stratus® CS STAT fluorometric analyzer [Siemens Medical Diagnostics, USA] (myoglobin, CK-MB, and cTnI), Roche cardiac reader [Roche Diagnostics, USA] (cTnT, myoglobin), RAMP™ cardiac marker testing [Response Biomedical Inc., Canada] (cTnI, CK-MM), and Cardiac STATus™ device [Nanogen Inc., USA] (myoglobin, CK-MB mass, and cTnI) (157). Table 3 lists the set of published work and commercial devices for microfluidic cardiac marker detection.

6.7. High Throughput and Multiplexed Diagnostic Screening

Microfluidic-based technology is ideal for developing highly parallel diagnostic assays that would allow high-throughput screening, but there has been limited success in this area. The lack of success is not due to problems with microfluidic devices; for example, drug screening requires high-throughput methods to find and test different drug candidates. Microfluidic high-throughput screening (HTS) techniques have been applied to drug discovery to perform thousands of tests in parallel with some success (162–164). As of now these methods haven't been applied in microfluidic diagnostics for several reasons. Current diagnostics are typically performed in large hospitals or reference labs. In these labs, most tests are

Table 3
Performance and detection methods of POC devices for cardiac marker diagnostics

Biomarkers	Detection method	Material/Device	Detection Limit	Reference
Published work				
CRP	Fluorescence	Thermoplastic (Zeonor™)	2.6 ng/ml	(147)
CRP	Fluorescence	PDMS	1 ng/ml	(148)
Myoglobin, BNPCRP cTnI	Fluorescence	Cyclic Olefin Copolymer (COC)	70 ng/ml 0.1 ng/ml700 ng/ ml0.7 ng/ml	(149)
CRP	Chemiluminescence	Thermoplastic (Zeonex™)	100 ng/ml	(150)
cTnI	Chemiluminescence	Polycarbonate	0.027 ng/ml	(151)
AFP	Chemiluminescence	PMMA	0.23 ng/ml	(152)
cTnI	Chemiluminescence	Glass/Polymer	Not reported	(153)
BNP	SPR	PDMS & Glass	5 pg/ml	(154)
Cardiac enzyme (myoglobin)	Electrochemical (impedimetric)	Pressure sensitive adhesive & PMMA	100 ng/ml	(155)
cTnICRP	Electrochemical (anodic stripping voltammetry)	PDMS	0.01–50 μg/l 0.5–200 μg/l	(158)
CK-MB CRPD-dimer	SAW	PDMS & PMMA	0.25 μg/ml 1 μg/ml5 μg/ml	(160)
Commercial devices				
Troponin ICK-MBBNP	Electrochemical	i-STAT®	0–50 ng/ml Not reportedNot reported	(142)
Troponin IMyoglobin CK-MB	Fluorescence	Triage®	Not reported	(142)
Troponin T Myoglobin NT-proBNP D-dimer CK-MB	Fluorescence	Roche cardiac reader	0.1–2.0 ng/ml 30–700 ng/ml 60–3,000 pg/ml 0.1–4.0 μg/ml 1.0–25 ng/ml	(142)
Troponin ICK-MB Myoglobin NT-proBNP	Fluorescence	RAMP™ 3.2	0.2 ng/ml 7.2 ng/ml 100.0 ng/ml Not reported	(142)
Troponin IMyoglobin CK-MB	Chemiluminescence	Cardiac STATus™	Not reported	(142)

batched and performed using robots in a highly parallel, high throughput approach. Replacing these robots by using microfluidics is unlikely in the short term due to the large infrastructure already developed. Essentially, a solution to this problem already exists, so adoption of microfluidics for these assays will only occur if there are compelling assay improvements. In addition, if an assay can be performed in a batch mode using microfluidics, it is likely to be able to be performed in the clinic or POC setting, and for nearly the same price. Thus, microfluidics is likely to be driven to the POC rather than to large reference laboratories.

The reverse of high-throughput screening (multiple samples with one target) is multiplexed screening, where one sample is tested for multiple agents or biomarkers. A few examples of multiplexed screening have already been provided, especially for cardiac biomarkers, but highly multiplexed diagnostics are still being developed. Multiplexed screening is likely to have a more significant impact on diagnostics than high throughput screening, especially with the move towards personalized medicine. Microfluidics has been combined with microarray technology, which is used regularly in genomics and proteomics, and which will likely have diagnostic applications in the future; however, this is beyond the scope of the chapter. More relevant are microfluidic devices that can diagnose multiple diseases simultaneously. A recently released product that uses "mesoscale" fluidics can simultaneously diagnose 15 respiratory diseases associated with viruses (165). A challenge with getting the device to commercialization is that regulatory agencies such as the FDA require individual validation of each assay, meaning that multiplexing must clear very challenging regulatory requirements, which will likely limit substantial multiplexing in the near future. Nevertheless, microfluidics will probably lead to highly multiplexed assays that can perform 100s or 1,000s of diagnostic assays on one sample.

7. Microfluidic Commercialization

About 1,200 patents related to microfluidics have been issued in the USA through 2010. In spite of immense academic interest in microfluidics and significant research investment directed towards both academic and industrial organizations, relatively few commercial products based on microfluidics have been introduced into the market (166, 167); however, the rate of introduction is increasing and many barriers are coming down. One of the reasons cited for lack of commercial success is the lack of a potential "blockbuster" end-user product that could generate billions of dollars in revenue. Until the industry can find a product with high volume demand, the fabrication costs due to lack of "economies of scale" are going to remain high. Existing materials like PDMS, which are hugely popular

Table 4
Leading POC diagnostic companies and products (168)

Company	Product	Application
Abbott Point of Care	i-STAT®	POC blood analyzer
Agilent	Agilent 2100 Bioanalyzer	Microfluidics-based platform for sizing, quantification and quality control of DNA, RNA, proteins and cells on a single platform, PCR/QPCR products
Beckman Coulter	AmpliSpeed	Thermal cycler, single-cell analysis platforms
Biosite	Triage® Cardiac Panel	POC diagnosis of acute myocardial infarction
Caliper Life Sciences	LabChip GX	Nucleic acid and protein separations system
Cepheid	GeneXpert	Integrated real-time PCR system
Cynvenio Biosystems	Under development	Integrated System for molecular analysis of circulating tumor cells
Daktari Diagnostics	Under development	POC CD4 cell counting system
Eksigent	NanoLC	Microfluidic flow control based nanospray mass spectrometry system
Fluidgm	BioMark™ HDFluidigm EP1	Microfluidic devices for molecular diagnostics and personalized medicine
LeukoDx	Under development	POC flow cytometry device
Microfluidic Systems		Microfluidic systems for detection, processing of biological samples and biodefense
Micronics	PanNAT™	Multiplexed nucleic acid amplification device
RainDance Technologies	RDT 1000	Microdroplet-based solutions for human health and disease research
Rheonix	CARD®	Disposable microfluidic chip technology for multiplexed endpoint analysis for diagnostic applications
Shimadzu Biotech	PPSQ-31A/33A	Technologies to aid the protein research work flow and drug discovery
Siloam Biosciences	Optimiser™	Diagnostic systems using microfluidic and microsensor technology
Veridex	CellSearch®	Commercializing microfluidic circulating tumor cell diagnostics

in research, have not succeeded in the industry due to issues with manufacturability and scaling (168). Most of the LOC products are still focused on the business-to-business segment and not the business-to-consumer (167). There needs to be more focused research on microfluidic product development including issues like manufacturability and cost dynamics and a simultaneous search for new application areas where microfluidics could be applied. Table 4 provides a sample of microfluidic companies and products in the market. More comprehensive lists are available (169).

8. Summary and Future Outlook

Microfluidic diagnostic devices have been developing at a rapid rate over the past few years. While the potential for these devices was first recognized more than 20 years ago, the realization of that potential has been slow, even though thousands of devices and methods have been published. The continuing development of applications and microfluidic manufacturing methods, including platform technologies that can be customized easily for each diagnostic test, will be the drivers of success. Very recent progress and an emphasis on global health has helped move the field towards POC devices that will likely become ubiquitous in the years ahead. While most microfluidic devices have one diagnostic target, devices capable of diagnosing 100s or 1,000s of diseases will likely be developed and commercialized in the next decade, making microfluidics a major driver of disease diagnostics.

Acknowledgments

The authors thank Keng-Min Lin for the schematic diagram in Fig. 1. The authors would like to thank the Nano Institute of Utah for funding this work through a nanotechnology training fellowship.

References

1. Yager P, Domingo GJ, Gerdes J (2008) Point-of-care diagnostics for global health. Annu Rev Biomed Eng 10:107–144
2. Kiechle FL, Holland CA (2009) Point- of-Care Testing and Molecular Diagnostics: Miniaturization Required. Clin Lab Med 29(3):555–560
3. UNAIDS (2010) UNAIDS Report on the global AIDS epidemic, UNAIDS.
4. Steven J, Zullo SS, Patrick Looney J, Barker PE (2010) Nanotechnology: Emerging Developments and Early Detection of Cancer, A Two-Day Workshop sponsored by the National Cancer Institute and the National Institute of Standards and Technology, National Cancer Institute, Gaithersburg.
5. Foster RS, Heffler SK (2009) Updated and Extended National Health Expenditure Projections, 2010–2019, Office of the Actuary. Centers for Medicare & Medicaid Services, Maryland
6. Cirino NM, Musser KA, Egan C (2004) Multiplex diagnostic platforms for detection of biothreat agents. Expert Rev Mol Diagn 4:841–857
7. (USDA) U S D o HHS H a A (ed) (2011) National select agent registry: overview. Department of Health & Human Service, Washington DC
8. Food and Drug Administration (2008) Recommendations: Clinical Laboratory Improvement Amendments of 1988 (CLIA Waiver Applications for Manufacturers of In Vitro Diagnostic Devices.
9. Sia SK, Kricka LJ (2008) Microfluidics and point-of-care testing. Lab Chip 8:1982–1983
10. Whitesides GM (2006) The origins and the future of microfluidics. Nature 442:368–373
11. Arora A et al (2010) Latest developments in micro total analysis systems. Anal Chem 82:4830–4847
12. Salieb-Beugelaar GB et al (2010) Latest developments in microfluidic cell biology and analysis systems. Anal Chem 82:4848–4864
13. West J et al (2008) Micro total analysis systems: latest achievements. Anal Chem 80:4403–4419

14. Trietsch SJ et al (2011) Lab-on-a-chip technologies for massive parallel data generation in the life sciences: A review Chemometr Intell Lab 108:64–75.
15. Weston AD, Hood L (2004) Systems biology, proteomics, and the future of health care: toward predictive, preventative, and personalized medicine. J Proteome Res 3:179–196
16. Reyes DR (2002) Micro Total Analysis Systems. 1. Introduction. Theory, and Technology 74(12):2623–2636
17. Jacobson SC, Moore AW, Ramsey JM (1995) Fused quartz substrates for microchip electrophoresis. Anal Chem 67:2059–2063
18. von Heeren F et al (1996) Micellar electrokinetic chromatography separations and analyses of biological samples on a cyclic planar microstructure. Anal Chem 68:2044–2053
19. Moore AW Jr, Jacobson SC, Ramsey JM (1995) Microchip separations of neutral species via micellar electrokinetic capillary chromatography. Anal Chem 67:4184–4189
20. Woolley AT, Mathies RA (1995) Ultra-high-speed DNA sequencing using capillary electrophoresis chips. Anal Chem 67:3676–3680
21. Shoffner MA et al (1996) Chip PCR. I. Surface passivation of microfabricated silicon-glass chips for PCR. Nucleic Acids Res 24:375–379
22. Koutny LB et al (1996) Microchip electrophoretic immunoassay for serum cortisol. Anal Chem 68:18–22
23. Hadd AG et al (1997) Microchip device for performing enzyme assays. Anal Chem 69:3407–3412
24. Colyer CL, Mangru SD, Harrison DJ (1997) Microchip-based capillary electrophoresis of human serum proteins. J Chromatogr A 781:271–276
25. Raymond DE, Manz A, Widmer HM (1996) Continuous separation of high molecular weight compounds using a microliter volume free-flow electrophoresis microstructure. Anal Chem 68:2515–2522
26. von Heeren F et al (1996) Characterization of electrophoretic sample injection and separation in a gel filled cyclic planar microstructure. J Microcolumn Separations 8:373–381
27. Woolley AT, Sensabaugh GF, Mathies RA (1997) High-speed DNA genotyping using microfabricated capillary array electrophoresis chips. Anal Chem 69:2181–2186
28. Delamarche E et al (1997) Patterned delivery of immunoglobulins to surfaces using microfluidic networks. Science 276:779
29. Freaney R et al (1997) Novel instrumentation for real-time monitoring using miniaturized flow systems with integrated biosensors. Ann Clin Biochem 34:291–302
30. Manz A, Graber N, Widmer HM (1990) Miniaturized total chemical analysis systems: a novel concept for chemical sensing. Sensors Actuators B: Chemical 1:244–248
31. Harrison DJ, Manz A, Glavina PG (1991) Electroosmotic pumping within a chemical sensor system integrated on silicon. In: Proc. IEEE Int. Conf. Solid-State Sensors and Actuators (Transducers '91), San Francisco, USA, pp 792–795
32. Seiler K, Harrison DJ, Manz A (1993) Planar glass chips for capillary electrophoresis: repetitive sample injection, quantitation, and separation efficiency. Anal Chem 65: 1481–1488
33. Sobek D et al. (1993) A microfabricated flow chamber for optical measurements in fluids. In: Proc. 6th IEEE MEMS, Fort Lauderdale, USA, pp 219–224
34. Northrup MA et al. (1993) DNA amplification with a microfabricated reaction chamber. In: Proc. of The 7th Int. Conf. on Solid-State Sensors and Actuators (Transducers '93), Yokohama, Japan, pp 924–926
35. Bousse L et al (1994) Micromachined multichannel systems for the measurement of cellular metabolism. Sensors Actuators B: Chemical 20:145–150
36. Jacobson SC et al (1994) Precolumn reactions with electrophoretic analysis integrated on a microchip. Anal Chem 66:4127–4132
37. Jacobson SC et al (1994) Microchip capillary electrophoresis with an integrated postcolumn reactor. Anal Chem 66:3472–3476
38. Jacobson SC et al (1994) High-speed separations on a microchip. Anal Chem 66: 1114–1118
39. Mensinger H et al (1995) Microreactor with integrated static mixer and analysis system. Kluwer Academic Publishers, The Netherlands, p 237
40. Effenhauser CS et al (1994) High-speed separation of antisense oligonucleotides on a micromachined capillary electrophoresis device. Anal Chem 66:2949–2953
41. Woolley AT, Mathies RA (1994) Ultra-high-speed DNA fragment separations using microfabricated capillary array electrophoresis chips. Proc Natl Acad Sci 91:11348
42. Fan ZH, Harrison DJ (1994) Micromachining of capillary electrophoresis injectors and separators on glass chips and evaluation of flow at capillary intersections. Anal Chem 66:177–184
43. Fuhr G, and Wagner B. (1994) Electric field mediated cell manipulation, characterisation

and cultivation in highly conductive media. Paper presented at the MicroTAS, University of Twente, Netherlands, 21–22 Nov 1994

44. Manz A et al (1994) Electroosmotic pumping and electrophoretic separations for miniaturized chemical analysis systems. J Micromech Microeng 4:257
45. Raymond DE, Manz A, Widmer HM (1994) Continuous sample pretreatment using a free-flow electrophoresis device integrated onto a silicon chip. Anal Chem 66:2858–2865
46. Verpoorte EMJ et al (1994) Three-dimensional micro flow manifolds for miniaturized chemical analysis systems. J Micromech Microeng 4:246
47. Seiler K et al (1994) Electroosmotic pumping and valveless control of fluid flow within a manifold of capillaries on a glass chip. Anal Chem 66:3485–3491
48. Jacobson SC et al (1994) Open channel electrochromatography on a microchip. Anal Chem 66:2369–2373
49. Feustel A, Muller J, Relling V (1995) A microsystem mass spectrometer. Springer, Berlin, p 299
50. Kim E, Xia Y, Whitesides GM (1995) Polymer microstructures formed by moulding in capillaries. Nature 376:581–584
51. Xia Y, Whitesides GM (1998) Soft lithography. Ann Rev Mat Sci 28:153–184
52. Whitesides GM et al (2001) Soft lithography in biology and biochemistry. Annu Rev Biomed Eng 3:335–373
53. Mrksich M, Whitesides GM (1995) Patterning self-assembled monolayers using microcontact printing: a new technology for biosensors? Trends Biotechnol 13:228–235
54. Zhao XM, Xia Y, Whitesides GM (1996) Fabrication of three dimensional micro structures: Microtransfer molding. Adv Mater 8:837–840
55. Bartholomeusz DA, Boutté RW, Gale BK (2009) Xurography: Microfluidic Prototyping with a Cutting Plotter. In: Herold K, Rasooly A (eds) Lab on a Chip Technology: Fabrication and Microfluidics. Caister Academic Press, United Kingdom, pp 65–82
56. Gale BK et al (2008) Fabrication and packaging: Low-cost MEMS technologies. In: Gianchandani YB, Tabata O, Zappe HP (eds) Comprehensive microsystems. Elsevier, Amsterdam, pp 341–378
57. Martynova L et al (1997) Fabrication of plastic microfluid channels by imprinting methods. Anal Chem 69:4783–4789
58. Gonzalez C, Collins SD, Smith RL (1998) Fluidic interconnects for modular assembly of chemical microsystems. Sens. Actuators B: Chemical 49:40–45
59. Johnson KS et al (1996) Using neutral metastable argon atoms and contamination lithography to form nanostructures in silicon, silicon dioxide, and gold. Appl Phys Lett 69: 2773–2775
60. Lorenz H et al (1997) SU-8: a low-cost negative resist for MEMS. J Micromech Microeng 7:121
61. Larsson O et al (1997) Silicon based replication technology of 3D-microstructures by conventional CD-injection molding techniques. IEEE 1412:1415–1418
62. Klaassen EH et al. (1996) Silicon fusion bonding and deep reactive ion etching; a new technology for microstructures. Sens. Actuators A: Physical 52:132–139
63. Juan WH, Pang SW (1995) A novel etch-diffusion process for fabricating high aspect ratio Si microstructures. In: Proc. 8th Int. Conf. Solid-State Sensors and Actuators (Transducers '95), Stockholm, Sweden, pp 560–563
64. Madou MJ (2002) Fundamentals of microfabrication: the science of miniaturization. CRC, Boca Raton, FL
65. Rolland JP et al (2004) High Resolution Soft Lithography: Enabling Materials for Nanotechnologies. Angewandte Chemie 116:5920–5923
66. Wasatch Microfluidics, LLC, www.microfl.com. 12 Nov 2012
67. Rivet C et al (2010) Microfluidics for medical diagnostics and biosensors. Chem Engine Sci 66:1490–1507
68. Crews N, Wittwer C, Gale B (2008) Continuous-flow thermal gradient PCR. Biomed Microdevices 10:187–195
69. Bartholomeusz DA, Boutté RW, Andrade JD (2005) Xurography: rapid prototyping of microstructures using a cutting plotter. Microelectromechanical Syst J 14: 1364–1374
70. Klank H, Kutter JP, Geschke O (2002) CO2-laser micromachining and back-end processing for rapid production of PMMA-based microfluidic systems. Lab Chip 2:242–246
71. Becker H, Gärtner C (2008) Polymer microfabrication technologies for microfluidic systems. Anal Bioanal Chem 390:89–111
72. Martinez AW, Phillips ST, Whitesides GM (2008) Three-dimensional microfluidic

devices fabricated in layered paper and tape. Proc Natl Acad Sci 105:19606–19611

73. Martinez AW et al (2009) Diagnostics for the developing world: microfluidic paper-based analytical devices. Anal Chem 82:3–10
74. Bhattacharyya A, Klapperich CM (2006) Thermoplastic microfluidic device for on-chip purification of nucleic acids for disposable diagnostics. Anal Chem 78:788–792
75. Yager P et al (2006) Microfluidic diagnostic technologies for global public health. Nature 442:412–418
76. Squires TM, Quake SR (2005) Microfluidics: Fluid physics at the nanoliter scale. Rev Mod Phys 77:977
77. Psaltis D, Quake SR, Yang C (2006) Developing optofluidic technology through the fusion of microfluidics and optics. Nature 442:381–386
78. Demello AJ (2006) Control and detection of chemical reactions in microfluidic systems. Nature 442:394–402
79. Myers FB, Lee LP (2008) Innovations in optical microfluidic technologies for point-of-care diagnostics. Lab Chip 8:2015–2031
80. Singer PA et al (2007) Grand challenges in global health: the ethical, social and cultural program. PLoS Med 4:e265
81. Chin CD, Linder V, Sia SK (2007) Lab-on-a-chip devices for global health: past studies and future opportunities. Lab Chip 7:41–57
82. World Health Organization (2010) World malaria report: 2010. World Health Organization, Geneva
83. World Health Organization (2010) 2010/2011 Tuberculosis Global Facts, Geneva.
84. United Nations (2000) *United Nations Millennium Declaration.*
85. World Health Organization. (2009–2010) Situation updates - Pandemic (H1N1) 2009, World Health Organization, Geneva.
86. Reed C et al (2009) Estimates of the prevalence of pandemic (H1N1) 2009, United States, April-July 2009. Emerg Infect Dis 15: 2004–2007
87. Ivnitski D et al (1999) Biosensors for detection of pathogenic bacteria. Biosens Bioelectron 14:599–624
88. Kim J et al. (2010) Sample to answer: a fully integrated nucleic acid identification system for bacteria monitoring. In: Becker H, Wang W (eds) Proc. SPIE, vol 7593, pp 75930S
89. Kim J et al (2009) Microfluidic sample preparation: cell lysis and nucleic acidpurification. Integr Biol 1:574–586
90. Sant HJ et al (2010) Integrated Microfluidics for Serotype Identification of Foot and Mouth Disease Virus, in Proceedings of The 14th International Conference on Miniaturized Systems for Chemistry and Life Sciences. Groningen, The Netherlands
91. Velusamy V et al (2010) An overview of food-borne pathogen detection: In the perspective of biosensors. Biotechnol Adv 28:232–254
92. Lazcka O, Campo F, Munoz FX (2007) Pathogen detection: A perspective of traditional methods and biosensors. Biosens Bioelectron 22:1205–1217
93. Xiang Q et al (2006) Miniaturized immunoassay microfluidic system with electrokinetic control. Biosens Bioelectron 21(10): 2006–2009
94. Gao Y et al (2008) Multiplexed high-throughput electrokinetically-controlled immunoassay for the detection of specific bacterial antibodies in human serum. Anal Chim Acta 606:98–107
95. Meagher RJ et al (2008) An integrated microfluidic platform for sensitive and rapid detection of biological toxins. Lab Chip 8:2046–2053
96. JinSeok H, Hua SZ (2009) An overview of recent strategies in pathogen sensing. Sensors 9:4483–4502
97. Boehm DA, Gottlieb PA, Hua SZ (2007) On-chip microfluidic biosensor for bacterial detection and identification. Sensors Actuators B: Chemical 126:508–514
98. Liao JC et al (2006) Use of electrochemical DNA biosensors for rapid molecular identification of uropathogens in clinical urine specimens. J Clin Microbiol 44:561–570
99. Stevens DY et al (2008) Enabling a microfluidic immunoassay for the developing world by integration of on-card dry reagent storage. Lab Chip 8:2038–2045
100. Cheng X, Chen G, Rodriguez WR (2009) Micro-and nanotechnology for viral detection. Anal Bioanal Chem 393:487–501
101. Kortepeter MG, Parker GW (1999) Potential biological weapons threats. Emerg Infect Dis 5:523–527
102. Desai D, Wu G, Zaman MH (2010) Tackling HIV through robust diagnostics in the developing world: current status and future opportunities. Lab Chip 11(2):194–211
103. Lee WG et al (2010) Nano/Microfluidics for diagnosis of infectious diseases in developing countries. Adv Drug Deliv Rev 62:449–457
104. Lee SH et al (2008) A polymer lab-on-a-chip for reverse transcription (RT)-PCR based

point-of-care clinical diagnostics. Lab Chip 8:2121–2127

105. Cheng X et al (2009) Enhancing the performance of a point-of-care CD4+ T-cell counting microchip through monocyte depletion for HIV/AIDS diagnostics. Lab Chip 9:1357–1364
106. Cheng X et al (2007) A microfluidic device for practical label-free CD4+ T cell counting of HIV-infected subjects. Lab Chip 7:170–178
107. Cheng X et al (2007) A microchip approach for practical label-free CD4+ T-cell counting of HIV-infected subjects in resource-poor settings. JAIDS J Acquired Immune Deficiency Syndrom 45:257–261
108. Moon SJ et al (2009) Integrating microfluidics and lensless imaging for point-of-care testing. Biosens Bioelectron 24:3208–3214
109. Cheng X et al (2007) Cell detection and counting through cell lysate impedance spectroscopy in microfluidic devices. Lab Chip 7:746–755
110. Gohring JT, Fan X (2010) Label Free Detection of CD4+ and CD8+ T Cells Using the Optofluidic Ring Resonator. Sensors 10:5798–5808
111. Wang JH et al (2011) An integrated microfluidic system for counting of CD4+/CD8+ T lymphocytes. Microfluid Nanofluid 10:531–541
112. Yamanaka K et al (2011) Rapid detection for primary screening of influenza A virus: microfluidic RT-PCR chip and electrochemical DNA sensor. Analyst 136(10):2064–2068
113. Neumann G, Noda T, Kawaoka Y (2009) Emergence and pandemic potential of swine-origin H1N1 influenza virus. Nature 459:931–939
114. Lien KY et al (2011) Rapid detection of influenza A virus infection utilizing an immunomagnetic bead-based microfluidic system. Biosens Bioelectron 26:3900–3907
115. Ferguson BS et al (2011) Genetic Analysis of H1N1 Influenza Virus from Throat Swab Samples in a Microfluidic System for Point-of-Care Diagnostics. J Am Chem Soc 133(23):9129–9135
116. Weiss VU et al (2007) Virus analysis by electrophoresis on a microfluidic chip. J Chromatogr B 860:173–179
117. Zhu H et al (2008) Opto-fluidic micro-ring resonator for sensitive label-free viral detection. Analyst 133:356–360
118. Huh YS et al (2008) Enhanced on-chip SERS based biomolecular detection using electrokinetically active microwells. Lab Chip 9:433–439
119. Weng CH et al (2011) A suction-type microfluidic immunosensing chip for rapid detection of the dengue virus. Biomed Microdevices 13(3):585–595
120. Lee BS et al (2009) A fully automated immunoassay from whole blood on a disc. Lab Chip 9:1548–1555
121. Heinze BC et al (2009) Microfluidic immunosensor for rapid and sensitive detection of bovine viral diarrhea virus. Sensors and Actuators B: Chemical 138:491–496
122. Wang C et al (2011) An integrated microfluidic loop-mediated-isothermal-amplification system for rapid sample pre-treatment and detection of viruses. Biosens Bioelectron 26(5):2045–2052
123. Tothill IE (2009) Biosensors for cancer markers diagnosis. Elsevier, Amsterdam, pp 55–62
124. Choi YE, Kwak JW, Park JW (2010) Nanotechnology for early cancer detection. Sensors 10:428–455
125. Kumar S, Mohan A, Guleria R (2006) Biomarkers in cancer screening, research and detection: present and future: a review. Biomarkers 11:385–405
126. Hanash SM, Pitteri SJ, Faca VM (2008) Mining the plasma proteome for cancer biomarkers. Nature 452:571–579
127. Saerens D et al (2008) Antibody fragments as probe in biosensor development. Sensors 8:4669–4686
128. D'Haeseleer P (2006) How does DNA sequence motif discovery work? Nat Biotechnol 24:959–961
129. Makarov DV et al (2009) Biomarkers for prostate cancer. Annu Rev Med 60:139–151
130. Wang J (2006) Electrochemical biosensors: towards point-of-care cancer diagnostics. Biosens Bioelectron 21:1887–1892
131. Legendre LA et al (2008) Toward a Simplified Microfluidic Device for Ultra-fast Genetic Analysis with Sample-In/Answer-Out Capability: Application to T-Cell Lymphoma Diagnosis. J Assoc Laboratory Automation 13:351–360
132. Diercks AH et al (2009) A microfluidic device for multiplexed protein detection in nano-liter volumes. Anal Biochem 386:30–35
133. Lin DH et al (2010) Internally calibrated quantification of VEGF in human plasma by fluorescence immunoassays in disposable elastomeric microfluidic devices. J Chromatogr B 878:258–263

134. Lee H et al (2008) Chip–NMR biosensor for detection and molecular analysis of cells. Nat Med 14:869–874
135. Lazar IM (2008) Microfluidic bioanalytical platforms with mass spectrometry detection for biomarker discovery and screening In: Severine le Gac, Albert van den Berg (eds) Miniaturization and Mass Spectrometry, Royal Society of Chemistry, 1st ed., pp 151–172
136. Zhang K et al (2010) A microfluidic system with surface modified piezoelectric sensor for trapping and detection of cancer cells. Biosens Bioelectron 26(2):935–939
137. von Muhlen MG et al (2010) Label-free biomarker sensing in undiluted serum with suspended microchannel resonators. Anal Chem 82:1905–1910
138. Zani A et al (2011) A New Electrochemical Multiplexed Assay for PSA Cancer Marker Detection. Electroanalysis 23:91–99
139. Fragoso A et al (2010) Integrated microfluidic platform for the electrochemical detection of breast cancer markers in patient serum samples. Lab Chip 11(4):625–631
140. Kellner C et al (2011) Automated microsystem for electrochemical detection of cancer markers. Electrophoresis 32:926–930
141. O'Sullivan A et al (2011) Cost Estimation of Cardiovascular Disease Events in the US. Pharmacoeconomics 29(8):693–704
142. McDonnell B et al (2009) Cardiac biomarkers and the case for point-of-care testing. Clin Biochem 42:549–561
143. Mohammed MI, Desmulliez MPY (2010) Lab-on-a-chip based immunosensor principles and technologies for the detection of cardiac biomarkers: a review. Lab Chip 11(4):569–595
144. Wu AHB et al (2004) Evaluation of a point-of-care assay for cardiac markers for patients suspected of acute myocardial infarction. Clin Chim Acta 346:211–219
145. Ordonez-Llanos J et al (2006) Risk stratification of chest pain patients by point-of-care cardiac troponin T and myoglobin measured in the emergency department. Clin Chim Acta 365:93–97
146. James SK et al (2004) A rapid troponin I assay is not optimal for determination of troponin status and prediction of subsequent cardiac events at suspicion of unstable coronary syndromes. Int J Cardiol 93:113–120
147. Cramer GE et al (2007) Lack of concordance between a rapid bedside and conventional laboratory method of cardiac troponin testing: impact on risk stratification of patients suspected of acute coronary syndrome. Clin Chim Acta 381:164–166
148. Jönsson C et al (2008) Silane–dextran chemistry on lateral flow polymer chips for immunoassays. Lab Chip 8:1191–1197
149. Gervais L, Delamarche E (2009) Toward one-step point-of-care immunodiagnostics using capillary-driven microfluidics and PDMS substrates. Lab Chip 9:3330–3337
150. Hong B et al (2008) Highly sensitive rapid, reliable, and automatic cardiovascular disease diagnosis with nanoparticle fluorescence enhancer and MEMS. Adv Exp Med Biol 614:265–273
151. Bhattacharyya A, Klapperich CM (2007) Design and testing of a disposable microfluidic chemiluminescent immunoassay for disease biomarkers in human serum samples. Biomed Microdevices 9:245–251
152. Cho IH et al (2009) Chemiluminometric enzyme-linked immunosorbent assays (ELISA)-on-a-chip biosensor based on cross-flow chromatography. Anal Chim Acta 632:247–255
153. Huang H et al (2009) Rapid analysis of alpha-fetoprotein by chemiluminescence microfluidic immunoassay system based on super-paramagnetic microbeads. Biomed Microdevices 11:213–216
154. Sista R et al (2008) Development of a digital microfluidic platform for point of care testing. Lab Chip 8:2091–2104
155. Kurita R et al (2006) On-chip enzyme immunoassay of a cardiac marker using a microfluidic device combined with a portable surface plasmon resonance system. Anal Chem 78:5525–5531
156. Tweedie M et al (2006) Fabrication of impedimetric sensors for label-free Point-of-Care immunoassay cardiac marker systems, with passive microfluidic delivery. IEEE 1: 4610–4614
157. Christenson RH, Azzazy HME (2009) Cardiac point of care testing: a focused review of current National Academy of Clinical Biochemistry guidelines and measurement platforms. Clin Biochem 42:150–157
158. Billah M, Hays HCW, Millner PA (2008) Development of a myoglobin impedimetric immunosensor based on mixed self-assembled monolayer onto gold. Microchimica Acta 160:447–454
159. Zhou F et al (2010) Electrochemical Immunosensor for Simultaneous Detection of Dual Cardiac Markers Based on a Poly (Dimethylsiloxane)-Gold Nanoparticles

Composite Microfluidic Chip: A Proof of Principle. Clin Chem 56:1701–1707

160. Chen X et al (2008) Electrochemical impedance immunosensor based on three-dimensionally ordered macroporous gold film. Anal Chem 80:2133–2140
161. Mitsakakis K, Gizeli E (2011) Detection of multiple cardiac markers with an integrated acoustic platform for cardiovascular risk assessment. Anal Chim Acta 699(1):1–5
162. Hong J, Edel JB, deMello AJ (2009) Micro- and nanofluidic systems for high-throughput biological screening. Drug Discov Today 14:134–146
163. Liu J et al (2009) In Situ Microarray Fabrication and Analysis Using a Microfluidic Flow Cell Array Integrated with Surface Plasmon Resonance Microscopy. Anal Chem 81:4296–4301
164. Caliper LifeSciences-Target ID/Validation Applications. http://www.perkinelmer.com/industries/lifesciencesresearch/default.xhtml. Accessed 26 May 2011
165. The FilmArray Respiratory Panel. http://www.biofiredx.com/FilmArray/RespiratoryTest.html. Accessed 12 Nov 2012
166. (2011) search issued patents for "microfluidic" in title or abstract, United States Patent and Trademark office.
167. Mark D et al (2010) Microfluidic lab-on-a-chip platforms: requirements, characteristics and applications. Chem Soc Rev 39:1153–1182
168. Becker H (2009) It's the economy. Lab Chip 9:2759–2762
169. Kim L (2011) FluidicMEMS.com's list of microfluidics/lab-on-a-chip companies. http://fluidicmems.com/list-of-microfluidics-lab-on-a-chip-and-biomems-companies. Accessed 12 Nov 2012

Chapter 21

Quantitative Heterogeneous Immunoassays in Protein Modified Polydimethylsiloxane Microfluidic Channels for Rapid Detection of Disease Biomarkers

Peng Li

Abstract

Conventional detection of disease biomarkers employs techniques such as lateral-flow assays or central laboratory-based enzyme-linked immunosorbent assays (ELISA). Miniaturization and performance improvement of such traditional immunoassays using microfluidic technologies has proved promising in producing rapid, sensitive and automated next-generation immunosensors for quantitative diagnoses in the point-of-care setting. In this article a poly(dimethylsiloxane) (PDMS)-based immunosensor is presented for rapid detection of C-reactive protein. PDMS is selected in part because of the vast popularity of using PDMS as a material for microfluidic devices and in part because of the challenge of obtaining a stable surface coating with PDMS for immunosensing applications. Practical procedures for fabrication, surface modification, and preservation of the microfluidic immuno-chips as well as detailed descriptions of performing the microfluidic heterogeneous assay are presented.

Key words: Microfluidic immunoassay, Poly(dimethylsiloxane), Surface functionalization, Fluorescence detection, C-reactive protein, Point-of-care diagnostics, Long-term chip storage

1. Introduction

The advancement of clinical diagnosis and biodefense necessitates detection of disease biomarkers and infectious agents in a prompt and sensitive manner (1). Lateral-flow assays (LFA) have proved to be a simple and effective approach for performing rapid tests, but only qualitative or semiquantitative results can be obtained. Most of today's critical diagnostic tests are still conducted in sophisticated clinical or reference laboratories with enzyme-linked immunosorbent assays (ELISA) by highly

Gareth Jenkins and Colin D. Mansfield (eds.), *Microfluidic Diagnostics: Methods and Protocols*, Methods in Molecular Biology, vol. 949, DOI 10.1007/978-1-62703-134-9_21, © Springer Science+Business Media, LLC 2013

trained personnel. Although ELISA provides a quantitative, specific, and sensitive detection approach for analyzing biological agents, it requires long incubation times and multistep fluid handling. The prospect of microfluidic technology as a candidate for clinical and immunological tests was given critical support by reports of detection improvements in microfluidic heterogeneous immunoassays (μHEI) (2). The reduction in analysis time with nearly no sacrifice to sensitivity strongly verifies the promise of this technology.

Researchers have experienced a dramatic reduction in analysis time for μHEI because the diffusion time required for molecules to adsorb onto the solid reaction surface is proportional to the square of the characteristic length of the channel (3). For reactions with a high intrinsic rate such as the high affinity antigen-antibody binding, the incubation times were dramatically reduced by shortening the diffusion length (4). However, for reactions with a low intrinsic rate, it is expected that the incubation time will not be dramatically influenced by simply reducing the characteristic length of the channel. In fact, kinetically controlled reactions in microchannels have been reported in the literature. The experimental data of a study performed by Lionello et al. (5) indicated that the immunoglobulin G (IgG) adsorption on the walls of a 200×50 μm poly(ethylene terephthalate) (PET) microchannel was under kinetic control. Van Dulm and Norde (6) found that the adsorption of negatively charged human plasma albumin on negatively charged polystyrene surfaces was also controlled by kinetics at pH 7.4. This kinetic controlled adsorption is analogous to the blocking step in a μHEI on poly(dimethylsiloxane) (PDMS) surfaces because bovine serum albumin (BSA) (7) (the major blocking buffer component) and PDMS (8) are both negatively charged at pH 7.4. In both cases the long-range repulsive electrostatic forces must be overcome before the albumin molecules can bind with the solid reaction surface. Such kinetically controlled reactions are actually the major hindrance towards developing truly rapid μHEIs because a reduction in size from macro to micro dimensions does not usually change the nature of molecular reactions (9).

In addition to reducing the analysis time, developing stable microfluidic immuno-chips that are capable of maintaining their biofunctionality over sustained storage remains a critical challenge, especially given the fact that many microfluidic immunosensors are targeting diagnostics at resource-limited settings (10). According to three recent excellent reviews on surface modifications of PDMS (11–13), although many surface modification methods have been developed, the long-term stability of these methods has rarely been documented. Most of the documented stability tests were conducted up to a period of 3 months at best. Almost no information is available with regard to the biostability of these surface modification methods for immunosensing applications.

Here we report the development of a quantitative μHEI for rapid (~23 min) detection of disease biomarkers. The kinetically controlled blocking process is eliminated from the μHEI by using protein A functionalized PDMS microchannels. The nonspecific binding of the assay is maintained around the chip background level by using a pair of antibodies with different affinities to protein A under optimized experimental conditions. The BSA-glutaraldehyde-protein A functionalized PDMS microfluidic immuno-chips demonstrate biofunctionality up to 14 months under wet-stored conditions at 4°C with a storage buffer. High purity C-reactive protein (CRP) from human serum is selected as a model biomarker. Many clinical studies suggest that CRP is quite specific for predicting cardiovascular events. Its risk prediction is superior to low-density lipoprotein cholesterol alone and adds substantial prognostic information to conventional risk assessment (14). Practical procedures for fabrication, bioactivation, and preservation of the microfluidic immuno-chips as well as detailed descriptions of performing the μHEI are presented.

2. Materials

2.1. Fabrication of PDMS Microfluidic Chips

1. Black plastic or chrome photomask with transparent patterns of the microchannels (see Note 1).
2. Negative photoresist SU-8 2050 (MicroChem Corp., Newton, MA, USA).
3. Adhesion promoting solution: isopropanol with 25% (v/v) titanium diisopropoxide bis(acetylacetonate) 4 in. Silicon wafer.
4. Spin coater.
5. Leveling table.
6. Programmable hotplate.
7. 365 nm collimated light source (i-line, Karl Suss MJB-3, Suss Microtech, Garching, Germany).
8. SU-8 developer solution: propylene glycol methyl ether acetate.
9. Orbit Shaker (capable of 300 rpm).
10. PDMS prepolymer and curing agent: Sylgard 184 silicone elastomer kit (Dow Corning, Midland, MI, USA).

11. Oxygen plasma system (Femto, Diener Electronic, Reading, PA, USA).

2.2. Bioactivation of PDMS Microfluidic Chips

1. World-to-chip fluidic connectors: polyether ether ketone (PEEK) tubes, barbed tube adaptors (McMaster-Carr, Robbinsville, NJ, USA), and Tygon tubing. These components are connected in the following order—syringe, barbed tube adaptor, Tygon tubing, PEEK tube, and then the fluidic port of the chip.
2. BSA solution: 1.5 mg/mL BSA in 1× PBS buffer
3. Protein A solution: 50 μg/mL Protein A in 1× PBS buffer.
4. Cross-linker: 0.4% (v/v) glutaraldehyde in distilled water.
5. Humidity chamber: a Petri dish lined with cleanroom wipes soaked in distilled water and sealed with Parafilm.
6. Storage buffer: 10 mM Tris, 0.05% BSA, 0.05% ProClin 300, and 5% glycerol.

2.3. Microfluidic Heterogeneous Immunoassay

1. Highly pure human CRP (>99%) (Catalog # 30R–AC067: Fitzgerald Industries International, Inc., Concord, MA, USA).
2. Polyclonal rabbit anti-CRP (capture antibody, cAb): Rabbit anti-CRP (catalog # ab31156: Abcam Inc., Cambridge, MA, USA) 200 μg/mL prepared in protein A/IgG binding buffer.
3. Fluorescein isothiocyanate (FITC) labeled polyclonal goat anti-CRP (detection antibody, dAb) (catalog # ab19174 Abcam Inc., Cambridge, MA, USA): 120 μg/mL in PBS.
4. Protein A/IgG binding buffer (Pierce Biotechnology, Inc., Rockford, IL, USA).
5. PBST buffer: 1× PBS (phosphate buffered saline), 0.05% Tween 20.
6. Programmable syringe pump (NE-1000, New Era Pump Systems, Inc., Wantagh, NY, USA).

2.4. Data Collection and Processing

1. Epifluorescence microscope (ZEISS Axioplan 2 Imaging System, Carl Zeiss MicroImaging, Inc., Thornwood, NY, USA) equipped with a 100 W mercury lamp (ebq 100 isolated), a FITC filter set, a 5× objective (Plan Neofluar, numerical aperture = 0.15), and a high-resolution color digital camera (ZEISS AxioCam).
2. Image analysis software: ImageJ 1.41o (National Institutes of Health, USA).
3. Data processing software: Origin 7.0 (OriginLab Corp., Northampton, MA, USA).

3. Methods

3.1. Fabrication of PDMS Microfluidic Chips

1. Spin-coat a thin film of the adhesion promoting solution on the silicon wafer at 300 rpm, 100 rpm/s for 10 s and 3,000 rpm, 300 rpm/s for 30 s with the spin coater.
2. Next spin-coat a layer of SU-8 2050 on the wafer at 300 rpm, 100 rpm/s for 45 s and 3,000 rpm, 300 rpm/s for 50 s (see Note 2).
3. Place the coated wafer onto the leveling table for 1 h to allow the SU-8 to level out.
4. Pre-exposure soft bake: ramp from 50°C to 95°C at 2°C/min and bake for 30 min at 95°C on the well-leveled hotplate. Cool the wafer to room temperature at 1°C/min (see Note 3).
5. Expose the SU-8 to 365 nm light for 90 s under the photomask at intensity of 3.5 mW/cm^2 (see Note 4).
6. Post-exposure baking: ramp heating from 45°C to 85°C at 2°C/min and held for 40 min at 85°C. Cool the wafer to room temperature at 1°C/min.
7. Place the wafer in SU-8 developer solution on the Orbit Shaker at 300 rpm for 10 min.
8. Casting of the PDMS device: pour mixed and degassed PDMS prepolymer base and curing agent (base:curing agent = 10:1, mass ratio) onto the mold and cure at 60°C for 2.5 h on the hotplate. Peel the cured PDMS off from the master, and punch 1.5 mm diameter access holes using a biopsy punch. Next place the PDMS part with the patterns and a second flat PDMS slab into the plasma system and oxidize for 3 min. Place PDMS slabs together to form an irreversible bond.

3.2. Bioactivation of PDMS Microfluidic Chips

1. Prior to surface functionalization, leave the fabricated microchips in air at room temperature (~25°C) for at least 24 h, allowing the PDMS surface to regain its hydrophobicity. Then, clean all PDMS microfluidic channels with ethanol for 10 min and subsequently wash for 5 min with 1× PBS buffer.
2. In the first step of the surface functionalization process, fill the microchannels with BSA solution (see Note 5 and Fig. 1a) and then incubate for 4 h (see Note 6).
3. Inject cross-linker solution (see Note 7 and Fig. 1b) into the BSA coated channels and incubate for 1 h.
4. To complete the functionalization, inject protein A solution (see Note 8 and Fig. 1c) into the microchannels and incubate for 1 h. This results in protein A covalently attaching to the glutaraldehyde. All of the above steps are carried out at room

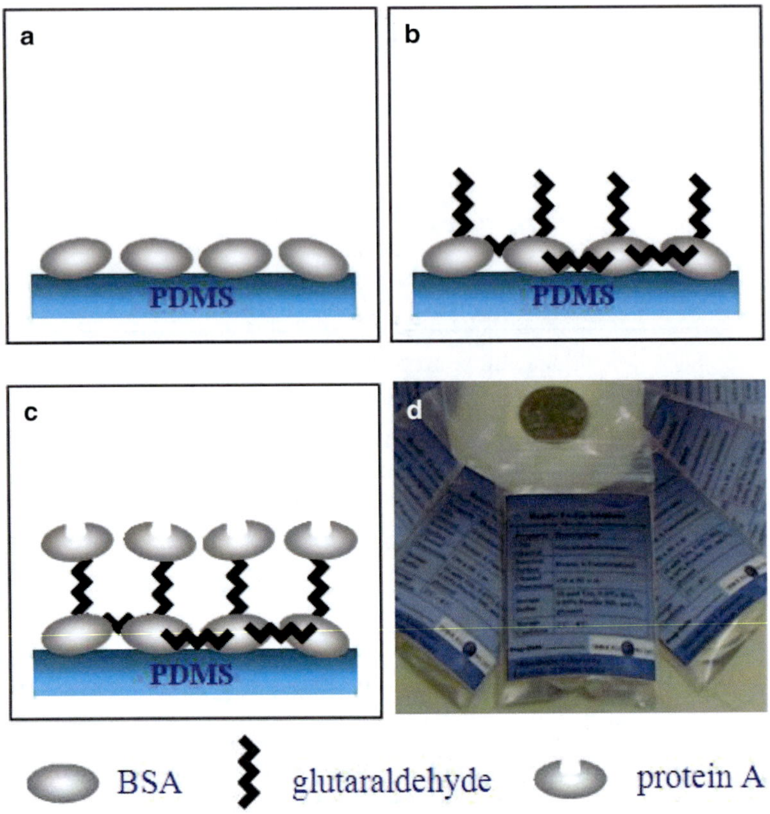

Fig. 1. Steps involved in preparing and preserving a bioactive PDMS surface for immunosensing: (**a**) BSA passivation, (**b**) glutaraldehyde cross-linking, (**c**) protein A coating, and (**d**) chip storage.

temperature inside the humidity chamber to suppress reagent losses in all microchannels due to the vapor permeable nature of the PDMS matrix.

5. Following surface functionalization, fill channels with the storage buffer and package each chip in a Ziploc bag which is also filled with the storage buffer (see Fig. 1d). The functionalized chips are stored at 4°C to preserve their biofunctionality (see Note 9).

3.3. Microfluidic Heterogeneous Immunoassay

1. To start the immunoassay, take a pre-functionalized PDMS chip out of the refrigerator and allow it to warm up to room temperature. Then remove it from its packaging and flush the microchannels with 1× PBS buffer.
2. Rabbit anti-CRP (see Note 10 and Fig. 2a) is then loaded into the microchannels and incubated for 7 min (see Note 6).
3. This is followed by injecting human CRP in PBS into the microchannels and incubating for 5 min (see Note 11 and see Fig. 2b).
4. Fill the microchannels with FITC labeled goat anti-CRP (see Note 12 and Fig. 2c) to detect the human CRP.

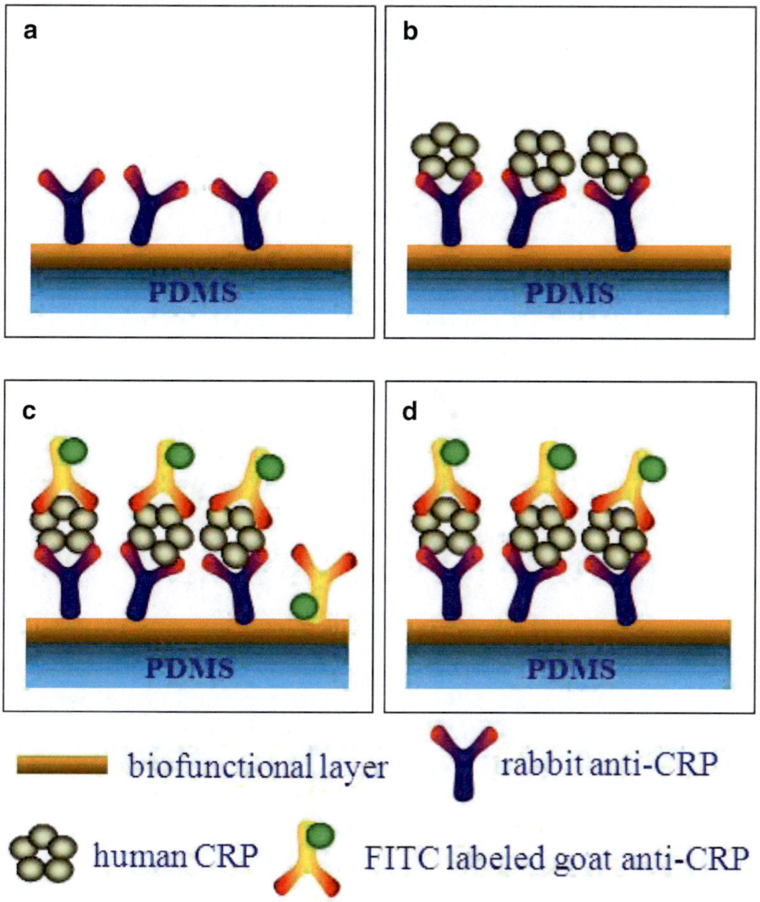

Fig. 2. The sandwich format fluorescence heterogeneous immunoassay in the functionalized microchannel for detecting human CRP: (**a**) immobilize cAb, (**b**) capture CRP, (**c**) react dAb, and (**d**) wash.

5. Following 5 min incubation, wash the microchannels by filling them with PBST buffer for 5 min (see Fig. 2d).
6. Perform negative control experiments without adding CRP to monitor the nonspecific binding on each chip (see Note 13).
7. The fluidic control of reagents in all steps is achieved with a programmable syringe pump. All incubations are accomplished under static conditions once the fluid has filled the microchannels (see Note 14). The microchannels are flushed with 1× PBS and emptied by aspiration between all steps in the assay. All the immunoassays are carried out at room temperature. Each microchip should be used only once (not recycled).

3.4. Data Collection and Processing

1. As soon as an immunoassay is complete acquire fluorescence images of the microchannels using an epifluorescence microscope. The optimal exposure time for imaging was found to be 2.5 s (see Note 15).

2. Obtain the average signal intensity within a selected area of interest (AOI) in each channel using the image analysis software (ImageJ 1.41o). The average signal intensity of each AOI is expressed in mean gray values (MGV) and exported to the data processing software (Origin 7.0). The local background intensity, taken from the areas in between the microchannels, is subtracted from the data to give the net fluorescence signal (NFS).
3. Dose–response curve: To generate dose–response curves, the assay should include CRP concentration in a broad range, such as 200, 100, 50, 10, 5, 1, 0.5 μg/mL and negative controls. After measuring the fluorescence intensity at each concentration (in triplicate using three different chips prepared at different times for improved reliability), the data are fitted with a four parameter logistic (4PL) function (see Note 16). The values of these parameters are obtained by performing curve fitting in Origin 7.0, with an estimated concentration of the sample determined directly from the obtained formula. The final curve should resemble the example in Fig. 3.
4. Limit of detection (LOD): The LOD is defined as the lowest human CRP concentration yielding an average fluorescence reading at least three standard deviations (99.5% confidence level) above the mean value recorded for the negative controls. Using the example 4PL function, the LOD for human CRP is calculated to be 360 ng/mL ($S/N = 3$).

4. Notes

1. Photomask: Photomasks can be designed with a CAD program such as Adobe Illustrator CS2 (Adobe, San Jose, CA). The patterns can be printed onto either a transparent Mylar film or transferred to a chrome coated glass slide.
2. Spin coating: This spin coating process generates a mold thickness of about 35 μm after baking. The mold thickness can be increased by reducing the spin speed. Starting the process at a low speed and a low acceleration is crucial to produce a uniform coating.
3. Soft baking: The time used for soft baking needs careful optimization based on film thickness to obtain a desirable solvent content in the final film, which is crucial to the success of the UV exposure and the post exposure baking process (15). The ramped heating and cooling is performed to reduce the internal stress of the photoresist film. The spin coating and soft baking processes can be repeated to build thick films (>300 μm).

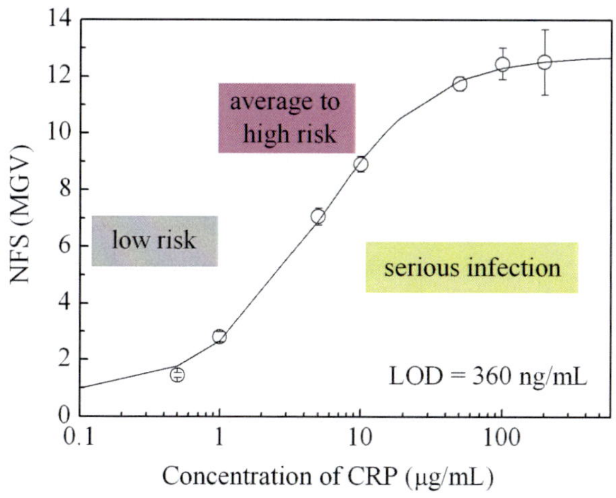

Fig. 3. A dose–response curve for human CRP. Error bars represent the standard error. According to the American Heart Association, three categories have been defined for cardiovascular disease risk assessment corresponding to CRP concentrations of <1 μg/mL (low), 1–3 μg/mL (intermediate), and >3 μg/mL (high). CRP level above 10 μg/mL is likely caused by acute disease (14).

4. UV exposure: Depending on the thickness of the film, the exposure time should also be optimized for best resolution and vertical sidewall profiles; however, it is generally favorable to over-expose the resist for better adhesion. For thick layers and for high exposure doses, the illumination should be done stepwise with sufficient intervals for the resist to cool down to room temperature. It is important to note that the optimal exposure energy relies on the nature of the substrate.
5. Surface coverage and Langmuir isotherm: PDMS is more hydrophobic than polystyrene and its high avidity for proteins results in high nonspecific binding in immunoassays. In the present protocol BSA is physically adsorbed to cover the hydrophobic sites and renders the PDMS surface amphiphilic. To obtain a desirable surface coverage ratio ($\Theta > 95\%$) in microchannels, the concentration of the BSA needs to be optimized. The obtained concentration-intensity data can be fitted to the Langmuir isotherm model to determine the optimal protein concentration. The Langmuir isotherm equation is $\Theta = K[P]/(1 + K[P])$, in which $[P]$ is the bulk concentration of the protein, and K is the equilibrium constant (16).
6. Protein adsorption kinetics: In addition to the protein concentration, the incubation time of each adsorption step is another important factor that affects the surface coverage. The incubation time-intensity data should be obtained and fitted to the protein adsorption kinetics equation, $[PS](t) = [S_0](1 - e^{-K_a[P_0]t})$ where the number of the bound

complexes, $[PS]$, depends upon the initial number of active sites on the surface ($[S_0]$), the initial bulk concentration of the protein ($[P_0]$), the association constant (K_a), and the reaction time (t) (17). A characteristic reaction time, $t_{\text{reac}} = 3/K_a[P_0]$ which allows 95% of the active sites on the surface to be covered, needs to be selected. For diffusion-controlled processes (e.g., antibody/antigen interaction), the reaction time should approximately equals to the characteristic diffusion time, $t_{\text{diff}} = L^2/2D$ where L is the diffusion length of the channel and D is the diffusion coefficient of the protein. For reaction-controlled processes (e.g., BSA/PDMS interaction), $t_{\text{reac}} >> t_{\text{diff}}$. The surface coverage and the adsorption kinetics are known to be affected by temperature, concentration, viscosity (18), buffer pH, buffer ionic strength, flow conditions (19), and surface to volume ratio (5). The influences of these parameters are reflected in the values of the association constant. The incubation times used in the example assay were selected based on the kinetics study for each step (20).

7. Glutaraldehyde cross-linking: In addition to serving as a cross-linker between BSA and protein A, as shown in Fig. 1b, the glutaraldehyde also cross-links individual BSA molecules to form a continuous surface layer so that the absorption or desorption of BSA molecules is sterically or entropically hindered. The cross-linked BSA molecules also restrict external protein access to the PDMS surface and therefore reduce nonspecific binding. It is worth noting that the glutaraldehyde cross-linking process also causes a weak fluorescence background in the bio-functional layer. This weak fluorescence is approximately 1 MGV higher than the background outside the functionalized channels (21).
8. Protein A for HEI: To increase the signal and reduce the sources of the nonspecific binding in this assay, it is desirable that the solid support phase only strongly binds to capture antibodies (cAb) but weakly or does not bind to detection antibodies (dAb) and the analytes. In this sense, protein A is an ideal candidate for being used as the supporting molecule in HEIs because it not only provides oriented immobilization of IgG (binds to Fc region) but also has selective binding characteristics to IgGs from different species (22). The characteristics of a protein A coated surface has been compared against those of the native PDMS surface elsewhere (21).
9. Long-term storage: Under wet storage at 4°C, the biofunctionality degradation of a protein modified PDMS surface is likely caused by: (1) the loss of the adsorbed proteins as a result of desorption or protein displacement (16), and (2) the denaturation of the adsorbed proteins due to the contamination of microorganisms and proteases. The biofunctionality

degradation caused by PDMS absorption will not be a concern because PDMS has so far only been reported to uptake small hydrophobic molecules with molecular weight (MW) up to ~300 Da (23) due to its porous natural. We have found that the BSA-glutaraldehyde-protein A functionalized PDMS microfluidic immuno-chips can maintain biofunctionality up to 14 months under wet-stored conditions at 4°C with our storage buffer (21). BSA is added to hinder desorption of coated BSA molecules from the surface. ProClin 300 is used as an antimicrobial reagent.

10. Effect of cAb concentration: Using low concentrations of the cAbs cannot provide enough binding sites for CRP. On the other hand, when the concentration of the cAbs is too high, the fluorescence signal of the assay also begins to decrease. This is caused by the formation of multiple layers of cAbs where some rabbit IgGs bind to other IgGs through protein–protein interaction instead of directly to the functionalized channel surface. CRPs that bind to rabbit anti-CRP antibodies that are not securely bound to the channel surface are removed during the washing step. By increasing the cAb concentration, the nonspecific binding level decreases and eventually levels off. This can be explained by the fact that using concentrated cAbs occupies more vacant binding sites on the protein A functionalized surface so that the free spaces left for the dAb adsorption are minimized.

11. Using blood sample: When the CRP test is performed from human blood samples (e.g., whole blood, plasma, and serum), an additional sample pretreatment chamber that is also functionalized with protein A should be added to the upstream of the immunoreaction zone to capture the human IgGs from the blood sample. Introducing human blood samples directly to the immunoassay area will compromise both the LOD and the dynamic range of the assay because the human IgGs in the blood can compete with the rabbit IgGs immobilized on the protein A surface.

12. Effect of dAb concentration: The fluorescence intensity increases sharply at low dAb concentrations and reaches a plateau as the available CRPs on the adsorbing surface start to be occupied and eventually saturated with dAbs. Increasing the dAb concentration does not increase the nonspecific binding level of the assay, in part because of the weak affinity between goat IgG (used for dAb) and protein A in PBS buffer, and in part because of a kinetically reduced dAb/antigen reaction time used here.

13. Nonspecific binding of the assay: Since there have been no reported affinities between protein A and CRP or between goat and rabbit antibodies, the nonspecific binding in this HEI

is primarily attributed to the adsorption of dAbs onto the solid reaction surface sites where no antigen is present. It is affected by several factors including surface biospecificity, antibody selection, buffer solutions, and incubation time. By employing IgGs with high affinity (from rabbit) to protein A as the cAb and IgGs with low affinity (from goat) to protein A as the dAb, the nonspecific adsorption of dAbs to unoccupied protein A is minimized in conjunction with using PBS buffer and a kinetically reduced antigen/dAb reaction time (20, 21). By using this scheme, our previous study revealed that the nonspecific binding of this assay can be maintained around the chip background level without adding a blocking step to the procedure (21).

14. Static incubation versus continuous flow: Under static situations, the adsorbing proteins (e.g., antibody or antigen) are likely to deplete due to the high surface-to-volume ratio in microchannels. The depletion leads to a decreased surface coverage on channel walls and therefore a low assay signal. If you have sufficient analyte-containing sample and you want to improve the assay signal at low analyte concentrations, the improvement can be achieved by using either a "stop-flow" or a continuous flow method, as proposed by Lionello et al. (24), to increase the adsorption level of the analyte on the channel walls.
15. The selected exposure time should be sufficient to distinguish the lowest signal from the background, but not make the highest signal saturate the camera.
16. 4PL function: The equation for the 4PL model is $y = \text{max} + (\text{min} - \text{max})/[1 + (x/EC_{50})]^{\alpha}$, where x is the concentration of the analyte (dose), y is the expected fluorescence signal (response), max is the signal for infinite analyte concentration, min corresponds to the signal with a zero dose of the analyte (the negative control), EC_{50} is the concentration of the analyte with an expected signal exactly halfway between max and min, and α is the slope factor.

References

1. Whitesides GM (2006) The origins and the future of microfluidics. Nature 442:368–373
2. Bange A, Halsall HB, Heineman WR (2005) Microfluidic immunosensor systems. Biosens Bioelectron 20:2488–2503
3. Squires TM, Quake SR (2005) Microfluidics: fluid physics at the nanoliter scale. Rev Modern Phys 77:977–1026
4. Rossier JS, Gokulrangan G, Girault HH, Svojanovsky S, Wilson GS (2000) Characterization of protein adsorption and immunosorption kinetics in photoablated polymer microchannels. Langmuir 16:8489–8494
5. Lionello A, Josserand J, Jensen H, Girault HH (2005) Protein adsorption in static microsystems: effect of the surface to volume ratio. Lab Chip 5:254–260
6. Van Dulm P, Norde W (1983) The adsorption of human plasma albumin on solid surfaces, with special attention to the kinetic aspects. J Colloid Interface Sci 91:248–255

7. Putnam FW (ed) (1975) The plasma proteins: structure, function and genetic control, vol 1. Academic, New York, pp 141–147
8. Kirby BJ, Hasselbrink EF Jr (2004) The zeta potential of microfluidic substrates. 2. Data for polymers. Electrophoresis 25:203–213
9. Dittrich PS, Manz A (2006) Lab-on-a-chip: microfluidics in drug discovery. Nat Rev Drug Discovery 5:210–218
10. Yager P, Domingo GJ, Gerdes J (2008) Point-of-care diagnostics for global health. Annu Rev Biomed Eng 10:107–144
11. Makamba H, Kim JH, Lim K, Park N, Hahn JH (2003) Surface modification of poly(dimethylsiloxane) microchannels. Electrophoresis 24:3607–3619
12. Wong I, Ho C-M (2009) Surface molecular property modifications for poly(dimethylsiloxane) (PDMS) based microfluidic devices. Microfluid Nanofluid 7:291–306
13. Zhou J, Ellis AV, Voelcker NH (2010) Recent developments in PDMS surface modification for microfluidic devices. Electrophoresis 31:2–16
14. Ridker PM (2004) High-sensitivity C-reactive protein, inflammation, and cardiovascular risk: from concept to clinical practice to clinical benefit. Am Heart J 148:S19–S26
15. Campo A, Greiner C (2007) SU-8: a photoresist for high-aspect-ratio and 3D submicron lithography. J Micromech Microeng 17:R81–R95
16. Holden MA, Cremer PS (2005) Microfluidic tools for studying the specific binding, adsorption, and displacement of proteins at the interfaces. Annu Rev Phys Chem 56:369–387
17. Dodge A, Fluri K, Verpoote E, Rooij NF (2001) Electrokinetically driven microfluidic chips with surface-modified chambers for heterogeneous immunoassays. Anal Chem 73:3400–3409
18. Schreiber G (2002) Kinetic studies of protein-protein interactions. Curr Opin Struct Bio 12:41–47
19. Salim M, O'Sullivan B, McArthur SL, Wright PC (2007) Characterization of fibrinogen adsorption onto glass microcapillary surfaces by ELISA. Lab Chip 7:64–70
20. Li P, Abolmaaty A, D'Amore C, Demming S, Anagnostopoulos C, Faghri M (2009) Development of an ultrafast quantitative heterogeneous immunoassay on pre-functionalized poly(dimethylsiloxane) microfluidic chips for the next-generation immunosensors. Microfluid Nanofluid 7:593–598
21. Li P, Sherry A, Cortex J, Anagnostopoulos C, Faghri M (2011) A blocking-free microfluidic fluorescence heterogeneous immunoassay for point-of-care diagnostics. Biomed Microdevices:. doi:10.1007/s10544-011-9515-9
22. Anderson GP, Jacoby MA, Ligler FS, King KD (1997) Effectiveness of protein A for antibody immobilization for a fiber optic biosensor. Biosens Bioelectron 12:329–336
23. Toepke MW, Beebe DJ (2006) PDMS absorption of small molecules and consequences in microfluidic applications. Lab Chip 6:1484–1486
24. Lionello A, Josserand J, Jensen H, Girault HH (2005) Dynamic protein adsorption in microchannels by "stop-flow" and continuous flow. Lab Chip 5:1096–1103

Chapter 22

Breast Cancer Diagnostics Using Microfluidic Multiplexed Immunohistochemistry

Minseok S. Kim, Seyong Kwon, and Je-Kyun Park

Abstract

A quantitative, reproducible, fast and inexpensive multiplexed immunohistochemistry (IHC) system might play a locomotive role in drug screening and personalized medicine. Currently, fully automated IHC machines and sequential multiplexed IHC methods based upon multiple color reagents have been developed, with the evolution of such methods having revealed novel biological findings over the conventional IHC method, which is time consuming and labor intensive. We describe a novel parallel multiplexed IHC method using a microfluidic multiplexed immunohistochemistry (MMIHC) device for quantitative pathological diagnosis of breast cancer. The key factors for success of parallel multiplexed IHC are the fabrication of a robust microfluidic device, the interface between the device and a tissue slide, and an accurate fluidic control for multiple IHC reagents. In order to apply conventional thin-section tissues into on-chip systems without any additional modification process, a tissue slide-compatible assembler was developed for optimal compatibility of conventional IHC methods. With this approach, a perfect fluid control for various solutions was demonstrated without any leakage, bubble formation or cross-contamination. The results presented in this chapter indicate that the microfluidic IHC protocol developed can provide the possibility of tailored cancer treatments as well as precise histopathological diagnosis using numerous specific biomarkers.

Key words: Breast cancer, Immunohistochemistry, Microfluidics, Multiplexing, Pathology, Quantification, Tissue.

1. Introduction

After Coons and his colleagues first developed an immunofluorescence technique for labeling proteins with a fluorescent dye (1), several enzyme-tagging methods to identify immunohistochemical reactions, including horseradish peroxidase (HRP) (2) and peroxidase–antiperoxidase techniques (3), were introduced in succession. Thereafter, immunohistochemistry (IHC) has been widely used for assessing therapeutic biomarkers and has become a major part of practical diagnosis in surgical pathology (4). Since IHC

Gareth Jenkins and Colin D. Mansfield (eds.), *Microfluidic Diagnostics: Methods and Protocols*, Methods in Molecular Biology, vol. 949, DOI 10.1007/978-1-62703-134-9_22, © Springer Science+Business Media, LLC 2013

supports not only identification of proteins of interest but also localization information of tissue morphology, tests for various biomarkers by IHC can be an ideal method to realize individualized tailored therapy with limited quantity of clinical samples (5). However, IHC is a labor intensive and time consuming process. Although fully automated machines can help address this issue, it is not necessarily feasible to use highly expensive instruments in general laboratories (6). Moreover, for a biological laboratory to study signal pathway and proteomics, it is more important to effectively study various biomarkers from a single sample rather than routinely run single biomarker tests for many samples.

Here, we introduce a novel microfluidic multiplexed immunohistochemistry (MMIHC) platform for quantitative pathological diagnosis (7, 8). Microfluidics enables construction of a well confined environment, precise manipulation of fluids, easy and fast fluidic control and use of small sample volumes. Therefore, such microfluidic technology is ideal to construct an inexpensive, fast, reproducible, cost-effective, and semiautomated IHC system. Moreover, the ability to fabricate numerous micron scale channels means that multiple biomarkers can be analyzed on a limited cancer core (biopsy sample) area. As the tumor size acquired from operational surgery has become smaller due to the development of cancer diagnostic technology and neoadjuvant therapy, the MMIHC platform is expected to enable examination of various predictive and prognostic biomarkers on a small sized tissue sample. Additionally, the MMIHC system is likely to also reduce process time, costs, and quantity of tissue sample required, as well as expand tests for prognostic biomarker candidates, and facilitate integration of genomic and proteomic approaches into the search for targets in oncology.

The protocol described in this chapter uses a poly(dimethylsiloxane) (PDMS) molding technique to fabricate a MMIHC device. Here, the device is composed of ten inlets and reaction channels for primary antibodies and one outlet (Fig. 1a). The antibodies are flowed through microchannels, which are reversibly sealed with a glass slide on to which the tissue (or cell) samples to be examined have been deposited (Fig. 1b). Hence, the area of reaction channels in the MMIHC device defines the staining area for a sample of interest. For homogeneous samples such as cell blocks, the area of reaction channels and alignment between the device and a sample is not a major concern. Therefore, it is possible to examine many biomarkers by designing a number of narrow reaction channels. For heterogeneous samples such as tissues, however, the area of reaction channels is critical to achieve an accurate staining result. Thus, the design of reaction channels should be carefully considered to maximize representativeness for a whole-section tissue slide.

Irregular flow rates among the channels might cause inconsistent staining intensities. Therefore, the same fluidic resistance should be guaranteed for equivalent flow rates in all reaction channels.

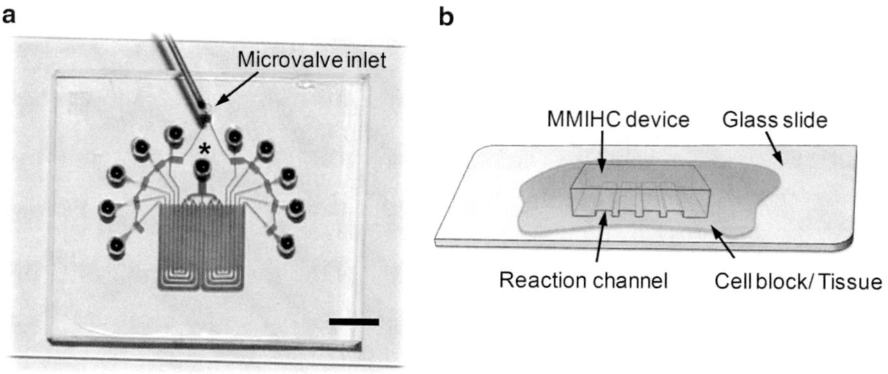

Fig. 1. Configuration of a MMIHC platform. (**a**) A fabricated MMIHC device having ten reaction channels. *Red line* (leading from microvalve inlet) is the microvalve and the asterisk indicates outlet of the device. Scale bar, 4 mm. Reproduced from ref. (8), with permission from Elsevier. (**b**) Schematic of the interface assembly between a MMIHC device and a tissue slide. Reproduced from ref. (7).

For rectangular channels, the channel flow rate is expressed as $Q = \Delta P/R$, where the ΔP is the pressure drop and R is the fluidic resistance. The resistance R is expressed as follows:

$$R = f\mathrm{Re}\frac{\mu L}{2D_h^2 A} \tag{1}$$

$$D_h = \frac{2wh}{w+h} \tag{2}$$

$$f\mathrm{Re} = 96(1 - 1.3553\alpha + 1.9467\alpha^2 - 1.7012\alpha^3 + 0.9564\alpha^4 - 0.2537\alpha^5) \tag{3}$$

where w, h, and L are the width, height, and characteristic length of a microfluidic channel, respectively; μ is the viscosity of a fluid; A is the cross-sectioned area of a microfluidic channel; and α is the aspect ratio (9). A withdrawal pumping mode for fluidic movement has the advantage to construct a simple MMIHC control system by connecting a syringe pump to an outlet. If the system is operated in this mode, it is desirable to equally apply the fluidic resistance between biomarker inlets and the outlet.

2. Materials

2.1. Fabrication and Preparation of MMIHC Device

1. Negative photoresist: SU-8 2025 (MicroChem Corp., Newton, MA).
2. SU-8 Developer: Propylene glycol monomethyl ether acetate (PGMEA).
3. Spin coater.
4. Silicon wafer.

5. Hotplate.
6. Collimated UV light source / mask aligner.
7. Photomask (three): one for rectangular reaction channels in a fluidic channel layer (defined using SU-8); one for rounded flow channels in a fluidic channel layer (defined using AZ photoresist); and one for pneumatic channels in a control channel layer (defined using SU-8).
8. Positive photoresist: AZ 9260 (MicroChemicals GmbH, Ulm, Germany).
9. Base and curing agent: PDMS.
10. O_2 plasma asher.
11. Tygon tube (inner diameter 0.02 in.).
12. Syringe pump with a gas-tight syringe.

2.2. Preparation of Samples

1. Breast cancer cell lines: AU-565, HCC70, MCF-7, and SK-BR-3 obtained from the American Type Culture Collection (ATCC, VA).
2. RPMI-1640 (Gibco, MD) supplemented with 10% fetal bovine serum (FBS, UT), 100 IU mL^{-1} penicillin, and 100 mg mL^{-1} streptomycin for AU-565, HCC70, and MCF-7.
3. Dulbecco's Modified Eagle's Medium (DMEM) (Gibco, MD) supplemented with 10% fetal bovine serum (FBS) (HyClone Laboratories, Logan, UT), 100 IU mL^{-1} penicillin, and 100 mg mL^{-1} streptomycin for SK-BR-3.
4. Incubation chamber maintained at 37°C and 5% CO_2.
5. Cell detachment: solution of trypsin (0.25%) and ethylenediamine tetraacetic acid (EDTA) (1 mM).
6. Phosphate buffered saline (PBS): pH 7.4, 1× stock (Invitrogen, USA).
7. 1% agarose, cell culture grade, in iso-osmotic PBS. Dissolve agarose in PBS by boiling. Check the amount of water lost as vapor during boiling and replace with distilled water. After preparation, cool to 50°C.
8. Tris-buffered saline (TBS) with Tween 20: pH 7.4, 1× stock (ScyTek, USA).
9. Formaldehyde.
10. Absolute ethanol, toluene, paraffin.
11. Microtome.
12. Positively charged microscope slides.

2.3. Off-Chip IHC

1. A cell block or a tissue slide (De-wax in xylene for 4 min twice before use).
2. Target retrieval solution: pH 9 (Dako, Denmark).

3. Washing buffer solution: 1× TBS with Tween 20, pH 7.4.
4. Hydrogen peroxide.
5. Blocking solution (see Note 1).
6. Syringe pump (with a withdrawal mode).
7. Primary antibodies (see Note 2).
8. Secondary antibodies (see Note 3).
9. Streptavidin-HRP.
10. 3,3′ diaminobenzidine tetrachloride (DAB).
11. Mayer's hematoxylin.
12. Canada balsam.
13. Food dye.

3. Methods

The MMIHC device is fabricated by a simple procedure consisting of: (1) preparation of the MMIHC device; (2) off-chip IHC process #1; (3) assembly of MMIHC device; and (4) off-chip IHC process #2. The overall procedure for MMIHC is described in Table 1. Staining intensity and quality are dependent on the concentration of antibodies, flow rate, and incubation time (see Note 4). Therefore, the condition of antibody treatment needs to be optimized by one's own strategy and the quality of staining confirmed via professional pathologists.

3.1. Fabrication of a MMIHC Device

1. The fluidic channel mold of the device can be fabricated by two-step multilayer soft lithography (8). For the rectangular reaction channels, a negative photoresist, SU-8 2025, is spin-coated at 500 rpm (10 s) and 1,500 rpm (30 s) to make a 60 μm thickness on a bare silicon wafer.
2. Allow the spin-coated wafer to rest on a level surface for 10 min.
3. Soft bake at 60°C for 3 min and then at 95°C for 6 min.
4. To pattern reaction channels on the silicon wafer, expose the wafer to ultraviolet (UV) light, through a photomask for 20 s at 20.3 mW/cm^2.
5. Post bake at 60°C for 2 min and then at 95°C for 6 min.
6. Develop the wafer with SU-8 developer for 3–5 min.
7. A positive photoresist, AZ 9260, is spincoated at 300 rpm (10 s) and 1,000 rpm (60 s) to make a 20 μm thickness on the wafer.
8. Soft bake at 110°C for 80 s.

Table 1
Procedure of the MMIHC methods

No.	Reagent	MMIHC method
1	Deparaffinization	4 min ×2
2	Rehydration process	1 min ×5
3	Washing using DI water	1 min ×4
4	Heat-induced epitope retrieval	20 min
5	Cooling	5 min
6	Hydrogen peroxide blocking	10 min
7	Washing using TBS tween buffer	1 min
8	Blocking solution	20 min
9	Primary antibody treatment	15 min
10	Washing using TBS Tween buffer	1 min
11	Secondary antibody treatment	20 min
12	Washing using TBS Tween buffer	1 min
13	Streptavidin-HRP	20 min
14	Washing using DI water	1 min
15	DAB	5 min
16	Washing using DI water	1 min
17	Mayer's hematoxylin	3 min
18	Washing using DI water	1 min
19	Dehydration	1 min ×5
20	Xylene	4 min ×2
	Total consumed time	154 min

In total, 154 min are required to stain ten biomarkers when using a MMIHC device. The gray area indicates on-chip process

9. Expose to UV light through the fluidic channel photomask for 20 s at 20.3 mW/cm^2.
10. Develop the wafer with an AZ photoresist developer for 15 min.
11. Reflow process by heating the wafer in the sequence of 70°C for 5 min, 85°C for 5 min, 95°C for 5 min, and finally 110°C for 30 min. Consequently, the fluidic channels, except the reaction channels (defined by the SU-8), are transformed into round shapes suitable for valving. Pour PDMS on the fluidic channel mold and wait for 5 min to fill the gap between reaction channels.

12. The fluidic channel mold is spincoated with PDMS at 500 rpm (10 s) and 2,000 rpm (30 s) to make a 40 μm-thick membrane and cured at 70°C for 1 h.
13. The control channel mold of the device is fabricated by conventional SU-8 photolithography (8). This mold defines channel structures which, when aligned over the top of the rounded fluidic channels, can allow effective valving of the fluidic channels by application of pressure.
14. The PDMS membrane containing the fluidic channels is aligned and bonded with the PDMS control channel layer using an O_2 plasma asher.

3.2. Preparation of Samples

The MMIHC device can be applied to both cell blocks and tissue sections, which have been widely used in conventional immunocytochemistry (ICC) and IHC. The device can cover any type of a cell block and a tissue section attached on a glass slide. The following protocol introduces a general sample preparation method for paraffin-embedded cell block samples.

1. Breast cancer cell lines: AU-565, HCC70, MCF-7 are maintained in RPMI-1640 supplemented with 10% FBS, 100 IU mL^{-1} penicillin and 100 mg mL^{-1} streptomycin for AU-565, HCC70 and MCF-7.
2. Breast cancer cell line SK-BR-3 is maintained in DMEM supplemented with 10% FBS, 100 IU mL^{-1} penicillin and 100 mg mL^{-1} streptomycin.
3. Cells are incubated in an incubation chamber maintained at 37°C and with 5% CO_2.
4. At 70–80% confluences, cells are detached using a solution of trypsin (0.25%) and EDTA (1 mM).
5. Prepare 1% agarose, cell culture grade, in iso-osmotic PBS. Dissolve agarose in PBS by boiling. Check for loss of water due to vapor during boiling and replace with distilled water.
6. Cool agarose to around 50°C and maintain this temperature to keep it in liquid form. The agarose will solidified below this point.
7. Add PBS to cells and harvest cells into a conical tube (1.5 mL).
8. Centrifuge the tube at 1,270 × *g* for 7 min, discarding the supernatant.
9. Fix the suspension of the specimen in formaldehyde at least for 1 h.
10. Discard the supernatant formaldehyde.
11. Resuspend the deposit in molten 1% agarose at 50°C and quickly place in a centrifuge again.

12. Spin at 1,270 × g for 2 min and then cool at 4°C for 30 min to harden the agar (agarose). Any excess agar is trimmed to leave a cell block.
13. Wrap the cell block in tissue paper and place in a plastic cassette for paraffin processing.
14. Transfer the blocks sequentially to 30, 50, 70, 80, 90, 95, and 100% alcohols for about 2 h each.
15. Place the blocks in a second 100% ethanol solution to ensure that all water is removed.
16. Move the blocks into a 50:50 mixture of absolute ethanol and toluene for 2 h.
17. Place the blocks into pure toluene.
18. Move the blocks into a 50:50 mixture of toluene and paraffin.
19. Place them in an oven at 56°C to–58°C.
20. Transfer the blocks to pure paraffin and place in the oven for 1 h.
21. Transfer the blocks to a second pot of melted paraffin and place in the oven for an additional 2–3 h.
22. Section the blocks using a microtome with 4 μm thickness.
23. Mount the sections on positively charged slides.
24. Dry for 1 h at room temperature and then 1 h in an incubator at 60°C.

3.3. Preparation of MMIHC Device

1. Prepare a tube by cutting a 30-cm length piece of Tygon tube (inner diameter 0.02 in.).
2. Insert one end of the tube into the outlet of a MMIHC device and attach the MMIHC device to the upper plate (Fig. 2a, b, see Note 5).
3. Fix the MMIHC device to the upper plate with an adhesive membrane (Fig. 2c, see Note 6).
4. Perform a plasma treatment on the bottom surface of the device for hydrophilicity.
5. After the plasma treatment, immerse bottom surface of the device into deionized (DI) water. Avoid bubbles on the surface of reaction channels.
6. Place a bottom plate on a flat surface.
7. A sample slide is loaded onto the bottom plate. Washing buffer or DI water is sufficiently poured on it.
8. Put a plasma-treated MMIHC device (attached under the upper plate) on a bare glass slide. In this process, the alignment rod must pass through the alignment hole of the upper plate (Fig. 2d, see Note 7).

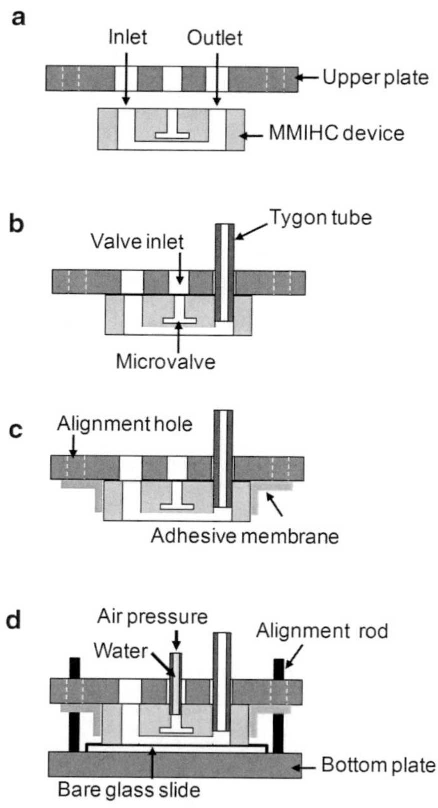

Fig. 2. Assembly process of the MMIHC device. (**a**) Alignment between an upper plate and a MMIHC device. (**b**) The device is attached to the upper plate. A tube is inserted into the outlet. (**c**) The MMIHC device is fixed to the upper plate using an adhesive membrane. (**d**) The device is loaded on the bare glass slide and a tube is inserted into the valve inlet. (**e**) A weight is loaded on the upper plate. (**f**) A sample slide is placed on the bottom plate. (**g**) The MMIHC device is aligned on the sample. (**h**) After mounting the weight on the upper plate, the sample slide is ready to be stained with the MMIHC device.

9. Connect a gas-tight syringe needle with one end of a tube, which is linked to outlet of the MMIHC device.
10. Install a syringe pump with a gas-tight syringe.
11. Prepare another tube for the microvalves.
12. Fill the tube with water using a syringe.
13. Insert the tube to the valve inlet (Fig. 2d).
14. Put a weight on the upper plate carefully through alignment holes and rods (Fig. 2e, see Note 8).
15. Apply air pressure to the valve inlet through the tube to fully fill water at the microvalve (Fig. 2e).
16. When the microvalves are filled with water completely, remove the air pressure (see Note 9).

3.4. Off-Chip IHC Process #1

1. De-wax a cell block or a tissue slide twice in xylene for 4 min.
2. Rehydrate through a graded series of ethanol solutions 100, 95, 90, 80, and 70% ethanol in series of 1 min.
3. Expose to DI water for 1 min, four times.
4. After the hydration process, target retrieval solution, pH 9 (Dako, Denmark) is preheated for 5 min.
5. When boiling, the samples are inserted into the solution and boiled for 20 min at 750 W using a microwave.
6. Cool the sample for 5 min in the target retrieval solution.
7. Take out the sample from the target retrieval solution and wash it with washing buffer solution at least three times. Wipe around the cell block or tissue section with tissue paper. Do not use a PAP pen (see Note 10).
8. Apply hydrogen peroxide on the area of the cell block or tissue section and incubate it for 10 min.
9. Wash the slide using washing buffer solution at least three times and wipe around the sample section with tissue paper.
10. Incubate the sample with blocking solution for 20 min. In this step, the sample is mounted on the bottom plate (Fig. 2f).

3.5. Assembly and Detachment Process of MMIHC System

1. Put a prepared MMIHC device on the sample and then aligned (Fig. 2g, see Note 11).
2. Mount a weight on the upper plate (Fig. 2h, see Note 12).
3. Apply air pressure to close the microvalve (40 kPa).
4. Remove liquid completely in all inlets using a syringe.
5. Inject 10 μL of antibodies into inlets using syringes (see Note 13).
6. After relieving the air pressure in the microvalves, operate the syringe pump in a withdrawal mode (i.e., 400 μL/h).
7. Primary antibodies are flowed along the reaction channels for 15 min.
8. Remove the weight from the assembly.
9. Immediately separate the upper plate and wash the sample with washing buffer.

3.6. Off-Chip IHC Process #2

1. Wipe around the cell block or tissue section with tissue paper (see Note 14).
2. Apply secondary antibody to the sample. Incubate it for 20 min.
3. Wash the sample using washing buffer.
4. Incubate the sample in streptavidin-HRP for 20 min.
5. Wash the sample using DI water.

6. Incubate the sample in DAB for 5 min.
7. Wash the sample using DI water.
8. Incubate the sample in Mayer's hematoxylin for 3 min.
9. Dehydrated the sample through a graded series of ethanol solutions 50, 80, 90, 95 and 100% ethanol in series of 1 min.
10. Expose to xylene treatment for 4 min, two times.
11. Mounting process with Canada balsam.

3.7. Characterization of MMIHC Device

It is important to check whether primary antibodies are properly flowed without any mixing. This can be characterized by injecting various colors of food dyes into inlets of primary antibodies (Fig. 3a). An important point to check is the confirmation of equal flow rates in individual reaction channels because different flow rates might cause different expressions of biomarkers (Fig. 3b). Another point is to confirm the maintenance of a rectangular cross section in the reaction channels (Fig. 3c, d, see Note 8).

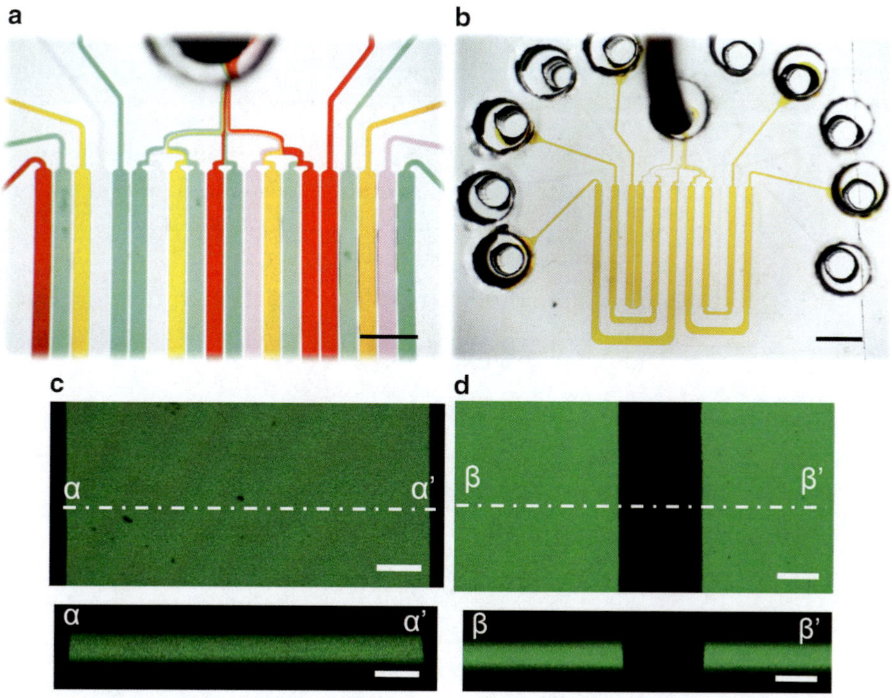

Fig. 3. Characterization of the MMIHC device. (**a**) Confirmation of fluid stream at reaction channels. The fluids are properly flowed without any cross-contamination. Scale bar, 1 mm. (**b**) Confirmation of equal flow rates at individual reaction channels. Liquids are simultaneously flowed out. Scale bar, 2 mm. Reproduced from ref. (8) with permission from Elsevier. (**c**) Plane and *z*-stacked confocal laser microscopy images of one reaction channel area under 8 kPa. Scale bar, 100 μm. (**d**) Plane and *z*-stacked confocal images of boundary area between reaction channels under 8 kPa. Reaction channels retain their original rectangular shape and individual reaction channel is completely separated. Scale bar, 100 μm. Reproduced from ref. (7).

1. Load a bare glass slide onto the bottom plate and pour washing buffer on it.
2. Place a prepared MMIHC device (described in Subheading 3.3) on to the slide.
3. Put a weight on the MMIHC device through alignment holes.
4. Apply air pressure to close the microvalves (40 kPa).
5. Aspirate solutions contained in the inlets with a syringe to remove the washing buffer.
6. Inject 10 μL of food dye into each inlet (Fig. 3b, see Note 15).
7. Relieve the air pressure in the microvalves.
8. Operate the syringe pump in a withdrawal mode (i.e., 600 μL/h).
9. Then place the assembled platform on the microscope to observe fluidic streams and flow rates of individual reaction channels.
10. Check for cross-contamination of liquids between the reaction channels via color change (Fig. 3a).
11. The consumption time of the solution in each inlet is measured and simultaneous removal of reagents in inlets is checked (Fig. 3b, see Note 16).

3.8. Image Analysis for Quantification of Biomarker Expressions

1. Place a cell block or tissue slide stained with the microfluidic system onto an inverted microscope.
2. Acquire images under 13,000 lux light intensity.
3. The acquired microscopic images are divided into three parts: staining part (SP) of a cell, non-staining part (NSP) of a cell, and background part (Fig. 4, see Note 17).
4. As one method to minimize variation in the image acquisition including cell density, the ratio of the staining area is calculated by SP over SP plus NSP.

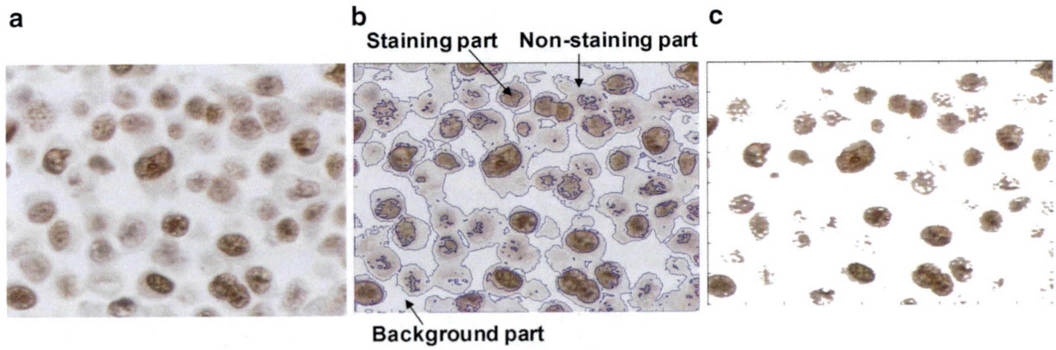

Fig. 4. Image analysis of biomarker expression level. (**a**) A microscopic image acquired via MMIHC. (**b**) The image is divided into three parts: the staining part (SP), the non-staining part (NSP), and the background. Only the cell area (SP and NSP) is considered to minimize the variation of expression level according to cell density. (**c**) Image after analysis. Only the brown-colored areas remained. Reproduced from ref. (7).

5. Staining intensity of the sample is averaged only with the SP.
6. "Expression level" is calculated by multiplying the staining ratio and the averaged intensity (see Note 18).

4. Notes

1. Blocking solution, secondary antibodies, streptavidin-HRP, and DAB are used, which are contained in the Cap-plus Detection kit (Zymed, San Francisco, CA). Blocking solution contains bovine serum albumin (BSA) and the serum from the same animal origin with the secondary antibodies. Usually, 2~3% of BSA and 5% of serum are mixed in 1× PBS.
2. Various primary antibodies can be used, which are produced from several species recognizing secondary antibodies. In this experiment, ER (SP1) antibody (Ventana, Tucson, AZ) and PR (1E2) antibody (Ventana) were used in conventional methods at 1× ready-to-use concentration, and in the microfluidic device at a dilution of 1:4. HER2 oncoprotein antibody (Dako) was used in conventional methods at a dilution of 1:500 and in the microfluidic device at a dilution of 1:2,000. Ki-67 (clone MIB-1) antibody (Dako) was used in conventional methods at a dilution of 1:50 and in the microfluidic device at a dilution of 1:200. CK14 (Novocastra, Newcastle-upon-Tyne, UK), CK5 (Novocastra), E-cadherin (BD Biosciences, Franklin Lakes, NJ), p53 (Ventana), p63 (Dako), and smooth muscle myosin heavy chain (S-H; Dako) were used in the microfluidic device at dilutions of 1:400, 1:200, 1:1000, 1:4, 1:200, and 1:800, respectively.
3. Secondary antibody supported in the Cap-plus Detection kit (Zymed, San Francisco, CA) was used. In multiplexed assay, it is noted that the combination between primary and secondary antibodies should be considered according to their specificity.
4. We found that dynamic incubation in microfluidic channels shows more effective staining than static incubation of antibody treatment. Therefore, the concentration of an antibody in MMIHC was normally 0.25 times diluted compared to conventional IHC staining.
5. The height that the tube is inserted should be less than half the height of the MMICH device to minimize leakage of fluids.
6. Alignment between the holes in the upper plate with the inlets and outlet in the MMIHC device is an important procedure for successful operation of the device. Avoid taping on the bottom surface of the device.

7. Avoid bubble formation in the reaction channel when the device is put into contact with the liquid on the glass slide.
8. The rectangular shape of the reaction channels should be maintained when a weight is loaded onto the MMIHC device. Deformation of reaction channels may occur owing to over-pressure. *Z*-stacked images via a confocal laser scanning microscope can be a good observation method for checking each device's geometry. It is recommended that the device works under 8 kPa of pressure.
9. Microvalves must be turned off (de-pressurized) before the weight from the MMIHC platform is removed through its disassembly. This is because if no external pressure is applied to the device by the platform, the applied air pressure can tear the membrane of the valve.
10. The use of a PAP pen prohibits stable contact between the microfluidic device and a sample slide, causing fluid leakages.
11. In application of tissue samples and needle biopsy sample, the device should be aligned at the area of highest cancer cell density, which is determined with the use of hematoxylin and eosin (H&E) staining slide (Fig. 5a). By the nature of heterogeneity of human tissues, staining intensity may not be homogeneous even in a reaction channel (Fig. 5b). Especially in needle biopsy samples, fatty tissue area should be avoided as an interested area. Biomarker expression in such regions might be inconsistent compared to other non-fatty areas.
12. When a weight is mounted on a device it is desirable to minimize the movement of the weight.
13. It is recommended that the antibody concentrations are fourfold lower than that of conventional IHC staining method. For antibody injection, the syringe needle is contacted at the bottom of an inlet to avoid bubble formation.
14. When the sample is separated from the device, the use of PAP pen is now allowable.
15. Adjacent inlets should be filled with different colors to estimate cross contamination between the reaction channels.
16. Simultaneous removal of liquids and bubble entrapment into each reaction channel mean that the same flow rate is achieved in all reaction channels.
17. The three parts are classified by a Bayesian classification. The Gaussian mixture model and expectation maximization method are used for the color distribution of each part in RGB color space (7).
18. At least three images are taken per reaction channel and the average expression level for the images is presented as a representative value of immunocytochemical staining for a

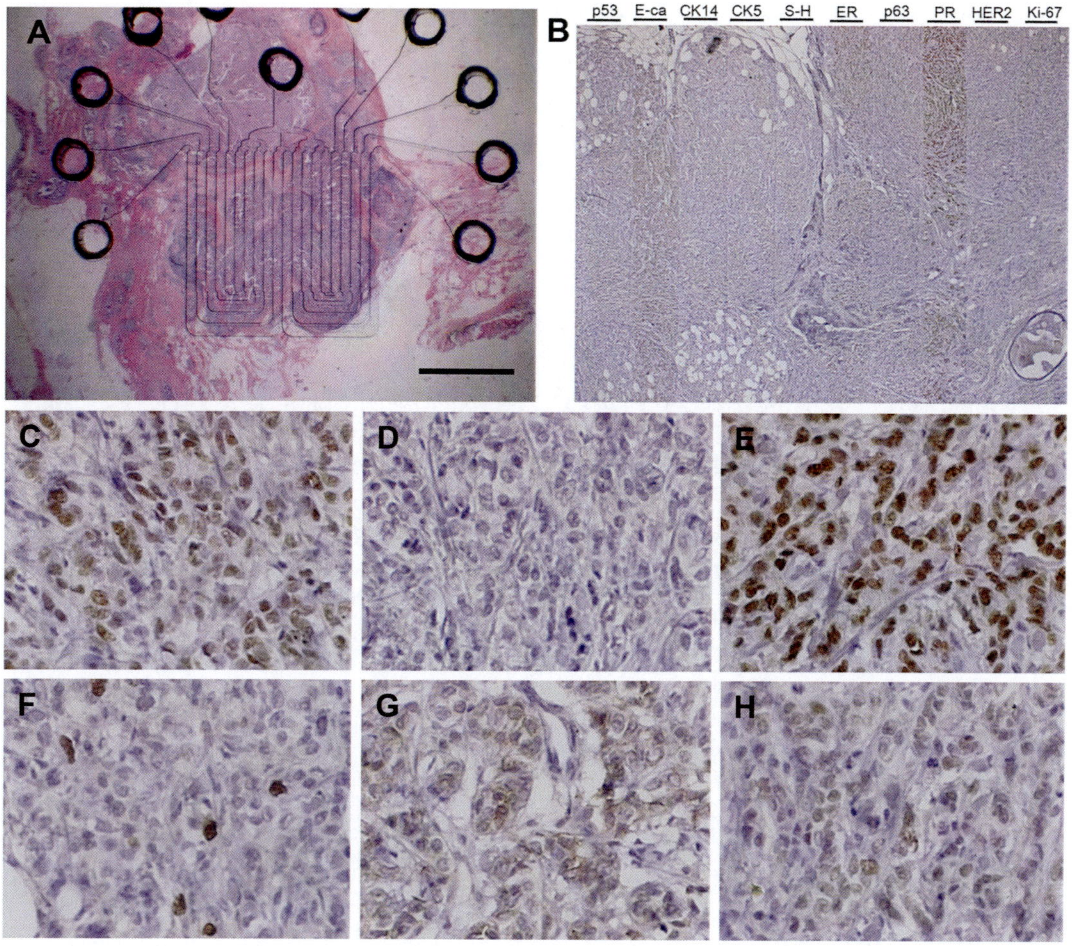

Fig. 5. MMIHC on a breast cancer tissue. (**a**) Alignment of a MMIHC device with an H&E-stained tissue sample. The device is aligned at a site showing numerous stained nuclei, indicating the area containing proliferating cancerous cells. Scale bar, 3 mm. (**b**) Result of multiplexed immunohistochemical staining for a human breast cancer tissue (×40). (**c–h**) Magnified images for expressed markers in Fig. 5b (×400). A tissue image stained with (**c**) ER, (**d**) HER2, (**e**) PR, (**f**) Ki-67, (**g**) E-Cadherin (E-ca), and (**h**) p63. Reproduced from ref. (8), with permission from Elsevier.

biomarker. Reaction channels having 300 μm width are enough to acquire staining images for a biomarker with ×400 magnification (Fig. 5c–h).

Acknowledgments

This research was supported by a National Leading Research Laboratory program (grant no. 2011-0018607). We acknowledge Eun Sook Lee, Chul Hwan Kim, Soim Kwon, and Bumi Kwon for clinical application, and Taemin Kim for image analysis.

References

1. Coons AH, Creech HJ, Jones RN, Berliner E (1941) The demonstration of pneumococcal antigen in tissues by the use of fluorescent antibody. J Immunol 45:159
2. Graham RC Jr, Karnovsky MJ (1966) The early stages of absorption of injected horseradish peroxidase in the proximal tubules of mouse kidney: ultrastructural cytochemistry by a new technique. J Histochem Cytochem 14:291–302
3. Sternberger LA, Hardy PH Jr, Cuculis JJ, Meyer HG (1970) The unlabeled antibody enzyme method of immunohistochemistry: preparation and properties of soluble antigen-antibody complex (horseradish peroxidase-antihorseradish peroxidase) and its use in identification of spirochetes. J Histochem Cytochem 18:315–333
4. Taylor CR, Levenson RM (2006) Quantification of immunohistochemistry–issues concerning methods, utility and semiquantitative assessment II. Histopathology 49:411–424
5. Lakhani SR, Ashworth A (2001) Microarray and histopathological analysis of tumours: the future and the past? Nat Rev Cancer 1:151–157
6. Ludwig JA, Weinstein JN (2005) Biomarkers in cancer staging, prognosis and treatment selection. Nat Rev Cancer 5:845–856
7. Kim MS, Kim T, Kong S-Y, Kwon S, Bae CY, Choi J et al (2010) Breast cancer diagnosis using a microfluidic multiplexed immunohistochemistry platform. PLoS One 5:e10441. doi:10.1371/journal.pone.0010441
8. Kim MS, Kwon S, Kim T, Lee ES, Park J-K (2011) Quantitative proteomic profiling of breast cancers using a multiplexed microfluidic platform for immunohistochemistry and immunocytochemistry. Biomaterials 32: 1396–1403
9. Vestad T, Marr DWM (2004) Flow resistance for microfluidic logic operations. Appl Phys Lett 84:5074–5075

Chapter 23

Charged-Coupled Device (CCD) Detectors for Lab-on-a Chip (LOC) Optical Analysis

Avraham Rasooly, Yordan Kostov, and Hugh A. Bruck

Abstract

A critical element of any Lab-on-a-Chip (LOC) is a detector; among the many detection approaches, optical detection is very widely used for biodetection. One challenge for advancing the development of LOC for biodetection has been to enhance the portability and lower the cost for Point-of-Care diagnostics, which has the potential to enhance the quality of healthcare delivery for underserved populations and for global health. We describe a simple and relatively low cost charged-coupled device (CCD)-based detector that can be integrated with a conventional microtiter plate or a portable LOC assay for various optical detection modalities including fluorescence, chemiluminescence, densitometry, and colorimetric assays. In general, the portable battery-operated CCD-based detection system consists of four modules: (1) a cooled CCD digital camera to monitor light emission, (2) a LOC or microtiter plate to perform assays, (3) a light source to illuminate the assay (such as electroluminescence (EL) or light emitting diode (LED)), and (4) a portable computer to acquire and analyze images. The configuration of the fluorescence detector presented here was designed to measure fluorogenic excitation at 490 nm and to monitor emission at 523 nm used for FITC detection.

The LOC used for this detection system was fabricated with laminated object manufacturing (LOM) technology, and was designed to detection activity of botulinum neurotoxin serotype A (BoNT-A) using a fluorogenic peptide substrate (SNAP-25) for botulinum neurotoxin serotype A (BoNT-A) labeled with FITC. The limit of detection (LOD) for the CCD detector is 0.5 nM (25 ng/ml). The portable system is small and is powered by a 12 V source. The modular detector was designed with easily interchangeable LEDs, ELs, filters, lenses, and LOC, and can be used and adapted for a wide variety of densitometry, florescence and colorimetric assays.

Key words: LED, Electroluminescence (EL), CCD, Fluorescence, Laminated object manufacturing (LOM), Fluorometer, Botulinum neurotoxin

1. Introduction

One key element for the development of Lab-on-a-Chips (LOC) for biodetection is the detection method, and many LOCs are utilizing optical detectors for analysis. There are many modalities for

Gareth Jenkins and Colin D. Mansfield (eds.), *Microfluidic Diagnostics: Methods and Protocols*, Methods in Molecular Biology, vol. 949, DOI 10.1007/978-1-62703-134-9_23,

optical detection, such as evanescence wave analysis, refractometry, colorimetry, densitometry, spectrometry, fluorometry, etc., making optical detection a very common approach for biological analysis.

In recent years, many optical detection methods have been developed for biodetection, including photodiodes (1–4), photomultipliers (5–10), and charged-coupled device (CCD) (10–18) based detection. Unlike most photodiodes and photomultiplier detectors, which are inherently point detectors for analysis of limited areas with high sensitivity, CCD-based detectors are spatial detectors which can monitor larger surface areas. This makes them ideal for LOC multichannel detection, because many channels can be analyzed simultaneously. In the last few years, complementary metal-oxide-semiconductor (CMOS) imagers have also emerged as a relatively simple, low cost, and sensitive alternative for optical detection to CCD imagers. While there are many scientific-grade CCD and CMOS imagers, their cost is typically high. Therefore, amateur astronomy digital imagers (or other compatible consumer electronics components) are very good alternatives providing sufficient sensitivity and image quality for biodetection at significantly lower cost. This is especially important for Point-of-Care diagnostics, where it is desirable to enhance the quality of healthcare delivery for underserved populations and for global health.

1.1. CMOS and CCD Technologies for Optical Detection

Charge-coupled devices (CCD) and Complementary metal-oxide-semiconductor (CMOS) are the two main semiconductor technologies used to produce imaging sensors. Both CMOS- and CCD-based imagers employ 2-D arrays of semiconductor pixels sensitive to light. Photons striking the semiconducting material generate electron–hole pairs and the resulting electric charge is processed into electronic signals. While both types of imagers collect and manipulate charge, in a CCD the charge from each pixel is sequentially transported across the chip to an analog-to-digital converter which converts the charge into a digital value. In most CMOS devices, each pixel has its own charge-to-voltage conversion and the sensor may include transistors at each pixel, noise-correction, and digitization circuits to convert the output signal to digital.

Today, both CCD and CMOS technologies are broadly used for imaging in variety of fields. While the boundaries between the technologies' capabilities are gradually blurring, CCD has been historically used for biodetection; therefore, in this chapter we addresses only CCD-based technology.

As discussed previously, the main advantage of CCD and CMOS imagers over photodiodes or photomultipliers for biodetection is their ability to analyze light from a relatively large surface area (e.g., from the entire surface of an LOC or plate array) (19–25). In contrast, photodiodes or photomultipliers are "point" detectors that measure light from small areas (such as one part of an LOC). To overcome the limitations of "point" detectors for the analysis of larger spaces, the measured surface (e.g., an ELISA microtiter plate) is normally raster scanned. Raster scanning requires mechanical movement of the LOC/plate, movement of the detector or a mirror imaging from the surface to the mirror, or the use of many "spot" detectors as pixels to cover the entire plate surface. Alternatively, CCD and CMOS imagers can visualize the entire plate at once, so they are much quicker and simpler to operate. In addition, CCD imagers are relatively low cost and are pre-configured for easy deployment and compatibility with LOC- and plate-based assays.

One notable example of the application of a CCD detector for biodetection is an array biosensor, where the CCD is used for multiplex detection of toxins and microbial pathogens and their toxins (11, 16, 26, 27). Sensitivity is a critical issue with biodetection, and CCDs in general are less sensitive than photomultipliers. However, with a long exposure time the CCD can be used for low level light detection provided the CCD has a low thermal noise level relative to the incoming photonic signal. This typically necessitates the use of a cooling system in order to minimize temperature and the subsequent thermal noise.

1.2. Optics for CCD-Based Detection

The optics for CCD-based detectors are relatively simple, consisting of a zoom lens and an extension tube which brings the lens closer to the imaged surface in order to capture more photons from a smaller surface area for macro imaging. The main advantage of zoom lenses is that their variable focal length enables imaging of both wide and narrow fields of view. Their main disadvantage is that they are slower, (depending on the focal length used—typically f/2.8—f/5.6 for a 1.2 lens), than a comparable fixed focal length lenses (e.g., f/1.2), which results in longer exposure times. A typical lens used for CCD detection is 4–12 mm diameter with f/1.2, which is a very wide angle lens that enables imaging of a wide field of view from a short distance (so the device can be compact). However, to achieve this wide field of view at such a short distance, the portion of the image captured along the edges of the lens ends up being curved and distorted.

Extension tubes are needed for greater magnification to achieve macro (i.e., close-up) imaging. The tube contains no optical elements and is placed between the CCD and the lens. It moves the lens farther from the CCD image plane and closer to the imaged surface. There is a trade-off when using extension tubes; the farther the lens is from the CCD, the greater the loss of light. Furthermore, extension tubes may not be the only optical element affecting light intensity. For fluorometry, emission and excitation filters are also needed that will further reduce the optical signal.

Therefore, the choice of lens, extension tube, and filters must all be compatible in order to image with enough sensitivity over the smaller surface areas of a typical LOC or microtiter plate.

1.3. Light Sources for Optical Detection

Light sources are critical for optical detection. Most current optical detectors utilize high intensity, high power, and bulky bench-top excitation sources including tungsten, mercury, or xenon lamps. These light sources are usually expensive and nondurable. They also require high voltage, which further limits the portability of the device. Lasers (11, 16, 26, 27) have been used for illumination sources since they are low cost, highly efficient, small, simple to operate, durable, and consume relatively little power. These characteristics make lasers ideal excitation light sources for portable detectors. However, lasers are inherently point light sources that can only be used effectively with photodiodes or photomultiplier detectors. When using lasers as point light sources to illuminate a larger surface area, mechanical translation stages can be used for raster scanning, but this complicates the operation of the imaging system and slows down the imaging. Alternatively, a line generator and evanescence light can be used with a laser (11, 16, 26, 27) to expand the light source and cover larger surface areas compatible with the spatial imaging of a CCD or CMOS. However, this may introduce light distribution and uniformity problems, and also complicates the design.

1.3.1. Electroluminescence (EL) Illumination

Electroluminescence (EL) illuminators are surfaces that provide spatial illumination, which is inherently more compatible with CCD and CMOS optical detectors (23, 24). EL illuminators are semiconductor surfaces that emit light in response to an alternating electric current. In EL, electrons that are excited into a higher state leave "holes" and when dropping back to the lower-energy ground state, the excited electrons release their energy as photons upon recombining with their "holes." The spectrum of the light emitted by EL depends on the electroluminescent materials. Although EL materials are widely used for signs and instrument dial illumination (e.g., indigo watches, clock radios, personal organizers, cockpit instrumentation, etc.), they are not commonly used for fluorescence excitation and biodetection. Nevertheless, they have great potential for biodetection applications given that the features of EL include spatial illumination (a great advantage for LOC detection), a range of wavelengths, low cost, high efficiency, small footprint, simple operation, durability, and low power consumption. However, EL generally emits fewer photons (i.e., less light) than other light sources used for fluorescence excitation, such as tungsten, mercury, and xenon lamps, LEDs, or lasers. Therefore, they require a longer exposure time. Another limitation is that they can require high voltage (e.g., 100 V) alternating electric current for operation to enhance photon emission.

1.3.2. Light-Emitting Diodes (LED) Illumination

Light-emitting diodes (LEDs) have been used as illumination sources in various optical detectors, such as for portable real-time PCR detection (28) and fluorometers (24, 25, 29). LEDs are semiconductor devices that emit light when an electric current passes through them. Similar to ELs, LEDs operate by electroluminescence, in which the emission of photons is caused by electronic excitation of a material. In the LEDs, two semiconductor materials of different composition, one negative, or n-type, and the other positive, or p-type, form a junction. Under an electric field, current flows across the p–n junction. The free electrons moving across a diode can fall into empty holes from the p-type layer. This involves a drop from the conduction band to a lower orbital, so the electrons release energy in the form of photons. The LED light is visible when the diodes are characterized by a wider gap between the conduction band and the lower electron orbit (i.e., wide bandgap). The size of this gap (which depends upon the semiconductor materials) determines the frequency of the photons (i.e., the wavelength the light).

LEDs have great potential use for optical sensor illumination (25) because they are available in a wide variety of wavelengths. Furthermore, some LEDs can emit light over a range of wavelengths, enabling multi-wavelength analysis. They are low cost, highly efficient, small, simple to operate, durable, and consume relatively little power. Their main advantage over ELs is they emit far more light than ELs and require low voltage DC for operation. Their main disadvantage is, unlike ELs, they are point illuminators like lasers. Therefore, surface illumination with LEDs requires optical systems (e.g., mirrors, diffusers, lenses, etc.) that may complicate the design and introduce light distribution and uniformity problems similar to Lasers.

1.4. Laminated Object Manufacturing (LOM) for LOC Fabrication

Laminated object manufacturing (LOM) in which layers of adhesive-coated laminates are successively glued together is an alternative prototyping technology for LOCs. In current LOM fabrication, the patterns and features of the device are cut through the thickness of thin sheets of polymer film, normally with a CO_2 laser cutter, and is followed by assembly and bonding to produce a three-dimensional object or device. In order to fabricate the layers, it is necessary to be able to create features at small length scales. For mass production, this must be done cheaply and rapidly. There are many approaches to manufacturing at these length scales, often referred to as "micromachining." These include lithography, etching, electrical discharge machining (EDM), fused deposition modeling (FDM), and laminated object manufacturing (LOM). The optimal manufacturing process is often dictated by the choice of material. For example, laminated object manufacturing (LOM) is based on the direct machining of polymer sheets; the patterns and features of the device are cut in thin sheets of polymer film, such as acrylic, polycarbonate or polyester, normally with a CO_2 laser cutter followed by assembly and bonding by successive lamination with an adhesive (30) or heat (31, 32) to produce a three-dimensional object or device. LOM is effective for rapid prototyping because all of the steps are relatively fast and inexpensive allowing for the design of complex features. For example, in a microfluidic device it is possible to cut simple microchannels for fluid flow, with more complex flow networks easily constructed by simply using more layers.

A critical element in the process is the laser cutter itself. Typically, the CO_2 laser is operated in a pulse mode. The intensity, duration, and frequency of the pulse will dictate the rapid heating characteristics that either melt or ablate the surface. If the heating is too intense, large areas are affected and can even burn. Furthermore, to cut surface features it is necessary to raster scan the laser, which means the raster speed can also affect the heating characteristics. Since each material absorbs the laser light at different rates, the exact conditions for laser cutting must be optimized for each material and the size of the desired features. Ultimately, the resolution of these features also depends on how small a spot size the laser light can be focused to, which is governed by the effective focal length of the laser lens. The more intense the focusing, the smaller the depth of focus (DOF) and the less parallel the cut surface features may be when cutting thicker materials. Typical spot sizes for most laser cutting systems are in the range of 200–400 μm. However, some systems can be configured to achieve spot sizes of less than 100 μm and even as low as 25 μm with properly sized expander/collimator and an aspheric lens. The minimum spot size is governed by the relation $\lambda f/d$, where λ is the wavelength of the light, f is the focal length of the lens, and d is the diameter of the collimated beam that enters the lens.

Fabrication of individual layers is followed by the layers' assembly and bonding by an adhesive; this precise alignment can be achieved by using a custom "guide pin" rig built to hold the machined layer in a precise position using "guide holes" in the machined layers. The "guide pins" are designed so that the PMMA layer will snap-fit into them for repeatable alignment.

1.5. Functional Elements of the Multichannel LOC System

In this chapter we describe an eight channel LOC system to mix the sample with reagents and enable assay detection for a BoNT cleavage assay (as described in our previous work (29)). Figure 1 shows the device design with Fig. 1a showing an enlarged single channel and Fig. 1b showing the complete design of the LOC. For the BoNT cleavage assay described in this chapter, each channel is designed to mix the sample with a FRET-labeled peptide and detect the signal. As shown in Fig. 1a, each channel is based on a V junction design that merges the sample and reagent channels. The sample reservoir (1) and the reagent reservoir (2) are joined via V junction (3), passing into the 1 mm joining channel (4). The BoNT

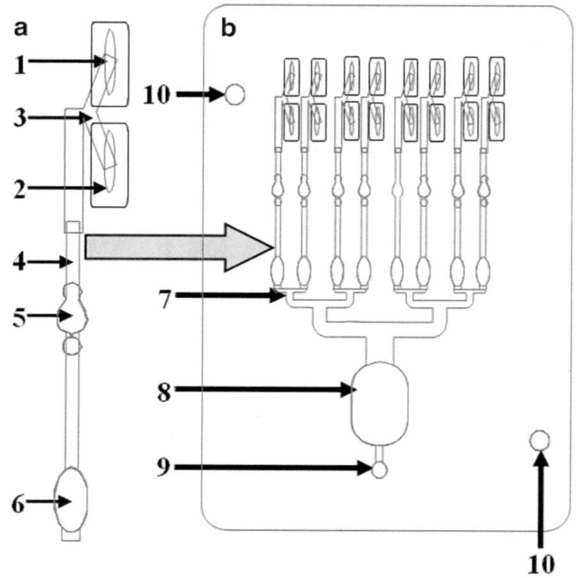

Fig. 1. Multichannel LOC for the detection of BoNT-A: An expanded single channel of the LOC (**a**) where the sample reservoir (1) and the reagent reservoir (2) are joined via V junction (3) to the joining channel (4). The cleavage reaction is monitored in the detection wells (5) designed with air traps to minimize air bubbles, followed by a mixing trap (6). The full schematic of eight-channel LOC chip (**b**), including the eight channel negative pressure distribution splitter (7) connected to a waste chamber (8) then to an outlet (9) for use with a pump or syringe. The alignment holes (10) were designed to simplify fabrication.

cleavage reaction is performed and monitored in the detection wells (5), designed with air traps on both side of the wells to capture air bubbles. The mixing of the toxin and the substrate peptide is done via flow mixing in the 1.85 mm diameter wells (not under laminar flow). At the end of the channel, a mixing trap chamber (6) prevents back flow and diffusion among the connected channels during incubation. It also serves as an indicator when the wells are filled.

Eight of these channel designs are connected to a negative pressure distribution splitter (7), which leads to a waste chamber buffer (8). The outlet of the device (9) is connected to a pump or syringe as the prime fluid mover. When a syringe is used, no electric power is needed. Another element of the chip is the set of two holes (10) for layer alignment of the layer during fabrication.

1.6. Botulinum Neurotoxins (BoNT) Detection

In order to demonstrate a typical application and show example results obtained with the use of CCD-based biodetection for LOC, we will briefly discuss a specific system we have previously developed for the detection of BoNT activity (23, 25, 29, 33). BoNT cleavage activity is detected by assaying the proteolytic activity of the toxin by measuring the cleavage product epitope for BoNT. The cleavage was analyzed using short peptide sequences fluorescently labeled with Förster resonance energy transfer (FRET) pairs in which toxin-mediated cleavage results in a change in the fluorescence signal that is measured via a variety of fluorescent sensors (34). Although full details of this assay are beyond the scope of this chapter, we will briefly present it as a suggested application for the CCD detection system described and use it to demonstrate typical results obtainable using the system.

2. Materials

2.1. CCD-Based Detector

1. CCD imager: Astronomy cooled black and white CCD imager SXVF-M7 equipped with Sony ICX-429ALL 752 × 582 pixel CCD (Adirondack Video Astronomy, Hudson Falls, NY) see Note 1 for details.

   Alternatively, another CCD imager which also uses the Sony ICX-429ALL is the Atik 16 (Adirondack Video Astronomy, Hudson Falls, NY).
2. 5 mm Pentax extension tube (Spytown, Utopia, NY).
3. Lens: Tamron manual zoom CCTV 4–12 mm, f1.2 lens. (Spytown, Utopia, NY).

   Alternatively, Pentax fixed focal length, 12 mm f1.2 lens (Spytown, Utopia, NY) with C adapter (see Note 2).
4. Green emission filter (EmF): D535/40 m (Chroma Technology Corp Rockingham, VT).
5. Blue excitation filter (ExF): HQ480/20× (Chroma Technology Corp, Rockingham, VT).
6. Blue Electroluminescence panel (Being Seen Technologies, Bridgewater, MA). The electroluminescence panels used as excitation sources (23, 24) are white and blue EL strips which are mounted on an acrylic card enabling it to be slid into the enclosure. The EL is powered by 110 V AC inverter utilizing 3 V DC. It is important to measure the illumination uniformity across the measured surface (see Note 3).
7. LED illumination box containing red, green, blue, and white LEDs was custom built by Luminousfilm (Shreveport, Louisiana, www.luminousfilm.com/led.htm) (see Note 4).
8. 110 V AC inverter (Being Seen Technologies, Bridgewater, MA).
9. Black acrylic enclosure designed to house detection setup.

2.2. Computer Control and Data Analysis

1. PC computer (laptop or desktop) with USB port.
2. Image analysis software: ImageJ-free NIH software, http://rsbweb.nih.gov/ij/download.html.

3. Data analysis software: Excel (Microsoft, Redmond, WA) and Sigma plot (Sigma plot, Ashburn, VA).

2.3. LOC and Assay Plate Fabrication

1. Black 3.2 mm thick acrylic (Piedmont Plastics, Beltsville, MD).
2. Clear 0.5 mm thick polycarbonate film (Piedmont Plastics, Beltsville, MD).
3. 3 M 9770 Adhesive transfer Tape (Piedmont Plastics, Beltsville, MD).
4. Epilog Legend CO_2 65 W laser cutter (Epilog, Golden, CO).

3. Methods

The basic configuration of the CCD detector platform is shown schematically in Fig. 2a along with an actual photograph of the detector Fig. 2b. The system consists of three main elements (described in more detail below): (a) the CCD detector module, (b) the illumination/excitation module, and (c) the assay chip module. All of these elements are contained within a black acrylic box designed for portability, reconfiguration, and to minimize extraneous signals from ambient light as well as light loss and internal light reflection.

3.1. Assembly of the CCD Detector Module

The assembly of the CCD-based fluorometer, which can use either EL or LED illumination, has been outlined in previous work (23–25) and is described as follows.

1. Place the SXVF-M7 cooled CCD imager (Fig. 2a-1) inside a light sealed (see Note 5) black acrylic enclosure (Fig. 2a-2).
2. Mount the camera in the enclosure with holes drilled in the top of the enclosure to enable air circulation for cooling the CCD.
3. Attach the optional extension tube (Fig. 2a-3) to the camera to enable focusing at short distances and over small fields of view.
4. Mount the Tamron manual zoom lens onto the extension tube (Fig. 2a-4) and then mount the green emission filter, (EmF), onto the lens (Fig. 2a-5). The green emission filter is to minimize the green light emission which would otherwise increase the noise and limit the detection sensitivity D535/40 m.
5. The LOC or detection plate (Fig. 2a-6), excitation filter (ExF) (Fig. 2a-7), and the EL card panel (Fig. 2a-8) are interchangeable and can slide on rails which should be glued on each side of the enclosure (t).

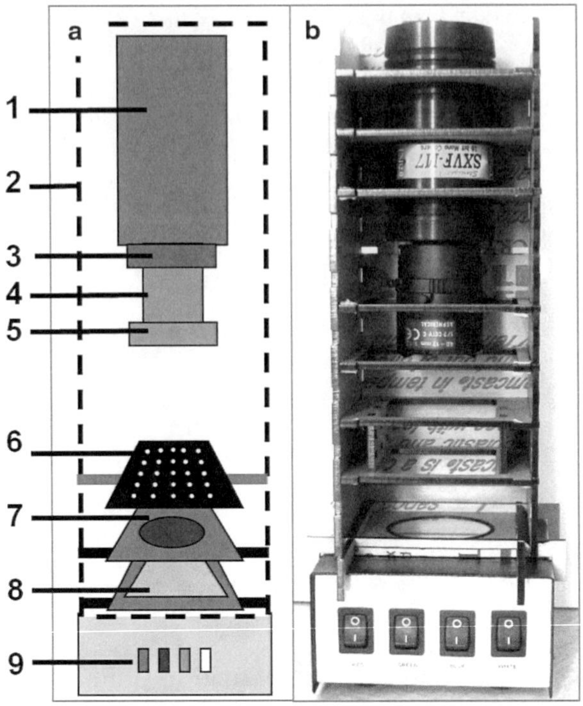

Fig. 2. LED-CCD multi-wavelength detector: (**a**) A schematic configuration of the multi-wavelength LED detector. (**b**) A digital camera image of the actual detector platform. The detector system components highlighted in the schematic (**a**) are (1) an SXVF-M7 CCD camera mounted in a homemade acrylic shelf box (2), which was designed to hold the filters and the sample chips. An optional extension tube (3) can enable the imager to be focused at a short distance and to cover small field. In this configuration the camera is equipped with a Tamron manual zoom CCTV 4–12 mm, f1.2 C-mount lens (4) with a green band pass emission filter (5) mounted on the end of the lens. The black acrylic sample chip (6) can slide on a shelf in the camera box above the blue band pass excitation filter (7) and the EL card (8) which also slides into the camera box. In the LED (9) configuration, the camera shelf box is placed on the top of the multi-wavelength LED illuminator or EL panel (7), the switches for the white and RBG LEDs are shown in (9).

6. To block the blue light emission to the CCD and to enable the measurement of the 523 nm light from the excited FITC, a blue excitation filter is used (see Note 6).
7. Two types of illumination modules can be used for the CCD detector (Fig. 3): (1) a multi-wavelength spatial LED illuminator (25) as shown in Fig. 3a or (2) an electroluminescence panel (23, 24) as shown in Fig. 3b.
8. If the LED illuminator is used, it is housed in a separate module (Fig. 2a-9).
9. The circular excitation filter (Fig. 2a-7) is placed in a round hole machined in an acrylic card, the placing of the filter in a card enables it to be slid into the enclosure above the EL illumination card (Fig. 2a-8) (see Note 5).
10. Similarly, the assay plates or LOC (Fig. 2a-6) can be slid onto the top rail of the enclosure. The design of the enclosure minimizes light loss and internal light reflection and enables the

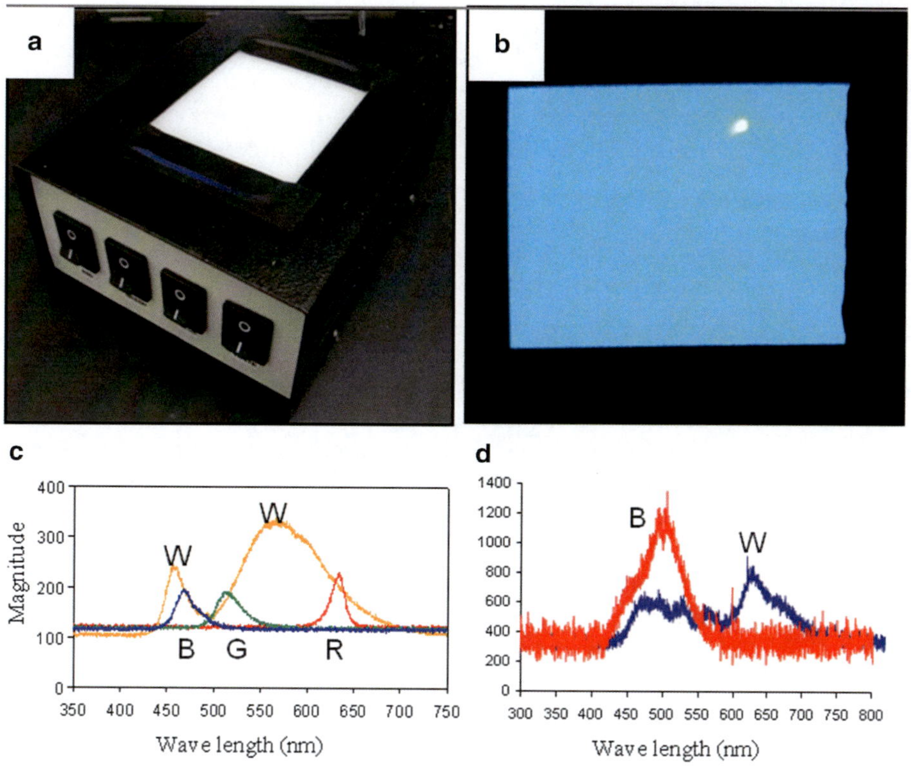

Fig. 3. Illumination modules for CCD detector: (**a**) A photograph of the multi-wavelength LED illuminator prior to attachment to the shelf box. (**b**) A photograph of EL illuminator prior to attachment to the shelf box (note that figures A and B are not to scale). (**c**) The spectra (measured by a spectrometer) was measured for the *white* (W), *blue* (B), *green* (G), and *red* (R) LED illumination (exposure time 100 ms) and compared with the (**d**) EL measurements for white EL (W) and blue EL (B) strips (exposure time 5 s).

components to slide easily into the rails each in a fixed location. For detecting low signals it is critical to reduce the noise consisting of low level green color from the blue LED or EL that limits the sensitivity of detection.

11. The assembly of the LED box (Fig. 2a-9) is also simple, consisting of housing the illuminator in a separate module on which the CCD enclosure rests.

3.2. Multi-Wavelength LED and EL Illuminators

Unlike the monochromatic light emitted from lasers or the limited spectrum light emitted from single LED- and EL-based illumination, the multi-wavelength LED illumination box provides illumination in the blue, green, red, and white ranges (Fig. 3c), covering a spectrum of 450–650 nm (red 610–650 nm, green 492–550 nm, and blue 450–495 nm) as described in our previous work (25). The only region of the spectrum not represented by the RGB LEDs strip is the range between 550 and 610 nm. The white LED (Fig. 3c) provides a broad illumination spectrum (440–680 nm with lower intensity in the range 472–510 nm). The multi-wavelength LED enables illumination in narrower spectra by using only one of the R,G,B, white LEDs, or combinations of them. Within each LED spectrum, interchangeable filters can be used to narrow the excitation band (see Note 7). The EL peak spectra of blue EL is 500 nm and for the white EL the spectral range is 425–700 nm (Fig. 3d). Longer exposure times were typically required to obtain spectra from the EL strips (exposure time 5 s) shown in Fig. 3c compared to LED illumination (exposure time 100 ms) shown in Fig. 3d, suggesting that LED illumination is approximately 11-fold higher than EL.

3.3. Image and Data Analysis

The exposure time for the signal will depend on the type of illumination and the light emitted, which is determined empirically (see Note 8). The CCD images can be analyzed using any standard image processing software. For our system, we chose the freeware ImageJ developed at NIH (see Note 9). It enables 2D linear image analysis of the intensity of each spot in a rows or columns (Fig. 4a), as well as 3D (Fig. 4b) spatial analysis (2D of the spatial position of the spots and the third dimension the intensity of the spots), to provide visual representation of the image, which is shown in Fig. 4b. For spot analysis the intensity value of every spot was exported to an Excel spreadsheet and to the scientific data analysis and graphing software Sigmaplot, which was used to plot the data. Several analyses were conducted, including subtracting baseline noise level and calculating the signal-to-noise ratio (S/N).

3.4. LED and EL Illumination for CCD Florescence Detection

To evaluate LED illumination (25) as compared to the EL (23, 24), the emission from nine concentrations ranging from 0.019 to 5 nM of a peptide (SNAP-25) labeled with FITC were excited with blue and white LEDs and blue EL strips illumination. The CCD analysis using the various types of excitation was reported in our previous work (25), as shown in Fig. 4a with the corresponding ImageJ 3D graph analysis of CCD images showing the pixel intensity Fig. 4b. In these experiments, the 9-well plate was loaded with three replicas of the dilution series (columns 1, 3, and 5), the other columns used as a control (no peptide) to measure background). The highest concentration of the peptide is in rows A and the concentration decreased through rows E. The signal from the blue LED excitation exposure of 2 s (Fig. 4-I) and exposure of 3 s for white LED (Fig. 4-II) were compared with the signal from illumination with blue EL (Fig. 4-III) excitation exposure of 60 and 300 s for blue EL (Fig. 4-IV), white EL was not tested because the relatively low emission in the blue spectra needed for FITC excitation.

Figure 4-A-I (CCD image) and Graph Fig. 4-B-I shows the corresponding signals for the blue LED compared with the white LED (panels-II), for the white LED, clearly the background (columns 2, 4, and 6) appears somewhat higher than the blue LED.

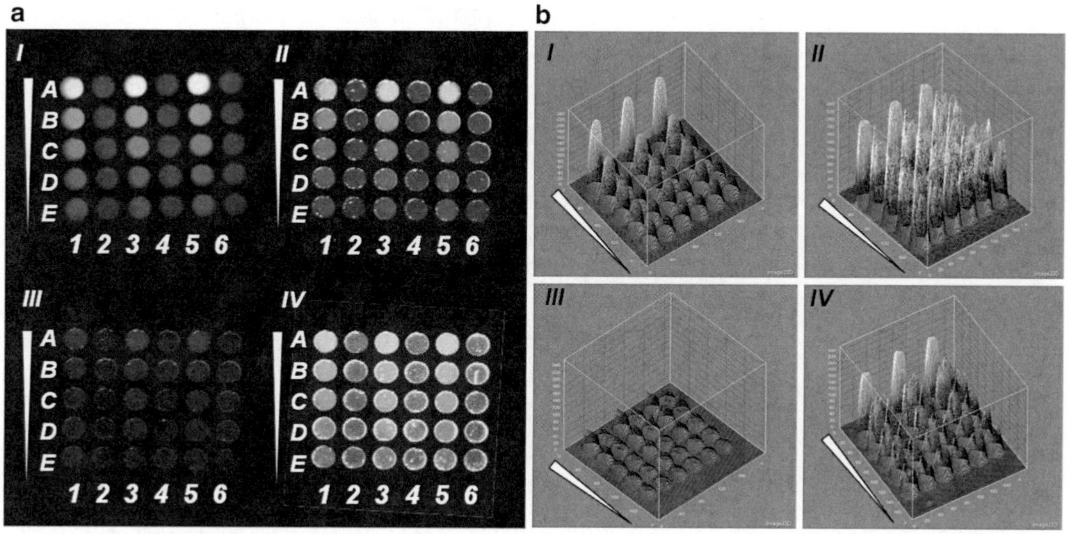

Fig. 4. CCD Images and 3D ImageJ analysis of EL and LED detection of unquenched FITC-SNAP peptide: The CCD images (**a**) and the corresponding 3D ImageJ analysis (**b**) of the CCD emission images of nine concentrations of a 50% dilution series of the unquenched SNAP-25 peptide ranging from 0.019 to 5 nM and a control. The plate was loaded with three replicas of the dilution series (columns 1, 3, and 5 and the other columns used as a control (no peptide) to measure background). The highest concentration of the peptide is in row A and the concentration decreased through rows E (direction indicated by the *triangles*). The illuminations are shown for (I) *blue* LED excitation exposure of 2 s and (II) *white* LED excitation exposure of 3 s, (III) *blue* EL excitation exposure of 60 s, and (IV) blue EL excitation exposure of 300 s.

The detailed 3D analysis (Fig. 4B Graph II) shows the variability of the white LED excitation clearly. While the shapes in the 3D graph of the blue LED spots are smooth cones (Fig. 4B-I), the white LED spots have an irregular rough surface (Fig. 4B-II) representing pixel–pixel variations). The signal from the blue EL excitation exposure of 60 s is low (Fig. 4A-III and 4-B-III) is very low, increasing the exposure time to 300 s improves the signal (Fig. 4A-IV and 4B-IV). These data suggest that the blue LED (Fig. 4-I) and the long exposure of blue EL (Fig. 4-IV) can be used for FITC detection.

3.5. LOC and Assay Plate Fabrication

The laminated LOC used to measure BoNT-A activity (29) is a multilayer, 3D microfluidic structure constructed with a rigid polymer core (e.g., 3.2 mm poly(methyl methacrylate) (PMMA) also known as acrylic) and laminated with layers of thin polymer (e.g., polycarbonate (PC) film) bonded with adhesive. The relatively thick core provides rigidity to the assembly and space for high volume fluidic reservoirs for purposes, such as liquid storage, waste, detection wells, and outlets. The thin polymer laminating layers encapsulate the chip, providing features such as microfluidic channels and fluid connections between layers and mixing chambers.

The assay plate is a single layer of black PMMA (acrylic) with nine holes, laminated onto a clear PC base plate. The polycarbonate material used for fabrication did not show detectable autofluorescence, and all the materials used to fabricate the device did not appear to inhibit enzymatic reactions. The fabrication method is as follows.

1. Laminate the clear polycarbonate channel layers with the adhesive tape as appropriate (some layers are without tape, others have tape laminated both sides—see step 4).
2. Cut the channel layer designs using the CO_2 laser (both the adhesive and the polycarbonate). Every second layer is a polymer (normally polycarbonate) layered with double-sided adhesive. The fluidics are made using 3.2 mm thick black acrylic, micromachined with a computer-controlled CO_2 laser system (see Notes 10 and 11). The well holes are 2 mm in diameter which allows analysis of 10 ml samples. Smaller holes make the loading of the sample less reproducible.
3. Assemble the pieces, layer-by-layer, by aligning them (using the alignment holes in a two pin device) with the adjoining layers and applying pressure.
4. The polycarbonate bottom was bonded to the acrylic with double sided pressure-sensitive adhesive transfer tape (see Note 12) which was bonded to the micromachined layer (so there is no contact between the adhesive and the fluids). Assembled devices were bonded together by firmly pressing layers together. An office laminating machine may be used to eliminate air bubbles and aid uniform bonding.
5. When designing the LOC, it is essential to minimize optical cross talk (i.e., light interference between wells) (see Note 13), air bubbles in the detection wells (see Note 14), and fluids mixing between channels (see Note 15). The layer numbers in Fig. 5 correspond to their order in the chip, with layer I on top. Each layer is described in more detail as follows.

   Layer I: The top layer, made from clear PC, seals the LOC upper fluidic channels while enabling loading of samples and reagents (2) and providing an outlet connection for a pump or syringe (9). Layer I carries no adhesive.

   Layer II: Clear PC which, in addition to the layer I features, includes joining channels (4), detection wells (5), mixing trap chambers (6), and a waste chamber (8). This layer carries adhesive on both sides.

   Layer III: This core layer is made from black PMMA, providing rigidity to the LOC and sufficient volume for sample and reagent reservoirs, detection chambers, and the waste chamber. An important element is a layer pass (3A, marked with an arrow) which connect the sample/reagent wells to the V junction (via connecting slits 1A and 2A in layer IV). The black PMMA is used for this layer minimize optical interference. The PMMA layer carries adhesive on the bottom side.

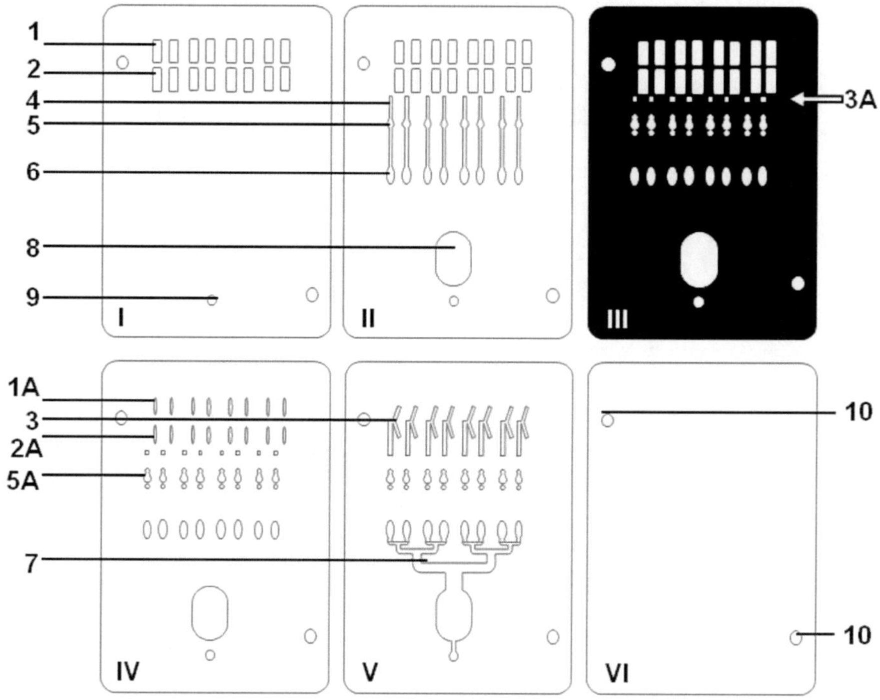

Fig. 5. Outline of the LOC's layers: The LOC's six layers are numbered in order from the top layer (I) down to the bottom layer (VI). All the layers are 250 μm clear PC expect the core layer III which is 3.2 mm black PMMA. The main elements shown are sample reservoirs (1), reagent reservoirs (2) connected to the V junction (3) via connecting slits (1A and 2A), through layer pass (3A marked with an *arrow*) to the mixing channel (4), detection wells (5) with air bubble traps (5A), mixing traps (6) horizontal merging channel (7) a waste chamber (8) to the outlet (9) for connection to a pump or syringe.

Layer IV: A clear PC forms a bottom seal for the core layer, sealing the various chambers of layer III. Layer IV also provides inlets for the fluid from the sample and reagent wells for the connecting element of the V junctions (which is located in layer V) via connecting slits (1A and 2A). These wide slits assure flow even if the device is not perfectly aligned during fabrication. Layer IV also contains outlets to the detection, waste chambers, mixing trap as well as an outlet for a pump or syringe. Layer IV carries no adhesive.

Layer V: A clear PC which forms the lower fluidic channels with the V junctions (3), negative pressure splitter (7), the detection and waste reservoir and outlet for a pump or syringe. This layer carries adhesive on both sides.

Layer VI: The bottom seal layer is made from clear PC with the alignment holes.

6. For the 9-well assay plate design, an array of 9, 2 mm diameter holes were cut into a single layer of 3.2 mm thick black acrylic and laminated with a PC base layer in a similar fashion as described for the LOC.

3.6. LOC Detection of Botulinum Neurotoxin-A Light Chain

A suggested application for the CCD LOC system described in this chapter is the Botulinum Toxin Activity FRET assay, in which the toxin cleaves the peptide substrate (SNAP-25), which is labeled with an FITC donor/DABCYL acceptor FRET pair. Because the DABCYL quenches the FITC emission, little light is emitted in its native state. The cleavage of the peptide sequence by the toxin disrupts the FRET and results in FITC fluorescence at 523 nm, which is measured using the CCD system. As described in our previous work (35), the FITC/DABCYL-SNAP-25 was used at a fixed 5 μM concentration and exposed to different concentrations of a twofold LcA serial dilution in the range 0.25–16 nM. The cleavage assay was carried out for 2 h at room temperature in the LOC (29) and in the 9-well plate as previously described (35), and was excited with blue LED. The actual CCD images for the LOC are shown in Fig. 6a and 9-well plate in Fig. 6b. Note the trapped air bubbles in plate assay (Fig. 6b) which complicate the measurement. The plot of the ImageJ quantization analysis of the results is shown in Fig. 6c (the 9-well plate (b) and the LOC (c)), where the S/N was calculated as the ratio between the value of the measurement to the value of the control sample (no LcA). As shown in Fig. 6c, the S/N increases with the increase in LcA concentration. The signal from the plate is higher than the signal from the LOC because the multilayer LOC absorbs light, see Note 16 for factors contributing to system sensitivity.

The LOD was defined as the toxin concentration that generates a signal three times the standard deviation above background, which is 0.5 nM (25 ng/ml) for both the LOC and plate format. This LOD is similar to our previous results using a plate assay (35) and is comparable to the measured level of LcA using peptide immobilized on beads and cleaved by the toxin, which releases fluorescent fragments into solution in a microfluidic system (36). The volume used for the assay is 10 μl and the detection level is below the adult human lethal oral dose of 70 μg of toxin (24, 33, 37).

The main advantages of the LOC are the ability to perform the assay in the field (e.g., no need for a microscope, incubator, etc.) and better protection and isolation of the user from the toxin. This ability to perform complex diagnostics assays without a laboratory is important for other medical applications, ranging from microbial detection to cancer diagnostics.

4. Notes

1. A black and white astronomy imager with cooled high sensitivity CCD imager SXVF-M7 (or Atik 16) was used as a CCD detector (Fig. 2-b). The imager employs a Sony ICX-429ALL CCD with 752 × 582 pixel resolution and a typical sensitivity of

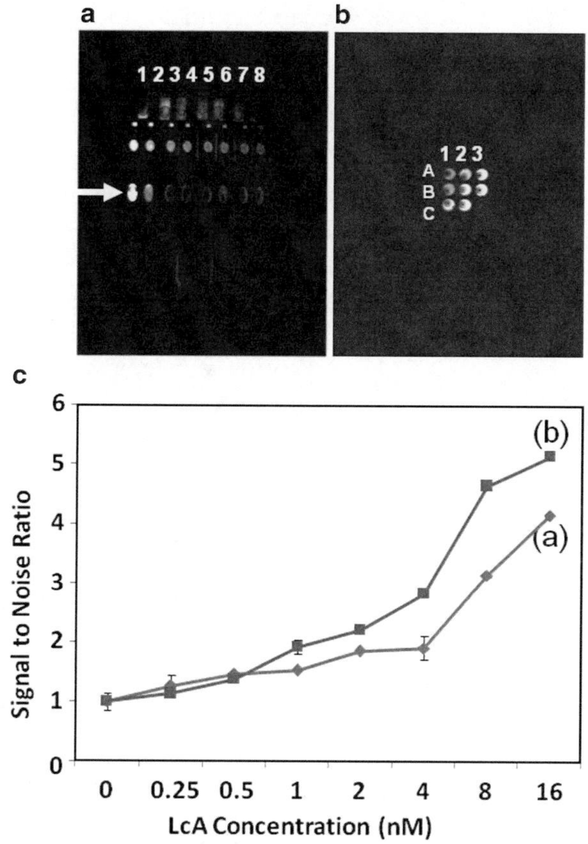

Fig. 6. CCD-based detectors for LOC in vitro activity detection of BoNT-A light chain: The SNAP-25 peptide substrate for BoNT-A is labeled with the FITC donor/DABCYL acceptor FRET pair. Interaction of the substrate with the toxin light chain LcA results in cleavage of the peptide sequence, disrupting the FRET and resulting in increased FITC donor emission as measured by CCD with a 500 ms exposure. FRET activity assay was performed in (**a**) eight-channel LOC with the detection wells marked with an *arrow*, and (**b**) a 9-well plate. Different concentrations of twofold LcA serial dilution in the range 0–16 nM (with 0 nM used as control) using four replicas loaded into the LOC from high concentration well #1 to low concentration well #8. For the plate, low concentration well #1A with increasing concentrations (wells 1B, 1C, 2A, 2B, etc.). The assay was carried out for 2 h at room temperature. The *S/N* of the FRET activity assay was calculated as the ratio between the value of the measurement to the value of the control (no LcA) and was plotted (**c**) for the LOC (a) and the 9-well plate (b).

5,500 mV (at shutter speed of 1/1,000 s). The imager is equipped with a 16 Bit Analog to Digital Converter enabling a dynamic range of 65,500 levels of grayscale. Peltier thermoelectric cooling is used to maintain the CCD at 25°C below ambient temperature so that much lower thermal noise is generated on the CCD.

2. The fixed focal length enables the use of a wider aperture. The fixed focal length lens was connected to the imager with a

C adapter, and a 5 mm extension tube that enables the imager to be focused at short distance (30 mm) and to cover small field (15 × 20 mm). The f1.2 lens permits operation at very little light.

3. To measure light uniformity, it is important to measure the light with a long enough exposure time (e.g., 60 s) to measure the CCD uniformity when all of the wells are loaded with the same FTCE fluorescence sample. If the lighting is not uniform, correction values for each well can be calculated and used for measurement composition.
4. The multi-wavelength spatial LED illuminator module (25) comprises a custom built multi-wavelength LED illumination box with two different types of LED; white which generates across the full optical spectrum, and RGB which generates at the red, blue, and green wavelengths. The LED illuminator contains four RGB LED strips and four white LED strips, both 18 LEDs per foot. Each LED emission color is controlled individually with a switch in the front panel. The top of the box consists of a diffusion panel (milky white plastic panel), which assures uniformity of the light and hence even illumination of the sample chip. The dimensions of the diffusion panel are 8.5 × 5.5 cm. The LEDs are powered by 3VDC.
5. The imager enclosure must be light sealed without any light leaks and using light reflecting material. A long exposure (e.g., 3 min) can be used to detect light leaks.
6. To measure the effectiveness of the filters, it is recommended to perform two long exposures (e.g., 3 min) without the assay plate, one with the EL on and one with the EL off. Ideally, the two measurements should be very similar (the blue filters pass only blue light which is blocked by the green filters). The difference between measurements may suggest that the blue filters do not block all green light and/or the green filters do not block all blue light.
7. Make sure the lens, filters, and sample plate are vertically centered and aligned and that the arrows on the coated filters are facing the CCD. Also, the filter holder should be leveled and at a height which allows proper focusing of the lens. The filters should slide freely into the EL enclosure, and always to the same position.
8. A fully open aperture (e.g., f1.2) will enable shorter exposure time, but focusing will be more limited and the image less sharp.
9. For data analysis, use the ImageJ high contrast visualization, which will not affect the values but will enable easy selection of the spots.

10. The laser power and speed for cutting polymers has to be determining empirically. It is recommended to use the minimum laser power to reduce overheating or burning the material.
11. The "holes" for the wells on the assay plate are 2 mm in diameter, which allows analysis of 20 μl samples. Smaller holes make the loading of the sample less reproducible because the fluid meniscus within the wells causes light diffraction, which complicate quantification.
12. For strong bonding, remove 1 cm of the adhesive tape cover, align the tape with acrylic surface, and attached the exposed tape to the acrylic surface. With a ruler, press the tape to the surface and slowly move the ruler across the tape with little pressure in order to prevent air bubbles. When bonding the tape, it should be aligned with the acrylic. For assembling of the assay plate, remove the protective cover from the other side of the double side adhesive (taped to the acrylic) and remove the protective cover from the polycarbonate. Then, align the two pieces, apply pressure, and run through the assay plate and attach the polycarbonate to the acrylic.
13. Controlling optical noise: Fluorescence emission and scattered excitation light can propagate through the chip, causing cross-talk between adjacent channels. This can become a major source of optical noise in the system (35, 38, 39) by increasing background noise, thus reducing the sensitivity of the measurements. To limit the effect of fluorescence background, PC, and not Mylar, which is the commonly used material for lamination-based fabrication, was used as the main fabrication material due to its lower fluorescence background (35). Using black material decreases the noise. Adding air gaps between channels did not reduce the background noise.
14. Avoiding air bubbles: Another issue with accurately measuring signals is occasionally air bubbles will be trapped in the detection chamber. A simple solution for this problem is to add small air traps to both sides of the detection chamber (Fig. 1a-5), trapping upstream and downstream bubbles. Also, placing the LOC at a slight angle decreases the trapping of air bubbles in the wells, especially for long incubation assays.
15. Avoiding cross-talk between channels: Diffusion of the analyte "upstream" into the channel from the detection port can cause significant dilution and reduce signal. To overcome this problem we use traps (Fig. 1a-6) which are not filled with liquids at the end of the channels to "separate" the fluidics of the connected channels. Longer channels are not effective to minimize diffusion because the incubation time for many biological assays is long (e.g., hours). Thus, longer channels (with smaller

volume) will not be as effective, and will increase the size of the device. In addition, when using a manually controlled device, the traps allow better monitoring of volume (the user can visually observe the trap filling).

16. Although CCDs and EL each inherently have less sensitivity compared with photomultipliers or avalanche photodiodes, the sensitivity of the CCD was increased to the level of a photomultiplier-based fluorometer through long exposure times and by optimizing several elements of the system. A sensitive Sony ICX-429ALL CCD-based imager is one key element of the system. It has thermoelectric cooling that enables the CCD temperature to be maintained at 25°C below ambient temperature, so the thermal noise is 16 times lower than at ambient and much longer exposures are possible. In general, the level of thermal noise will drop by ½ for every 6° of cooling. An f1.2 lens was used to maximize the amount of light transmitted to the CCD. Finally, as discussed previously, the type of EL, the filter arrangement for reducing light noise, a light absorbing polycarbonate material, and the compact configuration of the EL-enclosure all combined to minimize light losses and crosstalk.

References

1. Capitan-Vallvey LF et al (2007) Oxygen-sensing film coated photodetectors for portable instrumentation. Anal Chim Acta 583:166–173
2. Mac Sweeney MM et al (2004) Characterization and optimization of an optical DNA hybridization sensor for the detection of multi-drug resistant tuberculosis. Conf Proc IEEE Eng Med Biol Soc 3:1960–1963
3. Claycomb RW, Delwiche MJ (1998) Biosensor for on-line measurement of bovine progesterone during milking. Biosens Bioelectron 13:1173–1180
4. Bruno AE et al (1997) All-solid-state miniaturized fluorescence sensor array for the determination of critical gases and electrolytes in blood. Anal Chem 69:507–513
5. Moehrs S et al (2006) A detector head design for small-animal PET with silicon photomultipliers (SiPM). Phys Med Biol 51:1113–1127
6. Takei M, Kida T, Suzuki K (2001) Sensitive measurement of positron emitters eluted from HPLC. Appl Radiat Isot 55:229–234
7. Ruiz-Martinez MC et al (1993) DNA sequencing by capillary electrophoresis with replaceable linear polyacrylamide and laser-induced fluorescence detection. Anal Chem 65:2851–2858
8. Tibbe AG et al (2001) Cell analysis system based on immunomagnetic cell selection and alignment followed by immunofluorescent analysis using compact disk technologies. Cytometry 43:31–37
9. Tsukagoshi K, Jinno N, Nakajima R (2005) Development of a micro total analysis system incorporating chemiluminescence detection and application to detection of cancer markers. Anal Chem 77:1684–1688
10. Roda A et al (2003) A rapid and sensitive 384-well microtitre format chemiluminescent enzyme immunoassay for 19-nortestosterone. Luminescence 18:72–78
11. Ligler FS et al (2003) Array biosensor for detection of toxins. Anal Bioanal Chem 377:469–477
12. Svitel J et al (2001) Functionalized surfaces for optical biosensors: applications to in vitro pesticide residual analysis. J Mater Sci Mater Med 12:1075–1078
13. Liu Y, Danielsson B (2007) Rapid high throughput assay for fluorimetric detection of doxorubicin–application of nucleic acid-dye bioprobe. Anal Chim Acta 587:47–51
14. Burkert K et al (2007) Automated preparation method for colloidal crystal arrays of monodisperse and binary colloid mixtures by contact

printing with a pintool plotter. Langmuir 23:3478–3484
15. Tohda K, Gratzl M (2006) Micro-miniature autonomous optical sensor array for monitoring ions and metabolites 2: color responses to pH, K+ and glucose. Anal Sci 22:937–941
16. Feldstein MJ et al (1999) Array biosensor: optical and fluidics systems. Biomed Microdevices 1:139–153
17. Sohn YS et al (2005) A microbead array chemical sensor using capillary-based sample introduction: toward the development of an "electronic tongue". Biosens Bioelectron 21:303–312
18. Knecht BG et al (2004) Automated microarray system for the simultaneous detection of antibiotics in milk. Anal Chem 76:646–654
19. Taitt CR, Anderson GP, Ligler FS (2005) Evanescent wave fluorescence biosensors. Biosens Bioelectron 20:2470–2487
20. Ngundi MM et al (2006) Detection of deoxynivalenol in foods and indoor air using an array biosensor. Environ Sci Technol 40:2352–2356
21. Moreno-Bondi MC et al (2006) Multiplexed measurement of serum antibodies using an array biosensor. Biosens Bioelectron 21:1880–1886
22. Ligler FS et al (2007) The array biosensor: portable, automated systems. Anal Sci 23:5–10
23. Kostov Y et al (2009) A simple portable electroluminescence illumination-based CCD detector. Methods Mol Biol 503:259–272
24. Sapsford KE et al (2008) A fluorescence detection platform using spatial electroluminescent excitation for measuring botulinum neurotoxin A activity. Biosens Bioelectron 24:618–625
25. Sun S et al (2010) Multi-wavelength Spatial LED illumination based detector for in vitro detection of Botulinum Neurotoxin A Activity. Sens Actuators B Chem 146:297–306
26. Sapsford KE et al (2005) Biosensor detection of botulinum toxoid A and staphylococcal enterotoxin B in food. Appl Environ Microbiol 71:5590–5592
27. Golden JP et al (2007) Target delivery in a microfluidic immunosensor. Biosens Bioelectron 22:2763–2767
28. Higgins JA et al (2003) A handheld real time thermal cycler for bacterial pathogen detection. Biosens Bioelectron 18:1115–1123
29. Sun S et al (2009) Lab-on-a-chip for botulinum neurotoxin a (BoNT-A) activity analysis. Lab Chip 9:3275–3281
30. Munson MS et al (2004) Suppression of non-specific adsorption using sheath flow. Lab Chip 4:438–445
31. Rossier JS et al (1999) Microchannel networks for electrophoretic separations. Electrophoresis 20:727–731
32. Rossier J, Reymond F, Michel PE (2002) Polymer microfluidic chips for electrochemical and biochemical analyses. Electrophoresis 23:858–867
33. Rasooly R et al (2008) Detection of botulinum neurotoxin-A activity in food by peptide cleavage assay. Int J Food Microbiol 126:135–139
34. Dong M et al (2004) Using fluorescent sensors to detect botulinum neurotoxin activity in vitro and in living cells. Proc Natl Acad Sci USA 101:14701–14706
35. Sapsford KE et al (2008) A fluorescence detection platform using spatial electroluminescent excitation for measuring botulinum neurotoxin a activity. Biosens Bioelectron 24(4):618–625
36. Frisk ML et al (2008) Bead-based microfluidic toxin sensor integrating evaporative signal amplification. Lab Chip 8:1793–1800
37. Rasooly R, Do PM (2008) Development of an in vitro activity assay as an alternative to the mouse bioassay for Clostridium botulinum neurotoxin type A. Appl Environ Microbiol 74:4309–4313
38. Irawan R et al (2005) Cross-talk problem on a fluorescence multi-channel microfluidic chip system. Biomed Microdevices 7:205–211
39. Hawkins KR, Yager P (2003) Nonlinear decrease of background fluorescence in polymer thin-films—a survey of materials and how they can complicate fluorescence detection in microTAS. Lab Chip 3:248–252

Chapter 24

Multilayer Microfluidic Poly(Ethylene Glycol) Diacrylate Hydrogels

Michael P. Cuchiara and Jennifer L. West

Abstract

Development of robust, in vivo like tissues in vitro holds the potential to create regenerative medicine-based therapeutics, provide more physiologically significant preclinical models and supply a pharmacological and toxicological screening platform that reflects in vivo systems in both complexity and function. This protocol describes a simple, robust, multilayer replica molding technique in which poly(dimethylsiloxane) (PDMS) and poly(ethylene glycol) diacrylate (PEGDA) are serially replica molded to develop microfluidic PEGDA hydrogel networks embedded within independently fabricated PDMS housings, using a combination of soft and photo-lithography. This work has direct applications toward the development of robust, complex, cell-laden hydrogels for in vitro diagnostics and regenerative medicine applications.

Key words: Hydrogel, Poly (ethylene glycol), Microfabrication, Microfluidics, Photolithography, Soft lithography

1. Introduction

Microscale systems are advantageous to the life science field due to their ability to reduce reagent volume, increase system throughput, and allow investigation of microscale cellular phenomena (1–3). In particular microfluidic systems provide picoscale fluid volume transfer, robust control of molecular gradients, and miniaturization of complex multicomponent systems (4, 5). However, a majority of microfabricated systems use 2D substrates (6, 7) or are based on silicone elastomers that are not favorable for cell culture and tissue engineering applications (8, 9). Integration of microfabrication technologies with hydrogel materials that are biocompatible and biomimetic enables more accurate modeling of living systems in vitro (10–12). Applications of microfabricated hydrogels range

Gareth Jenkins and Colin D. Mansfield (eds.), *Microfluidic Diagnostics: Methods and Protocols*, Methods in Molecular Biology, vol. 949, DOI 10.1007/978-1-62703-134-9_24, © Springer Science+Business Media, LLC 2013

from regenerative medicine therapeutics (13) to more clinically relevant tissue models for drug screening applications (14, 15). Previously we have shown the ability to fabricate multilayer polydimethylsiloxane/poly(ethylene glycol) diacrylate (PDMS/PEGDA) hydrogels (16). Herein we describe methods to fabricate PEGDA hydrogels within PDMS housings using a combination of soft and photo-lithography.

The methods employed to build microfluidic or other microfabricated hydrogels vary widely depending on the application. The choice of hydrogel material and the mechanism that drives cross linking from a pre-polymer solution to a gel will alter procedures significantly (12, 16, 17). We have fabricated poly(ethylene glycol) (PEG) microfluidic networks with polymer concentrations ranging from 2.5 to 50% (w/v) at molecular weights ranging from 575 to 20,000 Da. The methods reported below will focus on the photopolymerizable derivative of PEG, poly(ethylene glycol) diacrylate (PEGDA), which is crosslinked using a photoinitiator that generates a free radical and initiates crosslinking upon exposure to ultraviolet (UV) light. Much care must be taken when choosing the most appropriate polymer, photoinitiator, and light sources for a given application. For a more detailed description of photoinitiators and photopolymerizable PEG derivatives please refer to West et al. (18).

Two hydrogel microfabrication approaches, using both soft and photo-lithographic techniques, will be described in detail. The first process describes the steps to independently fabricate a microfluidic hydrogel network within a robust polydimethylsiloxane (PDMS) perfusion housing using soft lithographic techniques. The second process describes the steps to photolithographically fabricate hydrogel microstructures within a PDMS housing followed by a soft lithographic step to independently fabricate microfluidic networks. This type of multimodal fabrication scheme will allow for the production of increasingly complex materials and highly biomimetic systems, which are important for the development of more relevant preclinical in vitro tissue models and for the production of complex, regenerative medicine-based therapeutics.

Depending on the chosen application and the required microfabrication tolerances, the use of a clean room facility may be required for the production of photoresist masters. Generally, we have found that use of clean room facilities produces robust, high fidelity, and long-lasting photoresist masters. However, much recent effort has been placed on adapting microfabrication technologies for out of clean room bench top applications (19–21). Herein, all photoresist masters were fabricated within a clean room but replica molded PDMS housings, PEGDA microchannels, and photolithographically fabricated PEGDA microstructures were produced on the bench top.

2. Materials

2.1. Photoresist Master Fabrication

1. Glass slides or other silicon-based substrates (see Note 1 on substrate resist adhesion effects).
2. Three glass slide staining dishes with one removable glass slide rack.
3. Piranha solution: concentrated sulfuric acid (18 M) and 30% hydrogen peroxide in a 7:3 v/v ratio (sulfuric acid:hydrogen peroxide) (see Note 2 on piranha handling and safety precautions).
4. SU-8 2100, SU-8 2050, and SU-8 developer (Microchem) protected from light and stored at room temperature.
5. Acetone (ACS grade), 2-propanol (ACS grade), and ultrapure water stored in standard 500 mL low density polyethylene wash bottles.
6. Temperature programmable 200°C hot plate (VWR).
7. Spin coater (PWM32-PS-R790, Headway Research).
8. Transparency photomask (25,000 dpi, CAD/ART Services).
9. Collimated mask aligner light source (MJB4 Mask Aligner, SUSS Microtec).
10. Filtered and dry nitrogen.
11. Sigmacote (Sigma Aldrich) protected from light and stored at 4°C in a flammables safe refrigerator.

2.2. Poly-dimethylsiloxane (PDMS) Housing Fabrication

1. Polydimethylsiloxane (PDMS, Sylgard184 kit, ESI, Inc.).
2. Round Petri dish (100×25 mm, polystyrene).
3. Disposable spatula or disposable serological pipette.
4. Vacuum pump and desiccator or oven with vacuum chamber.
5. 1 mm biopsy punch (see Note 3 on compression fitting).
6. Razor blade or scalpel.
7. Acetophenone (2,2-dimethoxy-2-phenyl acetophenone, Sigma Aldrich).
8. NVP (*n*-vinylpyrrolidone, Sigma Aldrich).
9. Ethanol (100%, ACS grade), and ultrapure water stored in standard 500-mL low density polyethylene wash bottles.
10. Filtered air.

2.3. Soft Lithographically Fabricated Poly(Ethylene Glycol) Diacrylate Hydrogel

1. Long wavelength UV lamp with spot bulb (365 nm, 10 mW/cm^2, Blak-Ray).
2. Poly(ethylene glycol) diacrylate (PEGDA) (see Note 4 on hydrogel content).
3. Phosphate-buffered saline (PBS, pH 7.4).

4. Irgacure 2959 photoinitiator (Ciba).
5. 1 mL syringe w/luer lock.
6. Polyethylene tubing (OD 1 mm ID 0.58 mm, see Note 3 on compression fitting).
7. Ethanol (100%, ACS grade).
8. Coverglass or glass slide.
9. Syringe pump with infusion and withdrawal mode (Harvard Apparatus Holliston, MA).

2.4. PEGDA Hydrogels Fabricated Using Photolithography and Soft Lithography

1. Long wavelength UV lamp with spot bulb (365 nm, 10 mW/cm^2, Blak-Ray).
2. Poly(ethylene glycol) diacrylate (PEGDA) (see Note 4 on hydrogel content).
3. Irgacure 2959 photoinitiator (Ciba).
4. 1 mL syringe w/luer lock.
5. Polyethylene tubing (OD 1 mm ID 0.58 mm).
6. Ethanol (100%, ACS grade).
7. Coverglass or glass slide.
8. Syringe pump with withdrawal mode (Harvard Apparatus).

3. Methods

The following methods describe multilayer soft lithographic and photolithographic fabrication methods to create PEGDA microfluidic hydrogels contained within a PDMS housing see Note 5 for design considerations.

3.1. Photoresist Master Fabrication

1. Load pre-cleaned glass slides into the staining rack and place them in a glass staining tray. Carefully add approximately 250 mL (or enough volume to cover the slides) of piranha solution into the staining tray. Allow the piranha solution to cleanse and etch the glass for 15 min with periodic, gentle agitation of the slide rack (see Note 2 on piranha handling and safety precautions).
2. Remove the rack of glass slides from the piranha solution and rinse in a staining tray of ultrapure water for 1 min while constantly and gently agitating. Next move the rack to a second tray of ultrapure water and allow the cleaned slides to soak for 5 min during which time the water will hyrdoxylate the glass surfaces.
3. Individually remove each slide from the water using tweezers, thoroughly dry with nitrogen and then place on a hot plate at 200°C for at least 10 min to dehydrate the slide surface.

4. After heating allow the slides to equilibrate to room temperature for at least 2 min.
5. Center a slide on the spin coater stage taking care to only handle it by the edges. Activate the spin coater vacuum to hold the slide in place. The dominant factors governing structure height are the choice of the resist and the speed and length of time chosen for the spin coating step. Please see Microchem's website (http://www.microchem.com/Prod-SU8_KMPR.htm) for parameter details (see Note 6 on proper protective equipment and spin coater operation).
6. To create the PDMS housing master, slowly pour the SU-8 2100 onto the substrate, making sure to coat the entire surface without trapping any bubbles. This is best achieved by a single, thick pour with the resist bottle spout almost touching the slide. Set the spin coater to a final spin speed of 1,000 rpm with a ramp speed of 300 rpm/s and a 40 s spin time (Fig. 1, Step 1).
7. Carefully remove the PDMS housing SU-8 2100 master from the spin coater stage and place it onto a hot plate that is at room temperature. Turn the hot plate to 65°C and bake for 30 min once the hot plate reaches temperature. Turn the temperature on the hot plate up to 95°C and bake for 120 min once the temperature reaches 95°C (see Note 7 on hot plates and the effects of thermal fluctuations).
8. To create the PEGDA microchannel master, repeat Subheading 3.1, Steps 5–7 with the following modifications. In Subheading 3.1, Step 6: replace SU-8 2100 with SU-8 2050. For the bake times in Subheading 3.1, Step 7: use 5 min at 65°C and 30 min at 95°C, taking care to gradually heat the substrate and allow the temperature to equilibrate before starting the timer.
9. Remove the substrate from the hot plate and cool to room temperature for 2 min. Place the coated substrate onto the mask aligner chuck or sliding stage. Mount the transparency mask with the ink side of the transparency in direct contact with the SU-8 surface on the slide (Fig. 1, Step 2) (see Note 8 on exposure light source).
10. The UV exposure time is dependent upon the type and thickness of the photoresist, the substrate used, and the energy output of the mask aligner lamp. Please see manufacturer recommendations for these materials and alter parameters as needed for optimal resist performance. A simple formula can be used to calculate the exposure time for a given exposure energy and lamp power.

$$\text{Exposure Time (s)} = \frac{\text{Exposure Energy}\left(\dfrac{\text{mJ}}{\text{cm}^2}\right)}{\text{Lamp Power Output}\left(\dfrac{\text{mW}}{\text{cm}^2}\right)}$$

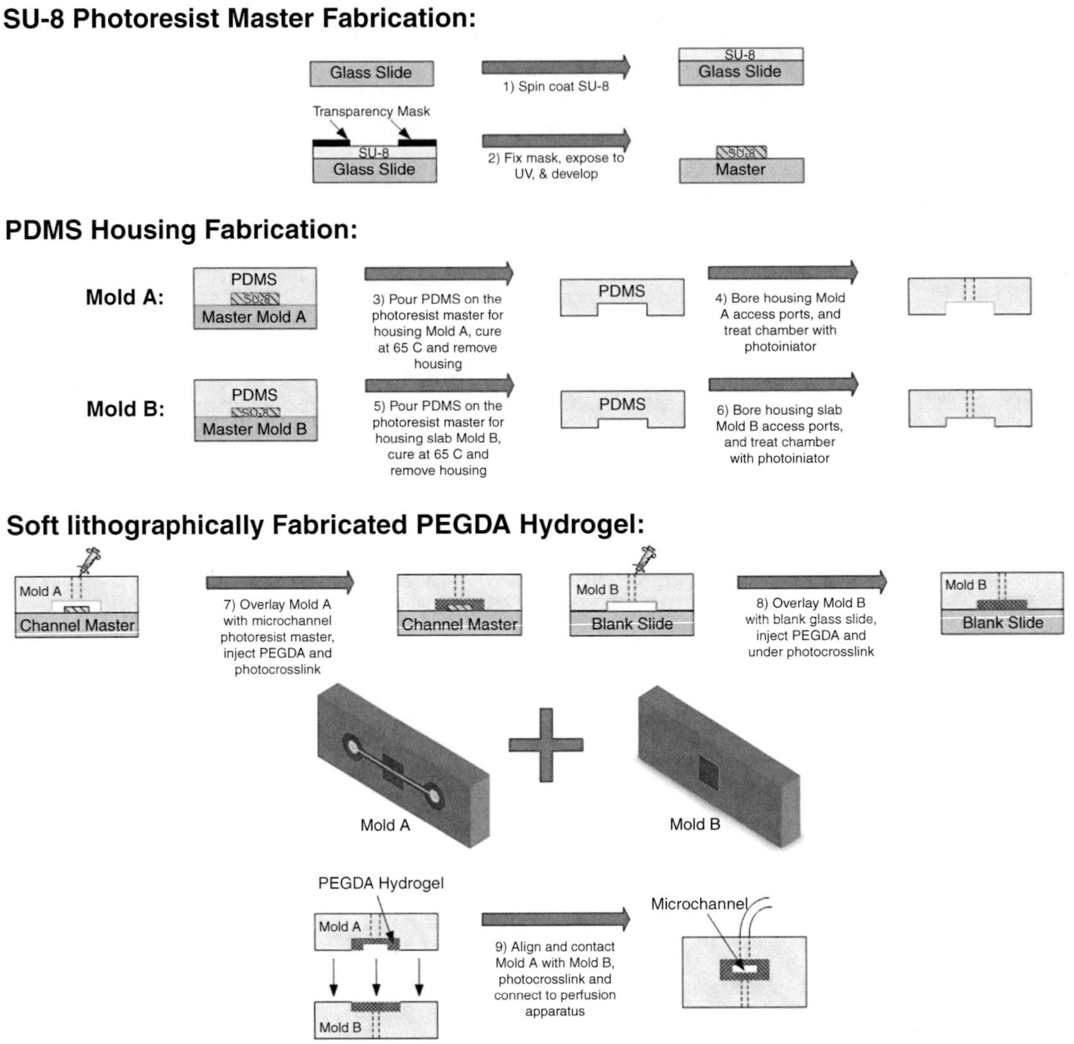

Fig. 1. Fabrication schematic of PDMS/PEGDA microfluidic networks. Serial replica molding of PDMS and PEGDA to photoresist masters produces a perfusable PEGDA microfluidic channel embedded within an exterior PDMS housing. The PDMS/PEGDA construct, Mold A, can be sealed to coverglass for imaging after Step 5 or crosslinked to a PEGDA blank slab, Mold B, to produce a 3D microchannel environment (Step 9).

11. After the resist is exposed, place the slide onto a hot plate at room temperature. For the PDMS housing SU-8 2100 master, set the hot plate to 65°C and bake for 5 min once the plate reaches this temperature. Next turn the hot plate up and bake for 20 min at 95°C. For the PEGDA microchannel SU-8 2050 master, the bake times are reduced to 65°C for 5 min and 95°C for 10 min. As before, take care to gradually heat the substrate and allow the temperature to equilibrate before beginning the bake time in each step.

12. Turn off the hot plate and allow the resist to cool to room temperature. Place the substrate into a "puddle" of SU-8 developer in a glass Petri dish and gently swirl or rock the dish intermittently while developing. Use a developer volume that is sufficient to fully cover the substrate.
13. Development times will vary depending on the thickness of the resist and the area of nonexposed photoresist that needs to be dissolved. For the PDMS housing SU-8 2100 master, the development time can be up to 60 min, with a recommended replacement of the developer solution after 30 min. The SU-8 2050 master is usually completely developed in 30 min and does not require a change of developer solution.
14. Remove the resist from the developer puddle and rinse with 2-propanol. White residue or precipitate that becomes visible upon washing is indicative of undeveloped resist and requires additional developing time. It may be necessary to repeat the development and 2-propanol rinse steps multiple times until no precipitate is formed (see Note 9 on developing and fine structure development).
15. Once the resist is fully developed, rinse it with ultrapure water and dry with filtered air or nitrogen.
16. This entire section (Steps 1–15) can be repeated with various combinations of PDMS housings and microchannel masters to create a wide variety of system geometries and configurations.

3.2. PDMS Housing Fabrication

1. Rinse a Petri dish and SU-8 2100 PDMS housing photoresist master with pure ethanol and dry with filtered air.
2. Mix PDMS at a 15:1 ratio of elastomer:curing agent using a disposable plastic spatula for at least 1 min or until trapped bubbles are homogenously dispersed within the mixture. Higher (20:1) and lower (10:1) ratios of elastomer:curing agent can be used to produce softer or stiffer housings, respectively. For more detailed information on silicone properties and reagents for silicone functionalization, see the commercial website http://www.gelest.com/gelest/forms/Home/home.aspx.
3. Place the PDMS housing master into the Petri dish and pour the PDMS mixture over the mold, making sure to recover all of the polymer (Fig. 1, Step 3) (See Note 10).
4. Next, degas the mixture in a vacuum oven (–30 psi) at room temperature for 1 h or until all bubbles are removed.
5. After degassing transfer the mold to a 60°C oven and cure for 3 h. A more rapid curing (1 h) can be accomplished at 100°C but this temperature is not suitable for the polystyrene Petri dishes.

6. Separate the PDMS housing from the photoresist master and the Petri dish using a razor blade. Bore access ports in the PDMS housing at the desired location using a 1-mm biopsy punch (Fig. 1, Step 4).
7. Finally, rinse the PDMS housing with ethanol and conformally seal it to a pre-cleaned glass slide until further use (see Note 11 on reversible conformal seals).
8. Repeat Steps 2–7 for each PDMS housing component that is required in the final design. Note that for a 3D hydrogel environment, as shown in Fig. 1, Mold A forms the template for the housing with the channel in relief (Fig. 1, Steps 3 and 4) while Mold B forms the blank slab for the device bottom (Fig. 1, Steps 5 and 6).

3.3. Soft Lithographically Fabricated Poly(Ethylene Glycol) Diacrylate (PEGDA) Hydrogel

1. Dissolve PEGDA in PBS at the desired concentrations and add the Irgacure 2959 photoinitiator (100 mg/mL in ethanol) to the hydrogel pre-polymer solution at 30 μL/mL (see Note 4 for choice of hydrogel composition).
2. Insert polyethylene tubing with an attached syringe into one of the PDMS housing access ports in Mold A, and inject the acetophenone photoinitiator (300 mg/mL in NVP) until the entire housing is filled. Allow the photoinitiator solution to coat the housing for 5 min.
3. To remove excess photoinitiator, peel the PDMS housing from the glass slide and rinse with ethanol and then ultrapure water before drying the device with filtered air.
4. Treat the SU-8 2050 photoresist master to be used as the PEGDA microchannel mold with Sigmacote for 1 min. Rinse with ethanol, and then dry with filtered air (see Note 12 on engineering interfaces).
5. Overlay and align the acetophenone coated PDMS housing with the SU-8 2050 microchannel photoresist master (Fig. 1, Step 7).
6. Inject the PEGDA pre-polymer solution with without cells (see Note 13) into the housing around the channel mold and expose the device to a long wavelength UV lamp (365 nm, 10 mW/cm^2) for 2.5 min (Fig. 1, Step 7), in order to form the 3D microfluidic network.
7. Next, conformally seal the acetophenone coated PDMS housing for the blank slab (Mold B) against a Sigmacoted glass slide. Inject the PEGDA pre-polymer solution with or without cells into the housing and crosslink using a short 1 min UV exposure time. This will form a hydrogel slab that will serve as the base of the hydrogel device (Fig. 1, Step 8).

8. Finally, overlay Mold A containing the PEGDA microchannel hydrogel with the under-crosslinked PEGDA slab (Mold B) and expose to UV for 2.5 min to graft the two components together (Fig. 1, Step 9). Alternatively, if a 3D hydrogel environment around the channels is not required, the PDMS housing containing PEGDA microchannels in relief can be conformally sealed to cover glass after Subheading 3.3, Step 6 to allow ease of high resolution imaging.
9. To activate the microfluidic device, first connect polyethylene (PE) perfusion tubes to the PDMS inlet/outlet access ports, perfusion reservoir, and syringe pump. Apply quick-dry 5 min epoxy at the PE tube/PDMS interface to help anchor the tubes and prevent pulling on the compression fit. Initiate perfusion at 600 μL/h using a syringe pump (Harvard Apparatus) in withdrawal mode. See the Chips and Tips section of the Lab on a Chip journal website for helpful and pragmatic tips on a wide variety of microfabrication techniques including alternative ways to interface tubing with chips (http://www.rsc.org/Publishing/Journals/lc/Chips_and_Tips/interfacing.asp).

3.4. PEGDA Hydrogels Fabricated Using Photolithography and Soft Lithography

Soft lithographic hydrogel fabrication methods described in Subheadings 3.2 and 3.3 can be combined with photolithographic fabrication methods to build multilayer and multicomponent PEGDA hydrogel structures within a PDMS housing.

1. Fabricate PDMS housing as described in Subheading 3.2 and treat with photoinitiator as described in Subheading 3.3, Steps 2 and 3.
2. Conformally seal the coated PDMS housing to a Sigmacoted glass slide. Insert a PE tube connected to a syringe into one of the punched access ports and inject the PEGDA solution to be used for the first set of structures into the housing (Fig. 2, Step 3, PEGDA Solution 1).
3. Align the transparency photomask with patterns for PEGDA Structure 1 with the PDMS housing and expose to UV light for 2.5 min (see Note 14 on alignment, light source, and alternative techniques).
4. After exposure, peel the PDMS housing with attached PEGDA structures from the Sigmacoted glass slide and gently rinse with ultrapure water (use buffer if cells are present). Gently dry the device with filtered air.
5. Figure 2, Steps 2–4 can be serially repeated with PEGDA hydrogel solutions of different content and photomasks with different structure geometries to produce multiple, independently fabricated PEGDA structures.

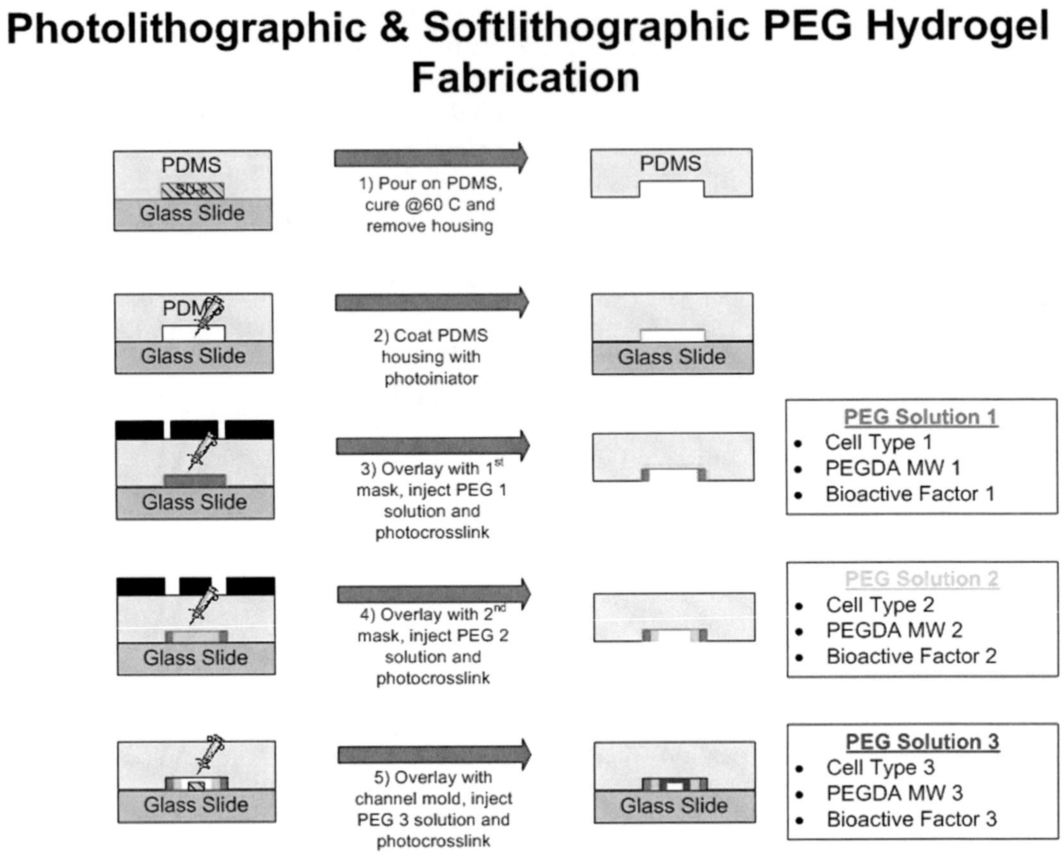

Fig. 2. Photolithographic and soft lithographic fabrication schematic of PDMS/PEGDA microfluidic networks. Serial mask-directed photopolymerization of PEGDA pre-polymers produces hydrogel structures with various physical and bioactive content (steps 3 and 4). Soft lithographic replica molding of PEGDA to a photoresist master produces a perfusable PEGDA microfluidic channel (Step 5).

6. Overlay the PDMS housing with attached PEGDA hydrogel structures with a Sigmacoted SU-8 2050 photoresist microchannel master (Fig. 2, Step 5).
7. Insert PE tubing with attached syringe into one of the punched PDMS housing access ports and inject the final PEGDA solution into the housing. Expose the device to a UV light source for 2.5 min.
8. Remove the PDMS housing with multiple PEGDA structures and PEGDA microchannels from the photoresist master and conformally seal to cover glass. Connect PE tubes and initiate perfusion. Example images of multilayer PEGDA Hydrogels are shown in Fig. 3.

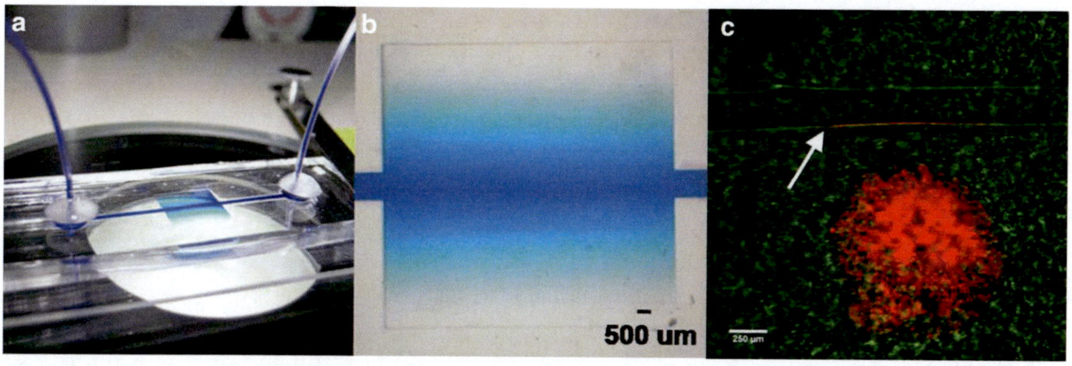

Fig. 3. (**a**) Multimodal fabricated PEGDA hydrogels. Isometric view of PDMS/PEGDA microchannel device perfused with *toluidine blue*. (**b**) Diffusion of *toluidine blue* into PEGDA diffusion chamber (10 × 10 mm) from a single 50 μm channel. Reproduced by permission from Elsevier. Cuchiara et al. (16). (**c**) Multimodal fabrication of murine hippocampal progenitor cell pillars (MHP 36, *red*) and human umbilical vein endothelial cells (HUVECs, *green*) replica molded to form a microfluidic channel (*arrow*).

4. Notes

1. Adhesion strength of SU-8 to various substrates can be found at Microchem's website (http://184.168.52.107/pdf/SU-8-table-of-properties.pdf). These effects will become more pronounced as resist structure aspect ratio (height/width) increases. The author has chosen to use glass because it is readily available and its performance is adequate for the given application.
2. Piranha is a very dangerous solution that is highly acidic and a strong oxidizer. Reaction of sulfuric acid and hydrogen peroxide is exothermic and the resulting solution can be explosive. Extreme caution must be employed to protect both the user and the user's workspace. Please thoroughly research piranha handling and disposal procedures before carrying out the protocols in this chapter.
3. In order to prevent leaking at the tube chip interface, a compression fit between the perfusion tubing and bored access port must be established. A compression fit requires that the diameter of the biopsy punch or access port be slightly less than or equal to the perfusion tubing outer diameter. The authors have found the a 1 mm biopsy punch combined with 1 mm OD PE tubing will hold a tight seal but can also be easily inserted.
4. Perhaps the most understated aspect of the protocols is the choice of PEGDA hydrogel system that satisfies the functional needs of the desired application. A rapidly growing body of PEGDA hydrogel research across disciplines has reported a wide variety of PEGDA hydrogel compositions that are suitable

for the photo and soft lithographic fabrication techniques described in this chapter. The reader is encouraged to start with a hydrogel composition that satisfies the desired application and work to design and optimize fabrication protocols around this target hydrogel composition. The reader is encouraged to reference book chapters by Miller et al. (22) and Bryant et al. (23) for more detailed information on photopolymerizable hydrogels.

5. Design of multilayer PDMS-PEGDA systems is dependent upon the differential geometry of the SU-8 photoresist master structures used to mold the PDMS housing and the SU-8 photoresist master structure used to mold the PEGDA channel (Fig. 1). Each user will have to define their structure geometry for a given application. The *x*–*y* plane geometry used will vary greatly from study to study and is easily modified by changes to the photomasks. The *z*-plane or structure height is determined by photoresist processing parameters and is not as readily tuned. The author has found that structure height of 350 μm for the PDMS housing mold and 100 μm for the PEGDA microchannel mold provides reliable perfusion and aspect ratios well within the manufacture recommended specifications. All process parameters mentioned herein are optimized from manufacturer recommended protocols. The reader is encouraged to reference Microchem's website for detailed information on adjustments to processing parameters and other helpful tips: (http://www.microchem.com/Prod-SU8_KMPR.htm).

6. Always operate the spin coater with a cover over the spinning chamber and use eye protection.

7. One of the most influential parameters on photoresist structure height, quality, and adhesion to the substrate is proper baking practices. Minimization of any rapid thermal fluctuations between baking steps and slowly ramped temperature baking protocols will go a long way to improving photoresist performance. In addition, the author has found that the temperature profile across a hot plate is not constant with the center of the hot plate likely being at the target temperature, while regions closest to the hot plate edges may show a decrease in temperature between 10 and 30% of the target temperature. Take care to make sure your substrate is experiencing the desired baking temperature. The author has found that the purchase of a standard infrared thermometer is a useful tool in validating bake temperatures and hot plate thermal profiles.

8. Various light sources can be used to expose photoresist. It is important that the spectral output of the light source corresponds to the photoresist's absorption spectrum. It is also important that the light source used is collimated. Mask aligner instrumentation with micromanipulators and imaging capability aid with multilayer alignment but are usually not necessary for single exposures.

9. The use of a standard light microscope periodically during photoresist development step may be helpful in determining when the photoresist structures are completely developed. The presence of a white film at the edge of a structure is indicative of undeveloped resist. For structures in close proximity, especially narrow "troughs" or "holes," more aggressive agitation, such as sonication, may help to remove undeveloped resist.
10. PDMS is very difficult to clean from glassware and other lab tools. It highly recommended that all materials used to prepare and cure PDMS be disposable.
11. For PDMS-PEGDA hydrogel microdevices, the PDMS-substrate seal must be reversible and methods for sealing the hydrogel laden PDMS chip must be cell compatible. The author has found that the only way to produce a PDMS-PEGDA microdevice that forms a reversible seal and does not leak is through a conformal seal. Conformal seal performance requires that the PDMS housing and substrate surfaces are very clean and free of particles. The best conformal seals form passively and do not require application of manual pressure. Conformal seals have been shown to hold a positive pressure of 5 psi (24) but the author has found that the most robust method for maintaining a conformal seal is the continuation of negative pressure via a syringe pump in withdrawal mode. This can be the same negative pressure used to drive flow through your device (16) or it can be a separate design that is only used to seal the interface and keep the chip from leaking (25).
12. Another understated aspect in this process is the need to selectively engineer adhesive and nonadhesive interfaces. In order to facilitate PEGDA polymerization at the PEGDA/PDMS interface and to promote PEGDA adhesion to the PDMS, the author recommends treating the PDMS housing with the photoinitiator acetophenone in the solvent NVP. It is thought that the NVP dissolves the PDMS polymer structure therefore facilitating PEG-PDMS chain–chain entanglement. Furthermore, the presence of locally high concentrations of acetophenone and the free radical acceptor NVP will counteract the free radical quenching effect of oxygen dissolved in the PDMS matrix. Other groups have recommended functionalizing the PDMS surface with acrylates (26) or coating with oxygen impermeable parylene c (27) to promote hydrogel polymerization at the PDMS interface. For a more complete review on PDMS surface functionalization please refer to Hu et al. (26). Conversely it is necessary to engineer a nonadhesive interface in order to avoid hydrogel tearing during the molding of PEGDA microstructures from SU-8 photoresist masters or blank glass slide substrates. The author has found that the commercially available Sigmacote provides a robust, easy-to-use, nonadhesive coating that improves the reliability of the PEGDA molding process.

13. All system components that are to contact cells or cell solutions must be sterilized prior to use. All flow loop components, including the PDMS chip, PE tubing for perfusion, valves for flow control, and reservoir bottles that are left in place during the duration of the cell culture are sterilized by either ethylene oxide exposure or autoclaving. Components that are only temporarily used during the fabrication process are sterilized under UV light in a laminar flow biosafety cabinet for at least 12 h. These components include PE tubing for solution injection, Sigmacoted glass slides, 200 proof ethanol and ultrapure water wash bottles, and UV lamp for polymerization. The PEGDA pre-polymer solution is sterile-filtered using a 0.22 μm PES syringe filter to remove debris and containments.
14. As with all multilayer lithography, multistructure fidelity is often limited by the ability to align the photomask with existing structures. The author has found that resolving >1 mm alignment tolerances is possible "by eye" with practice. However, when aligning multiple structures with micron-level precision, it is necessary to use a micromanipulator or a common microscope stage controller coupled with a microscope. The author has found that the use of a mask projection photolithography technique (28) allows for micron-level manipulation and alignment of the substrate with a single wavelength of light followed by exposure of the pre-polymer solution using the same light path with a different wavelength of light. This technique allows for precise alignment and is compatible with newer three-dimensional laser scanning lithography fabrication techniques (29).

Acknowledgments

The author would like to thank Professor Lisa Biswal and her research group, especially Gautam Kini, for technical assistance and Melissa McHale for her helpful insight proofreading and improving this chapter. This work was supported by the NIH Biotechnology Training Grant (T32 GM008362-18) and the NIH Quantum Grant (1 P20 EB007076 01).

References

1. Whitesides GM, Ostuni E, Takayama S, Jiang X, Ingber DE (2001) Soft lithography in biology and biochemistry. Annu Rev Biomed Eng 3:335–373
2. Weibel DB, Diluzio WR, Whitesides GM (2007) Microfabrication meets microbiology., Nature reviews. Microbiology 5:209–218
3. Khademhosseini A, Langer R, Borenstein J, Vacanti JP (2006) Microscale technologies for tissue engineering and biology. Proc Natl Acad Sci USA 103:2480–2487
4. Gómez-Sjöberg R, Leyrat AA, Pirone DM, Chen CS, Quake SR (2007) Versatile, fully

automated, microfluidic cell culture system. Anal Chem 79:8557–8563

5. Maerkl SJ, Quake SR (2007) A systems approach to measuring the binding energy landscapes of transcription factors. Science (New York, NY) 315:233–237
6. Chen CS (1997) Geometric Control of Cell Life and Death. Science 276:1425–1428
7. Quist AP, Pavlovic E, Oscarsson S (2005) Recent advances in microcontact printing. Anal Bioanal Chem 381:591–600
8. Hoganson DM, Anderson JL, Weinberg EF, Swart EJ, Orrick BK, Borenstein JT, Vacanti JP (2010) Branched vascular network architecture: a new approach to lung assist device technology. J Thorac Cardiovasc Surg 140:990–995
9. Carraro A, Hsu W-M, Kulig KM, Cheung WS, Miller ML, Weinberg EJ, Swart EF, Kaazempur-Mofrad M, Borenstein JT, Vacanti JP, Neville C (2008) In vitro analysis of a hepatic device with intrinsic microvascular-based channels. Biomed Microdevices 10:795–805
10. Ling Y, Rubin J, Deng Y, Huang C, Demirci U, Karp JM, Khademhosseini A (2007) A cell-laden microfluidic hydrogel. Lab Chip 7:756–762
11. Golden AP, Tien J (2007) Fabrication of microfluidic hydrogels using molded gelatin as a sacrificial element. Lab Chip 7:720–725
12. Choi NW, Cabodi M, Held B, Gleghorn JP, Bonassar LJ, Stroock AD (2007) Microfluidic scaffolds for tissue engineering. Nat Mater 6:908–915
13. Tsang VL, Chen AA, Cho LM, Jadin KD, Sah RL, DeLong S, West JL, Bhatia SN (2007) Fabrication of 3D hepatic tissues by additive photopatterning of cellular hydrogels. FASEB J 21:790–801
14. Chin VI, Taupin P, Sanga S, Scheel J, Gage FH, Bhatia SN (2004) Microfabricated platform for studying stem cell fates. Biotechnol Bioeng 88:399–415
15. Albrecht DR, Tsang VL, Sah RL, Bhatia SN (2005) Photo- and electropatterning of hydrogel-encapsulated living cell arrays. Lab Chip 5:111–118
16. Cuchiara MP, Allen ACB, Chen TM, Miller JS, West JL (2010) Multilayer microfluidic PEGDA hydrogels. Biomaterials 31:5491–5497, Elsevier Ltd
17. King KR, Wang CCJ, Kaazempur-Mofrad MR, Vacanti JP, Borenstein JT (2004) Biodegradable Microfluidics. Adv Mater 16:2007–2012
18. Nguyen KT, West JL (2002) Photopolymerizable hydrogels for tissue engineering applications. Biomaterials 23:4307–4314
19. Martinez AW, Phillips ST, Wiley BJ, Gupta M, Whitesides GM (2008) FLASH: a rapid method for prototyping paper-based microfluidic devices. Lab Chip 8:2146–2150
20. Martinez AW, Phillips ST, Whitesides, GM (2008) Three-dimensional microfluidic devices fabricated in layered paper and tape 2008.
21. Grimes A, Breslauer DN, Long M, Pegan J, Lee LP, Khine M (2008) Shrinky-Dink microfluidics: rapid generation of deep and rounded patterns. Lab Chip 8:170–172
22. Miller J, West J (2008) Biomimetic Hydrogels to Support and Guide Tissue Formation. In: Khademhosseini A, Borenstein J, Toner M, Takayama S (eds) Micro and Nanoengineering of the Cell Microenvironment, 1st edn. Artech House, Boston, pp 101–120
23. Bryant SJ, Anseth KS (2006) Photopolymerization of Hydrogel Scaffolds. In: Ma PX, Elisseeff JH (eds) Scaffolds in Tissue Engineering, 1st edn. Taylor and Francis, New York, pp 71–90
24. Mcdonald JC, Duffy DC, Anderson JR, Chiu DT (2000) Review General Fabrication of microfluidic systems in poly (dimethylsiloxane), Review Literature And Arts Of The Americas.
25. Wong AP, Perez-castillejos R, Love JC, Whitesides GM (2008) Partitioning microfluidic channels with hydrogel to construct tunable 3-D cellular microenvironments. Cell 29:1853–1861
26. Hu S, Ren X, Bachman M, Sims CE, Li GP, Allbritton NL (2004) Tailoring the Surface Properties of Poly (dimethylsiloxane) Microfluidic Devices. Society 20(13):5569–5574
27. Moraes C, Wang G, Sun Y, Simmons CA (2010) A microfabricated platform for high-throughput unconfined compression of micropatterned biomaterial arrays. Biomaterials 31:577–584, Elsevier Ltd
28. Love JC, Wolfe DB, Jacobs HO, Whitesides GM (2001) Microscope Projection Photolithography for Rapid Prototyping of Masters with Micron-Scale Features for Use in Soft Lithography. Langmuir 17:6005–6012
29. Hahn MS, Miller JS, West JL (2005) Laser Scanning Lithography for Surface Micropatterning on Hydrogels. Adv Mater 17:2939–2942

Chapter 25

Purification of DNA/RNA in a Microfluidic Device

Andy Fan, Samantha Byrnes, and Catherine Klapperich

Abstract

Often, modern diagnostic techniques require the isolation and purification of nucleic acids directly from patient samples such as blood or stool. Many diagnostic tests are being miniaturized onto micro-sized platforms and integrated into microfluidic devices due to the economies resulting from smaller sample and reagent volumes. Often, these devices perform sample preparation in series with the diagnostic tests. The sample preparation steps are vital in order to purify the desired genetic material from potential inhibitors that can interfere with the outcome of the test. There are various techniques used to selectively capture the nucleic acids while washing away potential contamination (proteins, enzymes, lipids, etc.). Two of the most common forms of selective capture are based on nucleic acid binding to silica surface or on the precipitation of nucleic acids with or without the presence of a carrier species. Each of these methods can be performed in liquid phase or in a solid support such as an extraction column. Here we discuss both methods and address microfluidic applications.

Key words: Sample preparation, Microfluidic device, Diagnostic tests, Micro-solid phase extraction, Purification of DNA/RNA, Blood lysis

1. Introduction

In a traditional medical lab setting, extraction and purification of nucleic acids (NA) from biologically complex samples, such as blood or organ tissues, often involves a common set of unit processes: (a) isolation of target cells from a heterogeneous matrix by fractionation, (b) lysis of the target cells, (c) isolation of NA from protein-laden lysate using additional fractionation steps, and (d) concentration of harvested NA. Before the advent of NA purification by solid-phase extraction with silica matrices (1, 2), the above tasks usually required cellular debris separations with centrifugation, handling of toxic substances such as phenol–chloroform, and time-consuming ice-cold precipitation of NA with alcohols. Even with modern advances in bench top solid-phase extraction (SPE) methods

Gareth Jenkins and Colin D. Mansfield (eds.), *Microfluidic Diagnostics: Methods and Protocols*, Methods in Molecular Biology, vol. 949, DOI 10.1007/978-1-62703-134-9_25,

offered by commercial entities such as Qiagen, Invitrogen, and others (3), a single limitation persists—the need for an accessible centrifuge, which implies access to a dedicated lab space with reliable power sources.

Recently, there has been a move toward miniaturization of diagnostics onto micro-sized platforms that include sample preparation steps (4–6). Besides the footprint advantage that microfluidic devices hold over dedicated lab equipment, the microscale platform allows the possibility of using a smaller patient sample size, automation of extraction (7) and most importantly, device portability that could potentially be the driving force toward future point-of-care diagnostics. To take advantage of the form factor while keeping the lysis chemistry simple and nontoxic, the majority of microfluidic NA extraction devices rely on the combination of a SPE column that captures and precipitates NAs and guanidine hydrochloride (GuHCl) or guanidine thiocyanate (GuSCN) as the lysis reagent (8, 9). In the presence of the guanidine cation, the most notable NA capturing technique is to utilize silica particles, which can be embedded within the column (10, 11) or kept as a separate layer from the column. An additional method of NA capture relies on a co-precipitant molecule, such as glycogen, which captures the NA and selectively precipitates it in the presence of a precipitating solvent, such as ethanol or isopropanol (12, 13) and small cation cofactor supplied by sodium chloride (NaCl), sodium acetate (NaOAc), ammonium acetate (NH_4OAc), or lithium chloride (LiCl) (8). Direct NA precipitation using this method, however, requires careful tailoring of the chemistry in order to avoid excess co-precipitation of cellular proteins and will not be considered in this chapter.

All of these sample preparation methods can take place on a fully integrated microfluidic chip or in small disposable systems. Additionally, these sample preparation methods do not differentiate between the type or source of NAs, resulting in broad-spectrum use of these techniques with various downstream applications such as PCR, ELISA, and ESI-MS (3, 14).

As will be detailed in Subheading 3.1, a simple, low-cost microfluidic chip can be constructed by thermo-compression bonding of two sheets of Zeonor 750R plastic. Briefly, channels are formed by patterning one sheet of Zeonor by hot-embossing with a noncompliant master mold made from SU8. The choice of Zeonor as the chip material was driven by two factors: the material is UV transparent and relatively chemically inert—both of which are advantageous attributes in most biological assays. This polymer is a cyclic polyolefin; similar materials with similar temperature and optical characteristics can be obtained from other suppliers. The bonded chips are filled with a pre-polymer solution and then exposed to UV light. After polymerization, the porogens and initiators are removed through washes with methanol and ethanol, and the resulting SPE matrix is equilibrated with lysis buffer prior to NA capture.

From a point-of-care diagnostics stance, there are cases where a very small form-factor microfluidic device is not feasible in the field. For instance, in resource-poor locations where electricity is at a premium, an electrically actuated fluidic pump operating at lower pressures can be replaced by a manual bicycle pump operating at an excess of 80 psi. As a result of favoring portability and laboratory independent operation over automation and flow accuracy, an existing chip design can be scaled up such that both the SPE column and the "microchannel" can be packaged together in a premade plastic vessel like a pipette tip. Preparation of such a non-chip-based extraction method is described in Subheading 3.2.

2. Materials

2.1. Microchip Components

1. Zeonor 750R, a thermoplastic polyolefin resin (Zeon Chemicals, Louisville, KY).
2. SU8 for making the master mold (Microchem, Newton, MA).
3. Metal sputtering system for titanium and aluminum coating of master mold.

2.2. Microchip Reagents for Bacterial DNA Recovery from Cell Culture or Blood

1. Test sample (e.g., infected blood or cell culture medium).
2. Resuspension buffer: 200 mM Tris–HCl, 2 mM EDTA, with sodium dodecyl sulfate (SDS) and Triton X-100.

   For gram-negative bacteria: Resuspension buffer should contain 0.03% SDS/3.6% Triton X-100.

   For gram-positive bacteria: Resuspension buffer should contain 0.3% SDS/7.2% Triton X-100.
3. Lysis reagent: 3 M GuSCN (from 6 M stock, Sigma Aldrich, St. Louis, MO).
4. Proteinase K (20 mg/mL stock in 50 mM Tris–HCl (pH 8.0), 10 mM $CaCl_2$); from Amresco (Solon, OH).
5. 70% Ethanol.
6. Nuclease-free water.

2.3. Reagents for Viral RNA Recovery from Plasma or Cell-Free Supernatants

1. Channel Buffer: 1.5 M GuSCN, 50% isopropanol, 1× RNASecure reagent. For 5 mL of channel buffer, mix 1.25 mL of 6 M GuSCN, 2.5 mL of 99% isopropanol, and 200 μL of 25× RNASecure reagent (stock is 25× from Applied Biosystems,Carlsbad, CA, Cat #AM7005) in a 15 mL tube. Use stock immediately.

2. Lysis buffer: 2 M GuSCN, 62.7% isopropanol, 1× RNASecure reagent. For 4.5 mL of lysis buffer, mix 1.5 mL of 6 M GuSCN, 3 mL of 99% isopropanol, and 180 μL of 25× RNASecure reagent in a 15 mL tube. Use stock immediately.
3. 70% Ethanol.
4. Nuclease-free water.

2.4. Acid-Treated Silica Slurry

1. Silica particles (Sigma Aldrich, St. Louis, MO).
2. Ultra-pure water.
3. Hydrochloric Acid Solution: hydrochloric acid (32 wt%/vol, 10.2 M).

2.5. UV Activated Micro-Solid Phase Extraction Column

1. Grafting Solution: ethylene diacrylate (90% EDA), methyl methacrylate (99%, MMA), benzophenone (99%). Mix a 1:1 solution of EDA and MMA with 3% benzophenone.
2. SPE column pre-polymer solution: porogenic solvents 1-dodecanol and cyclohexanol, monomer solution butyl methacrylate (99% BuMA) and ethylene dimethacrylate (98% EDMA). Mix 24 wt% of BuMA, 16 wt% EDMA, 42 wt% 1-dodecanol, 18 wt% cyclohexanol. Add ~0.4 wt% of the UV-Sensitive Free-Radical Initiator: 2,2-dimethoxy-2-phenylacetophenone (99%, DMPAP). The DMPAP should be 1 wt% with respect to the monomers.
3. 0.7 μm Silica microspheres (Polysciences, Inc., Warrington, PA).
4. Methanol.
5. Absolute ethanol.

3. Methods

3.1. Preparation of Microfluidic Chip with Embedded Micro-Solid Phase Extraction Column Applied to Bacterial DNA Extraction from Blood

A schematic of the chip-making process is shown in Fig. 1.

1. For preparation of the master mold: pre-clean 100 mm diameter p-type (100) Si wafers with piranha solution (1:3 vol ratio of 30 wt% hydrogen peroxide and 98 wt% sulfuric acid) and spin-coat 100 μm of SU8 onto the wafer surface. After a 30 min prebake at 95°C, the appropriate channel features are formed using contact mask lithography. In general, the SPE columns are fabricated inside of straight microfluidic channels. The channels do not need to have a specific type of sidewall (i.e., both isotropic and anisotropic techniques can be used). After pattern development, post-bake the wafers for 1.5 h at 175°C (see Note 1).
2. To aid the subsequent removal of the hot-embossed plastics from the SU8 mold after embossing (see Note 2), a bilayer of

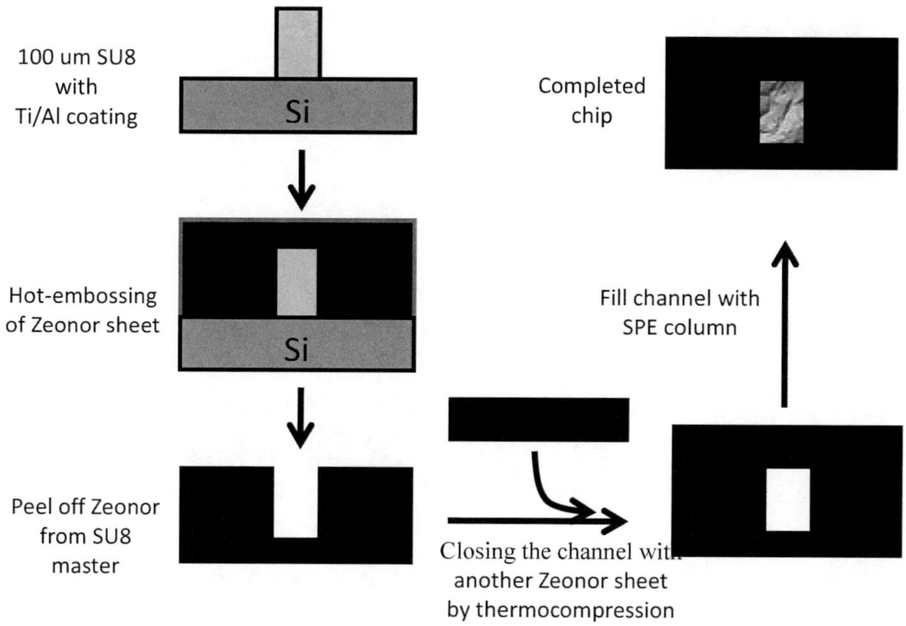

Fig. 1. Fabrication process of the microfluidic chip with embedded SPE column.

500 Å titanium/1,000 Å of aluminum should be sputtered onto the SU8 surface. The completed SU8 mold can be reused 2–5 times before structural degradation occurs from repeated hot-embossing.

3. To construct Zeonor microchannels, the Zeonor sheets to be used for subsequent hot-embossing can be made by melting raw Zeonor pellets in a heat press at 90°C (20°C above the glass transition temperature of Zeonor 750R) against two flat surfaces. Shims (spacers) can be used to create sheets or plaques of the desired thickness; otherwise, extruded sheet material can also be purchased directly through vendors.
4. Starting from a sheet of Zeonor, microfluidic channels with 100 μm depth and either 100 or 150 μm in width can be created by hot-embossing the Zeonor against the SU8 master using a heat press at 90°C and 250 psi for 5 min. The Zeonor-SU8 duplex is then cooled for 2 min at room temperature prior to manual separation (see Note 3).
5. Following separation, drill two, 1.5-mm diameter wells at each end of the microchannel to form the input and output ports.
6. To encapsulate the microchannel, an identically sized piece of Zeonor was thermally bonded to the hot-embossed Zeonor sheet on a hot-press at 70°C, 250 psi for 2 min.
7. In order to make the intra-channel solid-phase matrix, the channel walls should be treated with grafting agent to increase

sidewall adhesion to the polymer monolith, due to the surface chemistry of the Zeonor plastic (15). This can be accomplished by filling the channels with the grafting agent, followed by UV-irradiation in an oven for 10 min at a wavelength of 254 nm and at 200 mJ/cm^2.

8. After UV-irradiation, remove the excess solution from the channels by washing with 100 μL of methanol; between 10 and 33 column volume-worth (CV), where 1 CV equates to the SPE monolith volume.

9. After grafting and washing, the SPE-proper can be made by filling the channels with the SPE column pre-polymer solution and 15 wt% of 0.7 μm silica microspheres (wet volume, see Note 4) and then UV-irradiated for 2 min at 200 mJ/cm^2. Typically, 3–10 μL of the SPE column pre-polymer solution is used to make a SPE.

10. After the second UV-irradiation step, wash out any excess porogen and SPE column pre-polymer solution with 100 μL (10–33 CV) of methanol followed by 200 μL (20–66 CV) of absolute ethanol.

11. The first step of using the on-chip SPE column for the application of bacterial DNA extraction from blood is sample collection and off-chip lysis. To simulate bacteremia, where bacteria are found in human whole blood, one must first take an overnight culture (or take log-phase cultures at $OD_{600} \sim 0.5$) and make serial dilutions of 10^1–10^5 CFU/mL. Then, spin down the 1 mL—worth of cells at 6,800 × *g* for 10 min, decant the supernatant, and resuspend each bacteria pellet in 100 μL of human whole blood.

12. By adding 150 μL of resuspension buffer to the blood–bacteria matrix, indiscriminant cellular lysis commences with both SDS and Triton-X denaturing the plasma membranes of cells and the bacteria cell walls.

13. Add 200 μL of the lysis reagent (3 M GuSCN), vortex to mix, and add enough Proteinase K to a final concentration of 0.8 mg/mL. At this stage, the final concentration of GuSCN is 1.33 M, which is around the upper limit where both Proteinase K and GuSCN can function synergistically as a protein denaturant and, at the same time, protein digestion can take place without proteinase K itself being denatured by the chaotropic agent. Furthermore, the guanidinium cations in the mixture also act as cofactors in NA-silica binding.

14. For optimal lysis and digestion, the mixture can be incubated at 50°C–60°C for 30–90 min.

15. Condition the on-chip SPE column with loading buffer for 5 min (around 30 μL total, or 10 CV) prior to introducing the

sample for extraction. For the description of sample loading, we will assume the volume of the SPE column to be 3 μL.

16. Pipette the blood–bacteria lysate into a 1 cm^3 syringe. Then, using a syringe pump, load the lysate into the microchannel at 10–15 μL/min. The flow-throughs can either be collected for analysis or discarded.
17. Wash the microchannel with approximately 300 μL of 70% ethanol (~10 CV) to remove excess proteins that may have adsorbed onto the column.
18. Finally, elute the DNA from the column using nuclease-free water. Typical elution volumes range from 25 to 100 μL (8–30 CVs).

3.2. Preparation of UV Activated SPE Column in Pipette Tips Applied to Viral RNA Extraction from Cell-Free Supernatants

1. To the solid-phase extraction (SPE) column pre-polymer solution, add 15 wt% of 0.7 μm silica microspheres (wet volume, see Note 4).
2. Aliquot the pre-polymer mixture in 50 μL portions into 250 μL non-filter pipette tips (see Note 5). To polymerize the solution, the pipette tips are placed in a 254 nm UV oven and irradiated for 2 min at 200 mJ/cm^2. During this time, the solution will turn from a clear liquid to a solid white monolith.
3. After the polymerization is complete, wash the monoliths with 100 μL methanol (2 CV) followed by 200 μL (4 CV) of absolute ethanol (see Note 6). For the sake of discussion we are now assuming that 1 CV equates to 50 μL worth of SPE matrix.
4. For performing the viral RNA extraction from cell-free supernatants using the SPE column the pipette tip is first equilibrated with the channel buffer; fluid flow can be driven by compressed nitrogen or air at a range between 40 and 80 psi depending on the desired flow rate. After the channel buffer wash, the pressure source is disconnected, the pipette tip opening exposed, and then loading of the viral sample can proceed (see Note 7).
5. Mix the virus sample at 1:1 with the lysis buffer, hand-invert the tube at least 20 times to mix, and commence lysis by incubating for 15 min at room temperature. Next, manually pipette the lysate into the SPE column, refit the compressed air source onto the SPE-tip complex and push the entire lysate through the SPE column. The flow-through fractions can either be collected for further analysis or discarded.
6. After the sample mixture passes completely over the column, wash the column with 500 μL (10 CV) of 70% ethanol solution to remove excess salts and proteins that may have precipitated onto the column.

7. After the wash, the column is dried with air for 2 min to remove excess solvent.
8. Finally, elute the RNA from the SPE column by washing with several CV-worth of nuclease-free water (see Note 8).

4. Notes

1. Following the post-bake period, the masters had a negative pattern of the channels and had glass-like mechanical properties (1).
2. The sputter coating for the master is an optional step but the aluminum coating seems to help with the removal of the master after embossing. Of course, electroplating is a much more robust, but also more expensive method of achieving the same effect.
3. The master/substrate were cooled on an aluminum plate and separated manually when the plastic was no longer soft.
4. The silica microspheres come suspended in water so the volume is measured and spun down in a centrifuge at 2,500 × *g* for 10 min. The supernatant was removed and the resulting pelleted silica was dried overnight at 115°C. The pellet was then broken up into a powder to be mixed with the porogen/monomer solution.
5. The SPE pre-polymer solution can be applied to various column geometries and volumes. Certain geometries require treatment to help prepare the surface so that the SPE column does not get pushed out of the channel under pressure. Surface preparation can be achieved through the use of a grafting solution or by mechanically scoring the inside surface of the channel (12).
6. The best way to store the SPE columns is in a dry, sealed bag with a desiccant packet.
7. The input sample can be virions in a protein solution or in some other biological matrix such as blood. This protocol can also be performed using a chip-scale SPE column.
8. The rate of elution should be kept slow and consistent (approximately one drop every 5 s) to ensure the captured RNA is exposed to water long enough to return into solution from its precipitated form on the SPE column.

References

1. Boom R, Sol CJA, Salimans MMM, Jansen CL et al (1990) Rapid and simple method for purification of nucleic acids. J Clin Microbiol 28:495–503
2. Milton ID, Carter MJ (1993) An Inexpensive and simple method for DNA purifications on silica particles. Nucleic Acids Res 21(4):1044
3. Zhao L, Fu ZF (2009) Isolation of viral RNA from cultures, chapter 3. In: Liu D (ed) Handbook of nucleic acid purification. CRC, Boca Raton, FL, pp 41–60
4. Bhattacharyya A, Klapperich CM (2006) Thermoplastic microfluidic device for on-chip purification of nucleic acids of disposable diagnostics. Anal Chem 78:788–792
5. Wolfe KA, Breadmore MC, Ferrance JP et al (2002) Toward a microchip-based solid-phase extraction method for isolation of nucleic acids. Electrophoresis 23:727–733
6. Tan A, Benetton S, Henion JD (2003) Chip-based solid-phase extraction pretreatment for direct electrospray mass spectrometry analysis using an array of monolithic columns in a polymeric substrate. Anal Chem 75: 5504–5511
7. Han S-I, Han K-H, Frazier AB, Ferrance JP, Landers JP (2009) An automated micro-solid phase extraction device involving integrated high-pressure microvalves for genetic sample preparation. Biomed Microdevices 11:935–942
8. Sambrook J, Russell DW (2001) Extraction, purification, and analysis of mRNA from eukaryotic cells, molecular cloning: a laboratory manual, chapter 7, 3rd edn. Cold Springs Harbor Laboratory Press, Cold Springs Harbor, NY
9. Hanson LA (2009) Isolation of viral DNA from cultures, chapter 2. In: Liu D (ed) Handbook of nucleic acid purification. Boca Raton, FL, CRC, pp 23–39
10. Breadmore MC, Wolfe KA, Arcibal IG et al (2003) Microchip-based purification of DNA from biological samples. Anal Chem 75: 1800–1886
11. Chatterjee A, Mirer PL, Santamaria EZ, Klapperich C, Sharon A, Sauer-Budge AF (2010) RNA isolation from mammalian cells using porous polymer monoliths: an approach for high-throughput automation. Anal Chem 82:4344–4356
12. Alexandrov A, Dutta K, Pascal SM (2001) Sensitive ribonuclease protection assay employing glycogen as a carrier and single inactivation/precipitation step. Biotechniques 30:1198–1204
13. Chomczynski P, Sacchi N (2006) The single-step method of RNA isolation by acid guanidinium thiocyanate-phenol-chloroform extraction: twenty-something years on. Nat Protoc 1:581–585
14. Giordano BC, Ferrance J, Swedberg S et al (2001) Polymerase chain reaction in polymeric microchips: DNA amplification in less than 240 seconds. Anal Biochem 291:124–132
15. Rohr T, Olgetree DF, Svec F, Frechet JM (2003) Surface functionalization of thermoplastic polymers for the fabrication of microfluidic devices by photoinitiated grafting. Adv Funct Mater 13:264–270

Chapter 26

Agarose Droplet Microfluidics for Highly Parallel and Efficient Single Molecule Emulsion PCR

Xuefei Leng and Chaoyong James Yang

Abstract

Agarose emulsion droplet microfluidic technology for single copy emulsion PCR (ePCR) is a suitable technique for the detection of single copy DNA molecules. It improves the traditional ePCR by employing agarose with low melting and low gelling temperatures, which is coupled with PCR forward primers using Schiff-base reaction. Highly uniform monodisperse nanoliter agarose droplets each containing PCR reagents and single copy template are produced with a microfabricated emulsion generator. Following PCR, the cooled droplets transform to microbeads carrying amplicons to maintain the monocolonity of each droplet, which can be further analyzed. This method allows high-throughput generation of uniform droplets and enables high PCR efficiency, making it a promising platform for many single copy genetic studies.

Key words: Emulsion PCR, Microfluidics, Bioconjugation, Droplet, Genetic analysis

1. Introduction

Compartmentalization of individual samples in aqueous droplets dispersed in an oil phase is becoming a powerful method for high throughput assays in chemistry and biology (1–3). For example, by segregating single DNA molecules in individual water-in-oil emulsion droplets, emulsion PCR (ePCR) offers the advantage of massively parallel colonal amplification of DNA templates, which allows the identification and quantification of rare mutant genes within large populations (4, 5) and enables a new generation of ultra-high throughput DNA sequencing technologies (6, 7).

ePCR for single DNA molecule amplification was first implemented by S. Katsura and M. Nakano in 2003 (8). In their study a PCR mixture containing statistically dilute templates was stirred with oil to generate droplets with or without single copy of template. After amplification, the aqueous phase was collected and analyzed

Gareth Jenkins and Colin D. Mansfield (eds.), *Microfluidic Diagnostics: Methods and Protocols*, Methods in Molecular Biology, vol. 949, DOI 10.1007/978-1-62703-134-9_26, © Springer Science+Business Media, LLC 2013

using gel electrophoresis. This method has the disadvantage of a slow and limited analytical assay, making it almost impossible for statistical analysis of a large range samples. The use of microbeads solved this problem with its advantage of downstream, high-throughput manipulation and analysis. To immobilize and enrich PCR product, D. Dressman et al. developed BEAMing (beads, emulsion, amplification, and magnetics) introducing primer-bound magnetic microbeads encapsulated with single DNA molecules in individual droplets (1). DNA template was amplified on the microbead in the same droplet, which prevented the possibility of mixing of DNA products from different droplets after lysis of emulsion droplets. After PCR, beads bound with amplified molecules were labeled by fluorescent probes complementary to regions of known allelic variations and analyzed via flow cytometry. This method had the advantages of high sensitivity and high throughput analysis, which found great application in the detection of minor variants in a DNA population.

The use of microbeads in BEAMing is critical for downstream, high-throughput manipulation and analysis. However the use of microbeads still leads to many problems including poor PCR efficiency and short product length because PCR on a solid surface suffers from problems such as steric hindrance effects and charge repulsion. Furthermore, by encapsulating diluted molecules and microbeads into the droplets at random, the resulting Poisson statistics leads to a large number of void droplets including empty droplets or droplets with only either molecule or bead, which is wasteful, thus negating the speed and efficiency afforded by microfluidics. Besides, introducing microbeads in a chip might block microchannels easily and requires demanding fabrication technology of micropumps, which increases microchip complexity.

The agarose emulsion droplet microfluidic method aims to address these challenging problems. A special kind of agarose that has a low gelling temperature is melted and acts as an aqueous phase to generate droplets containing PCR mixture and single copy of template in a microfluidic chip. After amplification, the agarose droplets are cooled and transformed into agarose microbeads carrying PCR products. Fluorescence microscopy and FACS technology provides fast and precise analysis of results. Compared with conventional ePCR, this method has the advantage of high PCR efficiency and a high yield of valid droplets by using agarose as a matrix to capture amplicons instead of microbeads, and high-throughput generation of uniform monodispersed droplets by utilizing microfluidics.

In this method a microfluidic glass chip with a 200 μm (w) × 100 μm (d) cross-channel serves as a droplet generator. The inner surface of the channel is hydrophobically treated to be adaptable to the oil sheath so that the water-in-oil droplets can remain stable.

The size of the generated droplets is closely related to the channel dimension. If the resultant agarose microbeads need to be analyzed by FACS, their diameters should be controlled to below 80 μm. Hence the dimension of cross-channels should be sized accordingly.

The role of agarose in the aqueous phase is to act as a capturing matrix to replace conventional primer functionalized microbeads to preserve the monoclonal nature of the DNA product in each droplet. Ultralow gelling temperature agarose (Sigma A2576) has a melting point of about 56°C and gelling point around 16°C. Once melted, this agarose will remain in liquid phase until the temperature drops below 16°C, which ensures easy generation of agarose droplets on chip under room temperature. In addition, at all PCR temperatures, agarose remains in liquid phase, where PCR can take place with high efficiency. On the other hand, after PCR amplification, the solution form of agarose droplets can be switched to a solid gel phase by simply cooling the solution below the gelling point. Once solidified, agarose beads will remain solid unless the temperature rises above 56°C. As a result, DNA products amplified in the droplet will maintain their monoclonality even after the oil phase is removed. The clonal beads can be used for downstream sequencing or genotyping applications.

There are two basic requirements to realize high throughput single copy DNA detection. First, each single molecule must be isolated from a large population. A microfluidic glass chip is used in this method to compartmentalize DNA sample by forming uniform monodispersed nanoliter water-in-oil agarose droplets containing PCR reagents and single copy of template. Secondly, a high throughput analytical method is necessary to detect the signal of isolated molecules. To analyze, the PCR product must be maintained and stable in porous agarose beads. Thus, forward primer is conjugated to the agarose matrix using Schiff-base reaction to prevent diffusion. Such primer conjugation is also described in our method.

2. Materials

2.1. Conjugation of PCR Primer to Agarose

1. Ultralow gelling temperature agarose (Sigma A2576).
2. 40 nmole 5′ end modified lambda DNA forward primer 5′-NH_2-$(C18)_2$-TAAGCACGAACTCAGCCAGAACGA-3′ is lyophilized in a vacuum centrifuge and stored at −20°C.
3. Conjugation buffer: 0.15 M NaCl, 0.1 M $NaHCO_3$, pH 8.5. Store at room temperature.
4. Sodium periodate ($NaIO_4$) and sodium cyanoborohydride ($NaCNBH_3$). $NaIO_4$ is dissolved in ultrapure water at 0.128 M,

and $NaCNBH_3$ in conjugation buffer at 0.1 M. Both of them should be freshly made before addition.

5. TAE buffer.
6. 2 mL PE tube.
7. Ultrapure water.

2.2. Agarose Droplet Generation and PCR

1. Microfluidic glass chip with 200 μm (w) × 100 μm (d) cross-channel droplet generator. The inner surface of the channel is hydrophobically treated as follows: After device bonding, microchannels were rinsed with isopropanol, acetone, piranha solution (H_2SO_4:H_2O_2, 3:1), and deionized water respectively and dried with nitrogen gas. Channels were hydrophobically treated with a 0.1% solution of octadecyltrichlorosilane in dry toluene for 5 min. The treated channels were rinsed using dry toluene, isopropanol, and deionized water respectively.
2. Oil sheath: Mix 4.0 g of 5225C, 3.0 g of 749 FLUID FORMULATION AID (Dow Corning) and 3.0 g of Silicone oil AR 20 (Sigma-Aldrich) together by stirring for 15 min, followed by 15 min ultrasonic degas.
3. Syringe pumps (for pumping aqueous and oil fluids into the chip).
4. PEEK tubing.
5. Fluorescence Microscope.
6. Ultralow gelling temperature agarose (Sigma A2576).
7. Lambda DNA as PCR template (Fermentas, Canada) is dissolved in ultrapure water at 0.1 pM and stored at 4°C.
8. Reverse primer 5′-CAAGCTTTGCCACACCACGGTATT-3′ for lambda DNA PCR is dissolved in ultrapure water at 10 μM and stored at 4°C.
9. 2.5 units/μL Hot-start Blend Taq plus DNA polymerase, 10× polymerase buffer, and 2.5 mM dNTPs solution (TOYOBO, Japan).
10. Agarose-conjugated forward primer (final product in Subheading 3.1).
11. Peltier thermocycler.
12. 0.6 mL PCR tubes.
13. 1.5 mL PE tube.
14. SYBR Green I dye.
15. 1× PBS buffer.
16. Solvents for oil removal: acetone, isopropanol, deionized water.
17. FACS Flow Cytometer: e.g., BD FACSAria™ II Flow Cytometer.

3. Methods

3.1. Conjugate PCR Primer to Agarose

1. Weigh out 0.01 g (0.8 μM) ultralow gelling temperature agarose into a 2.0 mL flat bottom PE tube. Add 125 μL of ultrapure water into the tube and mix it on a vortex mixer briefly to distribute the agarose powder in the water. Incubate the tube in a dry bath at 70°C for 10 min. Mix the solution to make sure that all the agarose powder is melted and the mix becomes transparent.
2. Add and mix 125 μL of 0.128 M $NaIO_4$ solution to the melted agarose. Then immediately put the tube in refrigerator at 4°C and wait at least 1 h until the agarose is thoroughly solidified to gel.
3. Carefully cut off the bottom of the tube to take out the agarose gel. Put the gel into another 2.0 mL flat bottom PE tube and wash in 1.5 mL of water four times for 30 min each to remove the remaining $NaIO_4$. During the washing process, the gel should be kept at a low temperature to retain its mechanical strength; otherwise it may be broken or dissolved to some extent. Centrifuge the tube at 1,700 rcf for 3 min at 4°C and discard all the fluid.
4. Dissolve 40 nmole 5′ end modified lambda DNA forward primer with 240 μL of conjugation buffer. Incubate the $NaIO_4$ oxidized agarose at 70°C for 10 min (no longer). This heating temperature and time must be controlled strictly as a longer incubation time at high temperature can cause the remaining $NaIO_4$ to break the chain of agarose molecule. When the agarose gel is melted, add 240 μL of the primer buffer solution and 10 μL of 0.1 M $NaCNBH_3$ buffer solution. Mix and allow reaction to continue at 37°C overnight while stirring.
5. After reaction, keep the tube at 4°C for at least 1 h until the agarose is thoroughly solidified to gel. Cut off both top and bottom of the tube to allow electrical connection to an electrophoresis chamber. Fix the tube using adhesive tape in a suitable orientation to an electrophoresis chamber containing 1× TAE buffer and apply electrophoresis to the gel for 30 min at 100 V to remove free primer.
6. Wash the gel with water and store at 4°C. The volume of final gel product is approximately 500 μL containing 2% agarose and 10 μM primer.

3.2. Agarose Droplet PCR

1. Place the microfluidic chip (see Fig. 1) on the stage of an inverted microscope (see Note 1).
2. Connect syringes for aqueous and oil phase fluids to chip using PEEK tubing and syringe pumps.

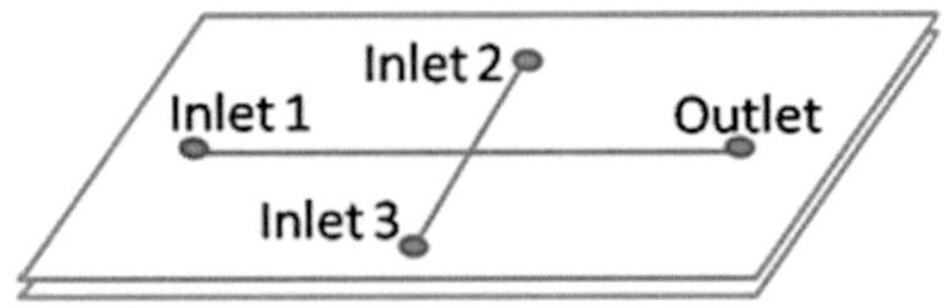

Fig. 1. Schematic of the microfluidic glass chip with a cross-channel as used in this method. *Inlet* 1 is for agarose PCR sample and *inlets* 2 and 3 are for oil sheath flow. Use a plastic three-port valve to divide oil fluid to two equal streams infusing into inlets 2 and 3. The generated emulsion is directly infused into a 0.6 mL PCR tube from the outlet.

3. Load syringes with heated oil and ultrapure water (at 65°C) and inject into the chip to pre-heat. Both flow rates are set to 1.0 mL/h. Meanwhile, prepare single-molecule agarose PCR reagents according to Table 1 (see Note 2).
4. Load prepared agarose PCR sample in the aqueous sample syringe slowly and carefully to avoid forming bubbles, which can damage the stability of fluid in chip. Set aqueous flow rate to 0.4 mL/h and oil flow rate to 4.0 mL/h.
5. Monitor the droplet generation process in the chip under the microscope. When PCR mix replaces the pre-heated water, a small quantity of air evolves between the two fluids and can be observed in the aqueous microchannel. When the PCR mix passes through the cross junction (after the air has passed through) and becomes stable, wait 10 s to let the preceding air and droplets run to waste before collecting sample agarose emulsion into a 0.6 mL PCR tube (see Note 3).
6. Stop collecting when all the agarose PCR sample is injected and keep the collected emulsion at 4°C. Repeat step 2 to wash the aqueous sample syringe and channel with 65°C water for 20 min and prepare the next sample.
7. After the last sample emulsion is generated and cooled to 4°C, wait 45 min until it is solidified. Transport sample emulsions into 0.6 mL PCR tubes at 90–100 μL volume each for PCR amplification. The thermal cycling conditions are as follows: 94°C for 3 min (initial denaturation), 25 cycles of 94°C for 30 s, 56°C for 30 s, and 72°C for 30 s, followed by a single final extension for 5 min at 72°C in a Peltier thermal cycler.
8. Keep the amplified emulsions at 4°C for 1 h to let droplets solidify into agarose beads. Combine each sample into one 1.5 mL PE tube.
9. Remove oil from emulsion by sequentially washing using acetone, isopropanol, and then deionized water in this order. Add 1 mL of each solvent to the emulsion in the tube and put the tube on a vortex mixer for 30 s. Separate beads by

Table 1
Agarose droplet PCR reaction compositions

	Forward primer	Reverse primer	dNTPs	Polymerase buffer	DNA polymerase	4% agarose	H_2O	Template
Blank	20	5	10	12.5	5	52.5	20	–
A	20	5	10	12.5	5	52.5	18.96	1.04
B	20	5	10	12.5	5	52.5	16.55	3.45
C	20	5	10	12.5	5	52.5	9.65	10.35

The unit of all volumes is μL. The template concentration of sample Blank, A, B, C is 0, 0.15, 0.5, 1.5 copy per droplet (65 μm diameter)

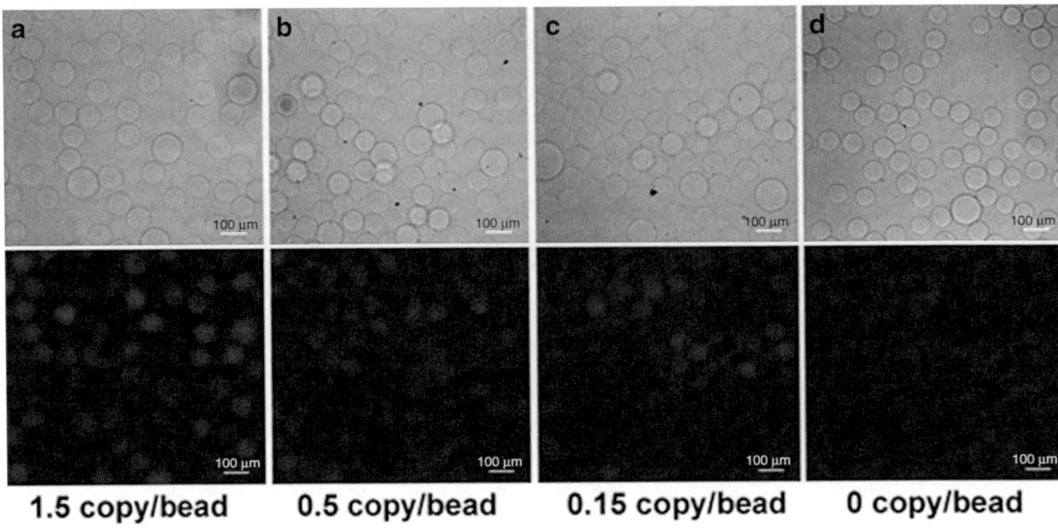

Fig. 2. Results of single copy agarose droplet PCR. Fluorescence microscope images of agarose beads after amplification from template concentration of (**a**) 1.5 copy/bead, (**b**) 0.5 copy/bead, (**c**) 0.15 copy/bead, and (**d**) 0 copy/bead.

centrifuging at 4°C for 4 min (30 rcf for organic solvent and 5,400 rcf for aqueous solvent) and throw the upper solution away (see Note 4).

10. Wrap the tubes with aluminum foil and add 250 µL of PBS buffer and 5 µL 20× SYBR Green I dye each. Incubate at 4°C for 1 h on a rotator to let the dye stain the agarose microbeads containing amplicons.
11. Examine each sample using a fluorescence microscope (see Note 3). Agarose beads containing amplicons will emit intense green fluorescence under blue light excitation. In contrast, beads with no DNA template are low in fluorescence. An example of the results produced is shown in Fig. 2.
12. Dilute the agarose beads of each sample into 1.0 mL volume with deionized water. Perform FACS analysis using flow cytometer with 100 µm nozzle. Figure 3 shows examples of the signals for each sample.

4. Notes

1. The environment temperature must be controlled strictly above 25°C during the droplet generation process. This prevents the agarose sample from curdling into small gel pieces that will severely affect the uniformity of droplet size. A domestic heater may be used to keep a warm environment around the sample and microfluidic chip if the room temperature is below 25°C.

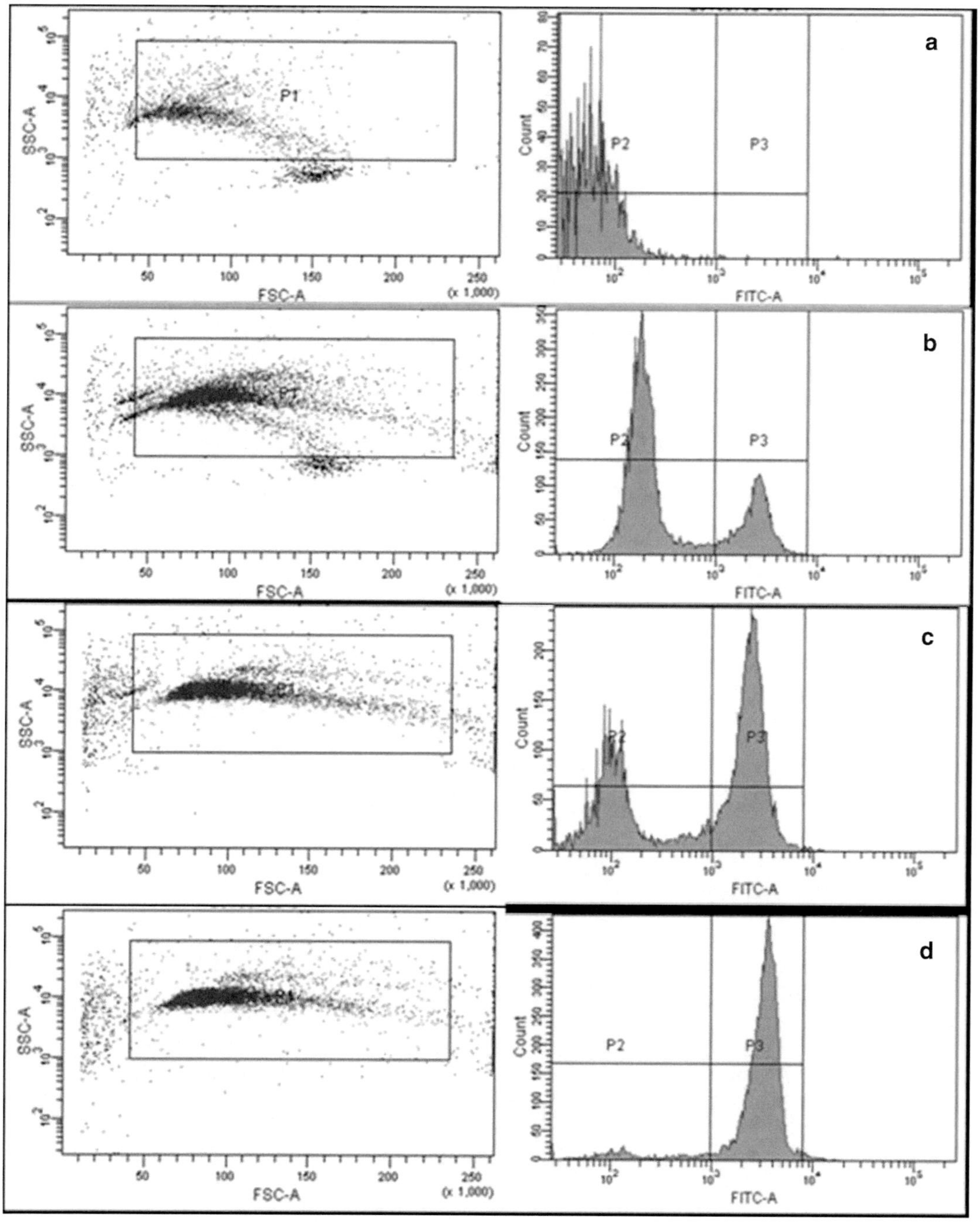

Fig. 3. FACS results of single copy agarose droplet PCR from template concentration of (**a**) 0 copy/bead, (**b**) 0.24 copy/bead, (**c**) 0.8 copy/bead, and (**d**) 2.4 copy/bead.

2. Agarose PCR reagents must be mixed sufficiently before loaded into the syringe. As 4% agarose is thicker than other solutions, it should be heated to 80°C when mixing into other reagents excluding Taq polymerase and template. After these reagents are mixed and cooled to 65°C, Taq polymerase and template are added and distributed to the whole solution by pipetting thoroughly. Bubbles created during mixing must be removed using centrifugation.
3. Try to keep the droplet generation system still to maintain uniformity of droplet size. If agarose beads larger than 100 μm are observed under the fluorescence microscope (step 11, Subheading 3.2), agarose microbeads must pass through a filter membrane with 80–100 μm grid size before performing FACS analysis.
4. Before the removing oil procedure, make sure the gel beads are solidified thoroughly. A temperature of 4°C for at least 1 h is necessary to maintain high mechanic strength of the gel beads.

Acknowledgments

This work was supported by National Scientific Foundation of China (20805038, 20620130427) and National Basic Research Program of China (2007CB935603, 2010CB732402).

References

1. Dressman D, Yan H, Traverso G, Kinzler KW, Vogelstein B (2003) Proc Natl Acad Sci USA 100:8817
2. Kelly BT, Baret JC, Taly V, Griffiths AD (2007) Chem Commun 1773.
3. Leamon JH, Link DR, Egholm M, Rothberg JM (2006) Nat Methods 3:541
4. Li M, Chen WD, Papadopoulos N, Goodman SN, Bjerregaard NC, Laurberg S, Levin B, Juhl H, Arber N, Moinova H, Durkee K, Schmidt K, He YP, Diehl F, Velculescu VE, Zhou SB, Diaz LA, Kinzler KW, Markowitz SD, Vogelstein B (2009) Nat Biotechnol 27:858
5. Li M, Diehl F, Dressman D, Vogelstein B, Kinzler KW (2006) Nat Methods 3:95
6. Wheeler DA, Srinivasan M, Egholm M et al (2008) Nature 452:872
7. Shendure J, Porreca GJ, Reppas NB, Lin XX, McCutcheon JP, Rosenbaum AM, Wang MD, Zhang K, Mitra RD, Church GM (2005) Science 309:1728
8. Nakano M, Komatsu J, Matsuura S, Takashima K, Katsura S, Mizuno A (2003) J Biotechnol 102:117

Chapter 27

Integrated Fluidic Circuits (IFCs) for Digital PCR

Ramesh Ramakrishnan, Jian Qin, Robert C. Jones, and L. Suzanne Weaver

Abstract

The Fluidigm Digital Array IFC is a nanofluidic biochip where digital PCR reactions can be performed with isolated individual DNA template molecules. This chip is part of a family of integrated fluidic circuits (IFC) and contains a network of fluid lines, NanoFlex™ valves and chambers. NanoFlex™ valves are made of an elastomeric material that deflects under pressure to create a tight seal and are used to regulate the flow of liquids in the IFC. Digital Arrays have enabled a different approach to digital PCR, by partitioning DNA molecules instead of diluting them. Single DNA molecules are randomly distributed into nanoliter volume reaction chambers and then PCR amplified in the presence of a fluorophore-containing probe. Positive fluorescent signal indicates the presence of a DNA molecule in a reaction chamber, while negative chambers are blank. IFC technology enables the delivery of very precise volumes of solutions in a simple, fast procedure, utilizing a minimum of sample and assay reagents. The development of the IFC technology and the Digital Array chip has revolutionized the field of biology, and has been utilized in gene copy number studies, absolute quantitation (molecule counting) of genomic DNA and cDNA, rare mutation detection, and digital haplotyping.

Key words: Integrated fluidic circuits, Digital arrays, Digital PCR

1. Introduction

The detection of single DNA molecules has been made possible by the use of what is known as a digital polymerase chain reaction (digital PCR) approach. Conventionally, the process of digital PCR has involved sequential limiting dilutions of target DNA followed by PCR amplification (1–3). In theory, when the target DNA is at limiting dilution levels, each of the dilutions contain a single copy of the target DNA template or less. PCR is then conducted with a positive amplification signal only observed if a DNA molecule is present.

The use of microfluidics has enabled a different approach to digital PCR by partitioning DNA molecules instead of diluting them (4–6). In this context, the original sample is effectively split into multiple partitions prior to PCR. The concentration of any

Gareth Jenkins and Colin D. Mansfield (eds.), *Microfluidic Diagnostics: Methods and Protocols*, Methods in Molecular Biology, vol. 949, DOI 10.1007/978-1-62703-134-9_27,

sequence in a DNA sample (copies/μl) can be calculated using the numbers of positive chambers that contain at least one copy of that sequence.

The combination of digital PCR and the Digital Array IFC has been used to detect differences in gene copy numbers (6), for absolute quantitation (molecule counting) of genomic DNA and cDNA (7), and digital haplotyping (8). Digital Array IFCs (Figs. 1 and 2) have also been shown to be between 50- and 200-fold more sensitive in detecting the Abl-T315I mutation in messenger RNA than conventional techniques (5).

This chapter focuses on the ability of the Digital Array IFC to identify and differentiate subtle copy number differences in two different DNA samples, obtained from Coriell Cell Repositories (Camden, NJ), which have previously been shown to have either four copies (NA19221) or five copies (NA19205) of the MRGPRX1 gene (9).

In particular, the protocol describes Specific Target Amplification (STA). We have previously shown that the use of Specific Target Amplification enables the separation of linked copies of a target gene (6). This ensures that multiple copies of a gene, which are closely linked on the same chromosome, are counted separately on the Digital Array IFC. STA is a simple PCR reaction with primers for both the reference gene and the gene of interest,

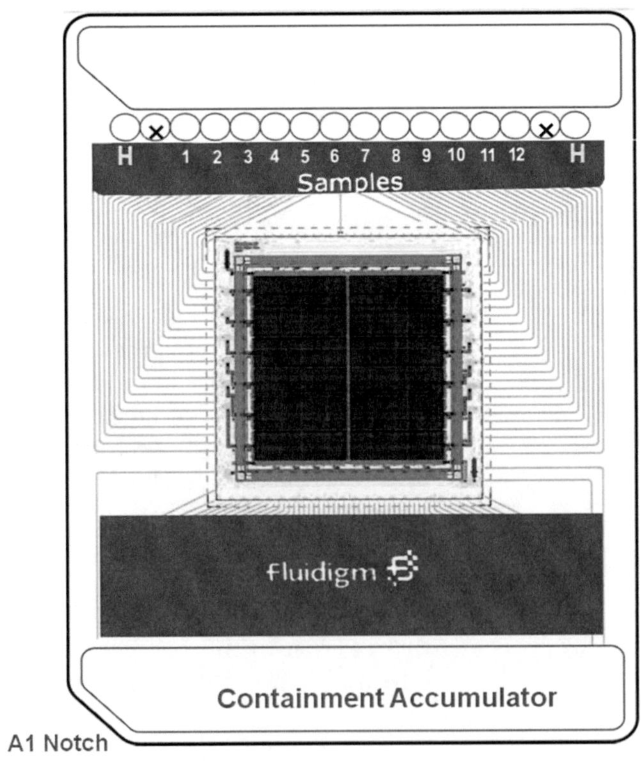

Fig. 1. A schematic of a digital array IFC.

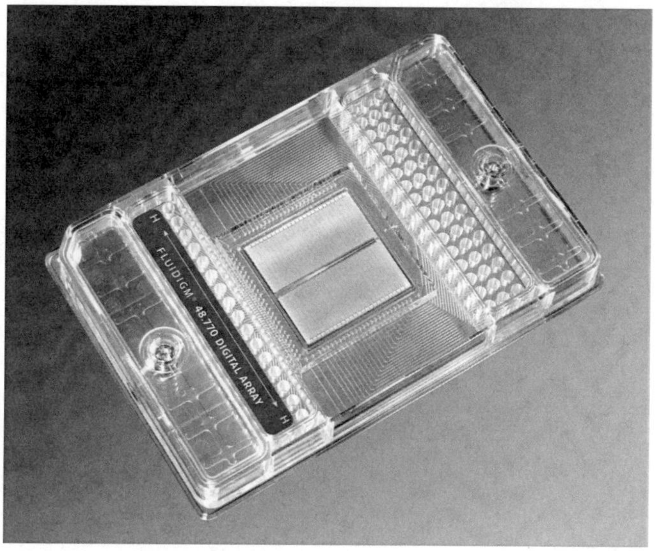

Fig. 2. A digital array IFC chip.

typically performed for a limited number of thermal cycles. The use of limited thermal cycles enables us to avoid having to compensate for different amplification efficiencies between the two sets of primers.

The operating mechanism is described as follows. Fluidigm's 48.770 Digital Array IFC has 48 individual panels with 770 reaction chamber per panel for digital PCR. Each inlet has 4 μl of sample-assay mix pipetted into it. The mixture fills the panel with the help of the IFC controller, which pressurizes the inlet with air, forcing the liquid into the array. The material used to manufacture the chips is permeable, enabling the air in the chip to be displaced by fluid. Each reaction chamber is approximately 0.8 nL in volume, and they are isolated from each other by pressurized valves that are controlled by an on-board pressurized accumulator. These valves partition the sample into hundreds of digital reactions (Figs. 1 and 2).

2. Materials

2.1. Equipment

1. PIPET-LITE LTS Pipettes from Rainin (#L-2PIPET-LITE LTS Pipet 0.1–2 μl; L-20PIPET-LITE LTS Pipet 2–20 μl; L-200PIPET-LITE LTS Pipet 20–200 μl; L-1000PIPET-LITE LTS Pipet 100–1000 μl).
2. PCR filtered pipette tips from Rainin (#RT-L10F, RT-L200F, RT-L100F).

3. Newly opened sample preparation tubes (0.5 and 1.7 mL).
4. Tube strips with attached bubble cap strips (VWR 82006-606).
5. Two biocontainment hoods to prevent DNA contamination of lab and samples.
6. Microfuge and vortex for mixing preps.
7. Standard Thermocycler; e.g., GeneAmp PCR 9700 thermal cycler (Applied Biosystems).
8. BioMark™ System (Fluidigm Corp., real-time thermocycler + reader).
9. IFC Controller (Fluidigm Corp.) (Fluidigm's IFC Controllers are designed to prime and load Digital Array™ IFCs).
10. Digital Array™ IFCs (Fluidigm Corp.)

2.2. DNA and Reagents

1. DNA samples NA19221 and NA19205 (Coriell Cell Repositories, Camden, NJ, USA).
2. 2× ABI TaqMan® PreAmp Master Mix (Applied Biosystems (Foster City, CA, USA), PN 4391128).
3. 2× ABI TaqMan® Gene Expression Master Mix (Applied Biosystems, PN 4369016).
4. 20× RNase P (VIC) Assay Mix (Applied Biosystems, PN 4316844) (Primer concentrations of this particular commercial assay are only 1/6 of the normal 20× assays).
5. MRGPRX1 Primer and probe sequences:

   Forward 5′-TTAAGCTTCATCAGTATCCCCCA

   Reverse 5′-CAAAGTAGGAAAACATCATCACAGGA

   Probe 5′ FAM-ACCATCTCTAAAATCCT-MGB

   100-nmol scale desalted primers were ordered from Integrated DNA Technologies (Coralville, IA, USA) and 6,000-mol scale HPLC-purified probe from Applied Biosystems (Foster City, CA, USA). 20× Assay mix contains 18 μM primers and 4 μM probe.
6. 20× GE Sample Loading Reagent (Fluidigm PN 85000735).
7. Deionized DNA-free, DNase-free, RNase-free water (Fluidigm PN W3330).

3. Methods

3.1. Specific Target Amplification (STA)

1. Combine the following for each 5 μl STA reaction in a tube of a tube strip.

   (a) 2.5 μl 2× TaqMan PreAmp Master Mix (Applied Biosystems).

(b) 1.25 μl assay mix for both RNase P and MRGPRX1 genes (6× RNase P (VIC) and 1× MRGPRX1. The concentration of all four primers is 900 nM in this assay mix and 225 nM in the final STA reaction).

(c) 1 μl 60 ng/μl genomic DNA.

(d) 0.25 μl Deionized DNA-free, DNase-free, RNase-free water.

2. STA is performed on a GeneAmp PCR 9700 thermal cycler. Thermocycling conditions are 95°C, 10 min hot start and five cycles of 95°C for 15 s and 60°C for 2 min.
3. 20 μl Deionized DNA-free, DNase-free, RNase-free water is added to each 5 μl reaction after the STA is completed.

3.2. Digital PCR on the BioMark™ system

1. Dilute the STA products 50-fold.
2. Combine the following for each 10 μl PCR mix for one panel of a Digital Array chip in a 0.5 mL tube:

(a) 5 μl 2× Gene Expression Master Mix (Applied Biosystems).

(b) 0.5 μl 20× RNase P (VIC) Assay Mix.

(c) 0.5 μl 20× MRGPRX1 (FAM) Assay Mix.

(d) 0.5 μl 20× GE Sample Loading Reagent.

(e) 3.5 μl diluted STA product.

3.3. Priming the Digital Array Chip (See Notes 1 and 2)

1. Add about 200 μl control line fluid to the containment accumulator.
2. Place the chip in the IFC controller.
3. Select the "115 Chip Prime" script and press "Start."
4. Once the chip has finished priming, inspect it under a top down microscope to verify that the interface valves are closed and that the containment line is primed.

3.4. Loading the PCR Mixes

1. Pipette 10 μl sample PCR mix into each of the middle top 12 wells of the carrier.
2. Pipette 10 μl water into each of the two hydration (H) wells of the carrier.
3. Place the chip in the IFC controller.
4. Select the "115× Load" Script, press "Start."
5. Once the chip has finished loading inspect the chip under a top down microscope for obvious loading defects.

3.5. Thermal Cycling on the BioMark Real-Time PCR System

1. Remove the blue IHS protector from the back of the chip.
2. Place the chip in the loading position of the BioMark instrument.

3. Select "Start" and "Load Chip"; the chip will be loaded into the instrument.
4. Select the HD location where you would like to store the experiment data.
5. Select "Application Type: Digital PCR."
6. Click "Select probes manually" and select "Two Probes."
7. Select probes "FAM_MGB" and "VIC-MGB."
8. Select protocol "Default-10 min-HotStart.pcl." Thermocycling conditions include a 95°C, 10 min hot start followed by 40 cycles of two-step PCR: 15 s at 95°C for denaturing and 1 min at 60°C for annealing and extension.
9. Select "Start Run."

3.6. Data Analysis (See Notes 3 and 4)

1. Launch the BioMark Digital PCR Analysis Software.
2. Click "Open a Chip Run" and double-click a chip run .bml file to open it in the software.
3. Click "Sample and Detector Setup" in the Chip Explorer pane.
4. Highlight the wells to name.
5. Click "Editor" in the Sample and Detector Setup pane.
6. Choose "Sample Type" from the drop-down menu in the Editor. Type a name for the sample.
7. Choose "Detector Type" from the drop-down menu in the Editor. Type a name for the detector.
8. Click "Update" to see the changes reflected in the highlighted wells.
9. Click "Panel Summary" in the Chip Explorer pane.
10. Click "Analyze" to process data.
11. Click "Panel Summary" or "Panel Details."
12. Choose a view from the drop-down menu: Results Table, Image View, and Heat Map View.
13. Save data and export as a .cvs file. The file can be opened in Microsoft Excel or Word.

    The ratio of the numbers of the MRGPRX1 molecules to the RNase P molecules and its 95% confidence interval for each panel can be found in the export file. The combined ratio and its 95% confidence interval from multiple panels can be calculated as described (10) (see Note 5).

4. Notes

1. Digital Array IFC (12.765) Layout and Pipetting Sequence (Fig. 1): Add 10 μl of deionized water to "H" wells and 10 μl of sample mixture to wells 1–12. Do not use "X" wells.
2. While priming the Digital Array IFC, take care not to drip control line fluid into or onto the chip.
3. The results from a representative run can be seen in Figs. 3 and 4.
4. While samples are read in digital mode (positive versus negative chambers), it is also possible to obtain real-time PCR curves from the same data (Fig. 5).
5. Typical results from such a run are shown in Fig. 6.

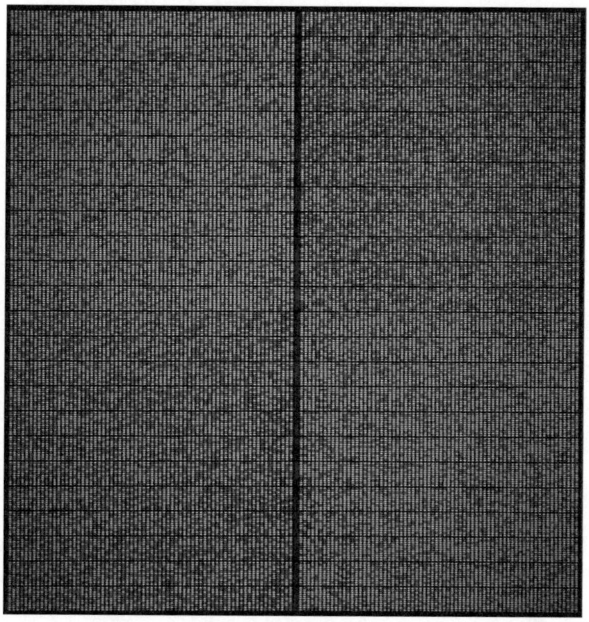

Fig. 3. A FAM-labeled image of a chip run comparing a 4-copy sample (*left*) with a 5-copy sample (*right*).

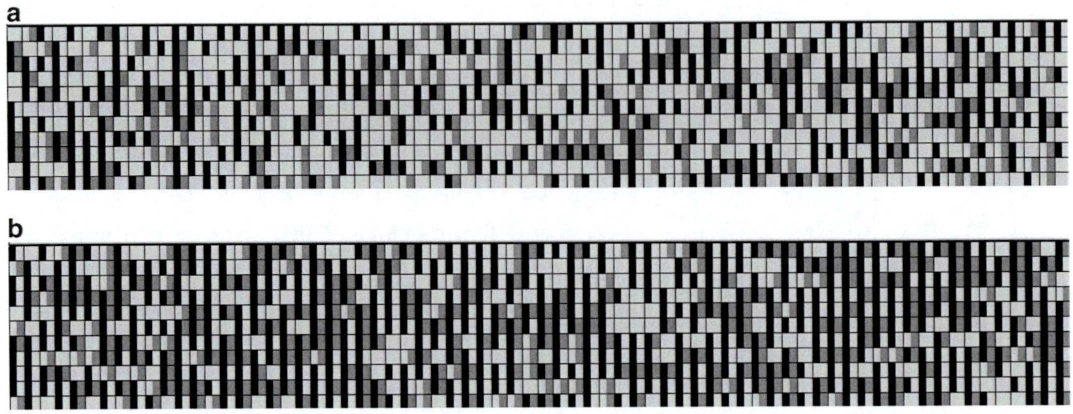

Fig. 4. A grayscale image of two panels (770 chambers each) from a chip run where light gray indicates a negative, dark gray indicates a positive hit for the VIC assay (reference), and black indicates a positive hit for the FAM assay (test), within a given chamber. Panel (**a**) shows the 4-copy sample, while panel (**b**) shows the 5-copy sample.

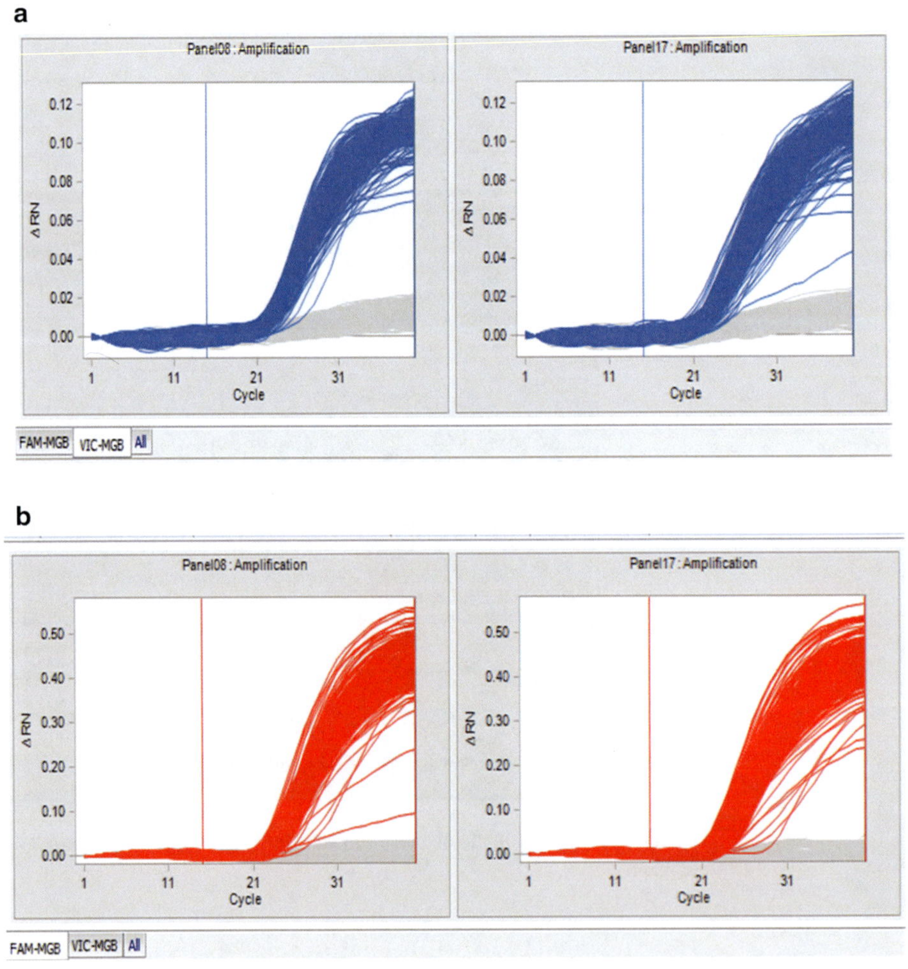

Fig. 5. Realtime PCR curves from a typical run. *Panel* (**a**) shows FAM-labeled assays, while *panel* (**b**) shows VIC-labeled results.

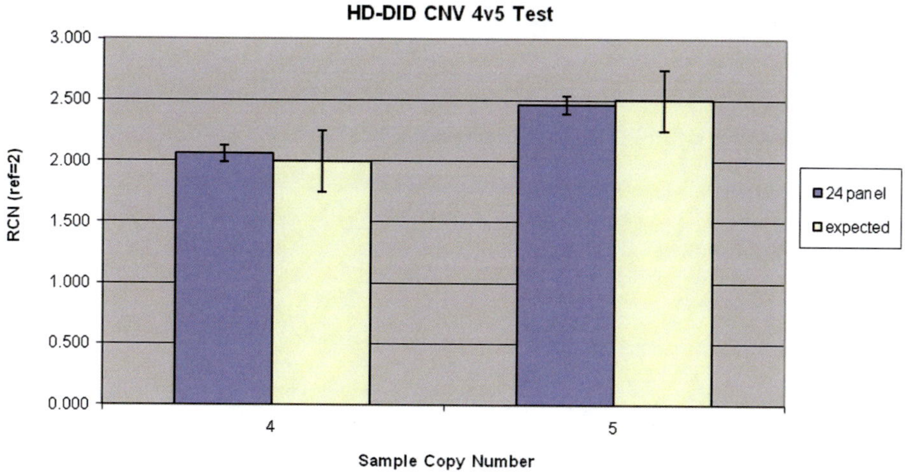

RCN Calculation

sample	24 panel	Rlow	rhigh
4	2.061	2.00	2.13
5	2.464	2.39	2.54

Fig. 6. Typical results obtained from a Digital Array IFC. RCN = relative copy number (test gene/reference gene).

References

1. Simmonds P, Balfe P, Peutherer JF, Ludlam CA, Bishop JO, Brown AJ (1990) Human immunodeficiency virus-infected individuals contain provirus in small numbers of peripheral mononuclear cells and at low copy numbers. J Virol 64:864–872
2. Sykes PJ, Neoh SH, Brisco MJ, Hughes E, Condon J, Morley AA (1992) Quantitation of targets for PCR by use of limiting dilution. Biotechniques 13:444–449
3. Vogelstein B, Kinzler KW (1999) Digital PCR. Proc Natl Acad Sci U S A 96:9236–9241
4. Ottesen EA, Hong JW, Quake SR, Leadbetter JR (2006) Microfluidic digital PCR enables multigene analysis of individual environmental bacteria. Science 314(5804):1464–1467
5. Oehler VG, Qin J, Ramakrishnan R, Facer G, Ananthnarayan S, Cummings C et al (2008) Absolute quantitative detection of ABL tyrosine kinase domain point mutations in chronic myeloid leukemia using a novel nanofluidic platform and mutation-specific PCR. Leukemia 23:396–399
6. Qin J, Jones RC, Ramakrishnan R (2008) Studying copy number variations using a nanofluidic platform. Nucleic Acids Res 36:e116
7. Bhat S, Herrmann J, Armishaw P, Corbisier P, Emslie KR (2009) Single molecule detection in nanofluidic Digital Array IFC enables accurate measurement of DNA copy number. Anal Bioanal Chem 394(2):457–467
8. Menzel S, Qin J, Vasavda N, Thein SL, Ramakrishnan R (2010) Experimental generation of SNP haplotype signatures in patients with sickle cell anaemia. PLoS One 5(9):e13004. doi:10.1371/journal.pone.0013004
9. Hosono N, Kubo M, Tsuchiya Y, Sato H, Kitamoto T, Saito S, Ohnishi Y, Nakamura Y (2008) Multiplex PCR-based real-time invader assay (mPCR-RETINA): a novel SNP-based method for detecting allelic asymmetries within copy number variation regions. Hum Mutat 29(1):182–189
10. Dube S, Qin J, Ramakrishnan R (2008) Mathematical analysis of copy number variation in a DNA sample using digital PCR on a nanofluidic device. PLoS One 3(8):e2876

Chapter 28

microFIND® Approach to Fluorescent in Situ Hybridization (FISH)

Andrea Zanardi, Emanuele Barborini, and Roberta Carbone

Abstract

FISH technology has gained increasing attention in the management of cancer disease, either for predictive or prognostic indications. Molecular cytogenetics has greatly improved diagnostic capability of classical cytogenetics analysis of metaphase-based chromosome for the identification of genetic aberrations. The availability of a large number of fluorescent probes, each specific for different genetic lesions, together with a robust protocol for interphase FISH, provide the pathologist with the essential tools for an accurate evaluation of patient's disease. Hemato-oncological and many of the solid tumors have been comprehensively characterized by peculiar genetic defects and are now routinely evaluated by interphase FISH. Despite the reliability of the method, which has undergone only minor changes since the 1970s, FISH assay is still hampered by reagents cost, preventing its adoption in large-scale oncological screening. In this chapter we describe a major improvement of interphase FISH assay for cytological samples through the description of the miniaturized device microFIND® that offers, besides reduction of cost per assay, a completely novel vision to the FISH technology, thanks to the perspective of full automation of FISH assay using a dedicated robotic platform for microFIND® handling, (not presently described in the chapter).

Key words: FISH, Screening, Miniaturization, Hemato-oncological cancer, microFIND®, Automation

1. Introduction

FISH technology (1, 2) relies on a series of manual procedures on biological samples that must be performed by skilled technicians in order to obtain high quality results for unambiguous data interpretations.

Among analytical assays FISH has not been substantially changed since its introduction (3): some procedures for automation have been proposed and implemented (4) to facilitate manual analysis; however the assay is costly and suffers the main drawback of not being easily scalable for disease screenings.

Gareth Jenkins and Colin D. Mansfield (eds.), *Microfluidic Diagnostics: Methods and Protocols*, Methods in Molecular Biology, vol. 949, DOI 10.1007/978-1-62703-134-9_28,

We have recently discovered a peculiar property of a nanotech coating (5, 6) that strongly immobilizes living and fixed, not otherwise spontaneously adhering, cells and by appropriate integration with a microfluidic element we have developed a diagnostic device, microFIND®, for FISH analysis on cytological samples.

The device is composed of a glass slide with a nanotech coating, and a polymeric microfluidic pad (7) with a microchannel and two wells for loading and aspiration of reagents. The microfluidic pad is coupled to the slide in order to have the slide constituting the bottom surface of the microchannel. Cells are loaded in the microchannel and allowed to adhere on bottom surface (of the coated slide). Due to the strong adhesion on the nanotech coating the analytical protocol can be carried out without loosening of cells, despite the significant shear stress caused by moving fluids inside the microchannel.

Following this approach, FISH can be performed in a miniaturized setting, strongly reducing sample requirement, reagents and probe volume. With dedicated automation, microFIND® provides an affordable and scalable method for genetic based-cancer screenings.

In this method we describe the device and a protocol for FISH on bone marrow and peripheral blood cells for the evaluation of a common type of leukemia, the chronic lymphocytic leukemia (CLL).

Specifically two genetic abnormalities have been tested for:

1. Deletion of locus D13S25 at 13q14.3 region that is one of the most common abnormalities in CLL. These patients have the best prognosis and most of them do not need therapy.
2. Trisomy 12, an additional chromosome 12, which is a relatively frequent finding, occurring in 20–25% of patients, indicating an intermediate prognosis.

2. Materials

2.1. microFIND® Device

1. microFIND® device is composed of two parts: a glass slide partially functionalized with a nanotech coating and a polymeric pad with microfluidic circuitry constituted by a straight channel acting as small volume reaction chamber and by two access wells (Tethis S.p.A., Milano, Italy). A larger diameter well within a cone-shaped pit is dedicated to sample/reagent loading (IN well), while smaller diameter well is dedicated to aspiration (OUT well).
2. Pads are contained in a plastic "assembler" designed to both store the pad in a proper way and prepare the final device (see Subheading 3 and Fig. 1).

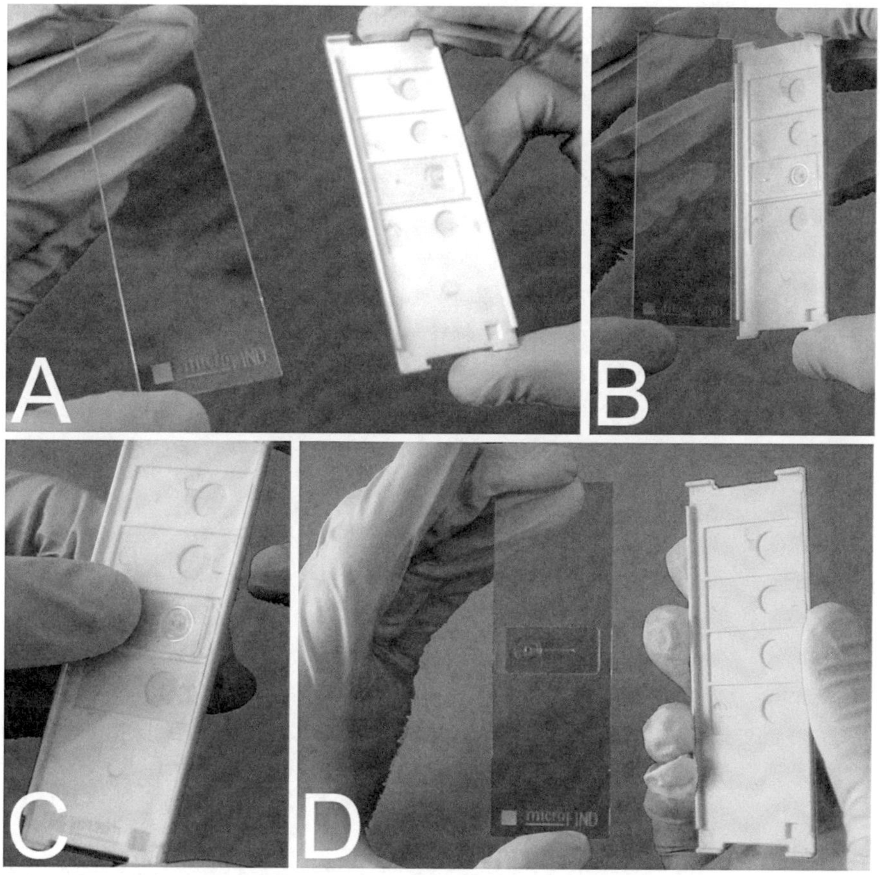

Fig. 1. Device assembly. (**a**) one coated slide and one assembler with a pad; (**b**) example of fitting the slide in the assembler; (**c**) example of pressing the slide in correspondence of the pad; (**d**) example of the device ready to be used.

3. Sample and reagents loading: standard pipettes with 50 and 10 μl tips.
4. Sample aspiration: a 10 ml syringe equipped with dedicated adaptor composed of 10 cm tube and sucker (Tethis S.p.A) is required for liquid removal.
5. As alternative for liquid aspiration the use of a motorized syringe equipped with a 20 ml syringe with a 1.5 mm tube is also possible (KDs 120, KD Scientific, Inc., Holliston, MA, USA).

2.2. Samples

1. Peripheral blood (PB) or bone marrow (BM) from a donor, in EDTA or Heparin stored at room temperature for not more than 48 h or at 4°C for not more than 4 days. Samples were kindly provided by Policlinico Ospedale Maggiore Ca' Granda Milan, Italy, according to hospital ethical regulations (see Note 1).

2.3. Reagents and Chemicals

1. Purified water (e.g., Milli-Q, Millipore, MA, USA) (see Note 2).
2. 1× DPBS Dulbecco phosphate buffer saline (Euroclone S.p.A. Italy).
3. TURK or Trypan blue for cell count (Sigma).
4. Mineral oil.
5. Mounting Media plus DNA counterstain ready to use (Vectashield, Vector Laboratories, Inc. CA, USA).
6. Immersion oil, type FF (Cargille Laboratories, USA).
7. FISH Probes (CE, IVD), LSI D13S25 Spectrum Orange™, LSI 13q34 Spectrum Green™; CEP®12 Spectrum Orange™ (Abbott Molecular, Illinois, USA).

2.4. Reagents Preparation (See Note 3)

1. Red blood lysis buffer (RBL) (for red blood cell lysis): 0.15 M NH_4Cl, 9.93 mM $KHCO_3$, and 0.13 mM EDTA (ethylenediaminetetraacetic acid) in purified H_2O.
2. Carnoy's fixative: mix 3 parts of methanol 100% and 1 part of Glacial acetic acid (Carlo Erba, Italy).
3. 2× SSC: mix 100 μl of 20× SSC (saline–sodium citrate buffer, Sigma) and 900 μl of purified H_2O.
4. Digestion buffer: mix 10 μl of 1 N HCl, 5 μl of 10% pepsin (Sigma, powder diluted in water at pH≤7.0), 989.5 μl of purified H_2O and bring it to 37°C (see Note 4).
5. Post fixative: mix 26 μl of 37% formaldehyde, 100 μl of 10× DPBS, 50 μl of $MgCl_2$ 1 M (powder diluted in water) in 824 μl of purified H_2O.
6. Denaturing solution: 70% Ultrapure formamide in 2× SSC: mix 700 μl of ultrapure formamide, 100 μl of 20× SSC (pH 5.3) and 200 μl of purified H_2O. Adjust pH to 7.0–8.0.
7. Ethanol solutions: dilute 100% ethanol in purified H_2O (700 μl in 300 μl for 70%, 850 μl in 150 μl for 85%).
8. Post-hybridization wash solution 1 (0.3% NP40 (Tergitol, Sigma) in 0.4× SSC): mix 155 ml of NP-40, 10 ml of 20× SSC in 488.5 ml of purified H_2O.
9. Post-hybridization wash solution 2 (0.1% NP40 in 2× SSC): mix 0.5 ml of NP-40, 50 ml of 20× SSC in 449.5 ml of purified H_2O.

2.5. Materials and Instruments

1. Coplin Jars (AGAR Scientific Ltd, Essex UK).
2. Glass Coverslips 18×18 mm (Zeuss super, Zeuss Padova, ITALY).
3. Microtiter Pipette and Tips P2, P10, P200, P1000 (Gilson, Inc., Middleton, WI, USA).

4. Slide hybridizer 37°C–75°C (Thermobrite™, Abbott Laboratories, Illinois, USA).
5. Cell count chamber (i.e., Burker chamber) (Brand GmbH, Germany).

2.6. Microscope Equipment and Filters

1. Microscope. For image acquisition and analysis, signals were evaluated using an Olympus BX61 microscope. A 100 W mercury lamp with a life maximum of about 200 h is recommended (see Note 5).
2. Camera. F-View II camera (Olympus, PA, USA).
3. Objectives. The objective has a profound influence on the brightness, resolution, and general quality of the image. A 40× objective in conjunction with 10× eyepieces is suitable both for scanning and for routine FISH analysis. When a higher magnification is needed, satisfactory results can be obtained with a 100× achromatic oil immersion objective (UPlanSApo 100× immersion objective, N.A. 1.40).
4. Immersion oil. Immersion oil used with immersion objectives should be formulated for low auto-fluorescence and specifically for fluorescence microscopy use (see Note 6).
5. Filters. Olympus U-MNIBA3 FITC for Spectrum Green probes, U-MWIG3 Cy3 for Spectrum Orange probes, U-MNUA2 DAPI.
6. Images were acquired using Cell^A software (Olympus) and elaborated with Adobe Photoshop 7.0 program (Adobe System Incorporated).

3. Methods

In this protocol we report the development of FISH starting from living not-fixed or fixed cells from standard cytogenetic pellet on microFIND® using specific probes for CLL, a common hemato-oncological disease, detecting chromosome 12 polisomy and the region of chromosome 13 13q14.3.

3.1. Device Assembly

1. microFIND® is operative once the microfluidic pad is attached to the coated side of the slide. For this purpose, the plastic package element hosting microfluidic pads (see Fig. 1a) acts as assembling tool.
2. The device is prepared by:
   (a) Holding the coated slide in your fingers from the short side and fitting it to the plastic element containing the microfluidic pad (Fig. 1a, b).

(b) Pressing gently the slide in order to let pad adhere to coated side (easily identified by microFIND® logo) (Fig. 1c).
(c) Removing the assembler (Fig. 1d).
(d) Pads will stay in contact with slides and will be lifted off from the assembler (see Note 7).

3. Once pad and slide are attached each other, microFIND® is ready for use.

3.2. Cellular Sample Preparation

microFIND® is designed to perform FISH directly on cytogenetic pellet, prepared according to standard procedures, or on living not-fixed cells, prepared as a cellular suspension in isotonic buffer from patient blood. Below we report the protocol for the preparation of living not-fixed cells from BM and PB: cytogenetic pellet can be prepared with available standard protocols.

3.2.1. Cellular Suspension of Living Not-Fixed Cells from BM and PB

1. Store BM and PB patient samples either in EDTA or Heparin at room temperature or 4°C until usage (up to 4 days from collection if stored at 4°C).
2. Protocols below are designed to process volumes from 1 ml down to 0.1 ml (respectively "Standard" or "Reduced" volume of patient samples).
3. Standard volume of patient samples:
   (a) Place 0.5 ml of BM or 1 ml of PB in 15 ml centrifuge tube.
   (b) Add RBL to a volume of 10 ml.
   (c) Keep 5 min at 4°C.
   (d) Centrifuge at 425 rcf (relative centrifugal force) for 5 min.
   (e) Discard supernatant.
   (f) Resuspend again cells in 10 ml of RBL buffer.
   (g) Centrifuge at 600 rcf for 5 min.
   (h) Resuspend cells in 1 ml of 1× DPBS.
   (i) Transfer to 1.5 ml tube.
   (j) Wash twice in 1× DPBS by centrifuging in a microfuge at 500 rcf for 5 min.
   (k) Before loading in microFIND® resuspend cells in 1× DPBS at a concentration between 8,000 and 15,000 cells/μl (optimal concentration of cell suspension is 10,000 cells/μl), after cell count with a cell count chamber (e.g., Burker chamber).
4. Reduced volume of patient samples: For samples from PB a volume down to 0.5 ml can be processed following the same

procedure as above. For samples derived from BM a volume down to 0.1 ml can be processed with the following protocol:

(a) Place 0.1 ml of BM in 1.5 ml centrifuge tube.
(b) Add RBL to a volume of 1.5 ml.
(c) Keep 5 min at 4°C.
(d) Centrifuge at 425 rcf for 5 min in a microfuge.
(e) Discard supernatant.
(f) Resuspend again cells in 1.5 ml of RBL buffer.
(g) Centrifuge at 600 rcf for 5 min in a microfuge.
(h) Resuspend cells in 1 ml of 1× DPBS.
(i) Wash twice in 1× DPBS by centrifuging in microfuge at 500 rcf for 5 min.
(j) Before loading in microFIND® resuspend cells in 1× DPBS at a concentration between 8,000 and 15,000 cells/μl (optimal concentration of cell suspension is 10,000 cells/μl) (see Note 8).

3.3. Cell Adhesion Protocol and Fixation

3.3.1. Adhesion and Fixation of Cells Suspension from Living Not-Fixed Cells

1. Before use, pre-incubate microFIND® at 37°C (using a hot plate, or a slide hybridizer), for at least 2 min.
2. Keeping microFIND® on the hot plate/hybridizer, pipette 1.5 μl of cell suspension into the IN well and slightly aspirate from the OUT well on other side of the microchannel with the manual syringe or with the motorized syringe (at 25 ml/h) for a maximum of 2 s, to favor the loading of cell suspension into the microchannel (see Note 9) (Fig. 2).
3. Incubate for 4 min at 37°C.
4. Do not empty the microchannel: pipette 20 μl of fixative Carnoy in the bottom of the IN well with a 10 μl tip.
5. Wait for 2 min at 37 ± 1°C, then add another 30 μl of fixative into the IN well and aspirate completely from the OUT well.
6. Cells are now tightly immobilized in microFIND®.
7. Observe the microchannel area under a Bright Field microscope. Examine the number of nuclei per field under a 10× objective. A minimum of approximately 150 cells per field is recommended for optimum assay results (see Fig. 3 for an example of optimal cellular density inside the microchannel).
8. In case of a suboptimal cell concentration per field it is suggested to repeat the procedure of adhesion and fixation on a new microFIND® (see Note 10); moreover in case of cellular suspension with low cellular number, it is recommended to decrease the volume of the suspension buffer down to few μl (not more than 2–3 μl) and load 1.5 μl on microFIND® to recover most of the cells.

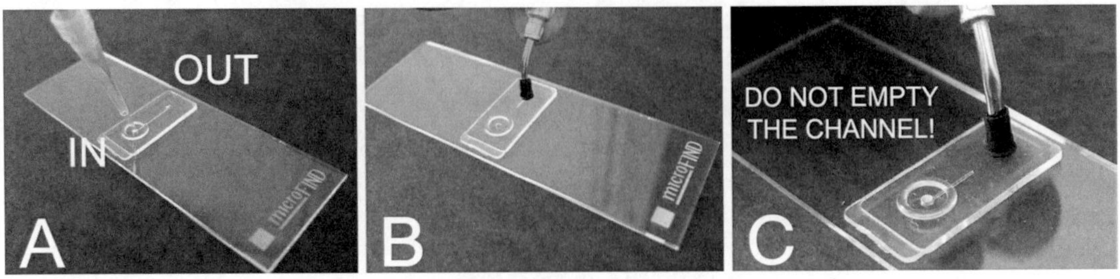

Fig. 2. Cell loading. (**a**) Example of loading of cell suspension in the "IN" well; (**b**) example of brief aspiration from the "OUT" well to allow cell entry inside the microchannel (only for living not-fixed samples) and subsequent adhesion; (**c**) example of wrong operation of excessive aspiration causing the loss of cells.

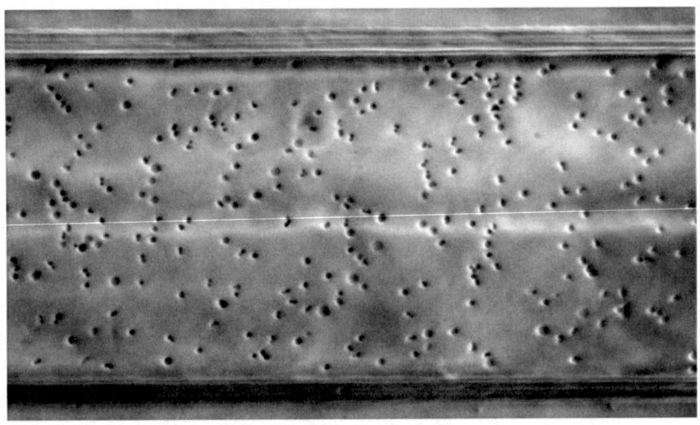

Fig. 3. Bright field image at 10× of a microFIND® to show optimal cell concentration after adhesion.

9. At this step microFIND® can be processed for FISH analysis or stored at room temperature in dry conditions, protected from light, for 24 h.
10. If patient cells must be reanalyzed it is recommended to prepare cytogenetic pellet from the remaining sample, keep it at –20°C, and perform the analysis on a new microFIND® following the protocol for standard cytogenetic pellets.

3.3.2. Loading and Adhesion of Standard Cytogenetic Pellet

1. Prepare a cytogenetic pellet according to the standard procedure.
2. Before cell loading in microFIND® it is recommended to evaluate cell density, to avoid cell clogging in the microchannel: after pellet preparation take 1 μl of cell suspension and spot on a standard glass slide; let the fixative evaporate and observe cell density under a Bright Field microscope at 10× magnification (see Fig. 4a as example of low cell density and Fig. 4b as optimal cell density).

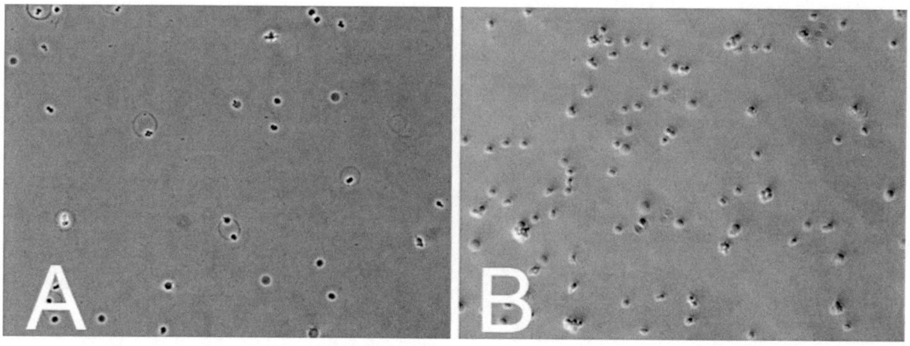

Fig. 4. Estimation of cytogenetic pellet concentration. (**a**) Bright field image at ×10 of low cell density; (**b**) bright field image at ×10 of optimal cell density.

3. If cell density is optimal, proceed with loading, otherwise dilute or concentrate the pellet with Carnoy fixative accordingly.
4. Load 1.5 μl of cytogenetic pellet into the IN well, at room temperature, and wait 2 min (see Note 11).
5. Put microFIND® at 37°C and wait until the complete evaporation of Carnoy's fixative (~5–10 min) (see Note 12).
6. Cells are now tightly immobilized in microFIND®.
7. Observe the microchannel area under a Bright Field microscope. Examine the number of nuclei per field under 10× objective. A minimum of approximately 150 cells per field is recommended for optimum assay results (see Fig. 3).
8. If an optimum cell density has not been reached repeat steps 1 and 2 and check again by microscopy. This operation can be repeated several times until an optimal cell concentration is reached in the microchannel (see Note 13).
9. At this step microFIND® can be processed for FISH analysis or stored at room temperature in dry conditions, protected from light for 24 h.

3.4. FISH Protocol

After loading of microFIND® with cell suspension (either from living not-fixed cells or from standard cytogenetic pellet) all the reagents for FISH analysis (see Subheading 3.4.1 below) except the probe will be used according to the following scheme, from A to D (see Fig. 5).

1. Reagent loading (Fig. 5a): Dispense a drop of reagent in the IN well with a pipette (volume is indicated below).
2. Partial aspiration from the OUT well (Fig. 5b): to let the reagent flow inside the microchannel: aspirate with the manual syringe as shown in the figure (or with motorized syringe at 50 ml/h) until the reagent drop on the IN well is reduced to a minimum (see Fig. 5b).

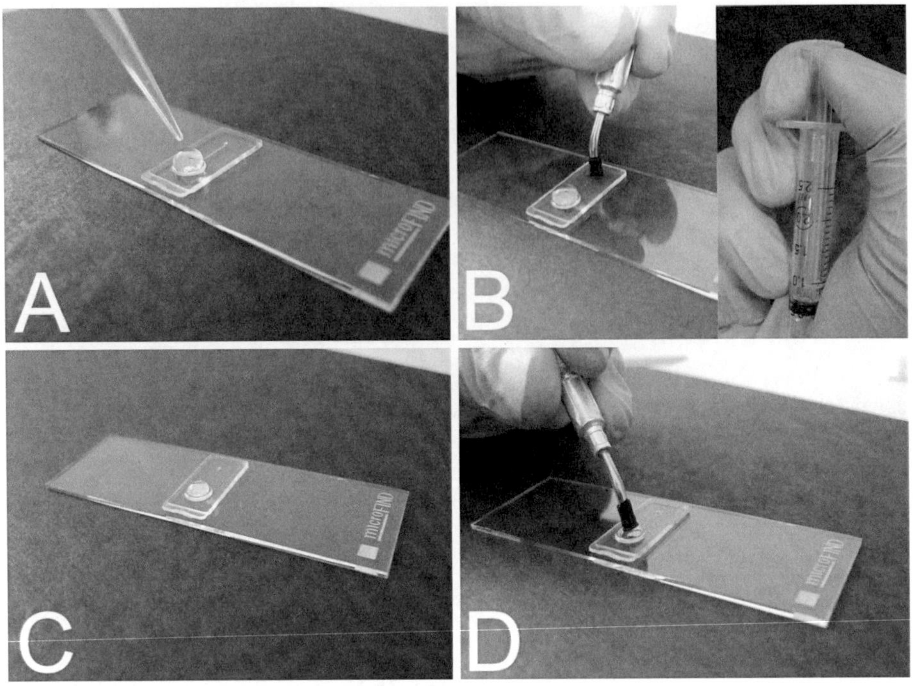

Fig. 5. FISH protocol. (**a**) Dispensing of reagents into the "IN" well; (**b**) aspirating from the "OUT" well; (**c**) aspiration from the OUT well and incubation for the indicated time; (**d**) removing of reagent excess from the "IN" well.

3. Incubation (Fig. 5c): follow the indicated Subheading 3.4.1 for incubation conditions.
4. Removal of reagent excess (Fig. 5d) (see Note 14): aspirate the residual drop over the IN well, touching it with the suction cup of the syringe.

At the end of step D, if required according to the protocol, start again from step A with the next reagent.

3.4.1. Reagents Volume, Incubation Time, and Temperature Conditions

FISH protocol consists of a series of reagent passes through the device as listed below. The hybridization with the fluorescent probe can be performed in two ways using the protocol for "Denaturation" or "Co-denaturation" (see Note 15). If using "Denaturation" for samples of living not-fixed cells, follow the complete protocol described below. For samples prepared from standard cytogenetic pellet start this protocol from step 5.

1. All the reagents must be used according to the above scheme (from A to D).
2. 50 μl of 2× SSC: 15 min at 37°C.
3. 50 μl of Digestion buffer: 10 min at 37°C.
4. 50 μl of 1× DPBS: 5 min at room temperature.
5. 50 μl of Post-fixative: 5 min at room temperature.

6. 50 µl of 1× DPBS: 5 min at room temperature.
7. 30 µl of EtOH 70%: 1 min at room temperature.
8. 30 µl of EtOH 85%: 1 min at room temperature.
9. 30 µl of EtOH 100%: 1 min at room temperature.
10. 50 µl of Denaturing solution: 3 min at 75°C (see Note 16).
11. 30 µl of EtOH 70%: 1 min at room temperature.
12. 30 µl of EtOH 85%: 1 min at room temperature.
13. 30 µl of EtOH 100%: 1 min at room temperature.

3.5. Probe Preparation

1. Most of commercial probes are provided already denatured, but if the probe is not denatured you will need to denature it separately following manufacturer instruction.
2. In the case of co-denaturation, the probe will be denatured with the sample DNA directly in the microchannel.
3. The amount of probe mix needed for each channel is 0.5 µl.

3.5.1. Hybridization with Denaturation (For Both Living or Fixed Cells)

1. After the last incubation with EtOH 100% (step 9, Subheading 3.4.1) empty the microchannel completely from the OUT well.
2. Place microFIND® at 60°C for 2 min to dry completely.
3. Load the probe by aspirating 0.5 µl of hybridization mix with a P2 pipette; to correctly fill the channel, direct the tip toward microchannel entry (see Fig. 6).
4. Wait a few seconds to let the probe fill the microchannel completely.

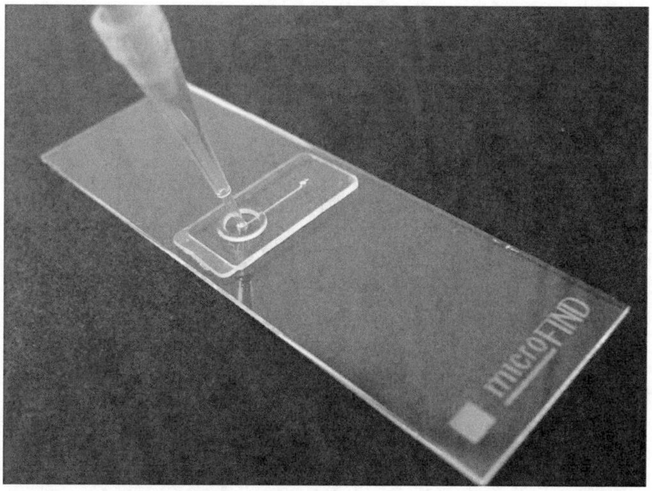

Fig. 6. Probe loading. Example of how to load the probe on the bottom of the "IN" well, by directing the tip toward the channel; the probe will fill the channel by capillarity.

5. Cover both wells with a drop (approximately 5–10 μl) of mineral oil to prevent evaporation of probe: use a 10 μl tip to pipette oil on the bottom of the IN well and on top of the OUT well.
6. Hybridize the samples on a slide hybridizer (ThermoBrite™ or similar) in controlled humidity conditions, however not above 75–80% relative humidity, following the probe manufacturer's suggested time and temperature. If you don't have a slide hybridizer, you can reach these conditions incubating microFIND® in a closed box with a vessel filled with saturated NaCl solution on the bottom, in a dry incubator (Note 17).

3.5.2. Hybridization with Co-denaturation

Co-denaturation is suitable and reproducible for CEP (Chromosome Enumeration Probes) LSI (Locus Specific Identifier) probes Although some times with lower signal to noise ratio.

For Samples Derived from Living Not-Fixed Cells

1. After the first ethanol scale (step 9, Subheading 3.4.1), empty the microchannel completely from the OUT well.
2. Place microFIND® at 60°C for 2 min to dry completely.
3. Load the probe, pre-heated at 75°C, by aspirating 0.5 μl of hybridization mix using a P2 pipette: to correctly fill the channel direct the tip toward microchannel entry (see Fig. 6).
4. Wait a few seconds to let the probe fill the microchannel completely.
5. Cover both wells with a drop (approximately 5–10 μl) of mineral oil to prevent evaporation of probe.
6. Co-denature the samples at 75°C for maximum 5 min.
7. Hybridize the samples following probe manufacturer's suggested time and temperature.

For Standard Cytogenetic Pellet

After the adhesion procedure (Subheading 3.3.2) leave microFIND® at 37°C for at least 5 min to ensure complete fixative evaporation, then follow the previous Subheading "For Samples Derived from Living Not-Fixed Cells" starting from step 3 (Load the probe), then proceed with co-denaturation, by directly loading the probe in microFIND®, as described above, incubating at 75°C for maximum 1.5 min, then hybridizing the samples following probe manufacturer's suggested time and temperature.

3.6. Post-hybridization Washes

1. After hybridization, remove the pad grabbing it by its flap and pulling it away from the slide (Fig. 7).
2. Perform post-hybridization washes by dipping the slide in coplin jars containing: Wash Solution 1 (pre-heated at 73°C) for 30 s and Wash Solution 2 at room temperature for 30 s.
3. Air dry the slide and mount coverslip with 4 μl of Vectashield per microchannel; the slide is then ready for the analysis.

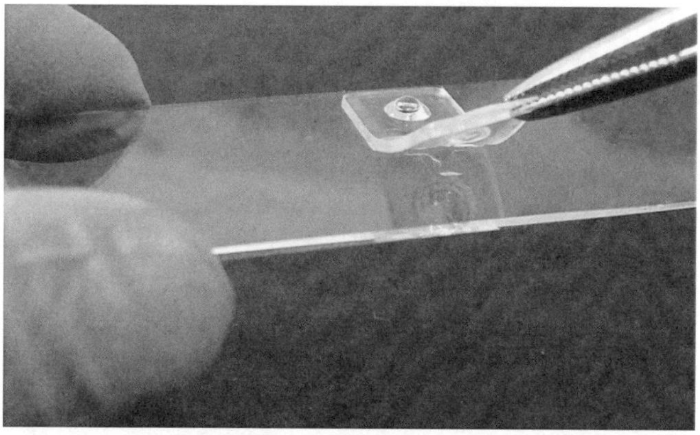

Fig. 7. Pad removal. After the overnight hybridization pad can be removed by pulling its flap.

Fig. 8. Example of staining of cells in microFIND®. (**a**) Low magnification images of DAPI staining (×4) showing the suggested analysis area of the microchannel; (**b**) ×20 magnification of a selected area of the microchannel for FISH analysis; *white dots* represent staining of chromosome 12 CEP probe.

3.7. Fluorescence Microscopy Analysis of FISH

1. Slides are analyzed by fluorescence microscopy.
2. Cells that are suitable for analysis and evaluation are located in the microchannel area (see Fig. 8a): do not analyze cells that are located in the correspondence of the IN and OUT wells if they appear as aggregates.
3. First the sample is analyzed at low magnifications (10×) to identify, through DAPI inspection, cells of interest (see Fig. 8b for a selected area of analysis).
4. A detailed evaluation of 200 cells at high magnification (100×) (scored per probe/per patient) showed that an average of 97.8% of cells were clearly positive for both probes.
5. Examples of FISH experiments on four CLL patients are reported in Fig. 9 (100× magnification). Three patients resulted

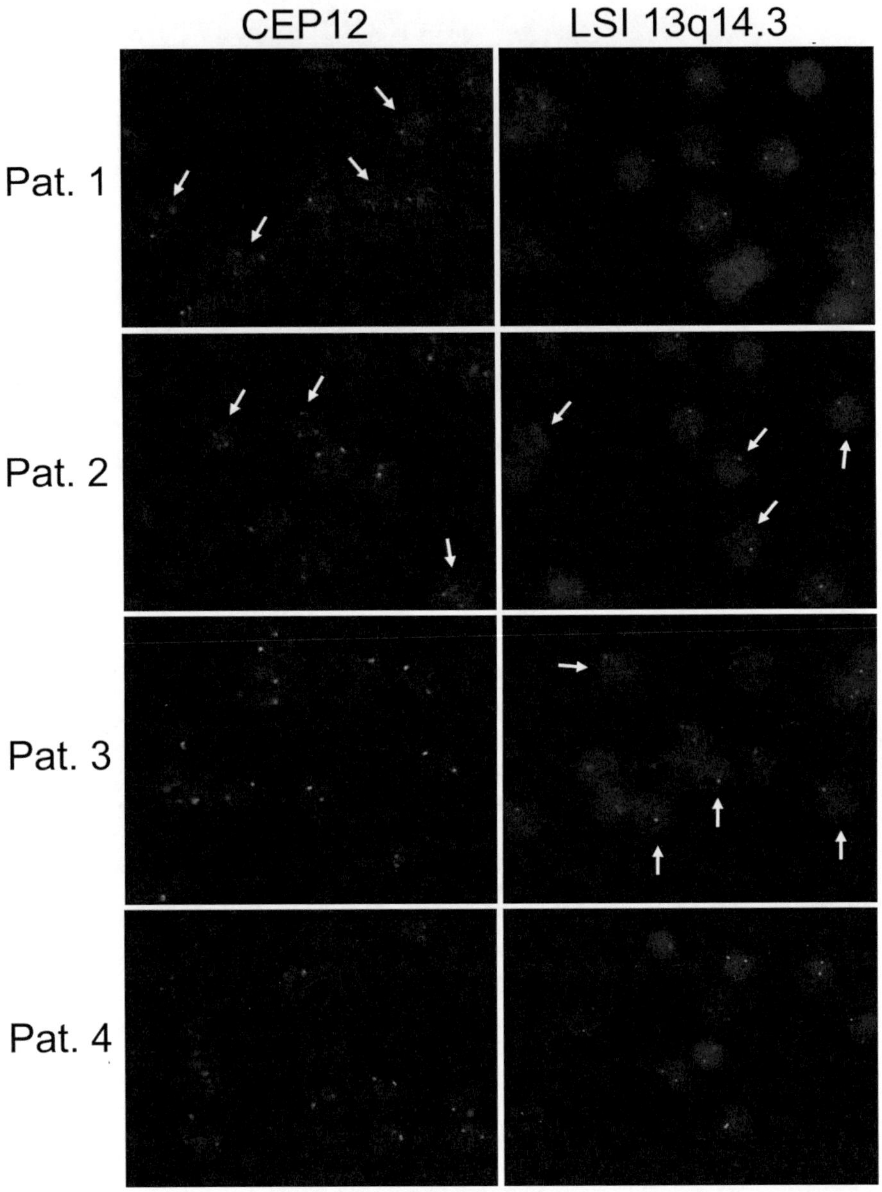

Fig. 9. FISH results on four CLL patients with CEP12 and LSI 13q14.3 probes. Patient (Pat.) 1 shows chromosome 12 trisomy (*see white arrows* indicating three *dots* per nucleus), Pat. 2 shows both chromosome 12 trisomy and LSI 13q14.3 deletion (*see white arrows* indicating single dot per nucleus), Pat. 3 shows LSI 13q14.3 deletion, and Pat. 4 shows a normal genetic asset.

positive and one negative for the genetic lesions tested; white arrows in Patient (Pat.) 1, 2, 3 indicate the presence of a specific lesion in the nucleus.

6. The percentage of positivity for each single lesion is reported in Table 1.

Table 1
Percentages of positivity of specific probes, with respect to the observed genetic aberrations

N=Negative P=Positive	+12		del 13q14.3	
Pat. 1	P	63%	N	100%
Pat. 2	P	53.5%	P	48.5%
Pat. 3	N	100%	P	76%
Pat. 4	N	99%	N	98.5%

+12 refers to trisomy of chromosome 12, del 13q14.3 to deletion of the specific region of chromosome 13

4. Notes

1. Patient samples consist of white blood cells isolated from total blood and resuspended in saline buffer. Different methods of purification are suitable, such as density gradient centrifugation (Ficoll method), hypotonic lysis buffer, or RBL buffer treatment, as described in this protocol.
2. Purified H_2O must have a controlled pH of 7.0. Uncontrolled pH in the water source can lead to degradation of fluorophores during post-hybridization washes.
3. It is recommended to prepare fresh solutions each time.
4. Prepare digestion buffer 15 min before use and keep at 37°C until use. It has been tested that Pepsin reaches optimal membrane digestion activity after a pre-heating time.
5. Routine microscope cleaning and periodic preventive maintenance by the manufacturer's technical representative are recommended; check the usage time of the microscopy mercury lamp: old lamps can result in a worse signal/noise ratio and therefore can produce inappropriate results. Use the indicated filters according to the information on probe's fluorophores; signal brightness can be improved using the correct filters.
6. Immersion oil contributes to two characteristics of the image viewed through the microscope: finer resolution and brightness. These features are most critical under high magnification: therefore use an immersion oil with refractive index suitable to your objective and matching the mounting media employed.
7. The polymeric material of the microfluidic pad slowly releases a minimal amount of volatile compounds that could modify the properties of the slide coating. This is the reason why slides and pads have separated packages. For this reason, the assem-

bling of microFIND® must be done when ready for the FISH protocol; the device must therefore be prepared and used within 1 h. During these operations, avoid touching the slide coating since this would result in degraded performance of the device.

8. Samples for FISH can also be derived by procedure of cell purification, like IMAC procedure (Immuno-affinity metal assisted chromatography) or FACS (fluorescence activated cell sorting); for instance, this is the case for the analysis of plasma cells deriving from multiple myeloma patients. In such case the amount of selected cells is often low and the preparation of a cytogenetic pellet for standard FISH can cause dramatic cellular loss: therefore cellular suspension can be prepared directly from living cells resuspended in a small volume of isotonic buffer. It is suggested to reduce the volume in not more than 5 μl of isotonic buffer and directly load the sample on the device to be able to recover all the cells present in the sample. Cell fixation and FISH analysis will occur inside microFIND®, avoiding cellular loss.
9. After loading the cell suspension of living not-fixed cells, do not completely aspirate the liquid from the microchannel. This would dramatically impair cell adhesion. It is important that no air bubbles pass through the channel before the Carnoy fixative has been loaded.
10. microFIND® is able to efficiently immobilize living not-fixed cells in isotonic buffer with a low amount of protein: salt and water-based buffers inactivate the surface after the first incubation with the cellular suspension; therefore it is not recommended to reload the cellular suspension after the first incubation and cell immobilization.
11. For an effective adhesion, be sure that your sample has been resuspended in fresh Carnoy fixative. Perform a check for cell density (see Fig. 4): if it is too concentrated proceed with a dilution of the pellet with fresh Carnoy, otherwise if too diluted, centrifuge the pellet again and resuspend in a lower volume. Repeat the check for cell density; a proper cellular dilution is crucial for an optimal result; refer to Fig. 4b as a correct reference.
12. Carnoy's fixative evaporation time depends on environmental factors like room temperature and relative humidity mainly. Evaporation will take a longer time in a highly humid environment, and vice versa.
13. Contrary to the living cell preparation, cytogenetic pellet can be repeatedly loaded in microFIND® to reach an optimal cellular number and distribution, particularly if the sample has a low cell density. In case of multiple loading, be sure that the

cytogenetic pellet is a uniform suspension, free of clumps. This is a critical point since cell aggregation will impair the correct loading of microFIND®. Between subsequent loading Carnoy fixative must be completely evaporated. It is possible to speed up this step by placing microFIND® at 75°C for few minutes.

14. To eliminate the liquid in excess on the IN well it is sufficient to place the suction cup over the residual volume while aspirating and the drop should disappear in an instant. Be sure that during all the incubations the reagent reservoir on the IN well DOES NOT DRY OUT: in dry environmental conditions (relative humidity below 20%) check the volume of the drop and, if it decreases rapidly, add an equal amount of reagent volume.
15. Recently the protocol of probe co-denaturation has been proposed as alternative to significantly reduce the protocol duration. However, using co-denaturation with microFIND®, reproducible results have been obtained only with CEP probes, while with LSI probes the results have not always been satisfactory, suggesting that the denaturation protocol must be recommended as first choice.
16. Check the heating block temperature before starting probe denaturation.
17. Do not incubate microFIND® in incubators for cell culture: the high humidity can cause condensation on the slide and consequent detachment of the polymeric microfluidic pad, leading to probe diffusion and incorrect hybridization conditions.

Acknowledgments

We sincerely want to acknowledge the participation and assistance of Dr. Sonia Fabris and Prof. Antonino Neri of Policlinico Ospedale Maggiore Ca' Granda Milan, Italy.

References

1. Tibiletti M (2007) Interphase FISH as a new tool in tumor pathology. Cytogenet Genome Res 118:229–236
2. Mitelman F, Johansson B, Mertens F (2007) The impact of translocations and gene fusions on cancer causation. Nat Rev Cancer 7:233–245
3. Lichter P et al (1990) High resolution mapping of human chromosomes 11 by in situ hybridization with cosmid clones. Science 247:64–69
4. Côté GB (2009) Nuclear FISH: automation, analysis, and interpretation. Diagn Mol Pathol 18:119–124
5. Carbone R et al (2006) Biocompatibility of cluster-assembled nanostructured TiO_2 with primary and cancer cells. Biomaterials 27:3221–3229
6. Barborini E et al (2003) Supersonic cluster beam deposition of nanostructured titania. Eur Phys J D 24:277–282
7. Zanardi A et al (2010) Miniaturized FISH for screening of onco-hematological malignancies. Biotechniques 49:497–504

Chapter 29

An ELISA Lab-on-a-Chip (ELISA-LOC)

Avraham Rasooly, Hugh A. Bruck, and Yordan Kostov

Abstract

Laminated object manufacturing (LOM) technology using polymer sheets is an easy and affordable method for rapid prototyping of Lab-on-a-Chip (LOC) systems. It has recently been used to fabricate a miniature 96 sample ELISA lab-on-a-chip (ELISA-LOC) by integrating the washing step directly into an ELISA plate. LOM has been shown to be capable of creating complex 3D microfluidics through the assembly of a stack of polymer sheets with features generated by laser micromachining and by bonding the sheets together with adhesive. A six layer ELISA-LOC was fabricated with an acrylic (poly(methyl methacrylate) (PMMA)) core and five polycarbonate layers micromachined by a CO_2 laser with simple microfluidic features including a miniature 96-well sample plate. Immunological assays can be carried out in several configurations (1×96 wells, 2×48 wells, or 4×24 wells). The system includes three main functional elements: (1) a reagent loading fluidics module, (2) an assay and detection wells plate, and (3) a reagent removal fluidics module. The ELISA-LOC system combines several biosensing elements: (1) carbon nanotube (CNT) technology to enhance primary antibody immobilization, (2) sensitive ECL (electrochemiluminescence) detection, and (3) a charge-coupled device (CCD) detector for measuring the light signal generated by ECL. Using a sandwich ELISA assay, the system detected Staphylococcal enterotoxin B (SEB) at concentrations as low as 0.1 ng/ml, a detection level similar to that reported for conventional ELISA. ELISA-LOC can be operated by a syringe and does not require power for operation. This simple point-of-care (POC) system is useful for carrying out various immunological assays and other complex medical assays without the laboratory required for conventional ELISA, and therefore may be more useful for global healthcare delivery.

Key words: ELISA, Lamination, Charge-coupled device, Micromachining, Microfluidics, Staphylococcal enterotoxins, Enhanced chemiluminescence, Carbon nanotubes, Point-of-care-settings, Food safety

1. Introduction

Enzyme-linked immunosorbent assay (ELISA) (1, 2) is one of the most common immunological methods used for medical diagnostic and research applications to identify and detect proteins based on their binding to immobilized antibodies. In an ELISA "sandwich

Gareth Jenkins and Colin D. Mansfield (eds.), *Microfluidic Diagnostics: Methods and Protocols*, Methods in Molecular Biology, vol. 949, DOI 10.1007/978-1-62703-134-9_29, © Springer Science+Business Media, LLC 2013

assay," a primary antibody is immobilized on a solid surface of a microtiter plate in order to bind with a target antigen. After washing the unbound material, a secondary labeled antibody (with an enzyme) is bound to the same target to detect the captured antigen. The binding of the secondary antibody to the target protein is quantified by measuring the activity of an enzyme tagged for the secondary antibody (e.g., by optical colorimetric or fluorescence reaction).

1.1. The Need for a New ELISA Format

Most ELISA assays are performed in 96-well plates which require a laboratory setting with dedicated instruments, including plate washers, and optical colorimetric, densitometry or florescent detectors. Normally the ELISA assay is slow (e.g., several hours) and requires large volumes of samples and reagents. In recent years, many LOC devices have been developed with the aim of simplifying and improving the performance of immunoassays. Although the majority of LOCs still utilize optical-based immunodetection (3–16), a few use other technologies including electrochemical detection in combination with nanoparticles (17), superparamagnetic beads (18), and a microcantilever transducer combined with an impedance analyzer (19). However, ELISA is still the gold standard and the most commonly used assay mainly because of the numerous assays and reagents already developed for ELISA during the last few decades.

Shifting to a new platform requires developing new assays and reagents which may not be practical unless the new format offers a great advantage. Moreover ELISA is simple, low cost and many samples can be assayed simultaneously, while the new detection methodologies are relatively complex, expensive, and require "exotic" technologies. More importantly, few LOC technologies can simultaneously test as many samples as the "classical" ELISA. One approach for resolving the limitations of ELISA is applying new technologies for the ELISA format which may enable retaining the ability to use current assays and reagents while broadening the use of the assay, especially in settings without a laboratory (i.e., Point-of-Care). The main limiting factors of using ELISA without a laboratory are the need for a detector and a washer. Developing low cost simple washers and detectors for Point-of-Care settings may enable broader use of the ELISA format in these settings.

1.2. Lab-on-a-Chip for ELISA

We recently developed a miniaturized 96-well ELISA plate (20), requiring as little as 5 μl of sample that can be combined with a CCD camera (20–24), to carry out low-cost immunodetection without a laboratory. However, the use of this miniature plate requires manual application of the various reagents needed for the assay. To eliminate the need for manual fluid handling, we developed an ELISA lab-on-a-chip (ELISA-LOC) which integrates the washing step directly into an ELISA plate (25). ELISA-LOC combines three functional elements: (1) reagent loading fluidics, (2) assay and detection wells plate, and (3) reagent removal fluidics. So ELISA-LOC enables the performance of ELISA assays without a plate washer since it is integrated directly into the plate assay.

1.3. LOM for ELISA-LOC Fabrication

The most commonly used technology used for LOC fabrication is polydimethylsiloxane (PDMS) replication (26, 27), mainly because it is relatively low cost, easy to fabricate, and easy to seal. As a material, PDMS is desirable because it is biocompatible, has low affinity for biomolecules and cells, has relatively high permeability to gases, and has a higher transmission of visible light. Some disadvantages of PDMS are the technology is not suitable for mass production, long time periods are needed for curing which slows prototype development, PDMS swells in some organic solvents making it difficult to control feature size, the mechanical strength is low enough to make it difficult to handle, and it is more difficult to fabricate multilayer 3D-configured devices.

Another method for rapidly and inexpensively prototyping LOC devices is LOM using direct machining of polymer sheets (28–32). In LOM, the patterns and features of the device are cut in thin sheets of polymer film, such as acrylic, polycarbonate or polyester, normally with a CO_2 laser cutter, followed by assembly and bonding by successive lamination with an adhesive (30) or heat (31, 32) to produce a three-dimensional object or device.

LOM is very effective for rapid prototyping because all of the steps are relatively fast and inexpensive. The microfluidics flow pattern can be rapidly designed using a CAD program in a very short time (depending on the complexity of the design). The laser cutter functions as a printer accessible directly from the CAD program, which enables rapid design, fabrication, and product realization. Finished multilayer LOC devices can be completed and ready for testing in as little as several hours. The method is reasonably versatile in allowing the design of complex fluidic structures, with more complex flow networks requiring more layers. For simpler designs, the technique is fast enough that a single machine can produce hundreds of units per day.

One critical element of LOM is the micromachining of features using techniques such as laser machining. Laser fabrication can work in either *vector* (line cutting) or *raster* (area etching) modes. Vector mode is most useful for cutting through the entire depth of the workpiece while in raster mode the laser passes back and forth across the surface, creating progressively advancing lines in a linear pattern that "fill out" the raster image. The distance of advance for each line is normally less than the actual dot-size of the laser, so the engraved lines overlap slightly to create a continuous and uniform engraving. Laser cutters utilize a computer for controlling engraving/cutting patterns made by the laser, in a manner similar to the function of a printer or plotter, where the printing head is replaced by the laser. Graphic programs, such as CorelDraw, are used to create the design that is converted by the laser cutter device driver into instruction for the engraving/cutting operation. These instructions enable the laser cutter to control the laser "printing" path, the speed at which the beam moves across the material and the laser intensity and pulse frequency, which determine the power delivered to the engraving surface. During laser micromachining, the workpiece material is oxidized or vaporized so that the fumes and debris must be removed by ventilation from the cabinet enclosure. The fumes and debris can also potentially contaminate the lens and mirror surfaces. To avoid such contamination, the cutting site is continuously flushed with a jet of ventilation air supplied by a compressor.

1.4. CCD-Based Detectors for ELISA

Most ELISA assays utilize optical detectors to measure the binding of the secondary antibody to the immobilized antigen. In most assays, such binding is measured and quantified optically by measuring the activity of the enzyme bound to the secondary antibody (e.g., optical colorimetric or through florescence reaction). Most of the optical detectors currently used for ELISA analysis utilize photodiodes or photomultipliers. As an alternative optical detector, charge-coupled device (CCD) cameras are a relatively simple low cost instrument for optical detection and CCDs have been used with several different types of array immunosensors (33–36). The main advantage of CCD cameras for ELISA is that as spatial detectors, CCDs can be used for analyzing light from a relatively large space (e.g., from the entire surface of an ELISA plate). This is in contrast to photo diodes or photo multipliers which are "point" detectors measuring light over a small area (e.g., from a single well on an ELISA plate). To overcome the limitations of point detectors for the analysis of larger spaces, the measured surface (e.g., ELISA plate) is normally scanned. Such scanning requires mechanical movement of the plate or the use of many detectors to cover the plate surface. Since a CCD camera can visualize the entire plate at once it does not require an additional scanning system and therefore they are relatively lower in cost, easier to operate, more reliable, and readily available.

1.5. Improving Sensitivity of ELISA-LOC

The ELISA-LOC system combines several biosensing elements: (1) Carbon nanotube (CNT) technology to enhance primary antibody immobilization, (2) sensitive ECL (electrochemiluminescence) detection, and (3) a charge-coupled device (CCD) detector for measuring the light signal generated by ECL. The large surface area of the CNTs (typically 400–1,000 m^2/g for HiPco) was used to increase the amount of primary antibody on the plate, and thus improved the sensitivity of ELISA assays to Staphylococcal enterotoxin B (SEB) from 1 to 0.01 ng/ml (37). While fluorescence or colorimetric detection is widely used for ELISA detection, fluorescence detection requires an illumination source, as well as excitation and emission filters which complicate the CCD detector and increase cost. Colorimetric assays are not as sensitive, so for the ELISA-LOC a sensitive enhanced chemiluminescence (ECL) assay (21, 23) can be used for detection with a simple but sensitive CCD imager (38, 39).

1.6. Staphylococcal Enterotoxins

We used ELISA-LOC for the detection of Staphylococcal enterotoxins (SEs) which are a group of 21 (known) heat labile toxins produced by *Staphylococcus aureus*. SEs are implicated in several illnesses including food-borne diseases resulting from the consumption of a broad variety of contaminated food (e.g., dairy products, unrefrigerated meats, salads, bakery products, etc.) (40–44) causing various gastrointestinal symptoms such as vomiting, nausea, and diarrhea even at low levels of exposure (e.g., total intake of SEA of approximately 20–100 ng per person) (45). In addition, SEs have also been implicated in other diseases such as atopic eczema (46–48), rheumatoid arthritis (49, 50), and toxic shock syndrome (51), and are also recognized as potential bioweapons (52–55). Many pathogenic *S. aureus* strains contain several SE genes (56–59). SEs are traditionally assayed immunologically mainly with ELISA. Rapid detection of SEs is important for public health, especially for managing SEs outbreaks.

2. Materials

2.1. Preparation of CNT Immunosensor

1. Single-walled carbon nanotubes (CNTs) (Carbon Solutions, Inc., Riverside, CA).
2. 3:1 v/v Mixture of concentrated (98%) sulfuric acid: 70% nitric acid for CNT oxidization and shortening.
3. Poly(diallyldimethylammonium chloride) polymer (PDDA) (Sigma-Aldrich, St. Louis, MO).
4. Rabbit anti-SEB affinity purified IgG, 0.01 mg/ml in 20 mM, pH 8.0 PBS buffer (Toxin Technology, Sarasota, FL).
5. Fisher (FS-14) Sonicator (Fisher Scientific, Pittsburgh, PA).
6. Beckman mini centrifuge (Beckman, Fullerton, CA).

2.2. Design and Fabrication of ELISA-LOC Using LOM

1. Clear 0.25 mm thick polycarbonate film and 1/8 in. (3.2 mm) black acrylic (Piedmont Plastics, Beltsville, MD).
2. Adhesive: 3M 9770 adhesive transfer double-sided tape (Piedmont Plastics, Beltsville, MD).
3. Epilog Legend CO_2 65W computer controlled laser cutter (Epilog, Golden, CO).
4. Corel Draw 11 (Corel Corp. Ontario, Canada).
5. Alignment plate with two pins for aligning multiple layers.

2.3. ELISA-LOC SEB Assay and Detection

1. Staphylococcal enterotoxin B (SEB) (Toxin Technology, Sarasota, FL).
2. Phosphate buffer (PBS) (20 mM, pH 8.0).
3. Washing buffer: 20 mM PBS, pH 7.4.
4. Blocking buffer: 1% Bovine serum albumin (BSA) w/v in PBS.
5. Horseradish peroxidase (HRP)-conjugated anti-SEB IgG, 0.01 mg/ml in 20 mM, pH 8.0 PBS buffer (Toxin Technology, Sarasota, FL).
6. Immun-Star HRP Chemiluminescence Kit (Bio-Rad, Hercules, CA).
7. HRP substrate: *o*-Phenylenediamine dihydrochloride (OPD) (Sigma-Aldrich, St. Louis, MO).
8. SXVF-M7 Cooled CCD camera (Adirondack Video Astronomy, Hudson Falls, NY).
9. Tamron manual zoom CCTV 4–12 mm, f1.2 lens, (Spytown, Utopia, NY).
10. Enclosure for CCD and ELISA-LOC setup.
11. LED light box optical filters (optional—for fluorescence measurements).
12. ImageJ software; developed and distributed freely by NIH (http://rsbweb.nih.gov/ij/download.html).

3. Methods

3.1. Preparation of CNT Immunosensor

1. Carbon nanotube preparation: CNTs were prepared as described in previous work (21, 23, 37) as follows.
2. 30 mg of CNTs were first shortened and oxidized by mixing with a concentrated sulfuric acid and nitric acid mixture (3:1 v/v) and sonicating for 6 h.
3. The CNTs were then extensively washed in water (100 ml) until neutralized (pH 7.0). using a centrifuge.
4. The CNTs were then dispersed in 100 ml 1 M NaOH solution for 5 min to achieve net negative charged carboxylic acid groups.
5. They were washed again with water (100 ml).
6. CNT-Antibody functionalization: A linker molecule to the carbon nanotube was used (60) to facilitate functionalization. Polydiallyldimethylammonium chloride (PDDA) is positively

charged and the SEB antibody is negatively charged, so antibodies are electrostatically adsorbed onto the carbon nanotube. This positively charged polycation was adsorbed by dispersing the CNTs in 50 ml of 1 mg/ml PDDA in 0.5 M NaCl for 30 min, followed by centrifugation (12,000 × *g*) for 15 min and washing with 100 ml of water.

7. The CNTs were functionalized by dispersing in rabbit anti-SEB IgG in phosphate buffer solution (20 mM, pH 8.0) at a concentration of 0.01 mg/ml for 1 h at room temperature to allow adsorption of the antibody onto the CNT surface.
8. After centrifugation (15 min at 12,000 × *g*) and washing extensively with water (10 ml), the modified CNTs were stored at 4°C in pH 8.0 phosphate buffer at a concentration of approximately 1 mg/ml for no more than 2 weeks before use.

3.2. Fabrication of Laminated LOC

1. Corel Draw computer aided drawing (CAD) software was used to design the patterns and features of the device.
2. Before cutting, layers of polycarbonate (PC) and acrylic (poly(methyl methacrylate) (PMMA)) were coated with 3M 9770 adhesive transfer double-sided tape.
3. Remove approximately 1 cm of the liner from one side of the double-sided adhesive tape roll.
4. Align and attach this portion of the tape with the polymer surface to be bonded (e.g., polycarbonate film or acrylic layer).
5. With a ruler, press the tape against the polymer surface and slowly unwind the tape and press it against the acrylic surface with the ruler to prevent air bubbles.
6. Cut the tape.
7. To create a polymer surface with adhesive on both sides, repeat for the other side of the polymer.
8. The cutting and engraving was done with a CO_2 laser cutter as described in previous work (20, 22, 24), which is designed to work with thin sheets of polymer film, such as acrylic, polycarbonate, and polyester.
9. Depending on the material being cut, different power levels and speed settings have to be used to ensure a clean cut. Cutting parameters need to be determined empirically (see Note 1).
10. After cutting, the edge of the PC, PMMA and double-sided tape needs to be cleaned using a toothbrush and a needle.
11. The layers were assembled and bonded by successive lamination with adhesive (30) to produce a three-dimensional device (28–32). For LOM, the liner is removed from the adhesive on each layer. Each layer is then aligned using guide holes machined

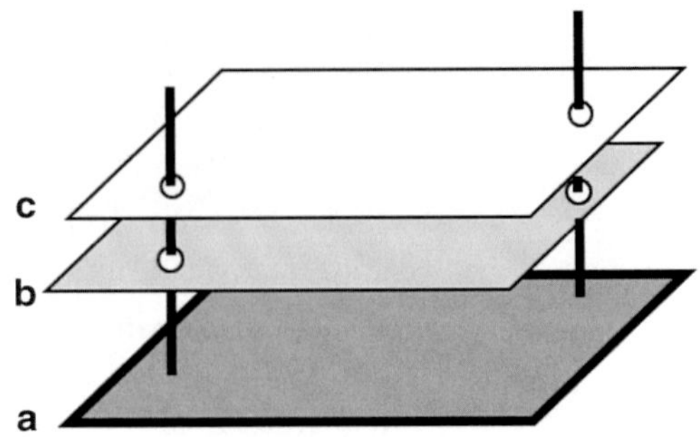

Fig. 1. Multilayer LOC assembly: An aligning plate (**a**) with two pins and two layers to be assembled (**b** and **c**).

in all layers. A simple plate with two pins is used to align the layers (Fig. 1) which are first joined together with gentle pressure.

12. The polycarbonate layers were bonded to the acrylic with double-sided pressure-sensitive adhesive transfer tape (see Notes 2–4). Since the transfer tape is bonded to the acrylic prior to machining, during machining the adhesive is removed in the area which will contact with the fluid. Thus, the sample has minimal contact with the adhesive during LOC operation.

3.3. Design and Fabrication of ELISA-LOC

Figure 2-I shows the multilayer, 3D microfluidic structure of the laminated LOC ELISA-LOC fabricated using the aforementioned LOM techniques. The device consists of (A) an interchangeable fluid delivery system (Fig. 2-I-A) which functions as the top of the plate and can be configured to perform assays in several different formats; (B) a miniature 96-well plate (Fig. 2-I-B) where the assays are carried out and detected; and (C) a fluid outlet system used to remove reagents from the wells (Fig. 2-I-C) attached to the bottom of the plate. The arrows in the figure show the direction of the flow.

The design overlay of the ELISA-LOC (Fig. 2-II) and pictures of the actual device (Fig. 2-III) show its major features, while the actual device is fabricated with black PMMA to minimize optical cross talk, Fig. 2-III shows a device fabricated with clear PMMA to show the elements of the device. To visualize the details, the device was loaded with fluorescein dye and photographed, after washing (Fig. 2-IV). The wells with the outlet holes (the bright spots in the center of the wells) and the outlet channels (the bright lines across the wells) are shown clearly in this image. The fluids are kept in the wells by surface tension. The diameter of the wells is 3 mm and the width of the channels and the hole in the bottom of the wells is approximately 0.2 mm.

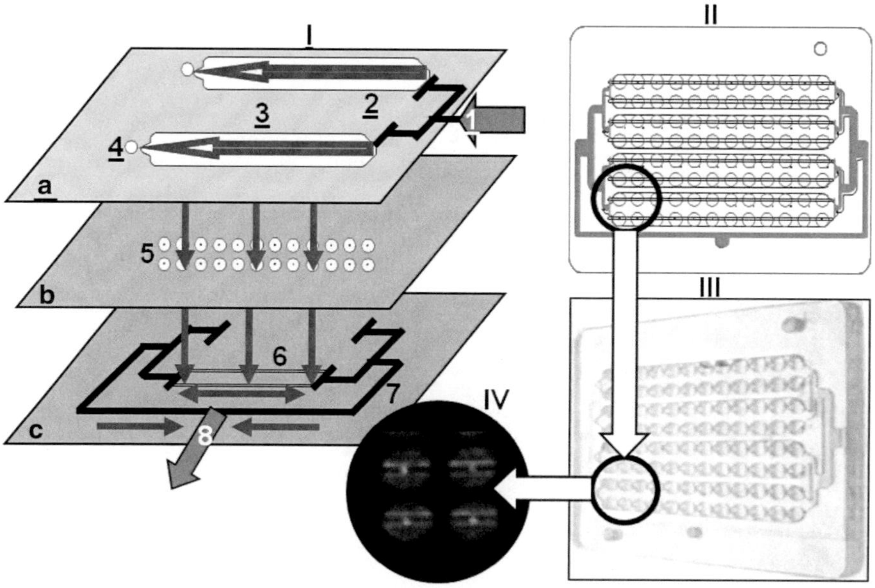

Fig. 2. Functional elements of ELISA-LOC: Diagram I shows an expanded 3D design of an ELISA-LOC, II shows the complete design overlay of the assembled 96-well device, and III shows the actual device and enlarged wells assembly is shown in IV. In the expanded diagram I, the three main layered elements of the LOC are shown: fluid delivery (I-A), which serves as the cover for the device, sample wells (I-B) for assay incubation and detection, and the outlet system (I-C) for fluid removal. The *arrows* in the figure show the direction of the flow.

3.3.1. Components of the ELISA-LOC

A. Fluid delivery system

The fluidic delivery system also forms the cover for the wells in the assay plate and is used to deliver assay reagents including HRP-labeled secondary antibodies, wash buffers, HRP substrate for enzymatic detection, and ECL reagents to generate the light from the HRP enzymatic reaction. The fluidics are arranged to create groups of wells ("nodes"), as described above, to divide the plate into different assays. A simplified fluidics with only two nodes is shown in Fig. 2-I-A. The fluid delivery system has an inlet port (1) which is connected to a syringe or pump for reagent delivery, an optional input fluid distribution splitter (2) to distribute reagents between nodes, and a fluid loading chamber (3) with a pressure equalization vent (4). As shown in Fig. 2-I, the inlet supplies fluid (Fig. 2-I-A) to four nodes (only two shown), each delivering reagents to two 12-well sample rows. The nodes can be made smaller (e.g., a single 12 sample row), or the nodes can be interconnected to carry out a single assay in all 96 wells.

B. 96-Well assay plate

The plate assay is fabricated with black PMMA, (Fig. 2-I-B), each of the 96 assay wells (5) has a volume of 13 μl with holes on the bottom, serving as an outlet for fluid removal (see Note 5).

An important and innovative aspect of ELISA-LOC is the use of a fluid tension passive valve to control flow. Valves to control liquid flow complicate fabrication of LOC, increase cost and can reduce reliability. We use a simple "surface tension" valve, which is an approximately 0.2 mm diameter hole in the bottom of the well, to control flow as shown in Fig. 2-IV. Fluid surface tension keeps the fluid in the wells (e.g., during binding antibody antigen). When a vacuum is applied through the outlet, fluid can be sucked through the holes and empty the wells to the outlet channels.

C. The outlet system for fluid removal
The bottom of the assay plate is attached to the outlet system (Fig. 2-I-C) and connected to a syringe used for fluid removal. The eight outlet channels (6) are positioned under the holes on the bottom of the wells. Groups of two channels are connected from both ends to the negative pressure distribution splitters (7), providing low pressure from both sides of the channels. The distribution splitters are connected to the outlet port (8) of a pump or syringe. When a vacuum is applied to the outlet port (8), fluid from the wells moves into the channels. A syringe may also be used, such that no electric power is required. The 3D fluidic path through the chip is in both the horizontal (within layers) and vertical (between layers) directions, as shown by the arrows in Fig. 2-I. Horizontal flow within the layers fills the fluid delivery system (3) and the outlet channels (6). Vertical flow between the three layers moves fluids between layers from the fluid delivery to the wells (5) and from the wells to the outlet system (6).

3.3.2. Fabrication of ELISA-LOC

Figure 3 shows the layers of the laminated ELISA-LOC for each of the three elements as shown in Fig. 2. These three elements are the fluid delivery system (Fig. 3a), the 96-well assay plate (Fig. 3b), and the outlet system for fluid removal (Fig. 3c). The LOC is constructed with PMMA cores laminated with five layers of thin polymer bonded with adhesive following the method described in Subheading 3.2. The relatively thick PMMA core provides rigidity to the assembly and space for high volume fluidic reservoirs (e.g., for the wells). In Fig. 3, the core well layer (III) and the bottom layer (VI) are both 3.2 mm PMMA, a rigid polymer. All the other layers are 250 μm polycarbonate (PC), except for the distribution layer II, which is 1.6 mm PC. Layers II, III, and V are laminated with the adhesive prior to CO_2 laser micromachining. The layers are assembled by aligning them with the adjoining layers using the alignment holes (elements 9A and 9B in Fig. 3) in a two-pin device (Fig. 1). These alignment holes are used again after the samples have been deposited into the wells to align the fluidic distribution

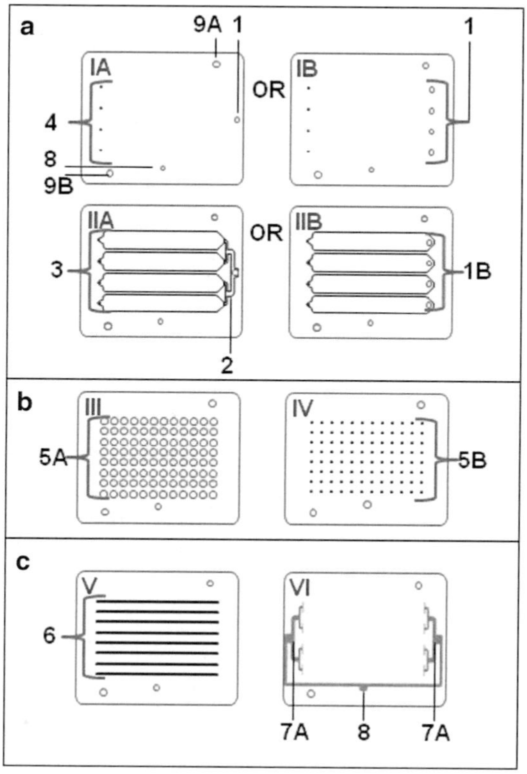

Fig. 3. Outline of the six layer ELISA-LOC: The three main elements of the six layers ELISA-LOC are shown as in Fig. 2: fluid delivery (**a**), samples wells (**b**), and the outlet system (**c**) for fluid removal. In the figure, layers are *numbered* in order from the top layer down (I to VI). All the layers are 250 μm PC, except the distribution layer II, which is 1.6 mm PC, and well layer III and bottom layer VI, which are 3.2 mm PMMA. Two configurations of the device are shown: four interconnected nodes (II-A) with its corresponding cover (I-A) and four independent nodes (II-B) with its corresponding cover (I-B). The main elements of the system are for fluid delivery (**a**) inlet (1) connected to a syringe for reagent delivery, which is a single port for interconnected nodes (I-A and II-A) or four ports for the independent nodes (I-B and II-B). The input fluid distribution splitter (2) distributes reagents between the nodes if the interconnected design is used (II-A); otherwise this element can be omitted. The fluid loading chamber (3) with its pressure vents (4) are the main components of the fluid delivery element (layers I and II). The assay plate (**b**) includes the PMMA wells in layer III for assay incubation and detection (5-A). Each well is fabricated with a small hole (5-B) in the bottom, which serves as fluid outlet (5-B) in layer IV. The outlet system (**c**) was designed with two outlet channels per node (6) in layer V, placed directly under the well holes. All outlet channels are interconnected via two negative pressure distribution splitters (7-A) engraved in the PMMA and connected to the outlet port (8) for a pump or syringe, which runs through all the layers. All the layers include two alignment holes (9-A and 9-B).

layer prior to bonding. The individual layers in the design shown in Fig. 3 are presented in a top-down view. The layer numbers correspond to their order in the chip, with layer I on top. The functional elements shown in Fig. 3 are labeled using the numbering scheme introduced in Fig. 2.

The fluid delivery system is shown in Fig. 3 panel a with two different configurations. I-A is one node configuration for loading all 96 wells with the same reagents and I-B is a four node configuration to load each of the four nodes, each with 24 wells, with the same reagents. In layer I, the system inlet (1) is connected to a syringe for reagent delivery either at a single port where all the nodes are interconnected to deliver the same reagents to all 96 wells (IA and IIA), or with four ports, each delivering reagents to a node with 24 wells (IB and IIB). Layer I also includes pressure vents (4) to let air out during filling (see Note 6).

Layer II is used for fluid distribution to the wells. An input fluid distribution splitter (2) distributes reagents between the nodes if the interconnected design is used (IIA). This element can be omitted when four independent nodes are used (IIIB). Reagents (buffers, secondary antibody, ECL reagent, etc.) are loaded for distribution into the wells in the fluid loading chamber (3). This layer is a double-sided adhesive layer. The volume of fluid distributed to the wells can be determined by varying the thickness of PC used.

The 96-well assay plate is shown in Fig. 3 panel b. Layer III is normally made of black PMMA and contains the wells for assay incubation and detection (5A). This layer has adhesive on the bottom. Layer IV forms the bottom of the wells, each with a small hole (5B) to serve as a fluid outlet when vacuum is applied.

Figure 3 panel c illustrates the outlet system, which removes fluid from the wells via the fluid outlet. Layer V is made of PC and has eight outlet channels placed directly under the well's holes of layer IV. It was designed with two outlet channels per node (6), each corresponding to a row of wells. This layer has adhesive on both sides. The bottom layer, layer VI is fabricated with PMMA and the elements are engraved in the PMMA (rather than cut all the way through). Layer VI allows all the outlet channels to be interconnected from both ends via two negative pressure distribution splitters (7A and 7B) engraved in the PMMA. The splitters are connected to the outlet port (8) for a pump or syringe to apply a vacuum, which drains the wells and moves the reagents in the channels.

As discussed above, the sample wells and the outlet (Layers III–VI) are preassembled, while the fluid delivery system is attached as a cover (Layers I–II) using the alignment holes after the samples are added to the wells.

3.3.3. Assembly of ELISA-LOC

In order to assemble the device, the assay plate, with its 96 wells (Fig. 2-I-B), and the outlet nodes (Fig. 2-I-C) are both preassembled. After loading samples into the wells, the fluid delivery system is used to cover the LOC. Fluids are delivered to the assembled LOC via the input port (Fig. 2-I-A-1) and the reagents are split among the four nodes. The immunological assay is carried out and monitored in the assay wells (5). The outlet of the device (8) is connected to a pump or syringe for fluid delivery.

3.4. Fluorescence Detection with ELISA-LOC Using LED-CCD-Based Detector

Whilst the ECL assay described in this chapter does not require illumination, fluorescence detection is also possible using an LED-based light box together with the CCD detector described in previous work (20, 22, 24). This consists of an enclosure, a custom LED-based light box with exchangeable filters for illumination, an SXVF-M7 camera equipped with a Sony ICX-429ALL CCD with 752 × 582 pixels, and a Tamron manual zoom CCTV 4–12 mm, f1.2 lens.

Fluorescence detection can be used to test that there are no unanticipated leaks or mixing between nodes. For this, the four independent nodes configuration (Fig. 3a) was used with two nodes (I and III) loaded with fluorescein and two nodes (II and IV) with buffer. In the four node 96-well fluidics configuration (Fig. 4b), seven of the eight rows of wells were loaded with fluorescein and the last row loaded with buffer to measure background. The fluorescein emission was detected with the CCD-based fluorescence detector (see Notes 7–9).

In both the single node and 4-node configurations, no fluorescence signal is seen from the buffer-only lanes, suggesting that there are no fluid leaks between the wells and that the nodes function independently. The 4 mm focal length of the CCD detector zoom lens enables wide angle imaging from a short distance (9 cm) which is suitable for signal detection from the broad array of assay formats, however as seen in Fig. 4, the intensity of the light at the edge of the image is lower than at the center and the image is curved. Such a "fish eye" image is typical for wide-angle lenses (e.g., 4 mm), especially with close-up photography. The distortion can be fixed by computer enhancement of the image or by increasing the focal length of the zoom lens. However, increasing the length of the lens will increase the size of the detector.

3.5. ELISA-LOC Assays

In an ELISA assay, there are several fluid handling steps: the primary antibody immobilization in the ELISA plate wells, the addition of the sample to the well, the washing away of unbound sample, addition of a labeled secondary antibody, and, following an additional washing step, the labeled antibody is assayed and measured. To perform these steps, current ELISA systems rely on manual pipetting or use separate equipment for reagent delivery and for washing. Fluid handling could be simplified by incorporating microfluidics into the ELISA plate to perform liquid handling steps with no additional equipment.

1. Preassemble the ELISA-LOC components described in Subheading 3.3, (the fluid delivery system, the 96-well assay plate, and the outlet system for fluid removal), using the LOM methods described in Subheading 3.2.
2. Functionalization of the plate: To prepare the plate assay, 15 μl of antibody functionalized CNT solution (prepared in

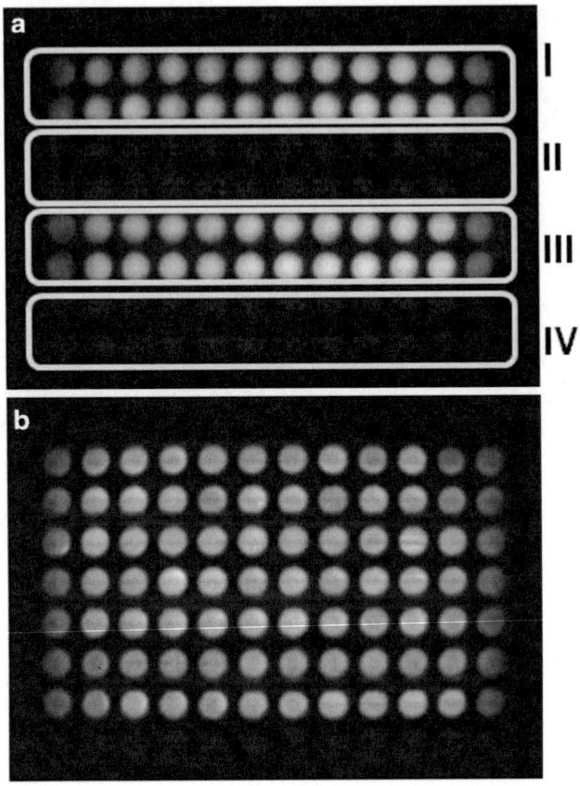

Fig. 4. Two configurations of fluid delivery system: (**a**), Configuration with four independent nodes (marked I–IV) and (**b**), Single node 96-well fluidics configuration. To demonstrate that there is no cross talk between nodes, in the single configuration, seven of the eight rows of wells were loaded with fluorescein, and for the four independent nodes configuration, nodes I and III were loaded with fluorescein and the nodes II and IV were loaded with buffer. Both chips were detected with a CCD camera. The ELISA-LOC was illuminated by blue LED with blue excitation filter and the signals of the detection wells were detected by the CCD camera, equipped with a green emission filter.

Subheading 3.1) was dropped into the wells of the 96-well sample chip.

3. This is then dried and rinsed with 2 ml washing buffer for 5 min to eliminate any loose or partially immobilized CNTs.
4. The CNT-antibody modified wells were then blocked with 1% BSA for 30 min.
5. Samples of various concentrations of SEB in PBS buffer solution are manually added to the ELISA-LOC wells (with the primary antibody–CNT complex) using 15 μl SEB sample per well. A range of concentrations of SEB in buffer (0.01–50 ng/ml) was used in this study.
6. After loading the samples, the fluid delivery system top is bonded onto the 96-well plate, sealing the plate. The entire assembly is then incubated for 60 min at room temperature.

7. The following steps were then performed by the ELISA-LOC fluidics system with reagents loaded sequentially into the loading fluid delivery system inlet (Fig. 2, I-a) using a syringe (connected to the outlet shown in Fig. 2-6) to move the fluids. In this configuration, no external power was needed to operate the device and cross mixing and diffusion between channels was not an issue because after the initial loading of the samples, the same reagents (wash buffer, secondary antibody, ECL reagents) are used for all steps.
   (a) The plate was washed with 2 ml of washing buffer delivered using a syringe via the fluid delivery system (while the device is slightly tilted to prevent air bubbles).
   (b) The buffer was removed via the fluid outlet system (bonded to the bottom of the plate).
   (c) Following washing, 2 ml of HRP conjugated anti-rabbit IgG (0.01 mg/ml) in PBS buffer was injected via the fluid delivery system and incubated for 1 h.
   (d) This was followed by three more cycles of washing with 2 ml of PBS washing buffer as described above.
   (e) The ECL assay was performed by adding 2 ml of ECL buffer (mixing the two solutions from Chemiluminescent Kit in a 1:1 volume ratio) and by measuring ECL intensity.
8. In the presence of HRP (bound to the secondary antibody), luminol in the ECL reagents reduces hydrogen peroxide, which emits light as it returns to its basal state and is detectable using the CCD detection setup. No illumination of the LOC is required because the ECL is light-emitting and is not a fluorescence assay.

3.6. ELISA-LOC Detection and Analysis of Staphylococcal Enterotoxin B

As seen in Fig. 5-I, the row with no SEB (row A) seems to be bright especially at the edge of the wells because of evanescent light passing through the PC (see Note 10) as shown in more detail in Fig. 5-II for rows A and B. A 3D analysis of panel 5-II is presented in 5-III showing a high signal at the edge of the well and a low signal (marked with an arrow) at the center. To minimize the effects of this light, only the center of the well (the dark portion) is measured (marked by a circle in Fig. 5-II). While the assay can resolve the 10× differences (e.g., between 0.01 and 0.1 ng/ml), it cannot discriminate 5× differences (e.g., between 0.1 and 0.5 ng/ml).

1. Signal intensities from the CCD images were quantified using ImageJ (see Note 11) and the dose response curve was plotted (Fig. 5-IV). The CCD image intensities (Fig. 5-I) show an increase in the signal corresponding to increasing concentration of SEB (Fig. 5-IV).

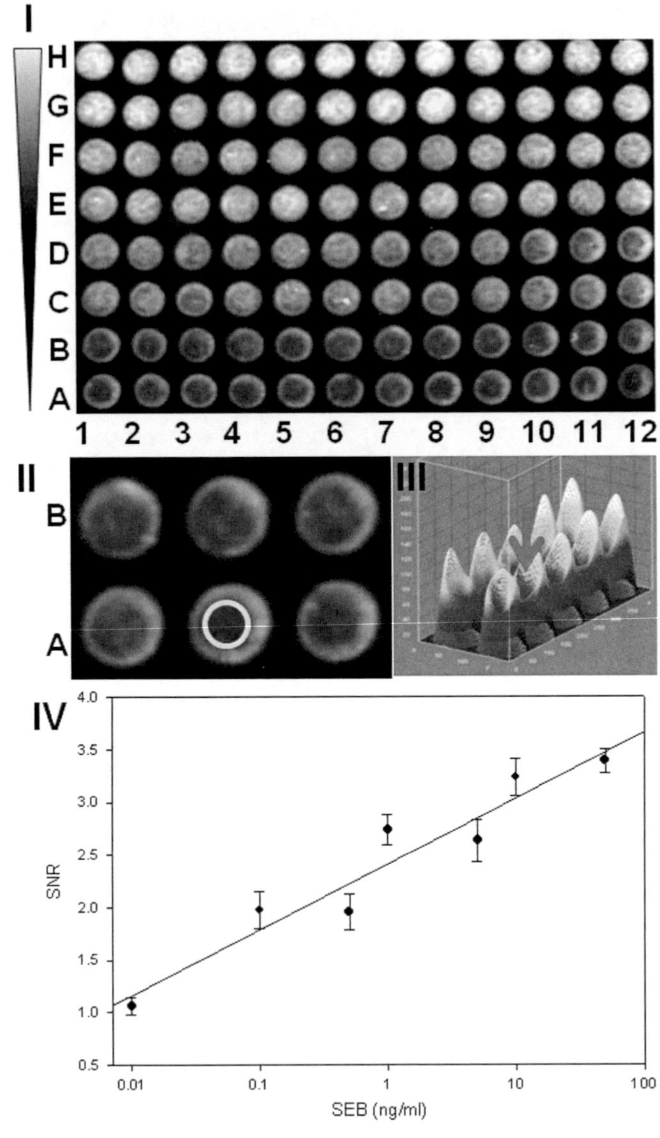

Fig. 5. ELISA-LOC detection of Staphylococcal enterotoxin B (SEB): Various concentrations of SEB were tested in a four node ELISA-LOC. I. The initial CCD image; II an enlarged image of rows A and B (the measured area marked with a *circle*), III. A 3D ImageJ analysis of *panel* II with an *arrow* pointing to the measured area and IV a plot of the signal/noise ratio (the noise is derived from row A with no SEB). The experiment was arranged so that each of the eight concentrations of SEB (0, 0.01, 0.1, 0.5, 1, 5, 10, 50 ng/ml) was loaded each in a row (represented by *letters*). The *side triangle* shows the order of loading the samples from high to low SEB concentrations, each of the eight concentrations (*rows*) were loaded with 12 replica columns (the *numbers*).

2. The ImageJ data were then imported into Microsoft Excel for further analysis. To measure the signal from each well, a circle (e.g., 20 pixels in radius) in ImageJ was used to measure the region of interest, which covers roughly 75% of the surface of each well on the plate.

3. The signal for the individual wells was calculated as the average of the intensity values of the respective pixels. The value obtained with a concentration of 0 ng/ml SEB was defined as the background. The signal-to-noise ratio (SNR) was further used to quantify the SEB concentration.

The limit of detection (LOD) represents a measured concentration that generates a signal three times the standard deviation above background, which is 0.1 ng/ml. This is comparable to the sensitivity of standard ELISA assays for Staphylococcal enterotoxins, which generally use optical detection, and have LOD ranges from 0.1 to 10 ng/ml (61). The typical LOD for commercial colorimetric ELISA systems is from 0.5 to 2 ng/g of food (62–67).

4. Notes

1. The laser power and speed for cutting polymers has to be determined empirically. It is recommended to use the minimum laser power to reduce overheating or burning the material. The appropriate settings are arrived at by trial and error. It is good practice to use the minimum necessary power for cutting to minimize burning of the material and localized heating of the workpiece. The power settings for the vector mode used for fabrication depend on the cutting objective. In most cases, vector cutting is designed to be through cutting which can be done at high power and high beam speed. It is essential to clean debris from the cuts after laser cutting as well to avoid blocking the channels by polymer and adhesive debris.
2. The type of adhesive used for LOC systems is seldom reported making it difficult to reproduce the designs of others. Several adhesive tapes can be used for bonding, including 3M 9770, 3M 9690, and 3M 501FL. 3M 9770 is an adhesive transfer tape with 2.0 mil (51 μm) hi-strength acrylic adhesive 330MP on a 4.2 mil (106 μm) polycoated Kraft liner. According to the manufacturer, the adhesive will hold securely after exposure to numerous chemicals including oil, mild acids, and alkalis. The tape can be used for short periods (up to an hour) at temperatures up to 120°C. The adhesive transfer tape is normally applied on one or both sides of a particular layer. To apply the tape, the liner is first removed from one side followed by bonding to the polymer surface (e.g., polycarbonate). Once the polymer/adhesive assembly is created, it can be cut in the laser cutter. Final assembly then involves removing the remaining liner, alignment of the layers and application of pressure to make the bond. The main challenge with this adhesive tape is

that the adhesive may flow when pressure is applied during bonding. When the design involves small channels, the flowing adhesive can block the small channels.

3. The double-coated tape 3M 9690 uses 330MP adhesive 2.8 mil (70 μm) on a clear 2 mil (50 μm) polyester carrier. The adhesive in this system may flow less than 3M 9770, resulting in less blocking of channels, although the potential still exists. The polyester carrier can be used as one of the layers for the microfluidics replacing a polycarbonate layer, thereby simplifying production. However, as with 3M 9770, channel blocking can still occur. Unfortunately the, polyester carrier used in 3M 9690 exhibits auto florescence so it is less useful when florescence detection is desired.
4. The 3M 501FL is called ultra-clean laminating adhesive transfer tape, and it comes without polyester backing. The 25 μm adhesive layer may reduce flow into the channels; it uses a 50 μm PET, silicone liner (3M 9770 uses a polycoated Kraft liner) which seems to produce cleaner cuts, especially when engraving is used.
5. The "holes" for the wells on the assay plate are 3 mm in diameter, which allows for analysis of 20 μl samples (only ~13 μl used). Smaller holes make the loading of the sample less reproducible because the fluid meniscus within the wells causes light diffraction, which complicate quantification.
6. *Loading the fluidics:* One issue with performing the assay is occasionally air bubbles will become trapped in the fluidic system. A simple solution for this problem is to place the LOC at a slight angle thus decreasing the trapping of air bubbles in the system. The air outlets (Fig. 3-A-4) enable air venting.
7. For the CCD detector, to measure light uniformity, it is important to measure the light with a long enough exposure time (e.g., 60 s) to measure the CCD uniformity when all of the wells are loaded with the same FITC fluorescence sample. If the lighting is not uniform, correction values for each well can be calculated and use for measurement composition.
8. For the CCD detector, to measure the effectiveness of the filters, it is recommended to perform two long exposures (e.g., 3 min) without the assay plate, one with the LEDs on and one with the LEDs off. Ideally, the two measurements should be very similar (the blue filters pass only blue light which is blocked by the green filters). The difference between measurements may suggest that the blue filters do not block all green light and/or the green filters do not block all blue light.
9. For taking images, a fully open aperture (e.g., f1.2) will enable shorter exposure times, but focusing will be more limited and the image less sharp.

10. *Controlling noise*: Fluorescence emission and scattered excitation light can propagate through the chip, causing cross-talk between adjacent channels. This can become a major source of optical noise in the system (22, 28, 68), which reduces the sensitivity of the measurements. To limit the effect of fluorescence background, PC, and not Mylar, which is a commonly used material for lamination-based fabrication, was used as the main fabrication material due to its lower fluorescence background (22). Using black material decreases the noise. Adding air gaps between channels did not reduce the background noise.

11. For data analysis, use the ImageJ high contrast visualization, which will not affect the values but will enable easy selection of the spots.

References

1. Engvall E, Perlmann P (1971) Enzyme-linked immunosorbent assay (ELISA). Quantitative assay of immunoglobulin G. Immunochemistry 8:871–874
2. Van Weemen BK, Schuurs AH (1971) Immunoassay using antigen-enzyme conjugates. FEBS Lett 15:232–236
3. Ihara M, Yoshikawa A, Wu Y, Takahashi H, Mawatari K, Shimura K, Sato K, Kitamori T, Ueda H (2010) Micro OS-ELISA: rapid noncompetitive detection of a small biomarker peptide by open-sandwich enzyme-linked immunosorbent assay (OS-ELISA) integrated into microfluidic device. Lab Chip 10:92–100
4. Gao Y, Sherman PM, Sun Y, Li D (2008) Multiplexed high-throughput electrokinetically-controlled immunoassay for the detection of specific bacterial antibodies in human serum. Anal Chim Acta 606:98–107
5. Kong J, Jiang L, Su X, Qin J, Du Y, Lin B (2009) Integrated microfluidic immunoassay for the rapid determination of clenbuterol. Lab Chip 9:1541–1547
6. Tseng YT, Yang CS, Tseng FG (2009) A perfusion-based micro opto-fluidic system (PMOFS) for continuously in-situ immune sensing. Lab Chip 9:2673–2682
7. Lee BS, Lee JN, Park JM, Lee JG, Kim S, Cho YK, Ko C (2009) A fully automated immunoassay from whole blood on a disc. Lab Chip 9:1548–1555
8. Javanmard M, Talasaz AH, Nemat-Gorgani M, Pease F, Ronaghi M, Davis RW (2009) Electrical detection of protein biomarkers using bioactivated microfluidic channels. Lab Chip 9:1429–1434
9. Tachi T, Kaji N, Tokeshi M, Baba Y (2009) Microchip-based homogeneous immunoassay using fluorescence polarization spectroscopy. Lab Chip 9:966–971
10. Liu C, Qiu X, Ongagna S, Chen D, Chen Z, Abrams WR, Malamud D, Corstjens PL, Bau HH (2009) A timer-actuated immunoassay cassette for detecting molecular markers in oral fluids. Lab Chip 9:768–776
11. Meagher RJ, Hatch AV, Renzi RF, Singh AK (2008) An integrated microfluidic platform for sensitive and rapid detection of biological toxins. Lab Chip 8:2046–2053
12. Reichmuth DS, Wang SK, Barrett LM, Throckmorton DJ, Einfeld W, Singh AK (2008) Rapid microchip-based electrophoretic immunoassays for the detection of swine influenza virus. Lab Chip 8:1319–1324
13. Luo Y, Yu F, Zare RN (2008) Microfluidic device for immunoassays based on surface plasmon resonance imaging. Lab Chip 8:694–700
14. Choi Y, Kang T, Lee LP (2009) Plasmon resonance energy transfer (PRET)-based molecular imaging of cytochrome c in living cells. Nano Lett 9:85–90
15. Lucas LJ, Chesler JN, Yoon JY (2007) Lab-on-a-chip immunoassay for multiple antibodies using microsphere light scattering and quantum dot emission. Biosens Bioelectron 23:675–681
16. Stevens DY, Petri CR, Osborn JL, Spicar-Mihalic P, McKenzie KG, Yager P (2008) Enabling a microfluidic immunoassay for the developing world by integration of on-card dry reagent storage. Lab Chip 8:2038–2045
17. Tang D, Tang J, Su B, Ren J, Chen G (2009) Simultaneous determination of five-type hepatitis virus antigens in 5 min using an integrated automatic electrochemical immunosensor array. Biosens Bioelectron 25(7):1658–1662

18. Mujika M, Arana S, Castano E, Tijero M, Vilares R, Ruano-Lopez JM, Cruz A, Sainz L, Berganza J (2009) Magnetoresistive immunosensor for the detection of Escherichia coli O157:H7 including a microfluidic network. Biosens Bioelectron 24:1253–1258
19. Lee SM, Hwang KS, Yoon HJ, Yoon DS, Kim SK, Lee YS, Kim TS (2009) Sensitivity enhancement of a dynamic mode microcantilever by stress inducer and mass inducer to detect PSA at low picogram levels. Lab Chip 9:2683–2690
20. Sapsford KE, Francis J, Sun S, Kostov Y, Rasooly A (2009) Miniaturized 96-well ELISA chips for staphylococcal enterotoxin B detection using portable colorimetric detector. Anal Bioanal Chem 394:499–505
21. Yang M, Kostov Y, Bruck HA, Rasooly A (2008) Carbon nanotubes with enhanced chemiluminescence immunoassay for CCD-based detection of Staphylococcal enterotoxin B in food. Anal Chem 80:8532–8537
22. Sapsford KE, Sun S, Francis J, Sharma S, Kostov Y, Rasooly A (2008) A fluorescence detection platform using spatial electroluminescent excitation for measuring botulinum neurotoxin A activity. Biosens Bioelectron 24:618–625
23. Yang M, Kostov Y, Bruck HA, Rasooly A (2009) Gold nanoparticle-based enhanced chemiluminescence immunosensor for detection of Staphylococcal Enterotoxin B (SEB) in food. Int J Food Microbiol 133:265–271
24. Sun S, Ossandon M, Kostov Y, Rasooly A (2009) Lab-on-a-chip for botulinum neurotoxin a (BoNT-A) activity analysis. Lab Chip 9:3275–3281
25. Sun S, Yang M, Kostov Y, Rasooly A (2010) ELISA-LOC: lab-on-a-chip for enzyme-linked immunodetection. Lab Chip 10:2093–2100
26. Xia Y, Kim E, Zhao XM, Rogers JA, Prentiss M, Whitesides GM (1996) Complex optical surfaces formed by replica molding against elastomeric masters. Science 273:347–349
27. Delamarche E, Bernard A, Schmid H, Michel B, Biebuyck H (1997) Patterned delivery of immunoglobulins to surfaces using microfluidic networks. Science 276:779–781
28. Irawan R, Tjin SC, Yager P, Zhang D (2005) Cross-talk problem on a fluorescence multichannel microfluidic chip system. Biomed Microdevices 7:205–211
29. Schilling EA, Kamholz AE, Yager P (2002) Cell lysis and protein extraction in a microfluidic device with detection by a fluorogenic enzyme assay. Anal Chem 74:1798–1804
30. Munson MS, Hasenbank MS, Fu E, Yager P (2004) Suppression of non-specific adsorption using sheath flow. Lab Chip 4:438–445
31. Rossier JS, Schwarz A, Reymond F, Ferrigno R, Bianchi F, Girault HH (1999) Microchannel networks for electrophoretic separations. Electrophoresis 20:727–731
32. Rossier J, Reymond F, Michel PE (2002) Polymer microfluidic chips for electrochemical and biochemical analyses. Electrophoresis 23:858–867
33. Taitt CR, Anderson GP, Ligler FS (2005) Evanescent wave fluorescence biosensors. Biosens Bioelectron 20:2470–2487
34. Ngundi MM, Qadri SA, Wallace EV, Moore MH, Lassman ME, Shriver-Lake LC, Ligler FS, Taitt CR (2006) Detection of deoxynivalenol in foods and indoor air using an array biosensor. Environ Sci Technol 40:2352–2356
35. Moreno-Bondi MC, Taitt CR, Shriver-Lake LC, Ligler FS (2006) Multiplexed measurement of serum antibodies using an array biosensor. Biosens Bioelectron 21:1880–1886
36. Ligler FS, Sapsford KE, Golden JP, Shriver-Lake LC, Taitt CR, Dyer MA, Barone S, Myatt CJ (2007) The array biosensor: portable, automated systems. Anal Sci 23:5–10
37. Yang M, Kostov Y, Rasooly A (2008) Carbon nanotubes based optical immunodetection of Staphylococcal Enterotoxin B (SEB) in food. Int J Food Microbiol 127:78–83
38. Hu D, Han H, Zhou R, Dong F, Bei W, Jia F, Chen H (2008) Gold(III) enhanced chemiluminescence immunoassay for detection of antibody against ApxIV of Actinobacillus pleuropneumoniae. Analyst 133:768–773
39. Rubtsova M, Kovba GV, Egorov AM (1998) Chemiluminescent biosensors based on porous supports with immobilized peroxidase. Biosens Bioelectron 13:75–85
40. Archer DL, Young FE (1988) Contemporary issues: diseases with a food vector. Clin Microbiol Rev 1:377–398
41. Olsen SJ, MacKinnon LC, Goulding JS, Bean NH, Slutsker L (2000) Surveillance for foodborne-disease outbreaks—United States, 1993–1997. MMWR CDC Surveill Summ 49:1–62
42. Bean NH, Goulding JS, Lao C, Angulo FJ (1996) Surveillance for foodborne-disease outbreaks—United States, 1988–1992. MMWR CDC Surveill Summ 45:1–66
43. Bunning VK, Lindsay JA, Archer DL (1997) Chronic health effects of microbial foodborne disease. World Health Stat Q 50:51–56
44. Garthright WE, Archer DL, Kvenberg JE (1988) Estimates of incidence and costs of intestinal infectious diseases in the United States. Public Health Rep 103:107–115
45. Asao T, Kumeda Y, Kawai T, Shibata T, Oda H, Haruki K, Nakazawa H, Kozaki S (2003) An extensive outbreak of staphylococcal food

poisoning due to low-fat milk in Japan: estimation of enterotoxin A in the incriminated milk and powdered skim milk. Epidemiol Infect 130:33–40
46. Breuer K, Wittmann M, Bosche B, Kapp A, Werfel T (2000) Severe atopic dermatitis is associated with sensitization to staphylococcal enterotoxin B (SEB). Allergy 55:551–555
47. Bunikowski R, Mielke M, Skarabis H, Herz U, Bergmann RL, Wahn U, Renz H (1999) Prevalence and role of serum IgE antibodies to the Staphylococcus aureus-derived superantigens SEA and SEB in children with atopic dermatitis. J Allergy Clin Immunol 103:119–124
48. Mempel M, Lina G, Hojka M, Schnopp C, Seidl HP, Schafer T, Ring J, Vandenesch F, Abeck D (2003) High prevalence of superantigens associated with the egc locus in Staphylococcus aureus isolates from patients with atopic eczema. Eur J Clin Microbiol Infect Dis 22:306–309
49. Howell MD, Diveley JP, Lundeen KA, Esty A, Winters ST, Carlo DJ, Brostoff SW (1991) Limited T-cell receptor beta-chain heterogeneity among interleukin 2 receptor-positive synovial T cells suggests a role for superantigen in rheumatoid arthritis. Proc Natl Acad Sci U S A 88:10921–10925
50. Uematsu Y, Wege H, Straus A, Ott M, Bannwarth W, Lanchbury J, Panayi G, Steinmetz M (1991) The T-cell-receptor repertoire in the synovial fluid of a patient with rheumatoid arthritis is polyclonal. Proc Natl Acad Sci U S A 88:8534–8538
51. Herz U, Bunikowski R, Mielke M, Renz H (1999) Contribution of bacterial superantigens to atopic dermatitis. Int Arch Allergy Immunol 118:240–241
52. Wiener SL (1996) Strategies for the prevention of a successful biological warfare aerosol attack. Mil Med 161:251–256
53. Ler SG, Lee FK, Gopalakrishnakone P (2006) Trends in detection of warfare agents. Detection methods for ricin, staphylococcal enterotoxin B and T-2 toxin. J Chromatogr A 1133:1–12
54. Henghold WB II (2004) Other biologic toxin bioweapons: ricin, staphylococcal enterotoxin B, and trichothecene mycotoxins. Dermatol Clin 22:257–262, v
55. Rosenbloom M, Leikin JB, Vogel SN, Chaudry ZA (2002) Biological and chemical agents: a brief synopsis. Am J Ther 9:5–14
56. Sergeev N, Volokhov D, Chizhikov V, Rasooly A (2004) Simultaneous analysis of multiple staphylococcal enterotoxin genes by an oligonucleotide microarray assay. J Clin Microbiol 42:2134–2143
57. Jarraud S, Peyrat MA, Lim A, Tristan A, Bes M, Mougel C, Etienne J, Vandenesch F, Bonneville M, Lina G (2001) egc, a highly prevalent operon of enterotoxin gene, forms a putative nursery of superantigens in Staphylococcus aureus. J Immunol 166:669–677
58. Jarraud S, Mougel C, Thioulouse J, Lina G, Meugnier H, Forey F, Nesme X, Etienne J, Vandenesch F (2002) Relationships between Staphylococcus aureus genetic background, virulence factors, agr groups (alleles), and human disease. Infect Immun 70:631–641
59. Omoe K, Ishikawa M, Shimoda Y, Hu DL, Ueda S, Shinagawa K (2002) Detection of seg, seh, and sei genes in Staphylococcus aureus isolates and determination of the enterotoxin productivities of S. aureus isolates Harboring seg, seh, or sei genes. J Clin Microbiol 40:857–862
60. Liu G, Lin Y (2006) Biosensor based on self-assembling acetylcholinesterase on carbon nanotubes for flow injection/amperometric detection of organophosphate pesticides and nerve agents. Anal Chem 78:835–843
61. Bennett RW (2005) Staphylococcal enterotoxin and its rapid identification in foods by enzyme-linked immunosorbent assay-based methodology. J Food Prot 68:1264–1270
62. Miyamoto T, Kamikado H, Kobayashi H, Honjoh K, Iio M (2003) Immunomagnetic flow cytometric detection of staphylococcal enterotoxin B in raw and dry milk. J Food Prot 66:1222–1226
63. Pan TM, Yu YL, Chiu SI, Lin SS (1996) [Comparison of immunoassay kits for detection of staphylococcal enterotoxins produced by Staphylococcus aureus]. Zhonghua Minguo wei sheng wu ji mian yi xue za zhi (Chinese journal of microbiology and immunology) 29:100–107
64. Park CE, Akhtar M, Rayman MK (1994) Evaluation of a commercial enzyme immunoassay kit (RIDASCREEN) for detection of staphylococcal enterotoxins A, B, C, D, and E in foods. Appl Environ Microbiol 60:677–681
65. Park CE, Warburton D, Laffey PJ (1996) A collaborative study on the detection of staphylococcal enterotoxins in foods by an enzyme immunoassay kit (RIDASCREEN). Int J Food Microbiol 29:281–295
66. Vernozy-Rozand C, Mazuy-Cruchaudet C, Bavai C, Richard Y (2004) Comparison of three immunological methods for detecting staphylococcal enterotoxins from food. Lett Appl Microbiol 39:490–494
67. Wieneke AA (1991) Comparison of four kits for the detection of staphylococcal enterotoxin in foods from outbreaks of food poisoning. Int J Food Microbiol 14:305–312
68. Hawkins KR, Yager P (2003) Nonlinear decrease of background fluorescence in polymer thin-films—a survey of materials and how they can complicate fluorescence detection in microTAS. Lab Chip 3:248–252

Chapter 30

Multiplexed Surface Plasmon Resonance Imaging for Protein Biomarker Analysis

Eric Ouellet, Louise Lund, and Eric T. Lagally

Abstract

The reliable detection of ligand and analyte binding is of significant importance for the field of medical diagnostics. Recent advances in proteomics and the rapid expansion in the number of identified protein biomarkers enhance the need for reliable techniques for their identification in complex samples. Surface plasmon resonance imaging (SPRi) provides label-free detection of this binding process in real-time. This chapter details the fabrication of an SPR imaging instrument and its use in analyzing molecular binding interactions with the use of a high-density microfluidic SPRi chip, capable of multiplexed analysis as well as various immobilization chemistries. Controlled recovery of bound biomarkers is demonstrated to enable their identification using mass spectrometry. Finally, activated leukocyte cell adhesion molecule (ALCAM), a protein biomarker associated with a variety of cancers, is identified from human crude cell lysates using the microfluidic surface plasmon resonance imaging (SPRi) instrument.

Key words: Surface plasmon resonance (SPR) imaging, Microfluidic arrays, Poly(dimethylsiloxane), PDMS, HeLa cells, Biomarkers, Mass spectrometry, Self-assembled monolayers

1. Introduction

Protein biomarker interactions are of fundamental importance in biology and medicine. With the publishing of the human genome sequence and the corresponding growth of systems biology as a method to discover individual markers of human disease, high-throughput techniques for measuring protein–protein interactions are increasingly required (1–5).

Surface plasmon resonance (SPR) (6, 7) is a widely used surface-sensitive method for biomolecular interactions analysis (BIA) capable of measuring molecular binding events at a metal surface. This is accomplished by coupling the photon energy from *p*-polarized incident light to the surface free-electrons at the metal–dielectric interface. The resulting oscillations of free-electron density are

Gareth Jenkins and Colin D. Mansfield (eds.), *Microfluidic Diagnostics: Methods and Protocols*, Methods in Molecular Biology, vol. 949, DOI 10.1007/978-1-62703-134-9_30,

called surface plasmons. A change in the local refractive index due to binding of molecules to the metal surface changes the propagation constant of these surface plasmons, resulting in a changing reflectivity of the incident light at the detector (8).

SPR provides real-time, label-free detection of binding kinetics by monitoring the reflectance as a function of time or by monitoring the angle of minimum reflectance as a function of time. Over the past decade, a growing number of publications using SPR as a detection tool have been reported (9). Surface plasmon resonance imaging (SPRi) is capable of sensing multiple interaction spots simultaneously by using a parallel detector such as a charge-coupled device (CCD) camera, rather than using point detection. Through its combination of the advantages of spectral SPR with parallel detection, SPRi is regarded as a method of choice for high-throughput binding affinity applications (1, 2, 10). In the past, this method has been used as a detection tool for affinity studies on DNA hybridization (11, 12), DNA–RNA binding (13), DNA–protein binding (3, 14), aptamer–protein binding (15), antibody–antigen binding (16, 17), detection of proteins from whole blood (18), and drug detection from saliva (19). However, commercial SPRi instruments are currently limited to a single target analyte stream across multiple immobilized ligand spots (20). Currently, commercial high-throughput SPR instruments employ either a cross-flow channel geometry, such as BioRad's ProteOn XPR (21, 22) or a flow cell-based system that have many ligand spots (up to 400 with Biacore's FlexChip) (23) but can only interrogate a single analyte in a single experiment.

Microfluidics has the potential to increase the throughput of SPRi by fabricating hundreds of individual reaction chambers in a small footprint, while simultaneously reducing the volumes of ligand and analyte required to perform a binding affinity assay. In addition, microfluidics has demonstrated the ability to accelerate reaction rates due to small volume requirements and reduction of diffusion distances (24, 25), to permit a better control over the reaction conditions because of the ability to quickly replenish reagents and/or change the local environment conditions (26), and to give greater reproducibility and precision due to automation via a computer-controlled manipulations (27).

In recent years, the integration of SPRi and microfluidics systems has emerged to provide a powerful platform for the identification of protein biomarkers related to disease diagnostics (10, 17, 28). In this chapter, a high-throughput microfluidic device capable of analyzing up to 264 different protein biomarkers from complex mixtures is presented, through the use of its element-addressable chambers (29). The protocol describes a complete process for the fabrication of microfluidic arrays and an SPR imager for the detection of proteins in a complex mixture. The immobilization of recognition elements is described using a carboxymethyl dextran hydrogel surface, widely popularized by GE Healthcare's Biacore platform. Composed of a 3-D matrix harboring activated carboxyl groups, this hydrogel improves the density of bound ligands compared to a planar metal surface, improving the overall detection of protein biomarkers. We demonstrate the methods presented here through an experiment to detect a representative biomarker from crude cell lysates. Activated leucocyte cell adhesion molecule (ALCAM, CD166) is a cell surface member of the immunoglobulin superfamily and is involved in cell–cell interactions through homophillic and heterophillic binding (30). Widely expressed in a variety of normal tissues, ALCAM was first identified on thymic epithelial cells and activated leucocytes, as well as in a wide variety of tissues and cells, including epithelia, lymphoid, myeloid cells, fibroblasts, neurones, hepatocytes, pancreas acinar and islet cells, and bone marrow (30, 31). Upregulation of ALCAM protein expression has been correlated with many cancers including prostate (32), breast (30), pancreatic (31, 33), ovarian (34, 35), and skin cancer (36). Finally, controlled recovery of binding species is presented for downstream mass spectrometry purification and identification.

2. Materials

2.1. Microfluidic Device Fabrication

1. Four inch (4″) silicon wafers are used for both the control and flow layer molds containing the desired features, and are designed using CAD software (ProgeCAD).
2. Poly(dimethylsiloxane) (PDMS) elastomer and curing agent (RTV615 A + B kit, GE Silicones).
3. Trimethylchlorosilane.
4. High-Resolution 20,000 DPI transparency mask (Cad/Art Services) of gold features created using CAD software.
5. SF-11 glass slides (3″ × 1″ × 1.1 mm, Schott Glass).
6. Positive photoresist (Shipley SPR220-7.0).
7. Developer: MF-24A (Microchem Corp.).
8. Gold Etchant: TFA (Transene Company, Inc.).
9. Chromium Etchant: 1020 (Transene Company, Inc.).
10. Piranha Solution: 5:1 (v/v) H_2SO_4:H_2O_2. CAUTION! Pour the sulfuric acid into the hydrogen peroxide. This will create a highly volatile exothermic reaction capable of degrading organic molecules.

2.2. SPR Imager Fabrication

1. Aluminum breadboard (12″ × 12″), mounting brackets and optical breadboard (Thorlabs).
2. Motorized rotation stages with mounting plates (URS Series) and motion controller (SMC100 Series, Newport).

3. Dovetail optical rails (12″), lens mounts (1″), rotation mount (1″), optical posts and post-holders (Thorlabs).
4. Near-IR LED (780 nm) with constant current power source.
5. Near-IR CCD Camera with Lens (Uniq).
6. SF-11 dispersion glass prism (30 × 30 mm) (Ealing cat no. 24-2966).
7. Collimation lens, 780 nm band-pass filter, linear polarizer (25.4 mm diameter, Edmund Optics).
8. Custom-machined aluminum plate with 23 × 25 mm hole (1/8″ thickness).

2.3. Microfluidic Connections

1. 23 Gauge blunt-end needles (BD Medical).
2. Tygon Micro-Bore Tubing, Formulation S-54-HL (0.020 ID; 0.060 OD) (Saint-Gobain).
3. 10 mm Solenoid 3-way valves with manifold and fittings (Clippard Minimatic).
4. Air supply connected to regulator with 50 PSI gauge capable of giving 10–15 PSI.

2.4. HeLa Cell Culture and Lysis

1. HeLa cells (cat. no. CCL-2, ATCC, Manassas, VA).
2. Dulbecco's modified Eagle's serum (DMEM) (Thermo Fisher Scientific), supplemented with 10% v/v qualified fetal bovine serum (FBS) (Invitrogen), an additional 2 mM L-glutamine (Thermo Fisher Scientific) and 0.1 U/L penicillin (Thermo Fisher Scientific).
3. 2-Mercaptoethanol.
4. 10 mM Phosphate-buffered saline (pH 7.4).
5. 1% NP 40 buffer: is made from NP-40 buffer (Sigma-Aldrich) supplemented with 50 mM Tris–HCl and 150 mM NaCl (pH 8.0) and 70% water (see Note 1).
6. Protease inhibitor tablets with EDTA (Complete, Roche).

2.5. SPR Imaging Array Preparation

1. 17.6 mM Cysteamine hydrochloride (Sigma-Aldrich) in 70% ethanol (see Note 2).
2. CM Dextran: 8 mg/mL carboxymethyl-dextran sodium salt in aqueous solution of 50 mM *N*-hydroxysuccinimide (NHS) and 200 mM 1-ethyl-3-(3-dimethylaminopropyl) carbodiimide.
3. 10 mM Phosphate-buffered saline (pH 7.4).
4. 25 μM Protein A from *Staphylococcus aureus* (Sigma-Aldrich).
5. 1 M Ethanolamine.
6. 10 mM Sodium acetate (pH 4.5).
7. 130 nM Recombinant human ALCAM protein (R&D Systems, cat no. 656-AL).

8. 1.67 μM Monoclonal anti-human ALCAM antibody (R&D Systems, clone 105901).
9. 50 mM Sodium hydroxide.

2.6. Sample Preparation for Mass Spectrometry

1. Digestion buffer: 1% sodium deoxycholate and 50 mM ammonium bicarbonate.
2. 0.5 μg/μL Dithiothreitol (DTT) (Sigma-Aldrich): Dissolve in water (see Note 3).
3. 13.5 mM Chloroacetamide (Sigma-Aldrich): Dissolve in water.
4. 0.5 μg/μL Trypsin: Sequencing Grade Modified Trypsin, Porcine (Promega).
5. 1× Sample Buffer: is prepared by mixing 12 mL acetonitrile with 4 mL trifluoroacetic acid, 2 mL acetic acid, and 400 mL water.
6. Buffer A: 0.5% acetic acid, 99.5% water (see Note 4).
7. Buffer B: 80% acetonitrile (High performance liquid chromatography (HPLC) grade), 0.5% acetic acid, 19.5% water (see Note 4).
8. Purification Tip: Empore Disk C18 (3M Company) packed into a 200 μL pipette tip (37).
9. HPLC Chip: Used with Agilent Technologies LC-MS/MS (ProtID-Chip-43 (II) G4240-62001).

3. Methods

3.1. SPR Imager Instrumentation

1. Attach mounting brackets to the backside of the optical breadboard and fasten the breadboard vertically on the optical table.
2. Attach one end of an optical rail onto a rotation stage, in line with the zero position. Mount the rotation stage at the upper end of the optical breadboard, with position 0 at the bottom.
3. Attach the second optical rail on the second rotation stage as described in step 1. Mount the second rotation stage on top of the other allowing enough clearance for motion. Both optical rails should be pointing downwards.
4. Connect the two rotation stages to the motion controller to control the motion. Adjust the position of the top rotation stage to 45 °, and −45 ° for the bottom stage. This step will require the use of the motion controller software (see Note 5).
5. Insert the linear polarizer into the manual rotation mount, and all other optics in lens mounts.
6. Attach the polarizer, the band pass filter and the CCD camera to the bottom rail. Attach the collimation lens assembly and the LED to the top optical rail.

7. Mount the prism on the custom machined aluminum plate by carefully inserting it from the top side, through the hole. When properly inserted, the prism should be sitting a few millimeters above the surface of the plate. Attach the plate to the optical post and mount on the optical table using a post-holder.
8. Connect the LED to a constant current power source and connect the CCD to a computer. Run the CCD software in live capture mode.
9. Align the surface of the prism to the center of the rotation stage by adjusting the height of the post. Align all other optics on the optical rails to have the LED light pass through the collimation lens assembly and the prism, and reflected back through the band pass filter, the polarizer, and to the CCD camera. Proper alignment will be accomplished when the CCD camera captures the light from the LED. Careful alignment is crucial to the successful operation of the SPR imager (see Note 6).
10. Enclose the SPR imager assembly to minimize stray light and dust from entering the system and the optics. A schematic of the imager with corresponding photographs is shown in Fig. 1.

3.2. Microfluidic Device Fabrication

3.2.1. PDMS Microchannels Fabrication

1. Clean both the flow and control molds. Remove small dust particles using a nitrogen stream. Never touch any part of the photoresist pattern and do not use any liquids.
2. Treat the molds with trimethylchlorosilane (TMCS) by placing 500 μL in a vial and incubating both molds for 1 h in a container in a fume hood. This step silanizes the molds for easy PDMS removal.
3. For the control layer, mix PDMS pre-polymers at a 1:5 ratio using 25 g Part A and 5 g Part B. Pour the liquids into a disposable cup and mix thoroughly. Place the control mold into an aluminum dish and add the mixture. You may need to press down on the mold to ensure uniformity (see Note 7).
4. Degas the PDMS in a dessicator to remove bubbles. If smaller bubbles are left on the surface, use a scalpel to pop them.
5. For the flow layer, mix PDMS pre-polymers at a 1:20 ratio using 15 g Part A and 0.75 g Part B. Degas the mixture in a dessicator.
6. Spin coat the PDMS mixture onto the flow layer mold at 2,500 rpm for 60 s to give a membrane thickness of 10 μm. When pouring the degassed PDMS mix, do so at a short distance from the mold and fast as to prevent the formation of bubbles (see Note 8).
7. Bake both molds in an oven at 80°C for 30 min.
8. Take both molds out of the oven and cut out the PDMS from the control layer at the edges of the wafer using a scalpel blade. Carefully peel the PDMS from the control mold.

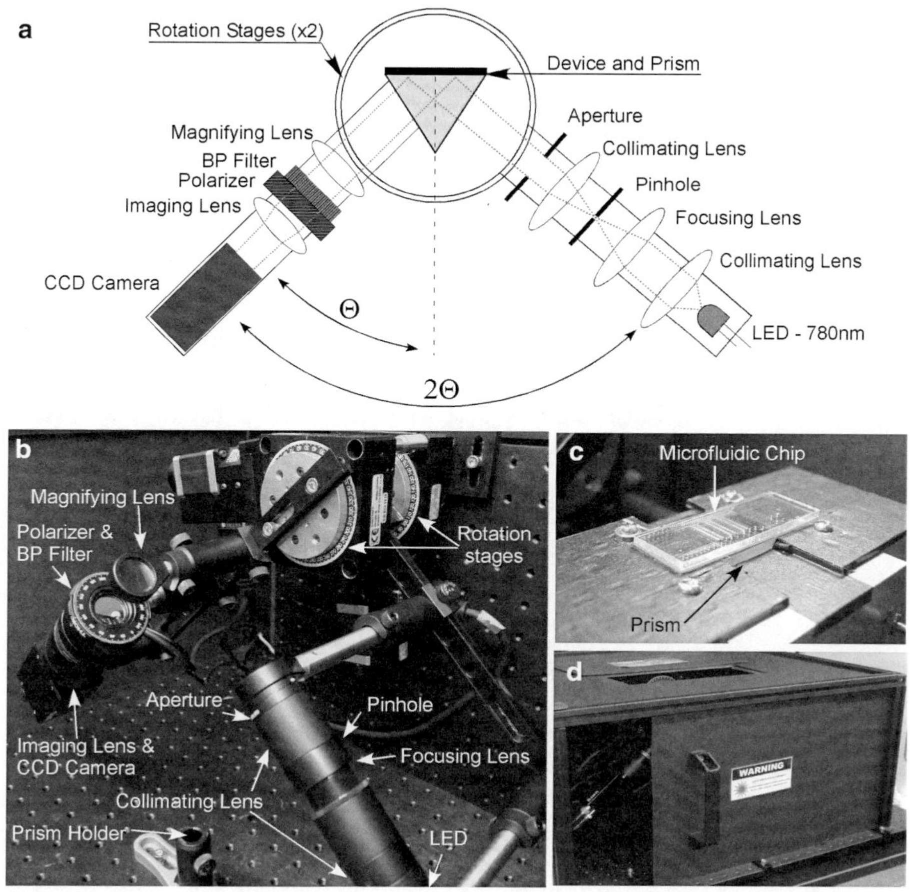

Fig. 1. Schematic of the surface plasmon resonance imager. (**a**) The incoming collimated Near-IR ($\lambda = 780$ nm) LED light beam transits the prism and illuminates the surface of the chip. The reflected light is then filtered and collected by a CCD camera. (**b**) The CCD camera and associated optical components are attached to one of two optical rails, controlled by two superimposed motorized rotation stages. The two rotation stages are driven by a stage controller that adjusts their positions *in-sync*. (**c**) The prism is fixed in place between the two optical rails using a custom prism mount. The microfluidic device is placed on a layer of refractive index matching fluid on the surface of the equilateral SF11 glass prism. (**d**) The entire setup is constructed in an upright position in order to facilitate the use of microfluidic chips and enclosed to protect from stray light and dust.

9. Immediately align the PDMS from the control mold to the PDMS from the flow layer using a stereo microscope. It helps to roughly align all control layers to the corresponding flow layers in order to avoid dust accumulation. Use the alignment marks on the corners to guide you. Do not try to peel off the PDMS from the flow layer. After final alignment, gently press on the PDMS to remove any trapped air between the two layers.
10. Place the aligned device in an oven at 80°C for 2 h.
11. Take the aligned device out of the oven and cut out the device from the mold using a scalpel blade. Trim off the edges of the device to fit a 3″ × 1″ gold slide.

12. Using a boring needle, punch out inlets and outlets of the device (see Note 9).
13. Remove any PDMS shards by using adhesive tape and store in a dust-free environment until needed. Be careful not to press on the device as the layers may stick together.

3.2.2. Gold Substrates Fabrication

1. Clean an SF-11 glass slide in piranha solution at 80°C for 15 min. CAUTION! Rinse thoroughly in flowing deionized (DI) water for 2 min before drying in a nitrogen stream.
2. Sputter 50 Å of chromium on the glass followed by 450 Å of gold.
3. Put the gold-coated slide on a hot plate for a dehydration bake at 150°C for 5 min. Remove the slide from the hotplate and let cool down to room temperature for 5 min.
4. Pipette positive photoresist along the length of the slide as to cover approximately half of its width. Spin coat at 500 rpm spread cycle for 5 s followed by 5,500 rpm for 40 s to give an approximate thickness of 6 μm.
5. Remove the coated gold slide from the spinner and pre-bake on a hot plate at 115°C for 2 min. Wait 10 min to let the photoresist rehydrate.
6. Place the coated gold slide on the platform of a mask aligner, followed by the transparency mask, followed by a 5″ × 5″ glass plate to ensure a hard contact between the mask and the photoresist. Expose the photoresist to 400 nm UV light for 65 s. Remove the slide from the mask aligner and wait 15 min for photoresist to relax before developing.
7. Develop the gold pattern in developer solution with gentle agitation for approximately 2 min. Rinse in flowing DI water immediately for at least 2 min and blow dry with nitrogen. This step reveals the pattern for the gold spots. Check the quality of the pattern under a stereo microscope before continuing. Repeat the development in 15 s increments to remove excess photoresist.
8. Post-bake the developed gold slide on a hot plate at 115°C for 2 min. Remove from the hot plate and let cool down to room temperature for 10 min.
9. Immerse gold slide in gold etchant at room temperature for 20 s. Immediately rinse the slide in flowing DI water for 2 min and dry in a stream of nitrogen. Immerse the slide in chromium etchant for 5 s. Immediately rinse the etched slide in flowing DI water for 2 min and dry the etched slide in a stream of nitrogen. This step etches the gold and chromium to reveal gold spots where the photoresist pattern was developed in step 7.

10. Strip the remaining photoresist in acetone for 2 min. Rinse the slide in flowing DI water and dry in a stream of nitrogen.
11. Clean the etched gold slide in piranha solution as described in step 1.

3.2.3. Bonding Device to Gold-Patterned Substrate

1. Place PDMS device (with microchannels facing up) and a gold-patterned glass slide (gold pattern facing up) in a UV-Ozone cleaner. Care must be taken to ensure there are no dust particles on either surface. Expose surfaces to UV-Ozone for 10 min (see Note 10).
2. Immediately pipette 50 μL of absolute ethanol on the gold patterned slide (gold pattern facing up) and bring the PDMS device (microchannels facing down) on top of the gold slide.
3. Align the gold spots with the PDMS chambers by gliding the PDMS device across the gold-pattern substrate until the gold spots register with the microchambers using a microscope and the alignment marks to help during the process.
4. Gently press down on the PDMS device to bring the microchannels into contact with the gold-patterned substrate. Care must be taken not to collapse the microchannels against the substrate.
5. Incubate the assembled device in an oven at 80°C for 2 min to evaporate the ethanol.
6. Take the assembled device out of the oven and gently press down in areas where air bubbles may be trapped. Care must be taken to ensure there are no trapped air bubbles between the PDMS and substrate.
7. Incubate the assembled device in an oven at 80°C for 1 h to allow bonding to occur. An overview of the fabrication steps involved is outlined in Fig. 2.

3.3. HeLa Cell Culture and Collection of HeLa Lysate

1. HeLa cells are maintained in DMEM.
2. Harvest cells at 90% confluency by placing the culture dish containing the cells on ice. Discard the media and gently wash three times with 15 mL of cold PBS.
3. Add 1 mL of PBS to the cells and remove them from the bottom of the culture dish using a cell scraper.
4. Collect the cells in a 15 mL tube and spin at 650 × *g* for 4 min at 4°C.
5. Carefully remove the supernatant without disturbing the cell pellet.
6. Resuspend the cell pellet in 1 mL of 1% NP40 buffer and ¼ protease inhibitor tablet. Add 1 μL of 2-mercaptoethanol and pipette to mix. Spin at 10,000 × *g* for 20 min (see Note 11).
7. Carefully collect the supernatant without disturbing cell debris and store at 4°C until needed (Subheading 3.4.2, step 17).

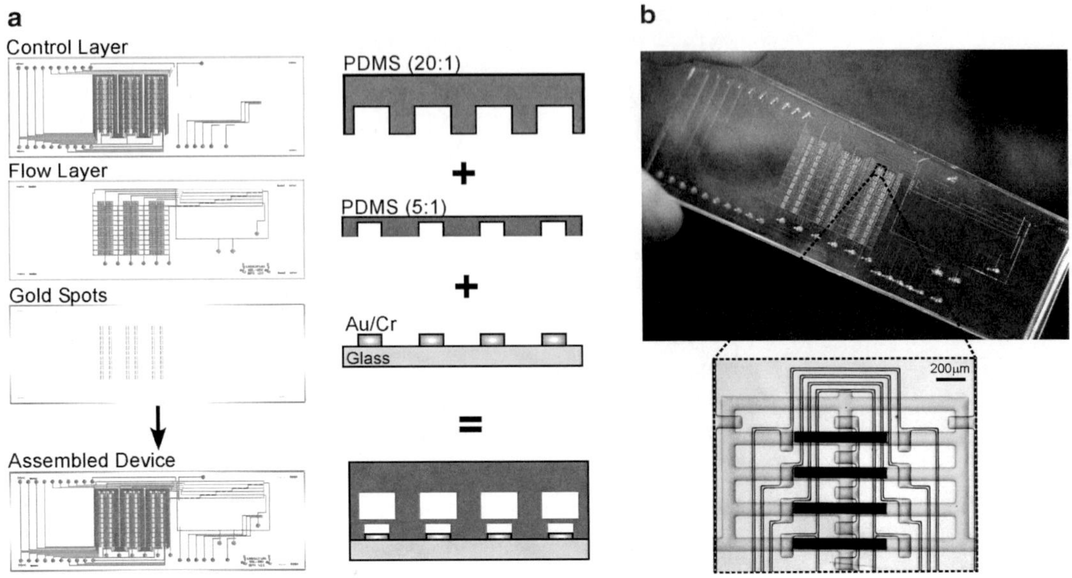

Fig. 2. Microfluidic surface plasmon resonance imaging chip. (**a**) Chip cross-sectional view fabricated using poly(dimethylsiloxane) (PDMS) soft lithography showing the control and the flow layer. The PDMS layers are irreversibly bonded to an SF11 glass substrate patterned with gold spots (47.5 nm gold with 5 nm chromium) for SPR sensing. (**b**) Photograph of the microfluidic device showing one of 66 chamber arrays.

3.4. SPR Imaging of ALCAM Biomarker Binding in Real-Time (See Note 12)

3.4.1. Microfluidic Device Pneumatic Connections

1. Using blunt-end needles, bend the end at a 90 ° angle using forceps. Cut off the Luer-Lok® fitting and reopen the collapsed end using forceps.
2. Connect the Tygon tubing to the longer end of the needle and fill the tubing with water using a syringe.
3. Connect the bent needle to a control inlet of the microfluidic device, and connect the tubing to the port fitting of a solenoid valve on the manifold. Connect the input port of the manifold to the air supply.
4. Actuate the solenoid valve to let 10–15 PSI of air enter the tubing. This will allow the water to enter the microchannel in the microfluidic device. Upon filling the microchannel, the pressure exerted on the water-filled microchannel will deflect the microvalve membrane, pinching off the flow layer below.
5. Repeat steps 1–4 for all remaining control inlets. These steps will allow for the actuation of the microfluidic valves in the control layer.
6. Connect the fluidic inlets and outlets to a needle connector (above) and place the outlet tubing into a suitable waste container (see Note 13).

3.4.2. Microfluidic Array Imaging and Recovery of Bound Species

1. Clean both the glass surface of the prism and the device with lens cleaner and lens paper. Ensure all surfaces of the prism are clean and free of any dust. Ensure proper placement of the prism on the SPR imager.
2. Place a small drop of refractive index matching fluid (1.7300, Cargille Series M) in the center of the surface of the prism. Place the microfluidic device on the prism ensuring that no air bubbles get trapped between the device and the prism. Should a bubble get trapped, remove the device and repeat from step 1.
3. To the sample inlet, add the cysteamine hydrochloride solution (17.7 mM in 70% ethanol, see Note 2). Actuate the solenoid valves for the microfluidic pump to introduce the solution in the device by peristaltic motion. Incubate for 1 h at room temperature. Figure 3 shows the details of the microfluidic device including the loading and washing modes of operation (29).
4. Rinse the device by pumping 70% ethanol at the sample inlet to remove the remaining cysteamine solution.
5. Purge the device with PBS injected from the wash inlet for 30 min at an equivalent flow rate of 50 μL/min.
6. From the CCD camera, select a region of interest (ROI) around a gold spot. Set the starting angle to 48 ° and record the pixel intensity (reflectivity) of the ROI at that angle. Repeat in 0.05 ° increments up to 54 °. The resulting plot of reflectivity versus angle will show a dip in reflectivity with a minimum. This angle is known as the surface plasmon resonance angle. Set the working angle below the minimum angle inside the linear region of the curve.
7. Select regions of interest (ROIs) for every gold spot and begin acquiring images from the CCD in real time. Plot the intensity from each ROI against time. A pixel intensity baseline value of 60–80 is typical.
8. Flow PBS in the device for 10 min to achieve a stable baseline (see Note 14).
9. Switch to the sample inlet and inject the CM dextran for 30 min at an equivalent flow rate of 5 μL/min (see Note 15).
10. Switch to the wash inlet and wash the surface with PBS for 10 min until a stable baseline is achieved.
11. Inject Protein A in sodium acetate buffer (pH 4.5) via the sample inlet for 15 min at 5 μL/min. Protein A will be positively charged at that pH and will react with the activated carboxyl groups on the surface. Protein A is used to orient the antibody immobilization for optimal results (see Note 16).
12. Switch to the wash inlet and wash the surface with PBS for 10 min until a stable baseline is achieved.

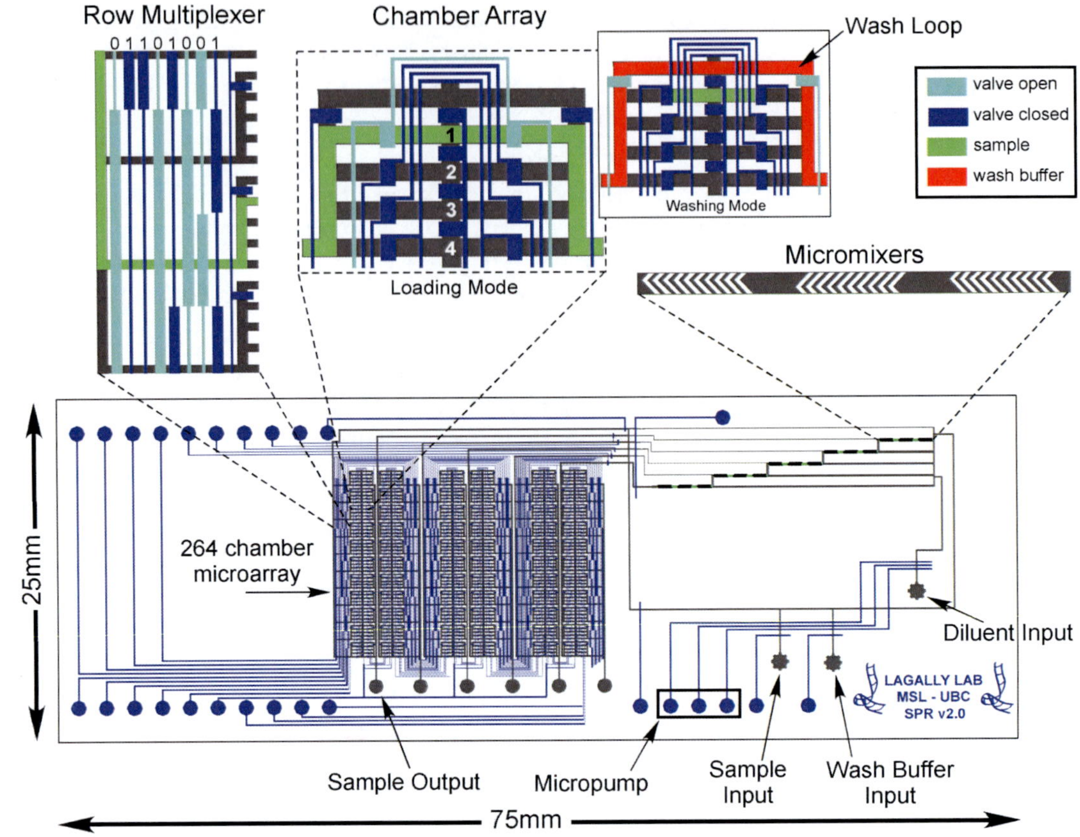

Fig. 3. Mask design of the microfluidic device. The chip consists of a top control layer and a bottom flow layer. The flow layer has microchannels of 100 μm width by 10 μm height. The control layer contains 1,132 microvalves (100 μm wide and 150 μm long) that are used to isolate and control the flow of fluids beneath. This layer also contains a micropump that can pump to the array of 264 individually addressable chambers, each with a volume of 700 pL. To control the loading of an individual chamber, a row multiplexer (*top left*) gates the entrance of fluid to a specific group of four chambers. A series of four chamber valves are then opened or closed to allow loading of a specific chamber within the group (*top middle*). Upon loading a ligand through the sample input port, the remaining sample can be flushed to the sample output port via the wash loop. A series of chaotic advection micromixers (*top right*) allow serial dilutions into six different concentrations by mixing the sample with diluent. (Reproduced from ref. 29 with permission from The Royal Society of Chemistry).

13. Inject the ethanolamine via the sample inlet for 30 min at 5 μL/min. This step is used to block unreacted ester groups on the surface to prevent any nonspecific binding.
14. Switch to the wash inlet and wash the surface with PBS for 10 min until a stable baseline is achieved.
15. Inject the ALCAM antibody via the sample inlet for 15 min at 5 μL/min.
16. Switch to the wash inlet and wash the surface with PBS for 10 min until a stable baseline is achieved.
17. Inject the HeLa cell lysate (that was prepared and collected in Subheading 3.3) through the sample inlet for 15 min at 5 μL/min (see Note 17).

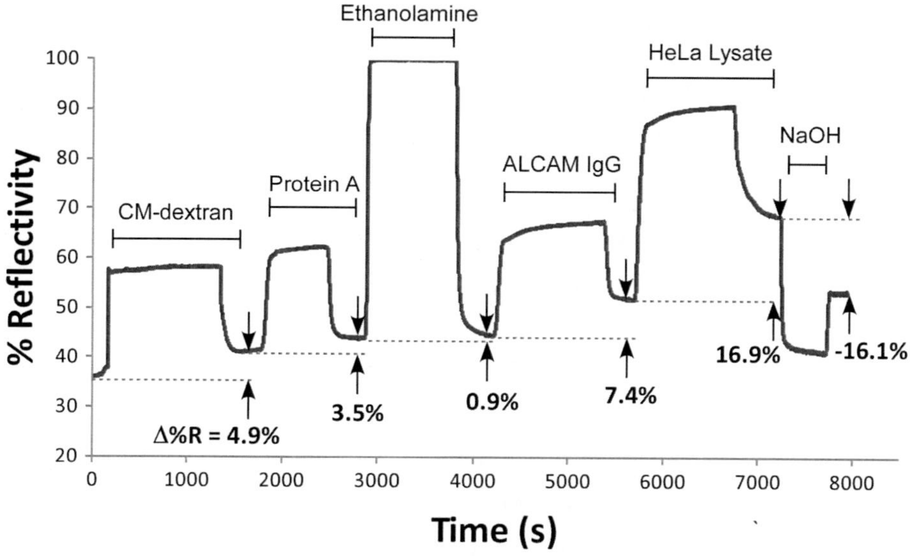

Fig. 4. SPRi sensorgram for ALCAM detection in crude HeLa cell lysate on the microfluidic chip. The binding was monitored at each step and the corresponding binding response is represented as the difference between baselines.

18. Switch to the wash inlet and wash the surface with PBS for 15 min at 100 μL/min until a stable baseline is achieved.
19. Remove the waste container and add a suitable container to collect the bound proteins.
20. Inject 50 mM sodium hydroxide through the sample inlet for 20 min at 100 μL/min.
21. Switch to the wash inlet and wash the surface with PBS for 10 min at 100 μL/min until a stable baseline is achieved. Figure 4 shows the resulting sensorgram with the binding response at each step.
22. Store collected samples at −20°C until needed. A typical sensorgram is shown in Figure 4.

3.5. Mass Spectrometry of Recovered Proteins

3.5.1. In-Solution Digestion

1. Dilute 10 μg of collected protein to 1 μg/μL in digestion buffer.
2. Add 1 μg of dithiothreitol (DTT) per 50 μg of sample protein and incubate for 30 min at 37°C.
3. Add 1 μg of chloroacetamide per 50 μg and incubate at 37°C for 20 min away from light.
4. Add 1 μg of Trypsin per 50 μg to proteins and incubate overnight at 37°C, away from light.
5. Following the trypsin digest, dilute the peptides 4× in sample buffer. Ensure that sample pH is below 3 (see Note 18).

3.5.2. Solid-Phase Purification of Peptides

1. Add 40 μL of methanol to the tip and spin at 750 × g for 2 min (see Note 19).

2. Add 40 μL of 1× Sample Buffer to the tip and spin at 750 × *g* for 2 min.
3. Load peptides to the tip and spin at 800 × *g* for 8 min.
4. Wash the tip with 20 μL of Buffer A by centrifugation at 800 × *g* for 5 min. Ensure that the entire buffer is out of the tip.
5. Elute the peptides from the tip into a microcentrifuge tube by adding 20 μL of Buffer B and spinning at 750 × *g* for 5 min.
6. Transfer the eluted sample to a 96-well plate.
7. Place samples in a dessicator and dry without heat for 30 min. Store dried peptides at 4°C until needed.

3.5.3. Mass Spectrometry

1. The following steps assume the use of a 6500 Series Accurate-Mass Quadrupole Time-of-Flight (Q-TOF) LC-MS/MS.
2. Add 5 μL of buffer B to each well of the 96-well plate to resuspend the dried peptides.
3. Load the 96-well plate into the autosampler of the machine.
4. Insert a proprietary HPLC chip into the unit.
5. Run the unit to begin peptide separation and identification. When complete, the MassHunter software will display the results. Figure 5 shows representative MS spectra.
6. Analyze the results from the LC-MS/MS MassHunter software using Spectrum Mill software to identify the proteins present in the sample. A typical MS spectrum is shown in Figure 5.

4. Notes

1. In all instances water refers to purified water at 18.2 MΩ cm resistivity.
2. Cysteamine hydrochloride is prepared in 70% ethanol to prevent the swelling of PDMS when injected in the microfluidic chip. Swelling will delaminate the PDMS from the substrate.
3. Dithiothreitol (DTT) is required to break the disulfide linkages between cysteine residues. To prevent the disulfide bonds from reforming, chloroacetamide is used to modify the reactive cysteine -SH groups, forming *s*-carboxylmethylated cysteines.
4. Prepare Buffers A and B in brown vials and protect from light.
5. The SMC100 controller software is used to register the rotation stages and control their motion. These require the use of a serial port on the computer. Alternatively, the motion control can be controlled using LabView software (National Instruments) with the libraries supplied by the motor manufacturer.

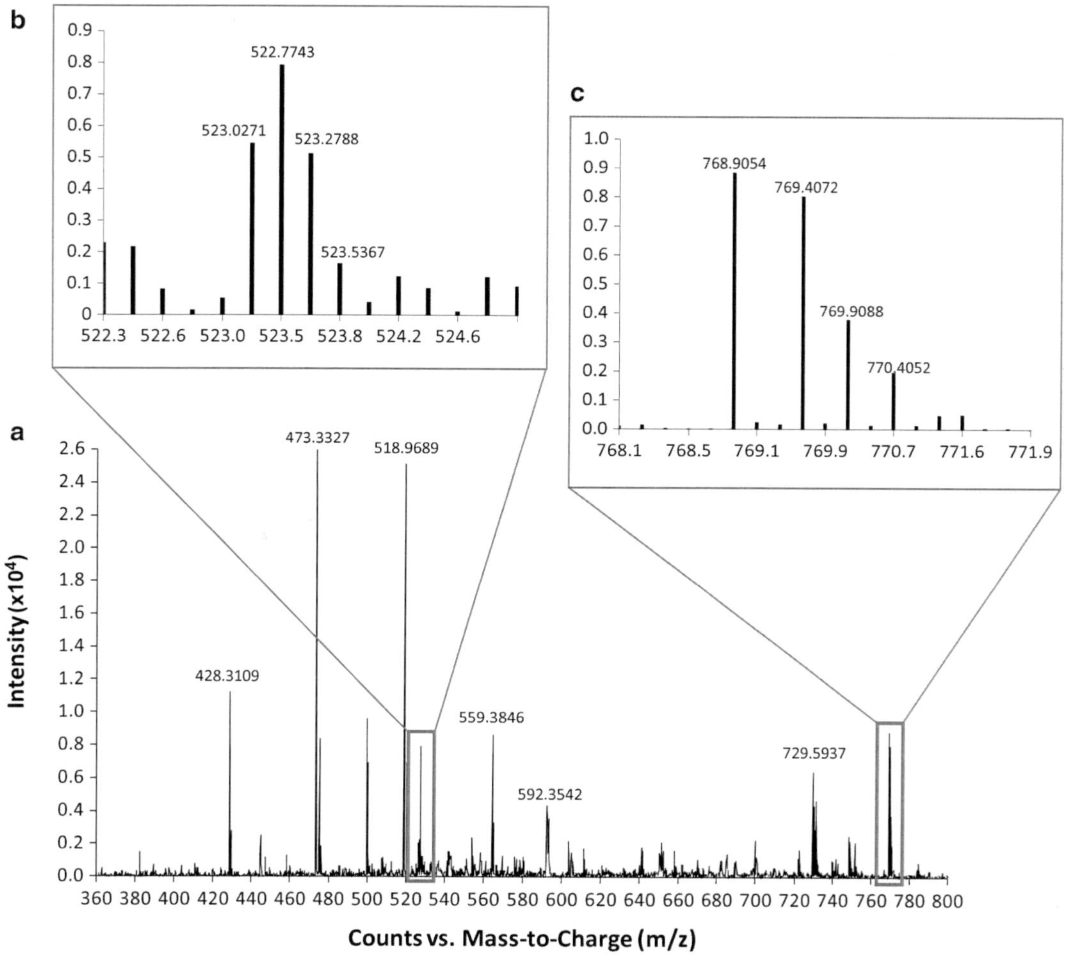

Fig. 5. (**a**) Representative MS spectrum for HeLa lysate from a Quadrupole time-of-flight mass spectrometer. (**b**) High-resolution spectra of peptide fragments from the M2 isoform of the cancer biomarker pyruvate kinase (PKM2). (**c**) High-resolution spectra of peptide fragments for the reporter heat shock protein (HSPA5).

6. During the operation and the alignment of the SPR imager, the angle on both the rotation stages should be changed *in-sync*.
7. Part A refers elastomer base and Part B refers to the curing agent.
8. Since the speed and time varies by spinner type, a calibration curve should be generated for best results. Unless otherwise stated, the text assumes the use of a Laurell WS-400-6NPP-LITE spinner.
9. For best reproducibility and performance, use an arbor press punching machine with a 23 gauge boring needle.
10. Alternatively, bonding can be achieved by using air or oxygen plasma.

11. This step is crucial in order to remove any cell debris which can block the microchannels on the microfluidic device.
12. The method here describes the detection of ALCAM, a protein biomarker from human crude cell lysate, and can easily be adapted for higher throughput screenings. Further expansion of this method to include identification of several other protein biomarkers can be performed in these microfluidic arrays.
13. Avoid disconnecting and reconnecting the needles at the PDMS inlet and outlet as this can damage the in/outlets over time.
14. Although PDMS is permeable to air, degassing of all solutions is recommended in order to avoid introducing bubbles which may affect the flow rate and introduce artifacts in the SPR sensorgrams.
15. Prepare CM dextran the night before and add NHS/EDC when ready to reduce reaction time required.
16. The protein A buffer solution should be approximately pH 4.5 to ensure that it is below the isoelectric point (p*I*) of the protein A. Reduced immobilization efficiencies have been observed at pH above the p*I*.
17. This step is crucial and care must be taken to prepare the HeLa lysate so as not to introduce bubbles or cell debris.
18. Use acetic acid to alter pH.
19. Ensure all the methanol has passed through the tip. Otherwise, replace with a new tip and recondition with methanol.

Acknowledgments

The authors would like to thank Dr. Leonard Foster for his advice and continued support, Dr. Leroy Hood and Christopher Lausted for their contributions to the microfluidic device design, Anders Riss Kristensen for his technical assistance with HeLa cell culture and Dr. Robert Parker for technical assistance with mass spectrometry analysis. Microfabrication was performed in the UBC Nanolab.

References

1. Piliarik M, Vaisocherová H, Homola J (2005) A new surface plasmon resonance sensor for high-throughput screening applications. Biosens Bioelectron 20:2104–2110
2. Wassaf D, Kuang G, Kopacz K, Wu Q-L, Nguyen Q, Toews M, Cosic J, Jacques J, Wiltshire S, Lambert J, Pazmany CC, Hogan S, Ladner RC, Nixon AE, Sexton DJ (2006) High-throughput affinity ranking of antibodies using surface plasmon resonance microarrays. Anal Biochem 351:241–253
3. Shumaker-Parry JS, Aebersold R, Campbell CT (2004) Parallel, quantitative measurement of protein binding to a 120-element double-stranded DNA array in real time using surface plasmon resonance microscopy. Anal Chem 76:2071–2082
4. Boozer C, Kim G, Cong S, Guan H, Londergan T (2006) Looking towards label-free biomolecular

interaction analysis in a high-throughput format: a review of new surface plasmon resonance technologies. Curr Opin Biotechnol 17: 400–405

5. Luo Y, Yu F, Zare RN (2008) Microfluidic device for immunoassays based on surface plasmon resonance imaging. Lab Chip 8:694–700
6. Rothenhausler B, Knoll W (1988) Surface–plasmon microscopy. Nature 332:615–617
7. Yeatman E, Ash EA (1987) Surface plasmon microscopy. Electron Lett 23:1091–1092
8. Campbell CT, Kim G (2007) SPR microscopy and its applications to high-throughput analyses of biomolecular binding events and their kinetics. Biomaterials 28:2380–2392
9. Homola J, Yee SS, Gauglitz G (1999) Surface plasmon resonance sensors: review. Sens Actuators B Chem 54:3–15
10. Shumaker-Parry JS, Zareie MH, Aebersold R, Campbell CT (2004) Microspotting streptavidin and double-stranded DNA arrays on gold for high-throughput studies of protein-DNA interactions by surface plasmon resonance microscopy. Anal Chem 76:918–929
11. Li Y, Wark AW, Lee HJ, Corn RM (2006) Single-nucleotide polymorphism genotyping by nanoparticle-enhanced surface plasmon resonance imaging measurements of surface ligation reactions. Anal Chem 78:3158–3164
12. Malic L, Veres T, Tabrizian M (2009) Biochip functionalization using electrowetting-on-dielectric digital microfluidics for surface plasmon resonance imaging detection of DNA hybridization. Biosens Bioelectron 24:2218–2224
13. Goodrich TT, Lee HJ, Corn RM (2004) Direct detection of genomic DNA by enzymatically amplified SPR imaging measurements of RNA microarrays. J Am Chem Soc 126:4086–4087
14. Wegner GJ, Lee HJ, Marriott G, Corn RM (2003) Fabrication of histidine-tagged fusion protein arrays for surface plasmon resonance imaging studies of protein–protein and protein–DNA interactions. Anal Chem 75: 4740–4746
15. Li Y, Lee HJ, Corn RM (2007) Detection of protein biomarkers using RNA aptamer microarrays and enzymatically amplified surface plasmon resonance imaging. Anal Chem 79: 1082–1088
16. Wegner GJ, Lee HJ, Corn RM (2002) Characterization and optimization of peptide arrays for the study of epitope–antibody interactions using surface plasmon resonance imaging. Anal Chem 74:5161–5168
17. Lee K-H, Su Y-D, Chen S-J, Tseng F-G, Lee G-B (2007) Microfluidic systems integrated with two-dimensional surface plasmon resonance phase imaging systems for microarray immunoassay. Biosens Bioelectron 23:466–472
18. Lausted CG, Hu Z, Hood LE (2008) Quantitative serum proteomics from surface plasmon resonance imaging. Mol Cell Proteomics 7(12):2464–2474
19. Fu E, Chinowsky T, Nelson K, Johnston K, Edwards T, Helton K, Grow M, Miller JW, Yager P (2007) SPR imaging-based salivary diagnostics system for the detection of small molecule analytes. Ann N Y Acad Sci 1098: 335–344
20. Rich RL, Myszka DG (2007) Higher-throughput, label-free, real-time molecular interaction analysis. Anal Biochem 361:1–6
21. Bravman T, Bronner V, Lavie K, Notcovich A, Papalia GA, Myszka DG (2006) Exploring "one-shot" kinetics and small molecule analysis using the ProteOn XPR36 array biosensor. Anal Biochem 358:281–288
22. Stroock AD, Dertinger SKW, Ajdari A, Mezic I, Stone HA, Whitesides GM (2002) Chaotic mixer for microchannels. Science 295: 647–651
23. Rich RL, Cannon MJ, Jenkins J, Pandian P, Sundaram S, Magyar R, Brockman J, Lambert J, Myszka DG (2008) Extracting kinetic rate constants from surface plasmon resonance array systems. Anal Biochem 373:112–120
24. Barry R, Ivanov D (2004) Microfluidics in biotechnology. J Nanobiotechnol 2:2
25. Bange A, Halsall HB, Heineman WR (2005) Microfluidic immunosensor systems. Biosens Bioelectron 20:2488–2503
26. deMello AJ (2006) Control and detection of chemical reactions in microfluidic systems. Nature 442:394–402
27. Khandurina J, Guttman A (2002) Bioanalysis in microfluidic devices. J Chromatogr A 943: 159–183
28. Wang Z, Wilkop T, Xu D, Dong Y, Ma G, Cheng Q (2007) Surface plasmon resonance imaging for affinity analysis of aptamer–protein interactions with PDMS microfluidic chips. Anal Bioanal Chem 389:819–825
29. Ouellet E, Lausted C, Lin T, Yang CWT, Hood L, Lagally ET (2010) Parallel microfluidic surface plasmon resonance imaging arrays. Lab Chip 10:581–588
30. Burkhardt M, Mayordomo E, Winzer K-J, Fritzsche F, Gansukh T, Pahl S, Weichert W, Denkert C, Guski H, Dietel M, Kristiansen G (2006) Cytoplasmic overexpression of ALCAM is prognostic of disease progression in breast cancer. J Clin Pathol 59:403–409
31. Vaisocherová H, Faca VM, Taylor AD, Hanash S, Jiang S (2009) Comparative study of SPR and

ELISA methods based on analysis of CD166/ ALCAM levels in cancer and control human sera. Biosens Bioelectron 24:2143–2148

32. Kristiansen G, Pilarsky C, Wissmann C, Stephan C, Weissbach L, Loy V, Loening S, Dietel M, Rosenthal A (2003) ALCAM/CD166 is up-regulated in low-grade prostate cancer and progressively lost in high-grade lesions. Prostate 54:34–43
33. Faca VM, Song KS, Wang H, Zhang Q, Krasnoselsky AL, Newcomb LF, Plentz RR, Gurumurthy S, Redston MS, Pitteri SJ, Pereira-Faca SR, Ireton RC, Katayama H, Glukhova V, Phanstiel D, Brenner DE, Anderson MA, Misek D, Scholler N, Urban ND, Barnett MJ, Edelstein C, Goodman GE, Thornquist MD, McIntosh MW, DePinho RA, Bardeesy N, Hanash SM (2008) A mouse to human search for plasma proteome changes associated with pancreatic tumor development. PLoS Med 5:e123
34. Pitteri SJ, JeBailey L, Faça VM, Thorpe JD, Silva MA, Ireton RC, Horton MB, Wang H, Pruitt LC, Zhang Q, Cheng KH, Urban N, Hanash SM, Dinulescu DM (2009) Integrated proteomic analysis of human cancer cells and plasma from tumor bearing mice for ovarian cancer biomarker discovery. PLoS One 4:e7916
35. Rosso O, Piazza T, Bongarzone I, Rossello A, Mezzanzanica D, Canevari S, Orengo AM, Puppo A, Ferrini S, Fabbi M (2007) The ALCAM shedding by the metalloprotease ADAM17/TACE is involved in motility of ovarian carcinoma cells. Mol Cancer Res 5: 1246–1253
36. van Kempen LCLT, van den Oord JJ, van Muijen GNP, Weidle UH, Bloemers HPJ, Swart GWM (2000) Activated leukocyte cell adhesion molecule/CD166, a marker of tumor progression in primary malignant melanoma of the skin. Am J Pathol 156:769–774
37. Rappsilber J, Ishihama Y, Mann M (2002) Stop and go extraction tips for matrix-assisted laser desorption/ionization, nanoelectrospray, and LC/MS sample pretreatment in proteomics. Anal Chem 75:663–670

Chapter 31

Surface Acoustic Wave (SAW) Biosensors: Coupling of Sensing Layers and Measurement

Kerstin Länge, Friederike J. Gruhl, and Michael Rapp

Abstract

Surface acoustic wave (SAW) devices based on horizontally polarized surface shear waves enable direct and label-free detection of proteins in real time. Signal response changes result mainly from mass increase and viscoelasticity changes on the device surface. With an appropriate sensor configuration all types of binding reactions can be detected by determining resonant frequency changes of an oscillator. To create a biosensor, SAW devices have to be coated with a sensing layer binding specifically to the analyte. Intermediate hydrogel layers used within the coating have been proven to be very suitable to easily immobilize capture molecules or ligands corresponding to the analyte. However, aside from mass increase due to analyte binding, the SAW signal response in a subsequent binding experiment strongly depends on the morphology of the sensing layer, as this may lead to different relative changes of viscoelasticity. Bearing these points in mind, we present two basic biosensor coating procedures, one with immobilized capture molecule and a second with immobilized ligand, allowing reliable SAW biosensor signal responses in subsequent binding assays.

Key words: Biosensor, Surface acoustic wave, SAW, SAW resonator, Protein coupling, Protein detection, Surface modification, Dextran, Polyethylene glycol, Carbodiimide chemistry

1. Introduction

1.1. SAW-Based Biosensors

Special types of surface acoustic wave (SAW) devices providing horizontally polarized surface shear waves allow the design of SAW-based biosensors for direct and label-free detection of biomolecules in real time (1–3). A SAW device typically consists of a piezoelectric substrate, such as lithium tantalate ($LiTaO_3$) or quartz, with interdigital transducers (IDT) as a planar electrode structure. The SAW is generated on the substrate by applying a high-frequency alternating voltage via the IDTs. For the detection of proteins the device surface has to be coated with capture

Gareth Jenkins and Colin D. Mansfield (eds.), *Microfluidic Diagnostics: Methods and Protocols*, Methods in Molecular Biology, vol. 949, DOI 10.1007/978-1-62703-134-9_31,

molecules or ligands binding specifically to the analyte of interest (3). Binding reactions on the biosensor surface are detected by determining changes in surface wave velocity caused mainly by mass adsorption or viscosity changes in the sensing layer (4–6). In the literature two different IDT designs are typically used: the SAW resonator (3, 7) and the SAW delay line design (3, 8, 9). The latter provides a free sensor surface without electrode structures allowing any coating including metallization. In the case of a gold coating, standard procedures based on alkane thiols (10) forming self-assembled monolayers (SAM) are typically used to compose the sensing layer. In contrast to this, the SAW resonator design does not permit any conductive layer because the sensor surface is completely covered with IDT electrodes. As we work exclusively with SAW resonators, the procedures described in this chapter use nonconductive coatings only. However, they are basically suitable for both types of SAW biosensor designs. In addition, these procedures can be applied on other acoustic sensing devices such as Lamb wave or Love wave devices and even on quartz crystal microbalances.

Generally, direct adsorption of the layer components on the SAW device surface would be the easiest way; however, as this method usually suffers from a lack of reproducibility, other immobilization methods, such as covalent coupling, are preferred (11). Prior to the application of biospecific layers the SAW resonators should be coated with a suitable polymer to chemically homogenize the device surface. A suitable material for this purpose is parylene C (poly-(2-chloro-*p*-xylylene)), which can be deposited by chemical vapor deposition methods (CVD) at room temperature (12). Parylene layers provide a chemically stable and homogeneous surface with almost no intrinsic mechanical stress and low intrinsic attenuation for the surface acoustic waves. Covalent binding on the inert parylene surface is possible by means of activation via plasma treatment and subsequent silanization of the parylene (13). However, for the detection of proteins it is useful to couple the corresponding capture molecules not directly on the parylene layer but via an intermediate hydrogel layer (7, 13). This approach enables mild reaction conditions so that capture molecules, e.g., antibodies, will keep their functionality (14). Furthermore, hydrogels have been shown to be good shielding layers against nonspecific interactions (15).

1.2. Effect of the Intermediate Hydrogel Layer on the SAW Signal Response

For our biosensor coatings we use two types of hydrogels, dextrans or polyethylene glycols (PEG), for coupling of capture molecules or ligands (6). Dextrans immobilized on the sensor surface typically represent a three-dimensional (3D) hydrogel scaffold, because functional groups occur not only on the top but also within the hydrogel layer. On the other hand, PEGs immobilized in an end-on configuration (i.e., the polymer chains are aligned perpendicularly to the device surface) represent a two-dimensional (2D) hydrogel scaffold since functional groups occur only on the top of the hydrogel layer. In principle, hydrogels based on 3D scaffolds should allow a greater number of binding sites in the sensing layer, implying a higher binding capacity than hydrogels based on 2D scaffolds.

Basically, any analyte binding in these hydrogel layers leads to mass loading and thus to a frequency shift of the sensor response. However, mass increase is not the only effect influencing the SAW biosensor signal. Another important parameter is the change of viscoelasticity of the sensing layer due to analyte binding. Changes of viscoelasticity are, amongst other factors, greatly influenced by the initial layer thickness with regard to the penetration depth of the SAW as well as changes of other viscoelastic properties based on the morphology of the complete layer setup (16). Hence, this effect may be promoted by sensing layers based on 3D scaffolds allowing analyte binding within the layer. A change of viscoelasticity during analyte binding might counteract or add to the effect of mass loading, i.e., leading to reduced or increased SAW signal responses (5, 6, 16, 17).

Morphology and thickness of the intermediate hydrogel layer influence the signal response of SAW biosensors, depending on the type of applied affinity assay and on the amount of binding sites in the sensing layer; the latter provided by capture molecules or ligands corresponding to the analyte. Only a limited amount of capture molecules can usually be immobilized in the sensing layer due to size and availability. In this case, a relatively thin 3D hydrogel layer such as dextran with a low molar mass implying low chain length, or a 2D hydrogel layer like PEG should be used. On the other hand, ligands, i.e., small molecules interacting specifically with the analyte, may allow a large number of binding sites in the sensing layer. In this case, a relatively thick 3D hydrogel layer such as dextran with a higher molar mass implying higher chain length should be used. In consequence, depending on the amount of the immobilized binding partner desired, the hydrogel has to be chosen carefully considering layer thickness and morphology to get an optimal signal response (5, 6).

1.3. Immobilization of Capture Molecule or Ligand

As previously mentioned, it is advantageous to couple capture molecules or ligands corresponding to the analyte via an intermediate hydrogel layer. In particular, hydrogels providing carboxyl groups allow the covalent coupling of capture molecules or linker proteins, as well as ligands, under mild reaction conditions by means of carbodiimide chemistry (Fig. 1). In this case, the carboxyl groups are activated by a carbodiimide, e.g., 1-ethyl-3-(3-dimethylaminopropyl) carbodiimide (EDC), and converted into an active ester, usually by means of *N*-hydroxysuccinimide (NHS). The NHS-ester reacts with the amino groups of proteins. The

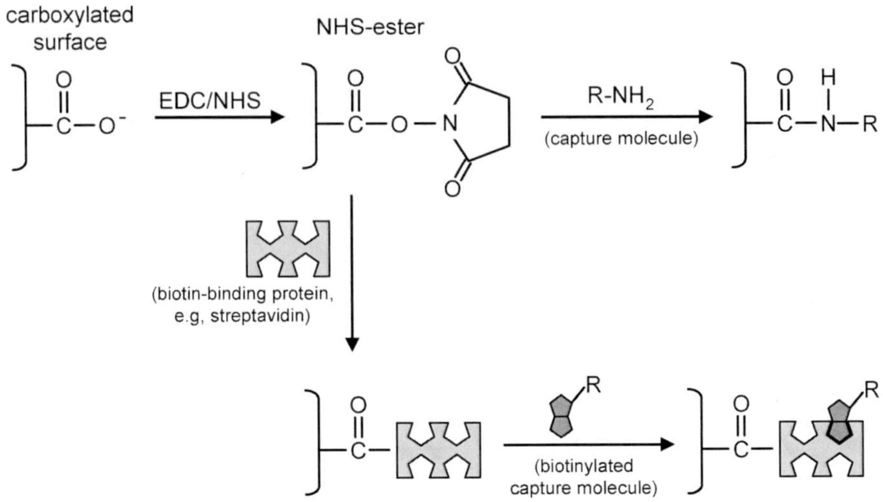

Fig. 1. Coupling procedures for capture molecules on carboxylated surfaces.

EDC/NHS coupling procedure is the most common and currently most used method (11). If the intermediate hydrogel provides amino groups, they can be converted into carboxyl groups by using dicarboxylic acid anhydrides to allow the same coupling scheme. However, the reaction conditions of the conversion reaction may also influence the morphology of the sensing layer and have to be considered carefully (18).

A disadvantage of the carbodiimide-based coupling procedure results from the randomized distribution of amino groups all over the protein molecule. This may lead to a blocking of binding sites after the coupling step, resulting in loss of signal response. For this reason a directed coupling of capture molecules via linker proteins may be preferred. This could be achieved, for example, by binding the corresponding antibody via Fc receptors such as protein A or G, or if biotinylated capture molecules are available by using biotin-binding proteins, such as streptavidin (11). However, it has to be taken into account that additionally introduced linker molecules may lead to a higher thickness of the sensing layer. This could be a problem for detection methods that imply a transduction mechanism near the surface, such as evanescent field techniques (11) or SAW biosensors (5, 6). In consequence, layer composition and concentrations of the single components have to be adjusted for each application to obtain optimal results in a subsequent assay, but approximate values valid for a group of sensing layers and applications can easily be given.

In this chapter we present general procedures to prepare SAW biosensors for affinity assays with either immobilized capture molecules or immobilized ligands corresponding to the analyte. Hydrogels providing carboxyl groups are used. Capture molecules are immobilized randomly oriented via a thin 3D dextran layer or via a 2D PEG layer. Additionally, capture molecules are immobilized in an oriented manner by means of linker molecules, which are also immobilized via a thin 3D dextran layer or via a 2D PEG layer. Ligands are immobilized via a thick 3D dextran layer. The decision on which of these basic procedures should be used for a given application, can be made with the considerations mentioned above.

2. Materials

2.1. SAW Biosensor Measurement Setup

1. SAW device: A shear horizontal SAW resonator based on a small (4×4 mm^2) 36° YX-$LiTaO_3$ device with gold transducers and a frequency of operation in aqueous media of 428.5 MHz (see Note 1) is used.
2. Electronic setup: SAW measurements are performed in an oscillator circuit developed in-house with the SAW resonator as the frequency-determining element. Output signals are obtained as difference frequencies relative to a permanently oscillating reference oscillator, featuring a constant frequency in the range of 434 MHz. Difference frequency changes (see Note 2) are monitored continuously with a time resolution of 1 s. The frequency resolution of the data acquisition is 1 Hz. The short-term noise is approximately 40 Hz.
3. SAW device integration: A flow cell was designed in-house, in which the SAW device is mounted upside down onto isolated contact pads of a circuit board and coupled capacitively. A flow channel located between the contact pads allows the fluid to pass along the SAW path. The dimensions of the flow channel at the sensor position are $0.6 \times 0.6 \times 4$ mm^3, corresponding to an effective sample volume of 1.44 µl. Stainless steel capillaries, inner diameter 0.8 mm, are used as connections to the fluidics.
4. Fluidic setup: A flow injection analysis (FIA) system equipped with two peristaltic pumps, Ismatec (Wertheim/Germany) and an injection valve, Besta-Technik (Wilhelmsfeld/Germany), is used. Polytetrafluoroethylene (PTFE) tubes serve as connections between single components and as sample loop. The volume of the sample loop (including connecting tubes) is 220 µl. Volumes of sample and reagent solutions are to be 250 µl each. A schematic of the FIA system and its settings is shown in Fig. 2.

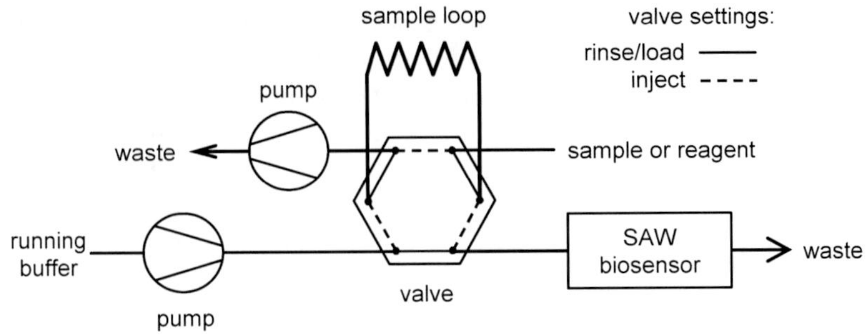

Fig. 2. Fluidic setup and system settings.

2.2. Surface Homogenization and Activation

1. Polymer: Parylene C dimer (di(2-chloro-*p*-xylylene)), Specialty Coating Systems (Indianapolis/USA).
2. Silane: (3-Glycidyloxypropyl)trimethoxysilane (GOPTS).
3. Rinsing solution: Acetone.

2.3. Coating with Hydrogel Layer

1. Hydrogel 1: Dicarboxy polyethylene glycol (DC-PEG), M_r 2,000, Rapp Polymere (Tübingen/Germany).
2. Hydrogel 2: Aminodextran (AMD), M_r 3,000, Life Technologies, Karlsruhe, Germany.
3. Hydrogel 3: AMD, M_r 70,000, Life Technologies, Karlsruhe, Germany.
4. Glutaric anhydride (GA).
5. Solvents: Dichloromethane (DCM), bidistilled water, dimethyl formamide (DMF).
6. Rinsing solutions: Bidistilled water, DMF, acetone.

2.4. Coupling of Proteins (i.e., Capture Molecule or Linker)

1. Running buffer: Phosphate buffered saline (PBS), pH 7.4, containing 10 mM phosphate buffer, 137 mM NaCl, and 2.7 mM KCl.
2. Immobilization buffer: Sodium acetate is diluted in bidistilled water to a concentration of 10 mM. The pH value of the solution is adjusted with KOH in such a way that the pH value of the resulting mixture containing capture molecule or linker will lie within the range of 5–6 (see Note 3). The immobilization buffer can be divided into working aliquots and stored at –20°C.
3. Activation solution: *N*-(3-dimethylaminopropyl)-*N*′-ethylcarbodiimide hydrochloride (EDC) is rapidly weighed and dissolved in water at 0.4 M. *N*-hydroxysuccinimide (NHS) is weighed and dissolved in water at 0.1 M. Freshly prepared solutions of EDC and NHS are mixed in a ratio of 1:1 immediately before use (see Note 4) to obtain a mixture containing 0.2 M EDC and 0.05 M NHS. Non-dissolved quantities of EDC or NHS can be stored at –20°C.

4. Capture molecule (e.g., antibody) binding specifically to the corresponding analyte (e.g., antigen) is dissolved in immobilization buffer. The concentration depends on the application (see Note 5). A reasonable concentration to start is 1.5 μM.
5. Linker (e.g., biotin-binding protein, such as streptavidin) binding specifically to the corresponding capture molecule (e.g., biotinylated antibody) is dissolved in immobilization buffer. For the subsequent coupling step capture molecule is dissolved in running buffer. Both concentrations depend on the application (see Note 5). A reasonable linker concentration to start is 1.8 μM. A reasonable capture molecule concentration to start is 0.05 μM.
6. Deactivation solution: Ethanolamine hydrochloride is diluted in bidistilled water at 1 M. The pH value of the solution is adjusted with HCl to a final pH value of 8.5. The solution can be divided into working aliquots and stored at −20°C.

2.5. Coupling of Small Molecules (i.e., Ligand)

1. Activation solution contains 1 M NHS and 1.2 M diisopropylcarbodiimide (DIC) dissolved in DMF. The solution must be prepared immediately before use (see Note 4).
2. Rinsing solutions: DMF, bidistilled water.
3. Ligand (small molecule) binding specifically to the corresponding analyte is dissolved in DMF. The concentration depends on the application (see Note 5). A reasonable concentration to start is 1 mM.

2.6. Measurements with the SAW Biosensor

1. Running buffer: PBS, pH 7.4 (see item 1 in Subheading 2.4).
2. Analyte buffer: PBS, pH 7.4 (see item 1 in Subheading 2.4, Note 6).
3. Blocking solution: Bovine serum albumin (BSA) is dissolved in running buffer at 1 mg/ml.
4. Rinsing solutions for cleaning sample loop in between the measurements: 2-propanol, bidistilled water.

3. Methods

3.1 Homogenization and Activation of the SAW Device Surface

1. Homogenization with parylene C:

   Coat all SAW devices used in binding experiments with 0.1 μm parylene C using a commercial parylene deposition system, Labcoater 1, PDS 2010, Speedline Technologies (Indianapolis/USA), to obtain a chemically homogeneous surface (see Note 7).
2. Activation of parylene surface:

Activate the parylene C-coated SAW devices by a two-step process: oxidation via plasma treatment and subsequent silanization. For this, treat the SAW devices 15 s in a plasma cleaner (PDC-32G, Harrick Plasma, NY, USA) at 0.2 mbar air pressure and 50 W power. Immediately after the oxidation step, apply GOPTS on the device surfaces (5 μl per SAW device). After 1 h silanization reaction, rinse the devices with dry acetone and dry them with nitrogen. The activated SAW devices have to be introduced to the next step (i.e., coating with hydrogel) immediately.

3.2. Coating of the SAW Device Surface with Hydrogel Layer

1. Coating with hydrogel 1, DC-PEG, $M_r = 2{,}000$:

   Apply a solution of 1 mM DC-PEG in DCM on the activated parylene surfaces (10 μl per SAW device). After evaporation of the solvent, heat the devices to 70°C for 20 h. After that, rinse the SAW devices thoroughly with hot bidistilled water, $T = 60$°C, and with room temperature bidistilled water (see Note 8) and dry them.

2. Coating with hydrogel 2, AMD, M_r 3,000, or hydrogel 3, AMD, M_r 70,000, and subsequent carboxylation:

   Apply an aqueous solution of AMD on the activated parylene surfaces (10 μl per SAW device). The concentration depends on the application (see Note 9). Reasonable concentrations to start are 1 mM for AMD, M_r 3,000, and 0.1 mM for AMD, M_r 70,000. After 20 h reaction, rinse the SAW devices thoroughly with bidistilled water and dry them. Convert the amino groups to carboxyl groups by means of GA in DMF, $c = 2$ mg/μl (10 μl per SAW device). After 30 h reaction, rinse the SAW devices with DMF and acetone and dry them.

3.3. Coupling of Proteins (i.e., Capture Molecule or Linker) with FIA System

1. Use a SAW device coated with DC-PEG, M_r 2,000, or a SAW device coated with carboxylated AMD, M_r 3,000 (see Subheading 3.2). Insert the SAW device in the flow cell and integrate it in the measurement setup as described in item 3 of Subheading 2.1.
2. Monitor the SAW sensor signal during the coupling procedure. An overview of the FIA system settings involved is given in Table 1.
3. Switch the FIA system to rinse/load mode and set the flow rate to 0.03 ml/min.
4. Rinse the carboxylated SAW device with running buffer until a stable baseline can be observed.
5. If prepared in advance, defrost vials of EDC, NHS, immobilization buffer and deactivation solution. Otherwise, use freshly prepared solutions.

Table 1
Coupling of proteins with FIA system: solutions and system settings

Time (min)	Mode	Reagents
Before start	Load	Activation solution: EDC/NHS
0–1	Flow rate: 0.03 ml/min rinse	Running buffer
1–9	Inject	Activation solution: EDC/NHS
9–17	Rinse	Running buffer
13–14 17–25	Load Inject	Protein solution: capture molecule or linker
25–33	Rinse	Running buffer
29–30 33–41	Load Inject	Deactivation solution: ethanolamine
45–60	Flow rate: 0.05 ml/min rinse	Running buffer
60	End of measurement	

6. Prepare the EDC/NHS activation solution immediately before use. Load the solution into the sample loop.
7. Activate the carboxylated SAW device surfaces by injecting the EDC/NHS mixture in the running buffer for 8 min (see Note 10).
8. Rinse with running buffer for 8 min.
9. During the rinsing step, load the solution of the capture molecule (e.g., antibody) or linker (e.g., biotin-binding protein, such as streptavidin) dissolved in immobilization buffer into the sample loop.
10. Couple the capture molecule or linker on the activated surface by injecting the mixture in the running buffer for 8 min.
11. Rinse with running buffer for 8 min.
12. During the rinsing step, load the deactivation solution into the sample loop.
13. Deactivate potentially remaining NHS-ester groups by injecting the deactivation solution in the running buffer for 8 min (see Note 11).
14. Rinse with running buffer for 4 min and switch the flow rate to 0.05 ml/min (see Note 10). Wash for further 15 min. Finish the measurement.

Table 2
Binding experiments with FIA system: solutions and system settings

Time (min)	Mode	Reagents
Before start	Load	Protein solution
0–1	Flow rate: 0.05 ml/min rinse	Running buffer
1–5	Inject	Protein solution
5–9	Rinse	Running buffer
9	End of measurement	

15. In the case of immobilized capture molecule, such as antibody, the SAW biosensor is ready for measurements.
16. In the case of immobilized linker (e.g., biotin-binding protein) a subsequent binding of capture molecule (e.g., biotinylated antibody) is required. Monitor the SAW biosensor signal during the binding procedure. An overview of the FIA system settings involved is given in Table 2. The flow rate is maintained at 0.05 ml/min. Load the solution of capture molecule dissolved in running buffer into the sample loop. Inject the solution in the running buffer for 4 min. Rinse with running buffer for 4 min. Finish the measurement. Now the SAW biosensor is ready for measurements.
17. Perform measurements with the SAW biosensor using the FIA system (see Subheadings 3.5 and 3.6) immediately after the SAW biosensor is ready.

3.4. Coupling of Small Molecules (Ligand)

1. Use SAW devices coated with carboxylated AMD, M_r 70,000 (see step 2 in Subheading 3.2). Insert the SAW devices in a glass chamber containing DMF vapor.
2. Prepare the activation solution as described in step 1 of Subheading 2.5 immediately before use.
3. Activate the carboxylated SAW device surfaces by applying the activation solution (15 μl per SAW device) and 2 h reaction at room temperature.
4. Rinse the SAW devices with DMF and dry them with nitrogen.
5. Couple the ligand on the activated surfaces by applying the ligand solution (15 μl per SAW device) and 20 h reaction at room temperature.
6. Rinse the SAW devices with DMF and bidistilled water and dry them.

7. SAW biosensor surfaces are now ready for measurements (see Note 12). Perform measurements with the SAW biosensor using the FIA system (see Subheadings 3.5 and 3.6).

3.5. Measurement with SAW Biosensor: Blocking Step/Quality Testing

1. Use a SAW biosensor with immobilized capture molecule or ligand corresponding to the analyte (see Subheadings 3.3 and 3.4). If the SAW biosensor is already inserted in the flow cell and integrated in the FIA system due to previous preparation steps, proceed with the next step. If not, insert the SAW biosensor in the flow cell and integrate it in the measurement setup as described in Subheading 2.1.
2. Monitor the SAW sensor signal during the testing procedure. An overview of the FIA system settings involved is given in Table 2.
3. Switch the FIA system to rinse/load mode and set the flow rate to 0.05 ml/min.
4. Rinse the SAW biosensor with running buffer until a stable baseline can be observed.
5. Load the blocking solution into the sample loop.
6. Inject the blocking solution in the running buffer for 4 min.
7. Rinse with running buffer for 4 min.
8. Finish the measurement. Signal heights can be determined by calculating the mean of the signal response curves in a defined range of 30 s after the injection interval (see Fig. 3). If the signal height obtained is 1 kHz or higher, the SAW biosensor is not suited for binding experiments and has to be discarded (see Note 13). If the signal height obtained is below 1 kHz, the SAW biosensor is ready for binding experiments. Perform

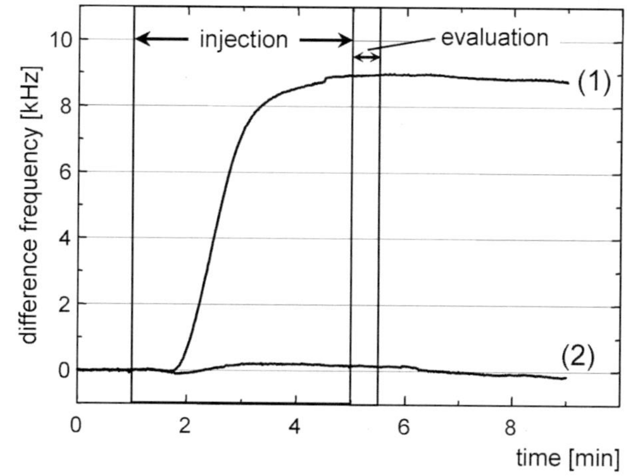

Fig. 3. SAW signal responses for analyte (1) and non-analyte (2) samples injected in the running buffer. Injection interval: 1–5 min, evaluation interval: 5.0–5.5 min.

binding experiments with the SAW biosensor using the FIA system (see Subheading 3.6) immediately after the SAW biosensor is ready.

3.6. Measurement with SAW Biosensor: Binding Experiment

1. Use a SAW biosensor with immobilized capture molecule or ligand corresponding to the analyte which passed the quality test (see Subheading 3.5). Due to the latter it is already inserted in the flow cell and integrated in the measurement setup.
2. Monitor the SAW sensor signal during the binding experiment. An overview of the FIA system settings involved is given in Table 2.
3. Switch the FIA system to rinse/load mode and set the flow rate to 0.05 ml/min.
4. Rinse the SAW biosensor with running buffer until a stable baseline can be observed.
5. Load the analyte solution into the sample loop.
6. Inject the analyte solution in the running buffer for 4 min.
7. Rinse with running buffer for 4 min.
8. Finish the measurement. Signal heights can be determined by calculating the mean of the signal response curves in a defined range of 30 s after the injection interval (see Fig. 3).
9. Clean the sample loop with rinsing solutions before starting a new binding experiment (see Note 14).

4. Notes

1. The operation frequency of the SAW device is one of the parameters determining the magnitude of the signal response. In principle, the higher the frequency, the higher the SAW signal response. However, the penetration depth of the surface acoustic wave decreases with increasing operation frequency, which could lead to reduced signal responses (see also Note 5). Therefore, the chosen frequency must neither be too high nor too low. We think best results should be obtained with a similar SAW resonator configuration as used here. However, if another operation frequency has to be used, then all coating-relevant data affecting the thickness of the sensing layer should be reconsidered. If a higher frequency is used it might be necessary to adjust all coating parameters towards a lower resulting layer thickness. Conversely, if a lower frequency is used it could be advantageous to adjust all coating parameters towards a higher resulting layer thickness to improve the sensor performance. In case of a delay line device all considerations concerning the

layer composition most likely must be reconsidered regardless of its operation frequency.

2. Difference frequencies below 20 MHz are more easily countable by a commercially available frequency counter than the absolute frequency of several hundreds of MHz of a SAW device. As difference frequencies (relative to the reference oscillator) were used as signal output, and as the frequency of the reference oscillator is higher than that of the SAW biosensor, processes leading to frequency decrease in the oscillator circuit, such as mass increase, result in increasing frequencies on the monitor. In some cases changes of viscoelasticity dominate, which may result in decreasing frequencies on the monitor. For the sake of clarity, signal response curves were plotted to start at 0 Hz (see Fig. 3).
3. The pH value of the coupling reaction has to be chosen regarding two facts. First it has to ensure that the amino groups of the protein molecules to be coupled are positively charged while maintaining a negative charge of the carboxyl groups on the surface to allow electrostatic attraction. Second, it has to allow the reaction with the active ester.
4. Rapid weighing of EDC is recommended due to the hygroscopic nature of EDC. Fresh preparation of carbodiimide mixtures is recommended due to the reduced stability of those solutions.
5. The optimal amount (and hence the concentration of the reagent solution) of capture molecule, linker or ligand depends, among others, on the morphology of the hydrogel layer. If, for instance, the thickness of the sensing layer, including immobilized capture molecules or ligands, is too high, the subsequent interaction of analyte and corresponding binding partner might not be detected due to the limited penetration depth of the surface acoustic wave, which would lead to a loss of signal response. As the coupling efficiency is one of the parameters determining the morphology of the sensing layer, the concentrations of the reagent solutions have to be optimized for each application and only approximate values can be given here.
6. Ideally, running buffer and analyte buffer are the same to prevent effects of the sample medium on the SAW signal response. If this is not possible, e.g., due to the sample medium in which the analyte is delivered, analyte signals still can be determined by the difference of the signal response before and after the injection interval, both monitored while rinsing with running buffer and thus avoiding medium effects.
7. The surface of the SAW resonator consists of two materials: $LiTaO_3$ and gold. As chemical coupling procedures can typically be optimized for one of the materials only, starting from

the device surface might result in inhomogenities of the subsequently prepared sensing layer. The initial coating with a polymer provides a chemically homogeneous surface resulting in a sensing layer without inhomogenities.

8. High temperature variations can damage the SAW device. Besides, it is easier to rinse off the remaining molten DC-PEG with hot water than with cold water.

9. The thickness (and hence the concentration of the reagent solution) of the 3D hydrogel AMD has to be chosen regarding two facts. First, the density of the AMD layer must be sufficient to prevent nonspecific adsorption of the sample medium in the subsequent assay. Second, in case of potential disadvantageous effects resulting from changes of viscoelasticity of the sensing layer, the thickness must not be too high, which in turn depends mainly on the morphology of the sensing layer. Therefore, the concentration of the hydrogel solution has to be optimized for each application and only approximate values can be given here.

10. The incubation time is one parameter to control the EDC/NHS coupling efficiency. It has been proven to be useful to chose the reaction times (and hence injection intervals) to be longer during the coupling reaction (8 min) than during the analyte binding (4 min). Therefore, in order to avoid increased reagent consumption, the flow rate is reduced to 0.03 ml/min during coupling procedure compared to 0.05 ml/min during the SAW biosensor binding experiments. When the EDC/NHS coupling procedure is finished, the flow rate is set to that of the binding experiments (0.05 ml/min) to increase the rinsing efficiency and to allow a first evaluation of the signal stability considering subsequent SAW biosensor measurements.

11. When performing the EDC/NHS coupling procedure, it has to be guaranteed that after the reaction there are no free NHS-ester groups remaining, because any protein used in a subsequent sample could couple to these active ester groups leading to false positive results. Therefore, a deactivation step is performed using a primary amine, such as ethanolamine, at a slightly alkaline pH to react with potentially remaining NHS-ester groups and prevent them from other coupling reactions in subsequent binding experiments.

12. The shelf life of SAW biosensors with immobilized ligands depends on the application. In our experiments, such biosensors could be stored up to 2 weeks at 4°C without loss of functionality.

13. It is not recommended to perform more blocking steps until the BSA signal meets the requirements, because proteins contained

in the sample of a subsequent binding experiment could bind nonspecifically to the BSA, leading to false positive results.

14. After each sample injection, the sample loop should be cleaned with the rinsing solutions to avoid carryover effects.

References

1. Flory CA, Baer RL (1987) Surface transverse wave mode analysis and coupling to interdigital transducers. IEEE proc Ultrason Symp:313–318
2. Shiokawa S, Moriizumi T (1988) Design of SAW sensor in liquid. Jpn J Appl Phys Suppl 27-1:142–144
3. Länge K, Rapp BE, Rapp M (2008) Surface acoustic wave biosensors: a review. Anal Bioanal Chem 391:1509–1519
4. Weiss W et al (1998) Viscoelastic behavior of antibody films on a shear horizontal acoustic surface wave sensor. Anal Chem 70:2881–2887
5. Länge K, Rapp M (2008) Influence of intermediate aminodextran layers on the signal response of surface acoustic wave biosensors. Anal Biochem 377:170–175
6. Länge K, Rapp M (2009) Influence of intermediate hydrogel layer and amount of binding sites on the signal response of surface acoustic wave biosensors. Sens Act B Chem 142:39–43
7. Länge K et al (2003) A surface acoustic wave biosensor concept with low flow cell volumes for label-free detection. Anal Chem 75:5561–5566
8. Gizeli E et al (1997) Antibody binding to a functionalized supported lipid layer: a direct acoustic immunosensor. Anal Chem 69:4808–4813
9. Gronewold TMA et al (2006) Discrimination of single mutations in cancer-related gene fragments with a surface acoustic wave sensor. Anal Chem 78:4865–4871
10. Masson JF et al (2004) Preparation of analyte-sensitive polymeric supports for biochemical sensors. Talanta 64:716–725
11. Gedig ET (2008) In: Schasfoort RBM, Tudos AJ (eds) Handbook of surface plasmon resonance, 1st edn. RSC, Cambridge
12. Bender F et al (2004) On-line monitoring of polymer deposition for tailoring the waveguide characteristics of Love-wave biosensors. Langmuir 20:2315–2319
13. Länge K, Grimm S, Rapp M (2007) Chemical modification of parylene C coatings for SAW biosensors. Sens Act B Chem 125:441–446
14. Löfas S, Johnsson B (1990) A novel hydrogel matrix on gold surfaces in surface plasmon resonance sensors for fast and efficient covalent immobilization of ligands. J Chem Soc Chem Commun:1526–1528
15. Österberg E et al (1995) Protein-rejecting ability of surface-bound dextran in end-on and side-on configurations: comparison to PEG. J Biomed Mater Res 29:741–747
16. McHale G et al (2000) Acoustic wave–liquid interactions. Mat Sci Eng C 12:17–22
17. Lucklum R, Hauptmann P (2006) Acoustic microsensors—the challenge behind microgravimetry. Anal Bioanal Chem 384:667–682
18. Länge K, Gruhl FJ, Rapp M (2009) Influence of preparative carboxylation steps on the analyte response of an acoustic biosensor. IEEE Sens J 9:2033–2034

Chapter 32

Microchip UV Absorbance Detection Applied to Isoelectric Focusing of Proteins

Junjie Ou and Carolyn L. Ren

Abstract

Isoelectric focusing (IEF) is considered as an attractive separation technique for biologically amphoteric compounds (e.g., proteins and peptides) based on their isoelectric point (pI). With the advancement in micromachining technology, microchip format IEF has attracted significant attention. Both single-point and whole column imaging detection (WCID) methods have been employed for analyzing the separation performance in a microchip. WCID is more favorable than single-point detection because the latter requires the focused bands to be mobilized and thus adds more complexity to the design and operation of such microchips. Fluorescence- and UV absorbance-based WCID have been successfully adapted in glass and PDMS microchips. We have developed polydimethylsiloxane (PDMS) microchips for IEF applications where UV-WCID is employed for evaluating the separation performance. The chips are designed for use in the iCE280 analyzer (Convergent Bioscience Inc., Toronto), for capillary-based IEF where UV-WCID is employed for analyzing the separation performance. Three kinds of microchips that have been successfully developed using standard soft lithography technology are described in detail.

Key words: Microfluidic chip, Isoelectric focusing, Protein separation, UV absorption detection, Whole channel imaging detection

1. Introduction

Isoelectric focusing (IEF) is a powerful and practical method for high-resolution separation of amphoteric molecules such as peptides, proteins, and components within complex biological mixtures. Capillary-based isoelectric focusing (CIEF) is a miniaturized approach to carry out IEF in capillaries, with fused silica being the most common material used in this technique (1–3). With the advancement in micromachining techniques, microchip format IEF has emerged and been successfully adopted for academic research and industrial processes over the past decade (4, 5). In microchip based IEF, similar to CIEF, a microchannel connecting two reservoirs is filled with a mixture of carrier ampholytes and samples.

Gareth Jenkins and Colin D. Mansfield (eds.), *Microfluidic Diagnostics: Methods and Protocols*, Methods in Molecular Biology, vol. 949, DOI 10.1007/978-1-62703-134-9_32, © Springer Science+Business Media, LLC 2013

The two reservoirs are filled with acidic buffer (anolyte) and basic buffer (catholyte), respectively. Upon applying an axial electric field, a linear pH gradient is formed in the microchannel, and the samples focus at the position where the pH becomes equal to their isoelectric points (pI).

For the method of single point detection, the focused zones need to be pumped passing through the detection point using chemical, hydrodynamic, or electroosmotic mobilization methods for detection purposes (4, 6). The single point detection method restricts IEF applications to a certain extent because the mobilization process, i.e., transporting the focused zones usually increases analysis time and lowers the resolution and repeatability of the separation (7). An excellent alternative is whole column imaging detection method (WCID) developed by Pawliszyn and coworkers (8), which does not require the mobilization process and thereby avoids the above-mentioned problems.

There are two commonly used detection methods for IEF applications, each of which has its pros and cons: laser-induced fluorescence detection (LIF), and ultraviolet (UV) absorbance detection. LIF has been widely used for microchip-based CE analysis due to its superior selectivity and sensitivity (9, 10). However, LIF detection normally requires the derivatization of proteins with a fluorescent label, which is time consuming and expensive. More importantly, the derivatization of proteins can alter their properties such as charge, molecular weight, hydrophobicity or can affect stability by rupture of disulfide linkages. In addition, labeling reactions result in a distribution of labeling products that often degrade separation efficiency (11). Although the UV absorbance detection method eliminates the need of derivatization, it has not been widely applied to microchip-based separations mainly due to the low detection sensitivity. The sensitivity of UV absorbance detection method is proportional to its light path, which in most applications is the channel height. The channel height is limited by the fabrication techniques. To fully utilize the advantages of UV absorbance detection, there has been growing interest in developing techniques to apply UV absorbance detection for microchip-based separations.

A CIEF instrument implemented with UV-WCID (iCE280 Analyzer) has been commercially available from Convergent Bioscience Inc. (acquired by Cell Biosciences in 2010) since 1998. Although this instrument has been widely adopted as a powerful tool for research and development as well as for quality control by many leading pharmaceutical companies (e.g., Pfizer, Amgen, Bristol-Myers Squibb, and Genentech), it has several disadvantages associated to the use of capillaries. These include, detection sensitivity limited by the inner diameter of the capillary used (100 μm), lens effect in detection due to the capillary's round cross-sectional area, and limited potential for multidimensional separation. To enhance the detection sensitivity, a metal slit with a 65 μm-wide opening must be manually glued on top of the capillary to block any stray light from entering the detector, which has proven to be challenging. Moreover, two hollow fibers must be glued to the separation capillary to separate the electrolytes in the reservoirs from the samples in the capillary. These manual processes also increase the cost of the cartridge. In this chapter, we focus on introducing the techniques we have recently developed for applying UV-WCID to microchip-based IEF aiming to address the limitations inherent to the cartridge used in the iCE280 analyzer.

2. Materials

2.1. Master Fabrication

1. 3-in. silicon wafers are used as a substrate to make masters for replica molding of polymer microchips (Wafer Reclaim Services LLC, San Jose, CA, USA). These wafers are cleaned by manufacturers prior to shipping and are dehydrated on a hot plate at 200°C for 30 min prior to use (see Note 1).
2. SU-8 2000 series photoresist (SU-8 2005, 2015, 2025 and 2075) and propylene-glycol-monoether-acetate (PGMEA) developer (Microchem, Newton, MA, USA) (see Note 2).
3. Photomasks with the designs created using a computer-aided design (CAD) program are made from Mylar films (CAD/Art Services Inc., Brandon, OR, USA).
4. Trichloromethylsilane (TCMS).
5. A combined spin-coating and heating system (Brewer Science, Rolla, MO) with a liquid dispensing unit.
6. An UV exposure system / mask aligner (Newport, Irvine, CA, USA).
7. Isopropanol.
8. Ultrapure water.
9. Glass vacuum desiccator.

2.2. Macroporous PDMS Membrane

1. Porogenic Solution: polystyrene of molecular weight (MW) 192 kDa dissolved at 1 g/mL in toluene.
2. Sylgard 184 PDMS kits with prepolymer base and curing agent are obtained from Dow Corning (Midland, MI, USA). Unless stated otherwise, the PDMS prepolymer mixture is freshly prepared by thoroughly mixing PDMS prepolymer base and curing agent at a weight ratio of 10:1 and degassing for 30 min under vacuum.
3. Standard microscope glass slide (1 × 3 in.).
4. Toluene, acetone and ethanol for rinsing cycle.
5. Air plasma cleaning system.

2.3. Microchip Fabrication

1. Nickel slit sheets (3 mm × 6.5 cm) with a 65 μm wide opening (StenTech, Mississauga, ON, Canada) (see Note 3). This metal slit is glued on the top of the microchannel to block the stray UV light from entering the detector for some applications.
2. Quartz slides (1 × 3 in.) (SPI Supplies, West Chester, PA, USA). These slides must be cleaned in a solution of H_2SO_4/H_2O_2 (3/1, v/v) and then dehydrated on a hot plate at 200°C for 30 min prior to use (see Note 4).
3. PMMA holders (1 × 3 in.) (Convergent Bioscience Inc., Toronto, Canada) with a 1 mm wide, 6 cm long slit (see Note 5), used to support the microchip and assist its best fit with the iCE280 analyzer.
4. Fused-silica capillaries (50 μm inner diameter × 100 μm outer diameter and 100 μm inner diameter × 168 μm outer diameter) (Polymicro Technologies, Phoenix, AZ, USA), used to connect the microchannel with the sample reservoir and waste reservoir for continuous operation.
5. Regenerated cellulose dialysis membrane with molecular weight cut-off (MWCO) of 12–15 kDa) (Spectrum Laboratories, Inc., Rancho Dominguez, CA, USA), used to replace the hollow fiber used in the commercial cartridge to separate the electrolytes in the reservoirs and the sample in the channel.
6. Vacuum oven.
7. Tool for punching holes in PDMS. Biopsy punches of various diameters may be used for this purpose.
8. Optical Microscope

2.4. Isoelectric Focusing with UV-WCID Evaluation

1. Poly(vinylpyrrolidone) (PVP; MW, 360 kDa), poly(vinyl alcohol) (PVA; MW, 13–23 kDa) and methyl cellulose solutions are dissolved in deionized (DI) water to the desired concentration, and then stored at room temperature (see Notes 6 and 7).
2. 0.2 μm pore size cellulose acetate filter (Sartorius, Gottingen, Germany).
3. Myoglobin and β-lactoglobulin are dissolved in DI water at a concentration of 5 mg/mL, respectively, and stored at 4°C.
4. Low molecular mass pI markers of 4.65, 5.12, 6.14, 7.05, 8.18, and 8.79 (Convergent Bioscience Inc., Toronto, ON, Canada) stored at 4°C.
5. Carrier ampholytes (Pharmalytes, pH range 3–10), stored at 4°C.
6. Solution of human hemoglobin control containing HbA, HbF, HbS, and HbC (Helena Laboratories, Beaumont, TX, USA), stored at 4°C.
7. The anolyte and catholyte are 0.1 mol/L phosphoric acid and 0.1 mol/L sodium hydroxide containing 1.5% PVP, respectively.

8. Sample solutions for IEF are prepared in DI water containing 1.5% PVP, 2% Pharmalytes (pH range 3–10) and different proteins or pI markers, stored at 4°C (see Note 8).
9. iCE280 analyzer (Convergent Bioscience, Toronto, ON, Canada), used to perform all IEF-WCID experiments (see Note 9).

3. Methods

3.1. Fabrication of Master and PDMS Mold Using Soft Lithography Technology

1. Spin-coating: To prepare a master, a silicon wafer is first spin-coated with SU-8 negative photoresist at 1,800 rpm for 35 s, giving rise to a 100 μm high microchannel (see Note 10).
2. Soft baking: The wafer coated with SU-8 is then baked at 65°C for 5 min and then 95°C for 15 min, and slowly cooled to room temperature (see Note 11).
3. UV-exposure: The photomask (see Note 12) is placed on the SU-8 layer of the soft baked wafer, and then the entire structure is put into the mask aligner. The designs on the mask are transferred to the coated photoresist through UV exposure at a dose of 800 mJ/cm^2 (see Note 13).
4. Post-exposure baking: After pattern transfer, the wafer coated with SU-8 is baked using the settings in the following order: 65°C for 5 min, 95°C for 10 min, 65°C for 5 min, and 50°C for 5 min. Afterwards, the wafer is slowly cooled down to room temperature (see Note 14).
5. Development: After post-exposure baking, the wafer coated with SU-8 photoresist is placed in a holder so that the wafer can be suspended in a jar filled with the SU-8 developer, PGMEA. Development time varies with the desired thickness. The developed wafer is then washed with isopropanol. If white solute is observed, then the wafer needs to be put back to the developer for another 30 s. This procedure is repeated until no white solute is observed, indicating that all of the non-cross-linked SU-8 has been removed. Finally, the wafer is washed with ultrapure water and blown dry with air.
6. Silanization: The wafer with the developed design is placed in a glass vacuum desiccator where a small plastic bottle of TCMS is left open. The wafer is left in the desiccator for approximately 25 min, and then stored in a wafer transport holder (see Note 15). This wafer is used as a master to replicate the design in PDMS.

3.2. Preparation of Macroporous PDMS Membrane

1. The solution of polystyrene (PS) in toluene is spin-coated on a microscope glass slide (1 × 3 in.) at 3,000 rpm for 30 s. Then the glass slide is baked at 60°C for 30 min to firmly deposit the PS film on the substrate.

2. A mixture of Slygard PDMS prepolymer base and curing agent is thoroughly mixed, and diluted with porogenic solution (toluene/polystyrene) (see Note 16). After degassing, the resulting suspension of diluted PDMS mixture is spin-coated on the top of the PS film deposited on the glass slide prepared in the previous step (step 1) and then soft baked on a hot plate at 60°C for 1 h. This PDMS layer is the membrane to be used for IEF separation.
3. The resulting glass slide is pretreated by air plasma and brought into contact with another PDMS layer (2 mm thick) containing two reservoirs (2 μm in diameter), which is also pretreated by air plasma to irreversibly bond them together (see Note 17). The distance between the two reservoirs is the desired separation channel length.
4. The bonded PDMS membrane and PDMS layer with two reservoirs are then peeled off from the glass slide. The resulting membrane is then immersed into toluene, acetone and ethanol, successively, to remove porogenic solvents, and then dried in an oven under vacuum at 50°C for 12 h (see Note 18).

3.3. Fabrication of the First Generation of PDMS Microchip

1. The mixture of PDMS prepolymer and curing agent is poured onto the master (see Note 19) and cured in the vacuum oven for 1 h at 80°C. After curing, the PDMS substrate with a thickness of 2 mm is peeled off from the master, and cut to a size of 1 × 3 in. Two 3 mm diameter holes are punched at the reservoir locations.
2. Another 2 mm thick PDMS substrate without any features is also prepared by simply pouring the PDMS prepolymer mixture onto a blank silicon wafer, curing it in the vacuum oven for 1 h at 80°C and then cutting it to a size of 1 × 3 in. (see Note 20).
3. The chip is made by bonding the PDMS substrate with the channel features (prepared in step 1) to the blank PDMS substrate (prepared in step 2) via air plasma treatment for 45 s at 29.6 W (see Note 21).
4. The PMMA holder is glued under the PDMS microchip serving as a supporting base to prevent chip deformation.
5. The nickel slit sheet with a 65 μm opening is glued onto the PDMS microchip under an optical microscope (see Notes 22 and 23) by aligning the opening with the microchannel.

3.4. Fabrication of the Second Generation of PDMS Microchip Sandwiched with a Dialysis Membrane

1. A PDMS layer with a 100 μm wide, 100 μm high, 56 mm long microchannel is prepared using standard soft lithography technology (12). This layer will serve as the bottom substrate of the final microchip.
2. The upper substrate of the microchip is prepared from another piece of PDMS layer fabricated with two 6 mm square, 40 μm

high recesses, which are separated with a distance of 57 mm (see Note 24) using standard soft lithography technology. Two reservoirs with 2 mm in diameter are punctured through the center of the recesses in the upper PDMS substrate.

3. The mixture of PDMS prepolymer and curing agent diluted with toluene (1:2 or 1:1, v/v) is spin-coated onto a clean glass microscope slide (see Note 25). The layer acts as "mortar" for binding the dialysis membrane in place. Sacrificial dialysis membranes with the desired size (6 mm square) are stamped onto the PDMS mortar for 1 min and then placed onto the recesses of the upper PDMS substrate for 1 min before being removed and discarded. Two fresh dialysis membranes are then put into the recesses to replace those membranes stamped with the PDMS mortar, respectively. The procedures are illustrated in Fig. 1.
4. The resulting PDMS substrate glued with the dialysis membranes is irreversibly bonded with the bottom PDMS substrate prepared in step 1 via air plasma. The combined PDMS chip is then placed in the vacuum oven to cure for 1 h at 65°C to harden the PDMS mortar.
5. A layer of glue is applied to the outer wall of two pieces of capillaries with an inner diameter of 50 μm and an outer diameter of 100 μm. These capillaries are used to connect the microchannel with the sample reservoir and the waste reservoir for continuous operation. Then, the capillaries are manually inserted into the two ends of the microchannel until the capillaries reach the edge of dialysis membrane (see Note 26). This step is done under an optical microscope.
6. The PDMS microchip sandwiched with the dialysis membranes and embedded with the injection and discharge capillaries is cut to the desired dimension and placed on the PMMA holder (see Note 5). A metal slit with a 65 μm wide, 50 mm long opening is carefully aligned with the microchannel and glued onto the top of the PDMS chip under an optical microscope (see Note 22). Two large reservoirs are glued on the top of the resulting microchips. A schematic microchip sandwiching a PDMS membrane is shown in Fig. 2.

3.5. Fabrication of the Third Generation of PDMS/SU8/Quartz Hybrid Microchip

1. A 50 μm thick SU-8 (SU-8 2025) layer is spin-coated onto a quartz substrate (see Note 4) at 500 rpm for 25 s followed by a spin-coating speed of 1,500 rpm for 30 s (see Note 2). The substrate is then soft baked on a hot plate at 65°C for 5 min and at 95°C for 7 min to remove the photoresist solvent and finally cooled to room temperature (see Note 11).
2. UV exposure is performed in a UV mask aligner at the wavelength of 365 nm for a dose of 300 mJ/cm^2 (see Notes 13 and 27). Post-exposure bake is performed on a hot plate at 65°C

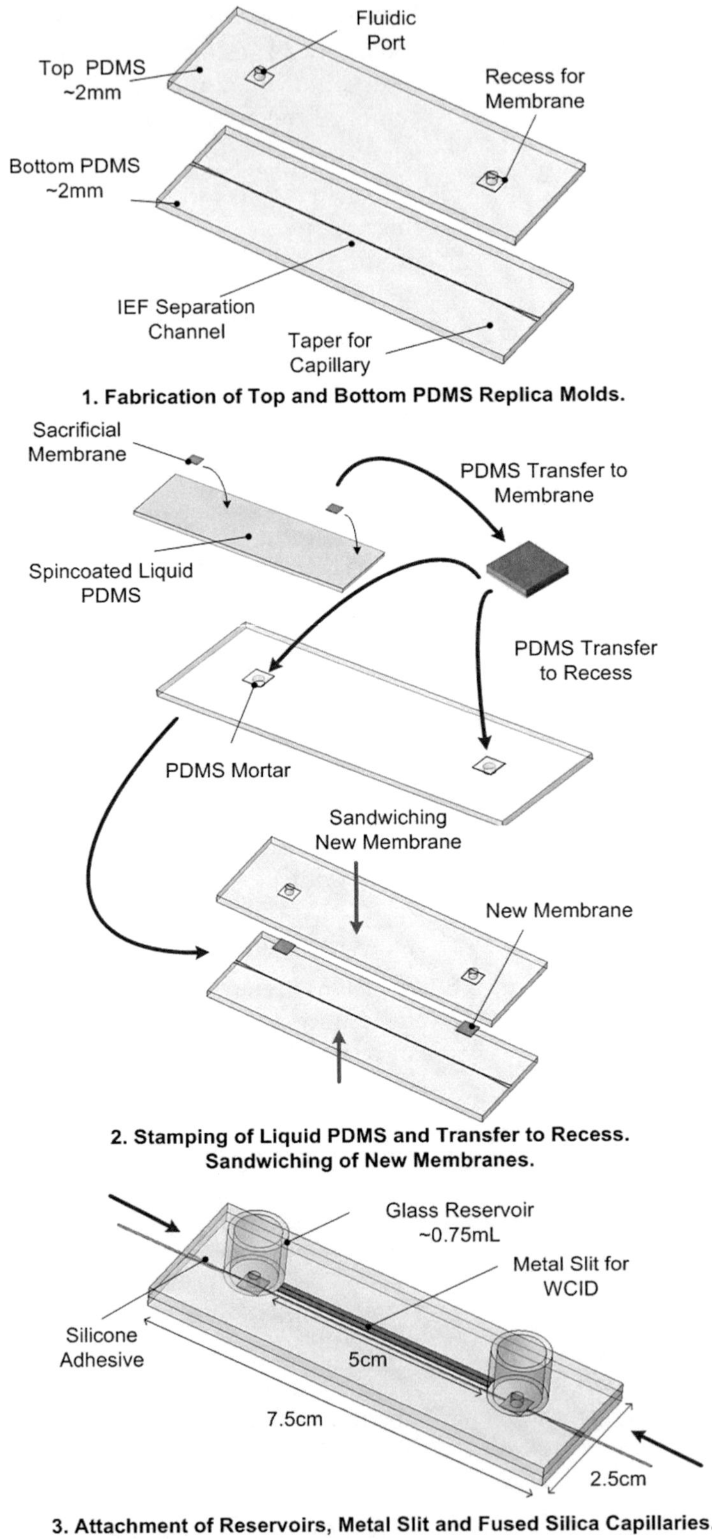

Fig. 1. Schematic fabrication procedures of PDMS microchip sandwiching dialysis membrane for IEF-WCID (reproduced from ref. 14 with permission from American Chemical Society).

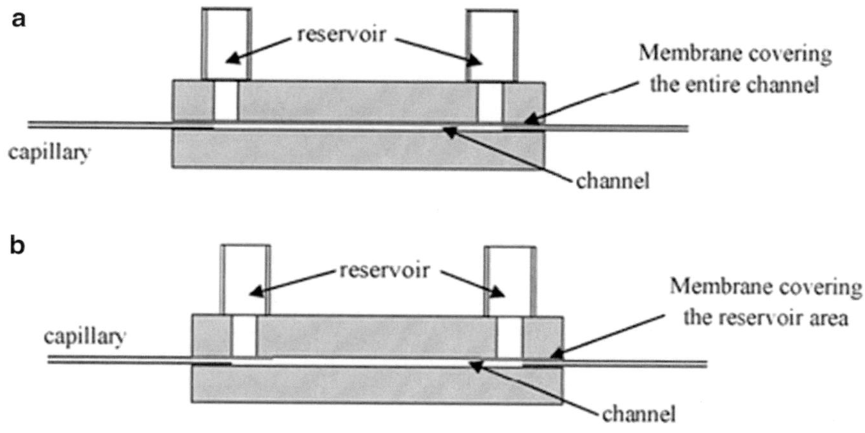

Fig. 2. Schematic depiction of the microfluidic chip (*side view*) sandwiching a macroporous PDMS membrane covering (**a**) whole channel and (**b**) two ends of channel (reproduced from ref. 13 with permission from Elsevier Science).

for 5 min and at 95°C for 6 min, successively. Afterwards, the substrate is cooled slowly to room temperature (see Note 14). The pattern is developed by dipping the substrate into the PGMEA developer for 7 min to dissolve the unexposed SU-8 (see Subheading 3.1, step 5).

3. The PDMS prepolymer mixture is poured onto a blank silicon wafer and cured for 1 h at 80°C to prepare a blank PDMS layer without any feature. After curing, the PDMS layer with a thickness of 2 mm is peeled off from the silicon wafer and cut to a size of 1 × 3 in. Two holes of 3 mm diameter are punched through the PDMS layer at the reservoir locations.
4. After pretreatment with air plasma at a dose of 29.6 W for 60 s, both the PDMS layer and the quartz substrate coated with SU-8 are brought into contact such that the PDMS contacts the SU-8 layer encloses the microchannel in the SU-8. The final chip is a hybrid PDMS/SU-8/Quartz chip, as shown in Fig. 3. Similarly, other hybrid chips with straight channels of 100 μm wide, 100 μm deep, 5 cm long, or 50 μm wide, 200 μm deep, 5 cm long, respectively, are fabricated according to the above-mentioned procedures with minor adjustments (particularly for the depth of SU-8—(see Notes 2, 10, 13 and 14).
5. The resulting hybrid microchip is positioned on the PMMA holder (see Note 5) for IEF-WCID analysis.

3.6. IEF-WCID Measurement of Proteins with the iCE280 Analyzer

1. Prior to measurement, the iCE280 analyzer is turned on according the guidelines from the manufacturer.
2. A volume of 20 ml PVP solution (1.5%, w/v) is added to one of the two reservoirs and then sucked into the microchannel using a vacuum pump connected to the other reservoir to condition

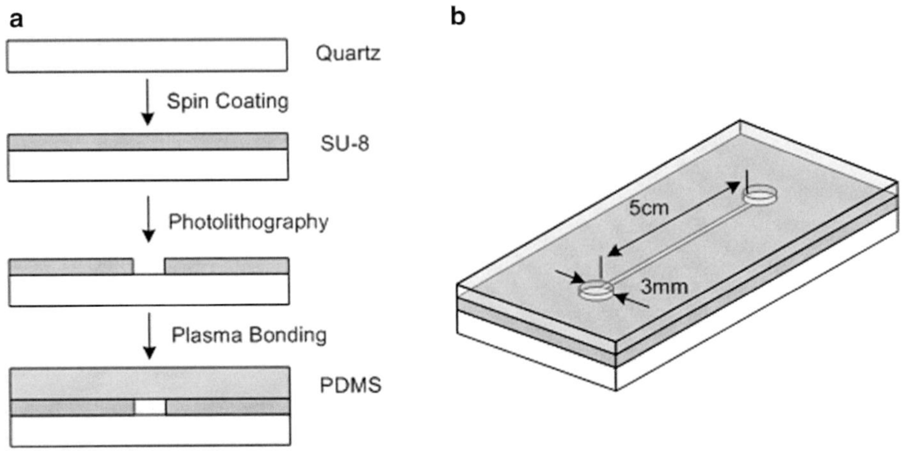

Fig. 3. (**a**) Diagram of the fabrication process and (**b**) side view of hybrid PDMS/SU-8/quartz chip (reproduced from ref. 15 with permission from Royal Society of Chemistry).

the microchannel. The microchannel is preconditioned by the PVP solution for 30 min (see Note 28). Afterwards, the solutions in both reservoirs are removed using a syringe.

3. The sample solution (20 μL) is filled into the whole channel using a vacuum pump.
4. Both anolyte and catholyte solutions (20 μL for each) are simultaneously added into the left reservoir and right reservoir of the microchip using two pipettes, respectively (see Note 29).
5. The microchip is quickly placed in the chamber of iCE280 analyzer for IEF. Focusing is performed by first applying a voltage difference of 1,500 V for 4 min and then maintaining it by applying a voltage difference of 3 kV until the focusing is finished. A typical electrophoretic profile for IEF-WCID analysis is shown in Fig. 4. Between different runs, the channel is washed with the PVP solution for 5 min.

4. Notes

1. The silicon wafer must be dehydrated and slowly cooled down to room temperature prior to use. The dehydration process lasts for about 4 h depending on the humidity of the room. Otherwise, the SU-8 feature may peel off from the wafer, resulting in a shorter lifetime of the master. It is necessary to remove the dust or dirt that may have accumulated on the wafer surface using nitrogen prior to spin-coating.
2. The thickness of the spin-coated SU-8 layer is influenced by the spin-coating speed and the viscosity of the SU-8 as recom-

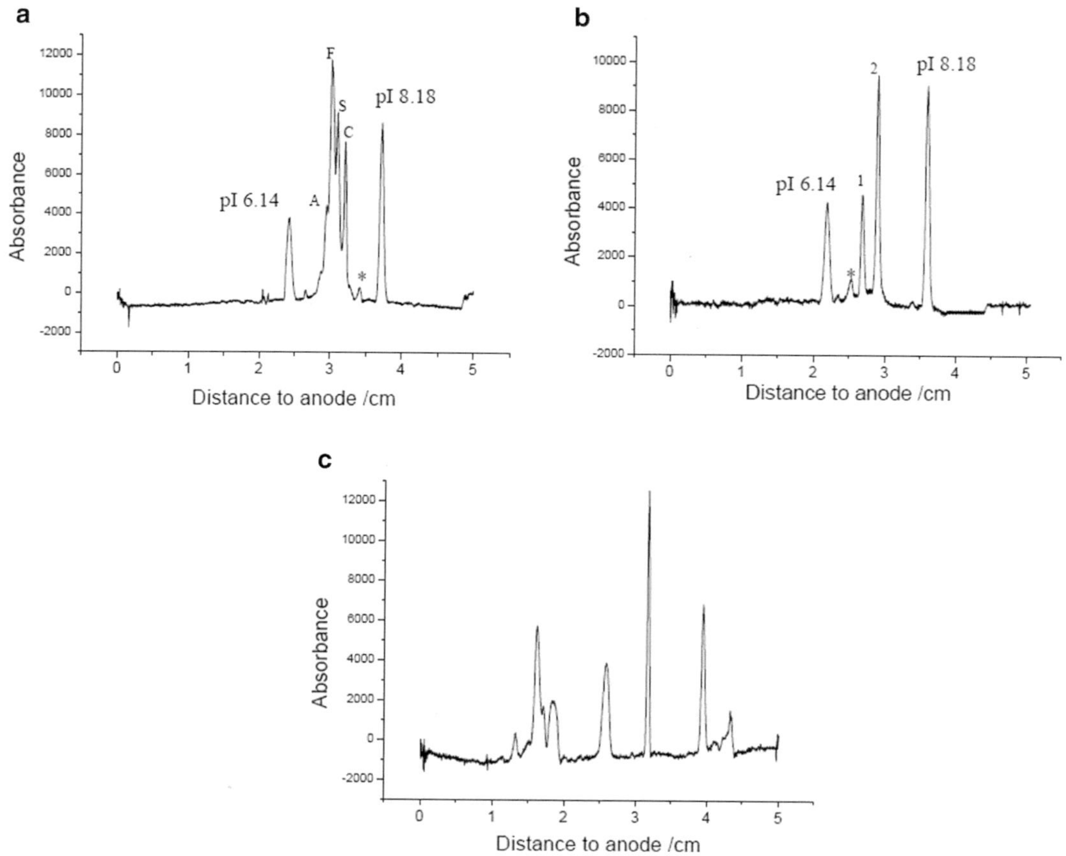

Fig. 4. IEF separation of (**a**) hemoglobin control AFSC, (**b**) myoglobin, and (**c**) the degenerative pI markers mixture on a PDMS microchip with a 100 μm high, 100 μm wide channel using IEF-WCID (reproduced from ref. 14 with permission from American Chemical Society).

mended by the manufacturer. It is highly recommended to develop protocols for a specific system.

3. Nickel slit sheets were custom-ordered. This slit sheet should be aligned with the microchannel molded in the PDMS chip to block the stray light from entering the detector, which in the IEF-WCID applications is a CCD camera. Care should be taken to align the slit sheet under an optical microscope. The slit must be collimated with the light beam that illuminates the channel.
4. This pretreatment for quartz slides is very important in order to obtain a satisfactory hybrid chip. Otherwise, the SU-8 layer may peel off from the quartz slide easily, resulting in a shorter lifetime of the fabricated chip.
5. The holder with a 1 mm wide, 6 cm long slit is used to support the PDMS chip because PDMS chips tend to deform due to their softness. Furthermore, the holder is designed to fit into the iCE280 analyzer and the slit is designed to allow the UV light to pass through the holder and the channel for detection.

6. Unless stated otherwise, all solutions should be prepared in water that is purified with an ultrapure system.
7. All solutions should be filtered using a 0.2 μm pore size cellulose acetate filter prior to use.
8. It is recommended to prepare the fresh sample of pI markers for IEF-WCID measurement because some pI markers might have been decomposed in the sample prepared previously even they are stored in a fridge at 4°C.
9. The commercial iCE280 analyzer (Convergent Bioscience) for IEF-WCID consists of a deuterium (D2) lamp as the light source and a whole-column optical absorption imaging detector operated at 280 nm. During the process of sample focusing, the light beam from the lamp is focused onto the separation channel by a bundle of optical fibers and a cylindrical lens. The UV absorption image of the whole channel is captured by a CCD camera.
10. The SU-8 master is fabricated on a silicon wafer. The thickness of the spin-coated SU-8 layer is dependent on the type of SU-8 and the spinning speed. Each type of SU-8 has a specific amount of solvent that determines its viscosity and thus its spin-coated thickness at a given spinning speed. Each SU-8 covers a specific range of thickness: SU-8 2005, 5–8 μm; SU-8 2015, 15–40 μm; SU-8 2025, 25–80 μm; and SU-8 2075, 75–180 μm. It is possible to create other thicknesses by multiple coatings with a combination of different types of SU-8 photoresist.
11. The soft bake removes the solvent from the SU-8 and solidifies it so that it will not stick to the mask during the exposure.
12. A series of designs, each of which contains a 5 cm long, 50 μm or 100 μm wide channel and two 6 mm square recesses, are commercially printed onto Mylar films which serve as the photomasks.
13. The UV exposure creates a chemical reaction in the exposed SU-8 that cross-links the material to make it harder. Areas that are not exposed will dissolve away in the developer. It is important to choose the right dose of the UV light for the desired thickness. Too low a dose will cause the SU-8 features to lift off since the cross-linking will not extend down to the substrate and too high will create features much larger than the designed. The procedures include (i) setting the auto-exposure dose; (ii) putting the wafer spin-coated with SU-8 photoresist in the mask aligner; (iii) placing the mask with the printed side down on the wafer; (iv) applying a vacuum pressure of 5 inHg (0.17 bar), which reduces the gap between the mask and SU-8 layer and thus reduces the diffraction of light; and (v) activating the UV exposure dose while ensuring the UV shield is in place for protection.

14. The post-exposure baking completes the cross-linking. During this procedure the exposed regions should become visible. Cooling down time should be increased for thicker photoresists. Cracking or lifting off of the SU-8 is likely caused by thermal stress created during the cooling down period.
15. After the SU-8 feature is developed, the wafer should be silanized to prevent PDMS from sticking to the SU-8 master and silicon wafer during the replica molding process. TCMS is a dangerous silanization chemical. Therefore, a vapor mask should be worn for processing and the entire assembly should be placed in a fume hood
16. The concentration of polystyrene in the suspension determines the pore diameter of the resulting PDMS membrane (13). The ratio of PDMS base to curing agent affects the membrane's mechanical strength. The effects of the curing temperature and time have been found trivial on the membrane morphology. The optimum preparation conditions are found to be a combination of the PDMS prepolymer (10/1, mass ratio of PDMS base to curing agent) and the same mass ratio of PS/toluene solution (10%, w/w). The thickness of the PDMS layer depends on the spin-coating speed and duration time. In our case, the resulting membrane is about 10 μm thick and the pore diameter is about 2 μm.
17. After bonding, baking on a hot plate at 60°C for 1 h would make the bonding stronger. Plasma treatment times will depend on the power and pressure of the plasma system used. Conditions should be optimized such that the PDMS becomes hydrophilic to ensure good bonding.
18. The resulting PDMS layer bonded with the macroporous PDMS membrane can be used to fabricate a PDMS microchip for IEF-WCID analysis, and the procedures are the same as described in Subheading 3.4).
19. The masters with positive SU-8 features are fabricated using standard soft lithography technology. The straight channels have the same length of 5.3 cm with different cross-sectional areas of either 100 × 100 μm or 200 × 50 μm.
20. As the blank PDMS substrate serves as the bottom layer of the resulting microchip, its homogeneous thickness is also very important for IEF-WCID analysis. This is because the variation in its thickness leads to the variation in the channel cross-sectional areas and results in hydrodynamic pressure flow in the microchannel, which in turn causes motion of the focused bands resulting in low repeatability of IEF-WCID analysis.
21. For storage of fabricated chips, the channel is kept wet by filling the reservoirs with water after bonding, and the reservoirs are also sealed with plastic film.

22. In the absence of this step it would take a longer time to align the slit to the microchannel as the opening section of the metal slit must perfectly overlap with the separation channel to enhance the sensitivity of UV absorption detection. Great care and experience are required to perform this procedure.

23. For the microchips with different aspect ratio channels, their layout is kept the same. The final microchip consists of three substrates (20 mm wide and 75 mm long): one supporting base and two PDMS substrates with the channel and reservoirs in the uppermost substrate. The metal slit is glued on the upper substrate with its opening aligned with the microchannel.

24. The thickness of the bottom and upper PDMS substrates is kept at 2 mm. The upper PDMS substrate has two 6 mm square, 40 μm high recesses, which would contain two pieces of 6 mm square dialysis membranes.

25. The thickness of the PDMS mortar layer is controlled by the weight ratio of the PDMS prepolymer to toluene, spin-coating speed and duration time, which play vitally important roles in successfully sandwiching the dialysis membrane into a PDMS microchip for IEF-WCID analysis (14). The optimal conditions are presented as follows: the PDMS mortar with a 1:1 weight ratio of PDMS prepolymer to toluene should be spin-coated at 6,000 rpm for 40 s to make a 5 μm thick mortar layer on a microscope glass slide. The gap surrounded the dialysis membrane is very small by using this mortar layer. Furthermore, the design of the recess facilitates excellent sealing between the two PDMS substrates.

26. The capillaries are incorporated into the chip for channel surface preconditioning and sample injection and discharge, which facilitate IEF-WCID experiment and shorten the analysis time.

27. A series of designs each of which contains a 5 cm long and 50 or 100 μm wide channel are commercially printed onto Mylar films to make photomasks.

28. PVP is selected to dynamically coat the channel wall of microchips, which is necessary to suppress EOF and prevent protein adsorption to channel walls for IEF-WCID analysis (15).

29. For the IEF-WCID evaluation of the first generation of microchip (16) and the third generation of hybrid microchip, the anolyte and catholyte solutions must be simultaneously filled into their respective reservoirs, immediately after the sample is filled into the channel. A voltage difference should be applied to the liquid through the electrodes in the reservoirs as soon as the reservoirs are filled to minimize the diffusion between the electrolytes and the samples. The uneven liquid height in the two reservoirs is likely accompanied with the meniscus difference, which results in the Laplace pressure and hydrodynamic flow through the channel. This undesired hydrodynamic flow

will affect the IEF separation performance. For the IEF-WCID evaluation of the second generation microchip, the integrated dialysis membrane allows the sample injection to be separated from the anolytes and catholytes through the injection and discharge capillaries which eliminates the induced hydrodynamic pressure between two reservoirs, as a result, repeatability of IEF-WCID analysis is enhanced.

Acknowledgments

This work was supported by the Strategic Project Grant from the Natural Sciences and Engineering Research Council of Canada (NSERC) awarded to Dr. Carolyn L. Ren and her collaborators.

References

1. Tang Q, Harrata AK, Lee CS (1996) High-resolution capillary isoelectric focusing-electrospray ionization mass spectrometry for hemoglobin variants analysis. Anal Chem 68:2482–2487
2. Shen Y, Berger SJ, Anderson GA, Smith RD (2000) High-efficiency capillary isoelectric focusing of peptides. Anal Chem 72: 2154–2159
3. Silvertand LHH, Torano JS, van Bennekom WP, de Jong GJ (2008) Recent developments in capillary isoelectric focusing. J Chromatogr A 1204:157–170
4. Hofmann O, Che D, Cruickshank KA, Muller UR (1999) Adaptation of capillary isoelectric focusing to microchannels on a glass chip. Anal Chem 71:678–686
5. Mao Q, Pawliszyn J (1999) Demonstration of isoelectric focusing on an etched quartz chip with UV absorption imaging detection. Analyst 124:637–641
6. Guillo C, Karlinsey JM, Landers JP (2007) On-chip pumping for pressure mobilization of the focused zones following microchip isoelectric focusing. Lab Chip 7:112–118
7. Vlckova M, Kalman F, Schwarz MA (2008) Pharmaceutical applications of isoelectric focusing on microchip with imaged UV detection. J Chromatogr A 1181:145–152
8. Wu X-Z, Huang T, Liu Z, Pawliszyn J (2005) Whole-column imaging-detection techniques and their analytical applications. Trends Anal Chem 24:369–382
9. Gotz S, Karst U (2007) Recent developments in optical detection methods for microchip separations. Anal Bioanal Chem 387:183–192
10. Peng Y, Pallandre A, Tran NT, Taverna M (2008) Recent innovations in protein separation on microchips by electrophoretic methods. Electrophoresis 29:157–178
11. Richards DP, Stathakis C, Polakowski R, Ahmadzadeh H, Dovichi NJ (1999) Labeling effects on the isoelectric point of green fluorescent protein. J Chromatogr A 853: 21–25
12. Duffy DC, McDonald JC, Schueller OJA, Whitesides GM (1998) Rapid prototyping of microfluidic systems in poly(dimethylsiloxane). Anal Chem 70:4974–4984
13. Ou J, Ren CL, Pawliszyn J (2010) A simple method for preparation of macroporous polydimethylsiloxane membrane for microfluidic chip-based isoelectric focusing applications. Anal Chim Acta 662:200–205
14. Ou J, Glawdel T, Samy R, Wang S, Liu Z, Ren CL, Pawliszyn J (2008) Integration of dialysis membranes into a poly(dimethylsiloxane) microfluidic chip for isoelectric focusing of proteins using whole-channel imaging detection. Anal Chem 80:7401–7407
15. Ou J, Glawdel T, Ren CL, Pawliszyn J (2009) Fabrication of a hybrid PDMS/SU-8/quartz microfluidic chip for enhancing UV absorption whole-channel imaging detection sensitivity and application for isoelectric focusing of proteins. Lab Chip 9:1926–1932
16. Liu Z, Ou J, Samy R, Glawdel T, Huang T, Ren CL, Pawliszyn J (2008) Side-by-side comparison of disposable microchips with commercial capillary cartridges for application in capillary isoelectric focusing with whole column imaging detection. Lab Chip 8: 1738–1741

Index

Gareth Jenkins and Colin D. Mansfield (eds.), *Microfluidic Diagnostics: Methods and Protocols*, Methods in Molecular Biology, vol. 949, DOI 10.1007/978-1-62703-134-9,